Hydrometeorology: Theory and Applications

Hydrometeorology: Theory and Applications

Edited by Frederick Green

MURPHY & MOORE
www.murphy-moorepublishing.com

Published by Murphy & Moore Publishing,
1 Rockefeller Plaza,
New York City, NY 10020, USA
www.murphy-moorepublishing.com

Hydrometeorology: Theory and Applications
Edited by Frederick Green

International Standard Book Number: 978-1-63987-318-0 (Hardback)

Cataloging-in-Publication Data

Hydrometeorology : theory and applications / edited by Frederick Green.
p. cm.
Includes bibliographical references and index.
ISBN 978-1-63987-318-0
1. Hydrometeorology. 2. Hydrology. 3. Meteorology. I. Green, Frederick.
GB2805 .H93 2022
551.57--dc23

TABLE OF CONTENTS

PREFACE

The branch of meteorology and hydrology which deals with the study of transfer of water and energy between the land surface and the lower atmosphere is referred to as hydrometeorology. The primary goal of this discipline is to forecast and mitigate the effects of hydrometeorological disasters. Such hazards include drought, floods, desertification and tropical cyclones. Hydrometeorology uses statistical techniques, numerical weather prediction and nowcasting to predict such phenomena. It also includes identifying threats caused by such hazards, analyzing their magnitude and taking the necessary steps to mitigate such risks. This book traces the progress of this field and highlights some of its key concepts and applications. It unfolds the innovative aspects of hydrometeorology which will be crucial for the progress of this field in the future. Coherent flow of topics, student-friendly language and extensive use of examples make this book an invaluable source of knowledge.

After months of intensive research and writing, this book is the end result of all who devoted their time and efforts in the initiation and progress of this book. It will surely be a source of reference in enhancing the required knowledge of the new developments in the area. During the course of developing this book, certain measures such as accuracy, authenticity and research focused analytical studies were given preference in order to produce a comprehensive book in the area of study.

This book would not have been possible without the efforts of the authors and the publisher. I extend my sincere thanks to them. Secondly, I express my gratitude to my family and well-wishers. And most importantly, I thank my students for constantly expressing their willingness and curiosity in enhancing their knowledge in the field, which encourages me to take up further research projects for the advancement of the area.

Editor

1

Improving the complementary methods to estimate evapotranspiration under diverse climatic and physical conditions

F. M. Anayah[1,*] **and J. J. Kaluarachchi**[2]

[1]Utah Water Research Laboratory, Utah State University, Logan, UT 84322-8200, USA
[2]College of Engineering, Utah State University, Logan, UT 84322-4100, USA
[*]currently at: Palestine Technical University, Kadoorie, Tulkarm, Palestine

Correspondence to: F. M. Anayah (fathi.anayah@aggiemail.usu.edu)

Abstract. Reliable estimation of evapotranspiration (ET) is important for the purpose of water resources planning and management. Complementary methods, including complementary relationship areal evapotranspiration (CRAE), advection aridity (AA) and Granger and Gray (GG), have been used to estimate ET because these methods are simple and practical in estimating regional ET using meteorological data only. However, prior studies have found limitations in these methods especially in contrasting climates. This study aims to develop a calibration-free universal method using the complementary relationships to compute regional ET in contrasting climatic and physical conditions with meteorological data only. The proposed methodology consists of a systematic sensitivity analysis using the existing complementary methods. This work used 34 global FLUXNET sites where eddy covariance (EC) fluxes of ET are available for validation. A total of 33 alternative model variations from the original complementary methods were proposed. Further analysis using statistical methods and simplified climatic class definitions produced one distinctly improved GG-model-based alternative. The proposed model produced a single-step ET formulation with results equal to or better than the recent studies using data-intensive, classical methods. Average root mean square error (RMSE), mean absolute bias (BIAS) and R^2 (coefficient of determination) across 34 global sites were 20.57 mm month^{-1}, 10.55 mm month^{-1} and 0.64, respectively. The proposed model showed a step forward toward predicting ET in large river basins with limited data and requiring no calibration.

1 Introduction

A reliable estimate of ET (evapotranspiration) in river basins is important for the purpose of water resources planning and management. ET represents a significant portion of rainfall in the water balance especially in semiarid regions where most rainfall is typically lost as ET (FAO, 1989). Therefore, the uncertainty in estimating ET can lead to the inaccurate prediction of water balance. A careful screening of available meteorological, land use/land class and related hydrologic data in typical rural river basins suggest that ET is more challenging to calculate given the limited data. Data limitations in most rural river basins highlighted the importance of using alternative methods as opposed to the classical methods using land use/land cover data. While remote sensing techniques are available to estimate ET, such methods are expensive and necessary data may not be readily available for verification (Jimenez et al., 2011). Complementary methods initially proposed by Bouchet (1963) and others are alternative methods that can be used to calculate ET using meteorological data such as relative humidity, temperature and sunshine hours.

There are several classical methods presently available to estimate potential ET whereas estimating actual ET requires detailed local data such as land cover/land use, crop pattern and growing cycle. Typically, these classical methods predict crop ET from crop covered areas during the growing season to manage agricultural water demands. Crop ET is nothing but the potential ET multiplied by an appropriate crop coefficient, which is sometimes called the two-step approach (Allen et al., 1998). However, the actual water loss from the land surface is not restricted to crop areas only; instead evaporation happens from open water bodies as well as from open land surfaces with minimal vegetation cover. In water resources planning, the important estimate is the total water loss from the land surface that may or may not include transpiration from crop areas.

For several decades, complementary methods, including CRAE (complementary relationship areal evapotranspiration; Morton, 1983), AA (advection aridity; Brutsaert and Stricker, 1979) and GG (Granger and Gray; Granger and Gray, 1989) methods, have been used to estimate ET or total water loss from the land surface independent of land cover. These methods are attractive due to simplicity and practicability in estimating ET, wet environment ET (ETW) and potential ET (ETP) at the regional scale using meteorological data only. Previous studies attempted to use the complementary methods with little success (Doyle, 1990; Hobbins et al., 2001; McMahon et al., 2013; Szilagyi and Kovacs, 2010; Xu and Singh, 2005) given the limited understanding of the methods and the conflicting definitions of different terms. Still the complementary methods offer a distinct advantage over the classical methods given the simplicity, ready availability of required data and the ability to estimate total water loss as opposed to crop ET only.

Any improvements to the complementary methods cannot be conducted without the use of actual ET measurements. As part of this study, it is important to use measured ET data for model validation. Currently, ET fluxes are directly measured using the eddy covariance (EC) method that uses surface energy fluxes for weather forecasting and hydrologic modeling. These fluxes include sensible heat (H) and latent heat (LE) fluxes. Compared to other methods such as lysimeters, an EC system produces minimal physical disturbance to the surrounding environment and captures the areal fluxes within the footprint area (Luo et al., 2010). Most importantly, EC data are freely accessible worldwide, for example, FLUXNET (http://fluxnet.ornl.gov/), which is a global network of micrometeorological sites that use the EC methods to measure land-atmosphere exchange of carbon dioxide, water vapor and energy fluxes (Baldocchi et al., 2001). FLUXNET comprises of free-access regional networks such as AmeriFlux, AsiaFlux, EuroFlux and CarboAfrica. Given the task of finding a large set of global data with different climatic conditions and physical conditions, this study used the FLUXNET sites similar to many other studies (Castellvi and Snyder, 2010; Huntington et al., 2011).

The major limitation of the EC method is the lack of energy balance closure (i.e., $H + \mathrm{LE} \neq R_n - G_{soil}$, where R_n is net radiation and G_{soil} is soil heat flux) that causes underestimation of ET (Wilson et al., 2002). Twine et al. (2000) and Wang et al. (2008) showed that underestimation of ET can be as high as 15 %, however, others, Castellvi et al. (2008), Huntington et al. (2011) and Wilson et al. (2002), found lower percentages within measurement uncertainty that can be < 5 %. These studies showed that the impact of energy imbalance in the EC method may not be significant as thought earlier (Castellvi and Snyder, 2010). Hence, the EC method is still attractive and served as the standard method for direct measurement of ET fluxes (Castellvi et al., 2008; Luo et al., 2010).

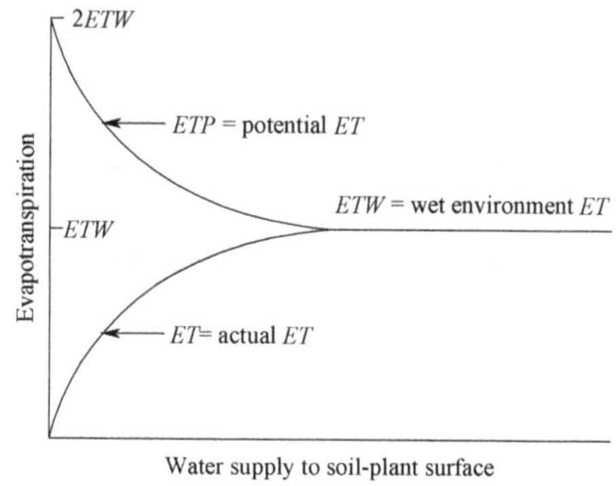

Figure 1. A schematic representation of the complementary relationship between ET, ETW and ETP (after Morton, 1983).

Hobbins et al. (2001) and Xu and Singh (2005) found limitations to the complementary methods in different physical and climatic conditions especially in arid settings. Some of these limitations lead to many unanswered questions such as how applicable are the complementary relationship to estimate ET? Are these methods only valid within humid climates? What are the limitations in the different complementary methods? Have complementary methods been compared to measured ET data under a variety of climatic and physical conditions? Given these unanswered questions, it is important to address the validity of the complementary methods in a scientifically justifiable manner.

It is found that there is no single study where the ET estimates from the complementary methods have been extensively predicted and evaluated using data from EC sites. To evaluate the applicability of the complementary methods and to propose suitable changes, the methods need to be evaluated under a variety of land cover/land use classes and climatic conditions. In addition, the three complementary methods, CRAE, AA and GG, have not been cross-compared and evaluated using measured ET data. Therefore the goals of this study are to investigate the applicability of the complementary methods in estimating ET in contrasting environments, perform necessary revisions to the existing methods to improve estimates if necessary and finally propose a universal model of estimating ET that is calibration-free, simple, robust and uses minimum data.

2 Complementary methods

2.1 Complementary relationship

Complementary methods describe the relationships between ET, ETW and ETP using the complementary relationship first introduced by Bouchet (1963). The theory states that a complementary relationship exists between ET and ETP as shown in Fig. 1 (see Davenport and Hudson, 1967; Pettijohn and Salvucci, 2009). ETW, however, is ET that would occur if the soil-plant surface is wet enough so that ET could approach its potential value, ETP (Granger, 1989). The development of the complementary relationships and the definitions of various terms are discussed in detail by Brutsaert and Stricker (1979), Granger and Gray (1989), Lhomme and Guilioni (2006), McMahon et al. (2013), Morton (1983) and Pettijohn and Salvucci (2009). The three definitions of ET are related as

$$\mathrm{ET} = 2\mathrm{ETW} - \mathrm{ETP}, \tag{1}$$

where ET, ETW and ETP are in millimeters per month ($\mathrm{mm\,month^{-1}}$). Equation (1), which is the Bouchet original expression, indicates that an increase in ET is accompanied by an equivalent decrease of ETP; i.e., $\delta\mathrm{ET} = -\delta\mathrm{ETP}$. In other words, as the surface dries, actual ET decreases causing a reduction in humidity and an increase in temperature of the surrounding air, and as a result ETP will increase. Once ETP and ETW are estimated, ET is subsequently derived.

In the literature, the complementarity relationship between ET and ETP shown in Eq. (1) is of controversy among scientists who claimed that many inherent assumptions of Bouchet's theory lack sufficient evidence (Granger, 1989; Lhomme and Guilioni, 2006). Recently, there have been several attempts to improve the complementary relationship and its predictive power of different ET definitions (see Brutsaert and Stricker, 1979; Granger and Gray, 1989; Morton, 1983). Han et al. (2012) developed a nonlinear approach to the complementary relationship but the results require further study and verification. Yet, Lhomme and Guilioni (2010) proposed a different model that can describe the complex relationship between ET and ETP based on the convective boundary layer.

2.2 CRAE method

ETP is estimated by solving the energy balance and vapor transfer equations iteratively (Morton, 1983). ETP is calculated by solving at equilibrium temperature (T_P in °C) at which the energy balance and vapor transfer equations for a moist surface are equivalent. The procedure describing the iterative solution is given by Morton (1983, Appendix C). The energy balance equation to estimate ETP is given as

$$\mathrm{ETP} = R_T - \lambda f_T (T_\mathrm{P} - T), \tag{2}$$

where R_T is net radiation for soil-plant surfaces ($\mathrm{mm\,month^{-1}}$) at air temperature T (°C), λ is the heat transfer coefficient (mbar °C) and f_T is the vapor transfer coefficient ($\mathrm{mm\,month^{-1}\,mbar^{-1}}$). To estimate ETW in Eq. (4), net radiation for soil-plant surfaces at T_P (R_TP) is first computed using Eq. (3).

$$R_\mathrm{TP} = \mathrm{ETP} + \gamma f_T (T_\mathrm{P} - T), \tag{3}$$

$$\mathrm{ETW} = b_1 + b_2 (1 + \gamma/\Delta_\mathrm{P})^{-1} R_\mathrm{TP}, \tag{4}$$

where γ is the psychrometric constant (mbar °C^{-1}), b_1 is a constant representing advection energy, b_2 is a constant and Δ_P is the rate of change of saturation vapor pressure with T at T_P (mbar °C^{-1}). Constants b_1 and b_2 were calibrated using climatic data from arid regions in North America and Africa (Morton, 1983). ETP from Eq. (2) and ETW from Eq. (4) are used in Eq. (1) to calculate ET of the CRAE method.

2.3 AA method

In the AA method, Penman's (1948) equation ($\mathrm{ET_{PEN}}$) is used to estimate ETP as shown in Eqs. (5) and (6).

$$\mathrm{ET_{PEN}} = \frac{\Delta}{\gamma + \Delta}(R_\mathrm{n} - G_\mathrm{soil}) + \frac{\gamma}{\gamma + \Delta} E_\mathrm{a}\,, \tag{5}$$

$$E_\mathrm{a} = 10.6 \times (\beta + 0.54U)(e_\mathrm{s} - e_\mathrm{a}), \tag{6}$$

where Δ is rate of change of saturation vapor pressure with T (mbar °C^{-1}), R_n is net radiation ($\mathrm{mm\,month^{-1}}$), G_soil is soil heat flux ($\mathrm{mm\,month^{-1}}$), E_a is drying power of air ($\mathrm{mm\,month^{-1}}$), β is a constant and usually equals to 1.0, U is wind speed at 2 m above ground level ($\mathrm{m\,s^{-1}}$), e_s is saturation vapor pressure at T (mm Hg) and e_a is vapor pressure of air (mm Hg). In the wind formulation of Penman (1956), β was updated to 0.5. Although both wind function formulae (when $\beta = 1$ or 0.5) are widely used in hydrology, Penman preferred a β value of 1 (see Brutsaert and Stricker, 1979; McMahon et al., 2013). Brutsaert and Stricker (1979) mentioned that their method is insensitive to the wind function. The first term of Eq. (5) is called equilibrium ET and the second is aerodynamic ET that is generated by large-scale advection effects (see Hobbins et al., 2001). When advection is minimal, the interactions of atmosphere with the soil-plant system will be completely developed and an equilibrium condition is approached (Brutsaert and Stricker, 1979).

ETW of the AA method is calculated using $\mathrm{ET_{PT}}$ of Priestley and Taylor (1972) in which minimal advection is assumed and given by Eq. (7).

$$\mathrm{ET_{PT}} = \alpha \frac{\Delta}{\gamma + \Delta}(R_\mathrm{n} - G_\mathrm{soil}), \tag{7}$$

where α is a coefficient that typically equals to 1.26 or 1.28 (Priestley and Taylor, 1972). The AA method in this study used the values α of 1.28 and β of 1. ETP from Eq. (5) and ETW from Eq. (7) are used in Eq. (1) to calculate ET of the AA method.

2.4 GG method

The complementary relationship given in Eq. (1) is primarily used by the CRAE and AA methods. In the GG method,

Granger and Gray (1989) used a modified version as shown in Eq. (8).

$$\mathrm{ET} = (1+\frac{\gamma}{\Delta})\mathrm{ETW} - \frac{\gamma}{\Delta}\mathrm{ETP} \quad (8)$$

Equation (8) is reduced to Eq. (1) only when $\gamma = \Delta$. In this method, two new concepts were proposed and empirically correlated together; relative drying power (D) and relative evaporation (G) shown in Eqs. (9) and (10), respectively.

$$D = \frac{E_a}{E_a + (R_n - G_{soil})}, \quad (9)$$

$$G = \frac{\mathrm{ET}}{\mathrm{ETP}}, \quad (10)$$

where D indicates surface dryness, i.e., D becomes larger as the surface becomes drier. G is ET that occurs under similar wind and humidity conditions from a saturated surface at the actual temperature (Granger and Gray, 1989).

In the original work, G was defined as G_1 through Eq. (11) where this equation was empirically derived using data from two stations in a semiarid region of Western Canada. Granger and Gray (1989) mentioned that G_1 is independent of land use.

$$G_1 = \frac{1}{c_1 + c_2 e^{c_3 D}}, \quad (11)$$

where $c_1 = 1.0$, $c_2 = 0.028$ and $c_3 = 8.045$. In the GG method, the selection of the function to calculate relative evaporation (G) has great impact on the actual ET estimates and any modification to this empirical formula may be significant in improving the predictability of the GG method. In essence, there is more research required in this effort. Thus, Eq. (11) was later modified by Granger (1998) to account for different surface conditions as shown in Eq. (12).

$$G_2 = \frac{1}{c_4 + c_5 e^{c_6 D}} + c_7 D, \quad (12)$$

where $c_4 = 0.793$, $c_5 = 0.2$, $c_6 = 4.902$ and $c_7 = 0.006$. Therefore G in Eq. (10) can be substituted by G_1 of Eq. (11) or G_2 of Eq. (12).

ETW required to solve Eq. (8) is obtained from Eq. (5), used earlier in the AA model. Thereafter G_1 is used in Eq. (10) together with Eq. (9) to solve for ET in Eq. (8). The final equation describing ET in the GG method is therefore given as

$$\mathrm{ET} = \frac{\Delta G}{\gamma + \Delta G}(R_n - G_{soil}) + \frac{\gamma G}{\gamma + \Delta G}E_a, \quad (13)$$

where ET, R_n, G_{soil} and E_a are in millimeters per month. Although the CRAE, AA and GG methods enable the direct prediction of ET without the need for surface parameters (temperature and vapor pressure), but the GG method is the only method that does not require a prior estimate of ETP (Granger, 1989).

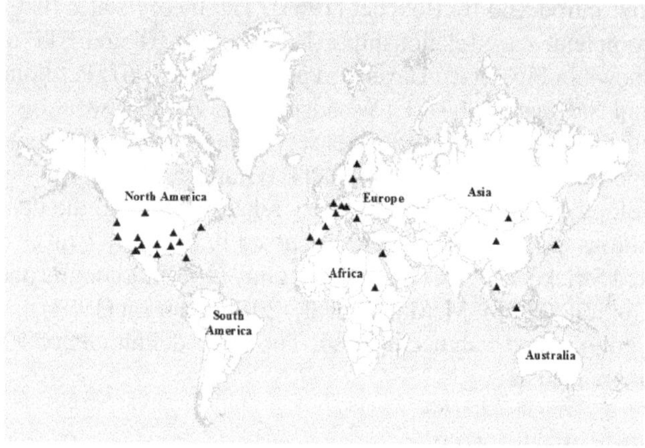

Figure 2. Map showing the locations of the 34 EC sites with measured ET flux data.

2.5 Alternative method (ASCE)

In the popular ASCE (American Society of Civil Engineers) method (Allen et al., 2005), input data to calculate net radiation (R_{ASCE}) are similar to those of the CRAE method. More specifically, the ASCE method requires minimum and maximum temperature data, which sometimes are not available. In such a case the procedure described by Allen et al. (2005, Eq. E.5) is followed. One major difference between the CRAE and ASCE methods is the albedo calculation. In the former, albedo is calculated using a set of equations whereas albedo is fixed at 0.23 in the latter. The ASCE method also requires wind speed measurements to calculate ETP while estimating crop ET requires detailed information of land cover/land use, crops, cropping pattern and the growing cycle. The ASCE method is specifically utilized in this study to compare R_{ASCE} with R_T and R_{TP}. The ASCE method is also used to calculate G_{soil} using monthly averages of temperature data.

3 Measured flux and meteorological data

3.1 Sites of EC data

In this study 34 global sites were selected with measured meteorological and flux data and these sites are distributed as follows: 17 from AmeriFlux sites, 11 from EuroFlux sites, 5 from AsiaFlux sites and 1 CarboAfrica site (see Fig. 2). Unfortunately, efforts to obtain data from other sites in CarboAfrica have not been successful. The selection of the 34 sites was based on data availability and climatic variability. The details of the sites and data collected are shown in Table 1 and Fig. 2.

The reason to select 34 sites is that prior studies have typically used a smaller number of sites and in most cases under similar climatic conditions. By using a variety of global sites in contrasting physical and climatic conditions with

Table 1. Characteristics of the 34 EC sites with measured ET data used in the study.

#	Site	Country	Lat	Long	Height	Data availability	ET_{EC}, mm month^{-1}			AI_M	Land cover
			°	°	m	from-to (# months)	min	mean	max	mm °C^{-1}	
Very humid											
1	Takayama	Japan	36.1	137.4	25	06–07 (24)	9.4	44.4	91.7	83.2	Deciduous forest
2	Walker Branch, TN	USA	36	−84.3	44	95–98 (48)	10.5	47.4	116.2	76.5	Deciduous forest
3	Qinghai	China	37.6	101.3	2.2	02–04 (36)	1.6	36.2	110.5	68.3	Alpine meadow
4	Palangkaraya	Indonesia	2.3	114	41.3	02–05 (47)	82.4	134.3	164	61.5	Tropical forest
5	Harvard Forest, MA	USA	42.5	−72.2	30	92–99 (96)	5.1	37.5	108.4	61.2	Mixed forest
6	Flakaliden	Sweden	64.2	19.8	15	96–98 (31)	−0.1	23	63.4	51.5	Coniferous forest
7	Bondville, IL	USA	40	−88.3	10	97–06 (120)	1.7	50.1	135.4	49.6	Cropland
8	Goodwin Creek, MS	USA	34.3	−89.9	4	03–06 (48)	2.4	55.5	138.7	47.9	Cropland/natural
9	Tharandt	Germany	51	13.6	42	96–99 (42)	6.5	39.2	95.9	47.1	Evergreen forest
10	Sarrebourg	France	48.7	7.1	22	96–99 (32)	−0.1	32.8	102.3	42.7	Deciduous forest
11	Kennedy Oak, FL	USA	28.6	−80.7	18	02–06 (48)	6	49.1	120.3	40.4	Woody savanna
12	Loobos	Netherland	52.2	5.7	24	96–98 (30)	7.4	32.4	63.1	39.7	Evergreen forest
13	Sakaerat	Thailand	14.5	101.9	45	01–03 (32)	37.7	63.8	109.5	36.8	Tropical forest
Humid											
14	Norunda	Sweden	60.1	17.5	103	96–98 (29)	1.3	30.9	80.8	34	Evergreen forest
15	Fort Peck, MT	USA	48.3	−105.1	4	00–06 (84)	1.3	26	164	33	Grassland
16	Freeman, TX	USA	29.9	−98	3	05–08 (48)	6	49.1	120.3	30.9	Grassland
17	Little Washita, OK	USA	35	−98	3	96–98 (32)	8.9	41.6	104.4	30.7	Grassland
18	Mehrstedt 2	Germany	51.3	10.7	n/a	04–06 (34)	0	27	95.3	29.6	Grassland
Subhumid											
19	Evora	Portugal	38.5	−8	28	05–05 (12)	−0.3	13.7	34.8	26.2	Savanna
20	Mauzac	France	43.4	1.3	3.5	05–07 (34)	8.3	37.2	91.4	25.5	Grassland
Mediterranean											
21	Bugac	Hungary	46.7	19.6	4	02–08 (72)	2.3	37.5	103.9	23.8	Cropland
22	Metolius, OR	USA	44.3	−121.6	12	04–08 (60)	2.3	30.3	71	22.8	Evergreen forest
23	Tonzi Ranch, CA	USA	38.4	−121	23	01–09 (80)	1.4	29.8	95.5	21	Woody savanna
24	Vaira Ranch, CA	USA	38.4	−121	2	01–09 (108)	−5.1	25.1	88	21	Woody savanna
Semiarid											
25	Kherlenbayan	Mongolia	47.2	108.7	3.5	03–10 (68)	−2.3	10.5	50.8	17.5	Grassland
26	Llano de los Juanes	Spain	36.9	−2.8	2.8	05–05 (12)	7.2	18.7	36.7	15.4	Closed shrubland
27	Audubon, AZ	USA	31.6	−110.5	4	02–09 (87)	2	24.4	92.5	13.5	Open shrubland
28	Kendall, AZ	USA	31.7	−109.9	6.4	04–09 (68)	2.2	20.2	72.4	13.2	Grassland
29	Santa Rita, AZ	USA	31.8	−110.9	6.4	04–07 (48)	4.3	26	91.1	10.7	Open shrubland
Arid											
30	Corral Pocket, UT	USA	38.1	−109.4	1.9	01–07 (39)	4.6	14.8	33.3	9.8	Grassland
31	Sevilleta grass, NM	USA	34.4	−106.7	3	07–08 (19)	4.5	22.2	69.7	9	Grassland
32	Sevilleta shrub, NM	USA	34.3	−106.7	3	07–08 (24)	3.3	23.5	74.7	9	Grassland
33	Demokeya	Sudan	13.3	30.5	12	97–98 (17)	6.1	38.1	106.3	8.9	Grassland
34	Yatir	Israel	31.3	35.1	14	01–09 (48)	5.7	17.8	57.3	8.6	Open shrubland

measured ET data, we will demonstrate the validity of the proposed complementary method in different land use/land class categories. While there are other global EC sites, these sites could not be considered due to the lack of diversity of land classes and climatic conditions required in this study. As mentioned earlier, data accessibility was also an issue in some cases.

To classify the climatic conditions prevailing at each site, a simple aridity index developed by De Martonne (1925), AI_M (in mm °C^{-1}), is chosen and given as

$$AI_M = \frac{P_{ann}}{T_{ann} + 10} \tag{14}$$

where P_{ann} is average annual precipitation in millimeters and T_{ann} is average annual T in degrees Celsius. Unlike other aridity indices, AI_M indicates the availability of both water and energy from readily available data. In effect, the sites were sorted to the following climatic classes; very humid ($AI_M \geq 35$), humid ($28 \leq AI_M < 35$), subhumid ($24 \leq AI_M < 28$), Mediterranean ($20 \leq AI_M < 24$), semiarid ($10 \leq AI_M < 20$) and arid ($AI_M < 10$).

As shown in Table 1, the 34 sites have different geographic and climatic conditions. The data set consists of 1657 monthly measurements across the 34 sites. The P_{ann} values range from 196 mm at site 25 to 2231 mm at site 4, and T_{ann} varies between −1.7 °C at site 3 and 26.3 °C at site 4. It is noticed that many sites fall within the very humid climatic class. The surface conditions also differ considerably from grasslands to forests. Data are available from 12 to 120 months from 1992 to 2010. At site 1, for example, data from 24 months are available in 2006 and 2007, while at site 4 there are no ET data in April 2003. Therefore, the total number of months included in the calculations from 2002 to 2005 is 47 instead of 48.

The EC tower heights vary from 2 m at site 24 to 103 m at site 14 with a median value of 10 m at site 7 and an average value of 17.1 m. The EC tower height reflects the vertical flux footprint that usually indicates the upwind area captured by the instruments mounted on the tower. Starting from very humid, humid, subhumid, Mediterranean, semiarid to arid climatic classes, the average EC tower heights are 24.8, 28.2, 15.8, 10.2, 4.6 and 6.8 m, respectively. It is no surprise that the tower heights are highest in the very humid sites where the land cover is dominated by forests of high canopy altitudes. However, low tower heights are required for arid and semiarid sites naturally characterized by grassland or shrubland covers. The high range of EC tower heights explains the suitability of selecting these particular 34 EC sites that have flux footprints of the scale of the complementary methods. This observation may lead to the conclusion that a perfect correlation between the EC and complementary methods may exist.

Compared to the lowest average-ET_{EC} flux (10.5 mm month^{-1}) that occurs at site 25, site 4 has the maximum of 134.3 mm month^{-1}. It is observed that site 4 has the highest ET_{EC} fluxes across the 34 sites because the site is located in tropical peat swamp forests where soil moisture is relatively high throughout the year (Hirano et al., 2005) and the site is also exposed to high energy demands. In general, the wide ranges of ET_{EC} fluxes and AI_M values reflect the diversity of hydrologic and climatic conditions present in this study.

3.2 Measured flux data from EC systems

In comparison to finer-resolution data, collecting data at a monthly scale is easier in rural and sparse areas, less problematic when data quality is poor and more appropriate for regional-scale studies. Thompson et al. (2011) examined model performance using different timescales from half hourly to interannual and found that a monthly time step is preferable. Data in this study were directly downloaded from its regional network website and sometimes obtained (or complemented) through personal communications. In cases where monthly data were not readily available, average monthly data were aggregated from finer-time-resolution data, e.g., daily or hourly. To keep minimal changes to the input data, only months of available data (50 % or more) were considered in the analysis.

Input data requirements are often the driver to select a specific method to estimate ET. Even in rural regions where data limitations are common, data to calculate R_n with the CRAE method (Morton, 1983) only requires monthly averages of temperature, humidity (or dew-point temperature) and sunshine hours (or solar radiation). Again, the CRAE method calculates two types of R_n, R_T and R_{TP}, at the same time. It is obvious that the CRAE method can also estimate ETP, ETW and ET using the same data. However, both AA and GG methods, similar to any classical method, need wind speed measurements to calculate ET (see Eq. 6).

The performance indicators used to assess the model predictions are root mean square error (RMSE), mean absolute bias (BIAS) and coefficient of determination (R^2). As the number of sites is large, the BIAS, which indicates the disparity of predicted and measured ET, is preferred over the mean bias value itself because negative values of mean bias cannot cancel positive values.

4 Model development and results

The approach used here is a systematic model sensitivity analysis across the three existing complementary methods to identify the major model components contributing to predicting ET compared to the EC observations. The findings from each step of the sensitivity analysis is later used to propose a universal model that is calibration-free and capable of predicting ET (or total water loss) independent of land cover/use. The proposed approach can be divided into four stages: (1) first, the three original complementary methods are applied across all 34 sites to identify the relative accuracy of each method, (2) using the results obtained from the first stage, a set of model variations representing the different model structures will be developed, (3) next the model variations with acceptable results will be selected for further analysis, and, finally, (4) a statistical analysis will be conducted to differentiate between the final model(s) to identify a universal model capable of predicting ET across all sites without calibration. To further test the proposed model, the results of this study will be compared with the results of recently published ET studies.

Table 2. Average values of RMSE, BIAS and R^2 of actual ET estimates from the different complementary methods, CRAE, AA and GG, for each climatic class.

Climatic class	RMSE (mm month^{-1})			BIAS (mm month^{-1})			R^2		
	CRAE	AA	GG	CRAE	AA	GG	CRAE	AA	GG
Very humid	27.6	29.0	22.6	15.8	12.2	10.6	0.73	0.71	0.73
Humid	31.2	35.2	27.1	19.2	16.5	14.3	0.77	0.73	0.75
Subhumid	46.6	54.7	45.0	31.9	28.7	26.5	0.39	0.33	0.41
Mediterranean	35.3	58.1	47.4	18.6	28.8	25.3	0.51	0.42	0.45
Semiarid	16.6	18.9	22.1	9.6	8.4	13.3	0.56	0.61	0.41
Arid	22.4	31.9	29.5	9.4	14.4	19.5	0.53	0.54	0.42
All classes	27.8	33.8	28.4	15.7	15.5	15.5	0.64	0.61	0.59

4.1 Comparison between original complementary methods

The ET estimates computed using the three original complementary methods were compared to the measurements from the EC sites (ET_{EC}) and the results are given in Table 2. It is no surprise that the subhumid climatic class has the poorest performance as there are only two sites in this class of which site 19 has the poorest values of RMSE, BIAS and R^2. For the CRAE method, the sites with arid climates have the lowest RMSE and BIAS values and sites with wet (very humid and humid) climates have the highest R^2 values. The AA method was developed for a watershed experiencing severe drought and, therefore, this method is expected to outperform the other two methods in arid climates. Hobbins et al. (2001) evaluated the CRAE and AA methods across 120 basins in the United States. They found that as aridity increases, the CRAE method tends to overestimate ET and the AA method tends to underestimate ET. Xu and Singh (2005) evaluated three sites of diverse climates and found that the predictive power of the methods increases with humidity. This conclusion contradicts the results in Table 2 as the CRAE and AA methods perform best in arid climates. In general, the three methods work relatively well under extreme climatic conditions, either arid or humid. Also the predictions of the GG method are slightly better in humid climates than arid as found by Xu and Singh (2005). Overall, the CRAE method is the best according to RMSE and R^2 while the GG method has the lowest BIAS. Still, the computed ET estimates are not close enough to the ET_{EC} measurements indicating that there is a need for improvements to the existing methods.

4.2 Development of alternative model variations

The prior estimates of ET are highly dependent on R_n. Net radiation computed by Morton (1983) is denoted as R_T, which is net radiation at T while R_{TP} is net radiation at T_P. Net radiation from Allen et al. (2005) is denoted as R_{ASCE}. When compared to the R_n measurements from the EC sites, the three estimates of net radiation perform better as humidity increases. Although detailed results are not shown here, the average R^2 values of R_T and R_{ASCE} estimates range from 88 to 98 % and from 92 to 98 %, respectively. While R_{ASCE} is the overall best estimator of R_n, R_T performs better in arid and semiarid regions. The results of this analysis clearly indicate that the net radiation prediction is dependent on the climatic class and therefore, any improvements should consider climate dependency.

Selecting the correct equations to calculate ETP, ETW and even ET may significantly influence the accuracy of the net radiation estimates. This work used the original model equations of the CRAE, AA and GG methods in different ways. This study is not meant to explore all possible relationships between ETP and ETW; instead the focus here is developing a reliable predictive model of actual ET that is applicable under a variety of climatic and physical conditions. Therefore, the relationships and model equations of the original methods were used here in a manner to preserve the physical processes controlling ET. Similarly, there are two formulae to describe the complementary relationship, namely Eqs. (1) and (8). It is true that there may be other possible formulae to simulate the complementary relationship between ET, ETW and ETP. The drawback of these approaches is the need for calibration for which the revised model will be applicable for a given site or region. This condition is against the original purpose of this study that attempts to develop a model that is widely applicable for many different climatic and physical conditions.

In stage 2, different combinations of model formulations are considered to develop a set of alternative model variations that may be better than the original methods. For instance, these alternative model variations can decide if R_T is a better estimator of net radiation compared to R_{ASCE} or not. Similarly another question is if the complementary relationships are adequately presented by Eq. (1) or Eq. (8) or if a different formulation is needed. In selecting these different alternative model variations, the criteria for the sensitivity analysis used are the method to calculate R_n, the representation of the complementary relationship, the value of

Table 3. Details of the 17 model variations developed based on the complementary relationships and the three original complementary methods (CR, complementary relationship).

Criteria	R_n		CR		α		β		G	
Equation			1	8					11	12
Value	R_T	R_{ASCE}			1.26	1.28	0.5	1.0		
CRAE	√		√							
CRAE1	√			√						
CRAE2		√	√							
CRAE3		√		√						
AA	√		√			√		√		
AA1	√			√		√		√		
AA2	√			√	√		√			
AA3		√	√			√		√		
AA4		√		√		√		√		
AA5		√	√		√		√			
AA6		√		√	√		√			
AA7	√		√		√		√			
GG	√			√				√	√	
GG1	√		√					√	√	
GG2	√			√				√		√
GG3	√		√					√		√
GG4		√		√				√	√	
GG5		√	√					√	√	
GG6		√		√				√		√
GG7		√	√					√		√

α in the ET_{PT} equation, the value of β in the wind function of the ET_{PEN} equation and the relative evaporation function (G) of the GG method. After studying the model structure of each complementary method, 17 different alternative model variations are proposed in Table 3 for subsequent analysis. As discussed earlier, this is a systematic parameter sensitivity exercise to identify the best alternative model variation. Although more model variations are possible, the 17 listed alternative model variations are adequate at this stage. For example, the AA and GG methods have four criteria each (R_n, complementary relationships, α and β) producing 16 model variations. An important consideration in the development of these model variations is the conclusions of others. For instance, Hobbins et al. (2001) found that changes to the AA method did not necessarily produce superior results, especially by perturbing β (see Brutsaert and Stricker, 1979).

The ET estimates produced by these 17 alternative model variations across the 34 sites were compared to the EC measurements and the results are shown in Fig. 3. It should be noted that Fig. 3 shows the anomalies from the original method for each model variation. In effect, the results are considered to show improvements if the anomaly of RMSE is negative. The same trend is valid for BIAS but opposite for R^2. It is observed that none of the CRAE- or AA-based alternative model variations improved RMSE and BIAS. Among the CRAE-based model variations, CRAE2 has the minimum deterioration of RMSE and BIAS while showing some improvement of R^2. A similar behavior is noticed with AA4 of the AA-based model variations. However, the GG-based model variations have obvious improvements across all three metrics. GG1, GG3, GG5 and GG7 model variations showed improved RMSE and BIAS values when compared with the original GG method. The only common feature among these four GG model variations is Eq. (1) representing the complementary relationship and not Eq. (8), which was used by the original GG method. This observation indicates that Eq. (1) is superior in representing the complementary relationships between ET, ETW and ETP. The deterioration of results in the GG-based model variations is deemed minor when compared to other model variations. The conclusion from stage 2 is that these GG model variations perform better than the CRAE and AA model variations.

Although ETP is usually given under saturated conditions in the equation of Penman (1948) as shown in the original AA method, the definition of ETW still has some ambiguity (Lhomme and Guilioni, 2006). One important difference of the original GG method compared to the other two methods is the equation describing ETW. ETW of the original CRAE and AA methods is derived from the ET_{PT} equation (Eq. 7) while the original GG method uses the ET_{PEN} equation or Eq. (5) (Brutsaert and Stricker, 1979; Granger and Gray, 1989; Morton, 1983). Given this departure of the GG model

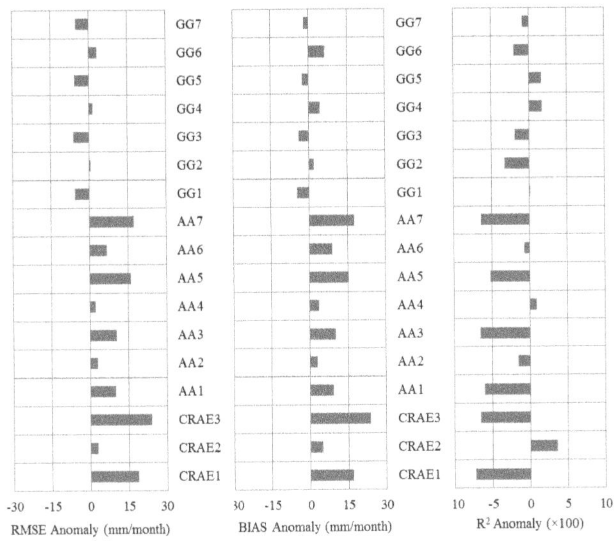

Figure 3. Anomalies of RMSE, BIAS and R^2 values for 17 model variations across the 34 sites. Here anomalies are computed based on the values computed with each corresponding complementary method.

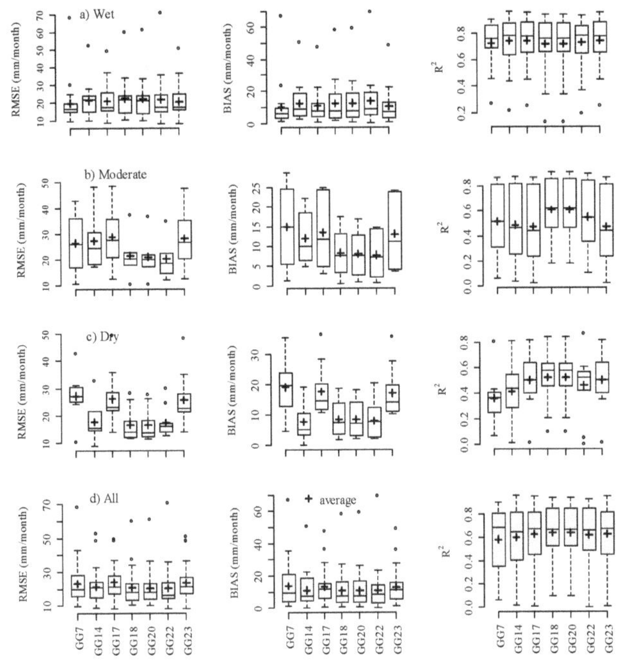

Figure 4. Box plots of RMSE, BIAS and R^2 metrics of the seven promising model variations for the simplified climatic classes.

from others, we further studied the GG model variations based on the model describing ETW. Accordingly, another set of alternative model variations from the GG model is possible. These variations consist of 16 models (GG8 through GG23) and the details are given in Table 4. In these variations, β is no longer changed while α in the ET_{PT} equation will be changed. ETW in all these variations will use the Priestley–Taylor equation (Table 4). In total, 24 GG model variations (GG1–GG23 from Tables 3 and 4) are now considered for the next stage.

4.3 Selection of best performing GG model variation

For the purpose of selecting the best GG model variation(s), each model from the latest 24 was run and the results were compared with EC observations (see Table 5). The performance metrics were used to identify the best GG model variation in each climatic and performance metric combination and the results are shown in Table 5. For example, GG3 was the best for RMSE, GG1 for BIAS and GG17 and GG23 for R^2 in the very humid class. In essence, 11 GG model variations became eligible from the 24 selected earlier from stage 2. It is also observed that GG20 is the best for six combinations of performance metric and climatic class combinations. In contrast, GG3 is the best only in RMSE for the very humid class. GG1, GG3, GG11 and GG13 are the best models each for one combination of metric and climatic class combination only. Therefore these GG model variations were rejected and the remaining seven (GG7, GG14, GG17, GG18, GG20, GG22 and GG23) were selected for further consideration.

There are other key observations made from the prior analysis. First, the original GG method uses the complementary relationship given by Eq. (8) (Granger, 1989), yet, five of the seven promising model variations selected earlier uses Eq. (1). In essence, this observation suggests that Eq. (1) is better in capturing the variability of ET compared to Eq. (8). Second, six of these seven promising GG model variations use the ET_{PT} equation to calculate ETW. Third, a comparison between R_T and R_{ASCE} shows that six of these promising GG model variations use R_{ASCE} to denote net radiation that supports the conclusion drawn earlier. Fourth, five of these GG model variations use Eq. (12) to calculate G. Lastly, changing the value of α in the ET_{PT} equation and varying the equation describing G did not alter the results.

The next step of the analysis will be to identify the best model variation of the seven selected earlier. Before proceeding to the next step, the six climatic classes are simplified to represent climatic variability using three simple classes; wet (from original very humid and humid), moderate (from original subhumid and Mediterranean) and dry (from original semiarid and arid). This revision shall not affect the results and will make the analyses and conclusions simple. Using these new definitions, the original 34 global sites are now reallocated as 18, 6 and 10 into wet, moderate and arid classes, respectively.

Figure 4 shows the results of performance metrics to these seven models using the simplified climatic classes of wet, moderate and dry. For all climatic classes, GG17 has the highest RMSE and GG7 has the highest BIAS values. GG7 performs well only in the wet climatic class, while it performs poor in the moderate and dry classes. The GG17 and

Table 4. Sixteen GG-based model variations developed given that ETW is calculated using the ET_{PT} equation.

Criteria	R_n		CR		α		G	
Equation			1	8			11	12
Value	R_T	R_{ASCE}			1.26	1.28		
GG8	√		√		√			√
GG9	√			√	√			√
GG10	√		√		√		√	
GG11	√			√	√		√	
GG12	√		√			√	√	
GG13	√			√		√	√	
GG14	√		√			√		√
GG15	√			√		√		√
GG16		√	√			√		√
GG17		√		√		√		√
GG18		√	√			√	√	
GG19		√		√		√	√	
GG20		√	√		√		√	
GG21		√		√	√		√	
GG22		√	√		√			√
GG23		√		√	√			√

Table 5. Results of the performance of different models in a given climatic class described through the best values of RMSE, BIAS and R^2.

Metric	Climatic class						All classes
	Very humid	Humid	Subhumid	Mediterranean	Semiarid	Arid	
RMSE	GG3	GG7	GG22	GG22	GG20	GG20	GG22
BIAS	GG1	GG7	GG20	GG22	GG14	GG14	GG18
R^2	GG17 & GG23	GG11 & GG13	GG18 & GG20	GG18 & GG20	GG17 & GG23	GG18 & GG20	GG18 & GG20

GG23 model variations have identical behaviors since these differ in the α value only. Both models fail in the moderate climatic class. It is also noticed that GG14 does not simulate ET well in the moderate climatic class.

Overall, GG22 has the lowest median and average values of RMSE that are 16.20 and 20.23 mm month^{-1}, respectively. These results indicate that GG22 has the potential to be the best model variation. Based on BIAS for all sites, the lowest average value is 10.55 mm month^{-1} for GG18, while the lowest median value is 7.45 mm month^{-1} for GG20. Comparing the three model variations, both GG18 and GG20 have same R^2 of 0.64 and GG22 produced 0.62. It is therefore reasonable to state that GG18, GG20 and GG22 are the best GG model variations for further consideration.

There is no evidence to suggest that a specific model variation from these three models is superior in a particular climatic class. The climatic class with poorest performance is the moderate class. The reason may be the low number of sites in this class and therefore extreme values such as those of site 24 can dramatically influence the results. In the moderate climatic class, GG22 has the lowest average RMSE and BIAS, however, GG18 and GG20 share the highest average R^2. It is also noted that all three model variations have the following similarities; net radiation is calculated by R_{ASCE}, the complementary relationship is represented by Eq. (1) and the ETW is computed by Eq. (7).

The performance metrics (RMSE, BIAS and R^2) for the three model variations can be compared with uncertainty associated with observed EC-based fluxes to assess the overall accuracy of the methods. For example, Mauder et al. (2007) showed that RMSE and bias of LE sensors normally range from 38 to 61 mm month^{-1} and from −29 to 30 mm month^{-1}, respectively. In another study, it was found that EC data are comparable to weighing lysimeter ET measurements (Castellvi and Snyder, 2010) when the RMSE was 26 mm month^{-1} and R^2 was 0.98. These results indicate the high efficiency of the three model variations, namely GG18, GG20 and GG22, in predicting the actual ET.

4.4 Statistical analysis

The applicability of the three GG model variations, GG18, GG20 and GG22, is further investigated using the analysis of variance (ANOVA) to assess if these three models are similar or not (Berthouex and Brown, 2002). The ANOVA test was used on the time series consisting of 1657 estimates of ET from each model variation and measured ET_{EC}. The average values of ET across the 34 sites are 35.9, 33.8, 33.2 and 32.0 mm month^{-1}, for GG18, GG20, GG22 and measured data, respectively. There is a tendency to underestimate average ET by all three model variations. The reason may be the similarity in structure of the three GG model variations. The ANOVA F test statistic ($F_{V1,V2,1-\mathrm{CI}}$) was computed for the four time series (three GG model variations simulated and ET_{EC} observations) at 95 % confidence level ($V1$ is number of models minus 1, $V2$ is number of measurements minus the number of models, and CI is the confidence interval) and compared to that of the F test of ANOVA. Simply, if the F test is smaller, methods are alike. In this case, $F_{3,1653,0.05}$ is found to be 2.60 (Berthouex and Brown, 2002, Table C in Appendix) while the F test is 4.58. Therefore, it is obvious at 95 % confidence, the averages of the four time series are not equal; however, the test cannot identify which model variation is different compared to the others.

For this purpose, Dunnet's method (Berthouex and Brown, 2002) was used to compare the three GG model variations to the measured ET_{EC} fluxes. Dunnet's method has the advantage to answer two questions: a confidence interval in which average values are alike and the direction of the difference. The results of Dunnet's method showed that at 95 % confidence interval, the average ET is between 32.3 and 39.4 mm month^{-1}. In other words, GG22 is statistically different while the difference in each of the other two model variations is likely to be insignificant. Figure 5 shows the average ET estimates across 33 sites according to the climatic class. At site 4, none of the models can simulate the elevated ET fluxes measured. In general, GG22 underestimates ET as humidity increases. However, the scatter of data around the 1:1 line for most climatic classes is more pronounced with GG18 and GG20. The similarity between GG18 and GG20 is visible because the only difference between the two models is α in the ET_{PT} equation that does not influence the results. In fact, GG18 has two advantages over the other two model variations: it has the closest average ET value to that of the ET_{EC} fluxes and closest to the 1:1 line (see Fig. 5). Hence, GG18 is deemed to be the best from the seven promising GG model variations.

In Fig. 6, the performance metrics of GG18 are shown for each site in the three climatic classes. The R^2 values have a minor increasing trend with humidity. The R^2 values at sites of the wet climatic class mostly lie above the average value and vice versa for the dry climatic class. There is no such trend with RMSE and BIAS. However, the RMSE and BIAS values at most sites of the dry climatic class are below the average value. Again, it is emphasized that site 4 has specific data issues that have to be further inspected. Generally, Fig. 6 demonstrates that GG18 is consistently predicting ET across these 34 sites that have diverse climatic and physical conditions. It also indicates that there is no evidence that the flux footprint (EC tower height) plays a major role or directly impacts the accuracy of the results.

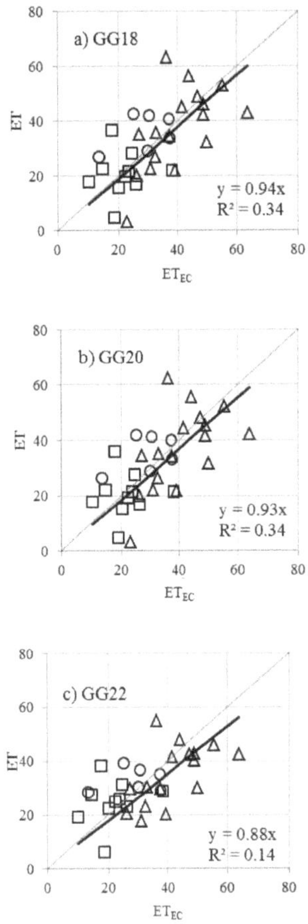

Figure 5. Scatter plots of average ET estimates (mm month^{-1}) for GG18, GG20 and GG22 model variations in comparison to measured ET_{EC} fluxes from 33 sites (all except site 4) in the wet (triangle), moderate (circle) and dry (square) climatic classes.

The average R^2 values of GG18 over the wet, moderate and dry classes are 0.72, 0.61 and 0.52, respectively. Since the ET fluxes differ between the wet and dry climates, the absolute values of RMSE may not be simply compared to each other. Instead, the RMSE value at each site is divided by the average ET_{EC} value shown in Table 1 such that the relative RMSE is computed and compared across all sites. The values of relative RMSE for GG18 range from 0.23 at site 11 to 1.59 at site 34 with an average of 0.69.

4.5 Comparison with recent studies

In this section, the results of the proposed modified complementary method, specifically GG18, are compared to the

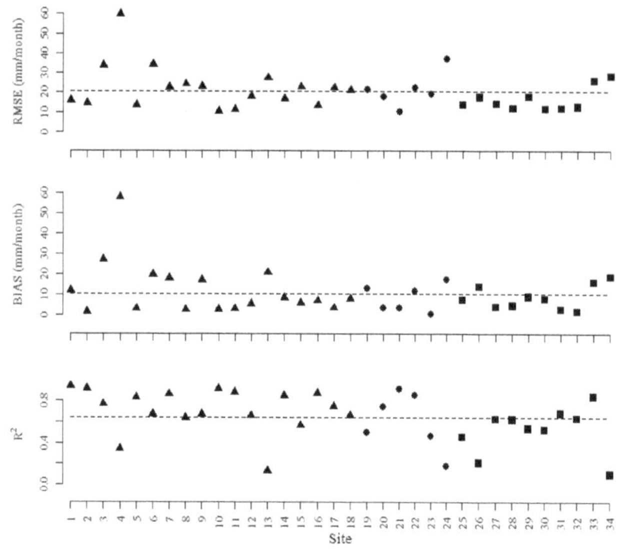

Figure 6. RMSE, BIAS and R^2 of the GG18 model variation at each site in the wet (triangle), moderate (circle) and dry (square) climatic classes and the dashed lines indicate the average values.

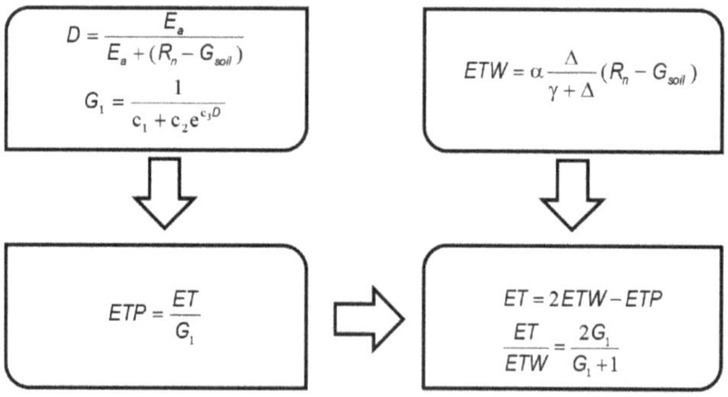

Figure 7. Schematic showing the structure of the proposed GG18 model.

results from recently published studies using the classical methods and original complementary methods.

Suleiman and Crago (2004) estimated hourly ET using radiometric surface temperatures in two grassland sites in Oklahoma and Kansas and validated using EC data. The results showed the RMSE values ranging from 32 to 53 mm month^{-1} while R^2 varied between 0.78 and 0.94. Mu et al. (2007) used data from 19 AmeriFlux EC sites to validate the estimates of a remotely sensed ET using a revised Penman–Monteith equation. The average RMSE, bias and R^2 were 29 mm month^{-1}, −6 mm month^{-1} and 0.76, respectively. When used with 46 AmeriFlux sites (Mu et al., 2011), the results showed average RMSE, absolute bias and R^2 of 26 mm month^{-1}, 10 mm month^{-1} and 0.65, respectively. Kuske (2009) estimated ET using Penman–Monteith and Priestley–Taylor equations and compared estimates to EC data. Both models were significantly overestimating the high ET fluxes and slightly underestimating the low ET fluxes. Thompson et al. (2011) tested ET "null" model that couples the Penman–Monteith equation to a soil moisture model at 14 AmeriFlux sites from which 8 sites are used in the present study. RMSE varied between 56 and 208 mm month^{-1} and therefore, changes were made to further improve the model to produce RMSEs of 34–175 mm month^{-1}.

However, complementary methods to predict ET have not been extensively compared with EC-based ET measurements. With the exception of Ali and Mawdsley (1987), researchers have recently started paying attention to the complementary methods. A monthly ET map using a modified Morton method was produced using MODIS imagery for Hungary (Szilagyi and Kovacs, 2010) and verified using three EC sites. At two sites, R^2 values were 0.79 and 0.80 and bias ranged between −19 and 21 mm month^{-1}. At the third site, however, the authors found a difference of 44 % with the EC measurements due to physical conditions at that particular EC tower (see Szilagyi and Kovacs, 2010). Shifa (2011) examined the wind function of the AA model using data under wet and dry conditions. With the original AA method, RMSE was 17 and 29 mm month^{-1} for the wet and dry conditions, respectively. The author found that the AA method performs best using calibrated wind function coefficients under wet conditions in which RMSE and R^2 were 12 mm month^{-1} and 0.7, respectively. Huntington et al. (2011) tested the AA method using data from arid shrublands at five EC sites in eastern Nevada. It was found that RMSE, R^2 and percent bias were 13 mm month^{-1}, 0.77 and 18 %, respectively. RMSE, R^2 and percent bias of a modified AA method were 11 mm month^{-1}, 0.71 and 1 %, respectively. Han et al. (2011) proposed an enhanced GG model at four sites under different land covers and compared it to the original GG method and EC-based ET data. The enhanced model was better than the original GG method at three sites and RMSE of the enhanced GG model ranged from 4 to 16 mm month^{-1}.

Table 6 shows the results from a set of the above-mentioned studies compared with the results of the proposed GG18 model variation. The comparison shows that the results of the GG18 model variation are equal or better and more reliable considering the wide range of physical and climatic conditions of the 34 global EC sites used in this study. More importantly, the ET estimates of GG18 outperform the estimates of ET of other studies given the minimal cost and data needed to compute reliable regional ET using meteorological data only. Furthermore, GG18 is a single-step method that does not require local calibration and therefore suitable to use in rural river basins with minimal data and monitoring while providing the total water loss from the land surface that is appropriate in water resources planning.

The GG18 model is close to a "universal model" and shows better behavior among the 34 sites and the results are more consistent across the spectrum of climatic classes as shown in Fig. 6. The ET estimates of the GG18 model for the moderate-climate sites are comparable to both wet

Table 6. Comparison of performance of GG18 to the most recently published ET studies.

Source	Method	# of sites	RMSE (mm month^{-1})			BIAS (mm month^{-1})			R^2		
			min	max	mean	min	max	mean	min	max	mean
Present study	GG18	34	10.3	59.9	20.6	0.5	58.1	10.6	0.01	0.94	0.64
Suleiman and Crago (2004)	Radiometric surface temperature	2	32.0	53.4					0.78	0.94	
Mu et al. (2007)	Revised remote sensing and Penman–Monteith	19	7.7	56.4	29.2	2.9	41.1	15.6	0.13	0.96	0.76
Szilagyi and Kovacs (2010)	CRAE method	3	2.6	39.7	15.3	0.0	21.0	8.4	0.79	0.95	0.85
Han et al. (2011)	Enhanced GG method	4	3.7	16.0	10.7				0.82	0.98	0.92
Huntington et al. (2011)	Modified AA method	5			11.0						0.71
Mu et al. (2011)	Modified remote sensing and Penman–Monteith	46	9.4	52.0	25.6	0.3	28.6	10.0	0.02	0.93	0.65
Thompson et al. (2011)	Penman–Monteith and soil moisture model	14	34.0	175.0	94.1						

or dry climatic classes (Fig. 6), and those of the most recent ET studies (Table 6). None of the original (CRAE, AA and GG) methods, however, succeeded to estimate ET under subhumid and Mediterranean climatic classes (see Table 2). The discrepancy is clear when compared to the more extreme conditions, i.e., dry and humid categories (Table 2). For example, one may argue that the average values of performance metrics of the GG18 model are slightly better than those of the original CRAE method that does not need wind measurements. The comparison cannot be made only between the overall average values given by the CRAE method and the GG18 model. There are other statistics (e.g., standard deviation) that show the accuracy (or distribution) of the ET estimates among the 34 sites. As discussed earlier, one major problem of the CRAE method is that it fails to estimate ET under subhumid and Mediterranean climatic classes (see Table 2). Under the diverse physical and climatic conditions, the GG18 model variation is quantitatively and qualitatively outperforming all original complementary methods. The model structure of the proposed GG18 model variation is given in Fig. 7.

One last concern is about the most proper temporal resolution of the GG18 model. It is known that the original AA and GG methods are usually used at daily timescales, while the original CRAE method is typically used at a monthly timescale. The goal of this study is to propose a universal ET model that can be successfully used for data deficit conditions under which daily data are missing or unavailable. It is believed that the regional estimates of ET entail monthly time resolution. Thus, the question now is whether applying the GG18 model at a monthly timescale will change the parameters of the model used in daily time steps or not. In order to answer this question, the proposed GG18 model was applied to a countrywide study of Ghana where daily data were available and climate varies from semiarid in the north to tropical humid in the south (Anayah et al., 2013). The predictions using monthly data from 2000 to 2005 were very much comparable to the daily estimates of the GG18 model. These results suggested that the GG18 model can accommodate both daily and monthly time steps to produce consistent results. The reader may refer to Anayah (2012) and Anayah et al. (2013) for further details.

5 Summary and conclusions

Complementary methods have the potential to predict regional ET using minimal meteorological data. However, prior studies used small data sets representing limited climatic variability and physical conditions that were not successful in improving the methods. A few of the successful studies used locally calibrated parameters that may not have the universal applicability simply due to the two-step approach required to compute ET. In addition, water resources studies require the total water loss from the land surface irrespective of the land use/land class. In this regard, complementary methods provide the distinct advantage over the classical methods that only provide crop ET using detailed input data such as land use/land class, cropping patterns and crop calendar. The state of the complementary methods is such that there is no single methodology consistently used over a wide variety of climatic and physical conditions. This study is aimed at developing a calibration-free universal model using the complementary relationship that requires meteorological data only to predict regional ET.

In this work, 34 global sites with measured ET data via the EC method are used to develop the proposed model using systematic sensitivity analysis conducted with the three original complementary methods. The sites have different climatic and physical conditions to ensure the universal application of the proposed model. The three original complementary methods consisting of CRAE, AA and GG are first evaluated and the need for improvement to all methods is determined. Based on the models' structures, 20 alternative model variations are proposed. The GG method was found to be the most attractive compared to the other two methods and therefore the GG method is further analyzed. The ETW that uses the Priestley–Taylor equation produced 16 GG model variations. Climates of the FLUXNET sites were initially sorted to six climatic classes based on the aridity index proposed by De Martonne (1925). The initial results identified seven promising model variations. Given the complexity of using six different climatic classes, the analysis later reduced this number to three distinct climatic classes consisting of wet, moderate and dry climates. This simplification identified three promising model variations from the earlier seven variations. Statistical analyses conducted via ANOVA testing and the Dunnet method showed that two of the model variations are similar while one GG model variation, GG18, clearly provided a different distribution and results. Therefore the GG18 model variation was considered the best. Also the comparison of results from recent studies showed that the GG18 model variation is capable of producing equal or better results while capturing a wide variety of physical and climatic conditions.

In the proposed GG18 model, net radiation R_n is computed using R_{ASCE} calculated by Allen et al. (2005), which outperforms R_T developed by Morton (1983). It is evident that the simple complementary relationships suggested by Eq. (1) can describe the behavior of ET fluxes better than the more generic complementary relationship of Eq. (8). Most importantly, the predictive power of the GG method (Granger and Gray, 1989) is improved when the ET_{PT} equation is used to calculate ETW. There is a strong indication that the proposed GG18 model can significantly enhance the accuracy of ETW using the GG method and consequently to predict regional ET using meteorological data only and without calibration. Furthermore, this one-step estimation method can reliably estimate ET regardless of the prevailing climatic conditions. Such an estimate will unequivocally lead to reliable predictions of water resources, in particular recharge estimation and impacts due to climate change.

Acknowledgements. This study was supported by the Utah Water Research Laboratory of Utah State University with additional funding available from the International Water Management Institute, Sri Lanka through the Rockefeller Foundation. We thank Drs. Vladimir Smakhtin and Paul Pavelic for their support and encouragement. The authors would like to acknowledge all scientists of FLUXNET agencies and Naama Raz-Yaseef from the University of California, Berkeley, for their willingness to share data. The assistance of Drs. David Stevens, Luis Bastidas and David Tarboton from Utah State University is appreciated. The English of the manuscript had been kindly edited by Naomi Higham.

Edited by: F. Tian

References

Ali, M. F. and Mawdsley, J. A.: Comparison of two recent models for estimating actual ET using only regularly recorded data, J. Hydrol., 93, 257–276, doi:10.1016/0022-1694(87)90099-0, 1987.

Allen, R. G., Pereira, L. S., Raes, D., and Smith, M. (Eds.): Crop Evapotranspiration: Guidelines for Computing Crop Water Requirements, FAO Irrig. Drain., Paper No. 56, Food and Agr. Orgn. of the United Nations, Rome, Italy, 1998.

Allen, R. G., Walter, I. A., Elliot, R., Howell, T., Itenfisu, D., and Jensen, M. (Eds.): The ASCE Standardized Reference Evapotranspiration Equation, Environment and Water Resources Institute of the Am. Soc. Civil Eng. (ASCE), Task Committee on Standardization of Reference Evapotranspiration, Final Rep., ASCE, Reston, VA, USA, 2005.

Anayah, F. M.: Improving complementary methods to predict evapotranspiration for data deficit conditions and global applications under climate change, Ph.D. dissertation, Utah State University, Logan, Utah, USA, 186 pp., 2012.

Anayah, F. M., Kaluarachchi, J. J., Pavelic, P., and Smakhtin, V.: Predicting groundwater recharge in Ghana by estimating evapotranspiration, Water Int., 38, 408–432, doi:10.1080/02508060.2013.821642, 2013.

Baldocchi, D., Falge, E., Gu, L., Olson, R., Hollinger, D., Running, S., Anthoni, P., Bernhofer, C., Davis, K., Evans, R., Fuentes, J., Goldstein, A., Katul, G., Law, B., Lee, X., Malhi, Y., Meyers, T., Munger, W., Oechel, W., Paw, K. T., Pilegaard, K., Schmid, H. P., Valentini, R., Verma, S., Vesala, T., Wilson, K., and Wofsy, S.: FLUXNET: A New Tool to Study the Temporal and Spatial Variability of Ecosystem-Scale Carbon Dioxide, Water Vapor, and Energy Flux Densities, B. Am. Meteorol. Soc., 82, 2415–2434, doi:10.1175/1520-0477(2001)082<2415:FANTTS>2.3.CO;2, 2001.

Berthouex, P. and Brown, L. (Eds.): Statistics for Environmental Engineers, 2nd Edn., Lewis publishers, CRC Press LLC, FL, USA, 2002.

Bouchet, R. J.: Evapotranspiration re'elle et potentielle, signification climatique, in: General Assembly Berkeley, Int. Assoc. Sci. Hydrol., Gentbrugge, Belgium, Publ. No. 62, 134–142, 1963.

Brutsaert, W. and Stricker, H.: An Advection Aridity Approach to Estimate Actual Regional Evaporation, Water Resour. Res., 15, 443–450, doi:10.1029/WR015i002p00443, 1979.

Castellvi, F. and Snyder, R. L.: A comparison between latent heat fluxes over grass using a weighing lysimeter and surface renewal analysis, J. Hydrol., 381, 213–220, doi:10.1016/j.jhydrol.2009.11.043, 2010.

Castellvi, F., Snyder, R. L., and Baldocchi, D. D.: Surface energy-balance closure over rangeland grass using the eddy covariance method and surface renewal analysis, Agr. Forest Meteorol., 148, 1147–1160, doi:10.1016/j.agrformet.2008.02.012, 2008.

Davenport, D. C. and Hudson, J. P.: Changes in evaporation rates along a 17-km transect in the Sudan Gezira, Agr. Meteorol., 4, 339–352, 1967.

De Martonne, E. (Ed.): Traité de Géographie Physique, tome 1, 4th Edn., A. Colin, Paris, 1925.

Doyle, P.: Modelling catchment evaporation: an objective comparison of the Penman and Morton approaches, J. Hydrol., 121, 257–276, 1990.

FAO (Eds.): Arid zone forestry: A guide for field technicians, FAO Conservation Guide 20, Food and Agr. Org. of the United Nations, Rome, Italy, 1989.

Granger, R. J.: A complementary relationship approach for evaporation from nonsaturated surfaces, J. Hydrol., 111, 31–38, doi:10.1016/0022-1694(89)90250-3, 1989.

Granger, R. J.: Partitioning of energy during the snow-free season in the Wolf Creek research basin, in: Proceedings of the Wolf Creek Research Basin – Hydrology, Ecology, Environment workshop, Whitehorse, Yukon, Canada, 5–7 March 1998, 33–44, 1998.

Granger, R. J. and Gray, D. M.: Evaporation from natural nonsaturated surfaces, J. Hydrol., 111, 21–29, doi:10.1016/0022-1694(89)90249-7, 1989.

Han, S., Hu, H., Yang, D., and Tian, F.: A complementary relationship evaporation model referring to the Granger model and the advection-aridity model, Hydrol. Process., 25, 2094–2101, doi:10.1002/hyp.7960, 2011.

Han, S., Hu, H., and Tian, F.: A nonlinear function approach for the normalized complementary relationship evaporation model, Hydrol. Process., 26, 3973–3981, doi:10.1002/hyp.8414, 2012.

Hirano, T., Segah, H., Limin, S., June, T., Tuah, S. J., Kusin, K., Hirata, R., and Osaki, M.: Energy balance of a tropical peat swamp forest in Central Kalimantan, Indonesia, Phyton, 45, 67–71, 2005.

Hobbins, M. T., Ramirez, J. A., Brown, T. C., and Claessens, L. H.: The complementary relationship in estimation of regional evapotranspiration: The Complementary Relationship Areal Evapotranspiration and Advection-Aridity models, Water Resour. Res., 37, 1367–1387, doi:10.1029/2000WR900358, 2001.

Huntington, J. L., Szilagyi, J., Tyler, S. W., and Pohll, G. M.: Evaluating the complementary relationship for estimating evapotranspiration from arid shrublands, Water Resour. Res., 47, W05533, doi:10.1029/2010WR009874, 2011.

Jimenez, C., Prigent, C., Mueller, B., Seneviratne, S. I., McCabe, M. F., Wood, E. F., Rossow, W. B., Balsamo, G., Betts, A. K., Dirmeyer, P. A., Fisher, J. B., Jung, M., Kanamitsu, M., Reichle, R. H., Reichstein, M., Rodell, M., Sheffield, J., Tu, K., and Wang, K.: Global inter-comparison of 12 land surface heat flux estimates, J. Geophys. Res., 116, D02102, doi:10.1029/2010JD014545, 2011.

Kuske, T. J.: Fluxes of energy and water vapour from grazed pasture on a mineral soil in the Waikato, M.Sc. thesis, The Univ. of Waikato, New Zealand, 152 pp., 2009.

Lhomme, J. P. and Guilioni, L.: Comments on some articles about the complementary relationship, J. Hydrol., 323, 1–3, doi:10.1016/j.jhydrol.2005.08.014, 2006.

Lhomme, J. P. and Guilioni, L.: On the link between potential evaporation and regional evaporation from a CBL perspective, Theor. Appl. Climatol., 101, 143–147, doi:10.1007/s00704-009-0211-0, 2010.

Luo, X., Wang, K., Jiang, H., Sun, J., Xu, J., Zhu, Q., and Li, Z.: Advances in research of land surface evapotranspiration at home and abroad, Sci. Cold Arid Reg., 2, 0104–0111, 2010.

Mauder, M., Oncley, S. P., Vogt, R., Weidinger, T., Ribeiro, L., Bernhofer, C., Foken, T., Kosiek, W., De Bruin, H. A. R., and Liu, H.: The energy balance experiment EBEX-2000, Part II: Intercomparison of eddy-covariance sensors and post-field data processing methods, Bound. Layer Meteorol., 123, 29–54, doi:10.1007/s10546-006-9139-4, 2007.

McMahon, T. A., Peel, M. C., Lowe, L., Srikanthan, R., and McVicar, T. R.: Estimating actual, potential, reference crop and pan evaporation using standard meteorological data: a pragmatic synthesis, Hydrol. Earth Syst. Sci., 17, 1331–1363, doi:10.5194/hess-17-1331-2013, 2013.

Morton, F. I.: Operational estimates of areal evapotranspiration and their significance to the science and practice of hydrology, J. Hydrol., 66, 1–76, doi:10.1016/0022-1694(83)90177-4, 1983.

Mu, Q., Heinsch, F. A., Zhao, M., and Running, S. W.: Development of a global evapotranspiration algorithm based on MODIS and global meteorological data, Remote Sens. Environ., 111, 519-536, doi:10.1016/j.rse.2007.04.015, 2007.

Mu, Q., Zhao, M., and Running, S. W.: Improvements to a MODIS global terrestrial evapotranspiration algorithm, Remote Sens. Environ., 115, 1781–1800, doi:10.1016/j.rse.2011.02.019, 2011.

Penman, H. L.: Natural evaporation from open water, bare soil, and grass, Roy. Soc. London, Ser. A, 193, 120–145, doi:10.1098/rspa.1948.0037, 1948.

Penman, H. L.: Evaporation: An introductory survey, Neth. J. Agr. Sci., 4, 9–29, 1956.

Pettijohn, J. C. and Salvucci, G. D.: A New Two-Dimensional Physical Basis for the Complementary Relation between Terrestrial and Pan Evaporation, J. Hydrometeorol., 10, 565–574, doi:10.1175/2008JHM1026.1, 2009.

Priestley, C. H. B. and Taylor, R. J.: On the assessment of surface heat flux and evaporation using large-scale parameters, Mon. Weather Rev., 100, 81–92, doi:10.1175/1520-0493(1972)100<0081:OTAOSH>2.3.CO;2, 1972.

Shifa, Y. B.: Estimation of evapotranspiration using advection aridity approach, M.Sc. thesis, University of Twente, The Netherlands, 52 pp., 2011.

Suleiman, A. and Crago, R.: Hourly and daytime ET from grassland using radiometric surface temperatures, Agron. J., 96, 384–390, 2004.

Szilagyi, J. and Kovacs, A.: Complementary-relationship-based evapotranspiration mapping (cremap) technique for Hungary, Periodica Polytech., 54, 95–100, doi:10.3311/pp.ci.2010-2.04, 2010.

Thompson, S. E., Harman, C. J., Konings, A. G., Sivapalan, M., Neal, A., and Troch, P. A.: Comparative hydrology across AmeriFlux sites: The variable roles of climate, vegetation, and groundwater, Water Resour. Res., 47, W00J07, doi:10.1029/2010WR009797, 2011.

Twine, T., Kustas, W. P., Norman, J. M., Cook, D. R., Houser, P. R., Meyers, T. P., Prueger, J. H., Starks, P. J., and Wesely, M. L.: Correcting eddy-covariance flux underestimates over a grassland, Agr. Forest Meteorol., 103, 279–300, doi:10.1016/S0168-1923(00)00123-4, 2000.

Wang, T., Sammis, W., and Miller, D. R.: Eddy covariance measurements of crop water uses: the energy closure problem and potential solutions, in: Agricultural Water Management Research Trends, Nova Science Publishers Inc., New York, USA, 1–7, 2008.

Wilson, K., Goldstein, A., Falge, E., Aubinet, M., Baldocchi, D., Berbigier, P., Bernhofer, C., Ceulemans, R., Dolman, H., Field, C., Grelle, A., Ibrom, A., Law, B. E., Kowalski, A., Meyers, T., Moncrieff, J., Monson, R., Oechel, W., Tenhumen, J., Valentini, R., and Verma, S.: Energy balance closure at FLUXNET sites, Agr. Forest Meteorol., 113, 223–243, doi:10.1016/S0168-1923(02)00109-0, 2002.

Xu, C. Y. and Singh, V. P.: Evaluation of three complementary relationship evapotranspiration models by water balance approach to estimate actual regional evapotranspiration in different climatic regions, J. Hydrol., 308, 105–121, doi:10.1016/j.jhydrol.2004.10.024, 2005.

2

Evaluation of multiple climate data sources for managing environmental resources in East Africa

Solomon Hailu Gebrechorkos[1,2], **Stephan Hülsmann**[1], **and Christian Bernhofer**[2]

[1]United Nations University Institute for Integrated Management of Material Fluxes and of Resources (UNU-FLORES), 01067 Dresden, Germany

[2]Faculty of Environmental Sciences, Institute of Hydrology and Meteorology, Technische Universität Dresden, 01062 Dresden, Germany

Correspondence: Solomon Hailu Gebrechorkos (gebrechorkos@unu.edu)

Abstract. Managing environmental resources under conditions of climate change and extreme climate events remains among the most challenging research tasks in the field of sustainable development. A particular challenge in many regions such as East Africa is often the lack of sufficiently long-term and spatially representative observed climate data. To overcome this data challenge we used a combination of accessible data sources based on station data, earth observations by remote sensing, and regional climate models. The accuracy of the Africa Rainfall Climatology version 2.0 (ARC2), Climate Hazards Group InfraRed Precipitation (CHIRP), CHIRP with Station data (CHIRPS), Observational-Reanalysis Hybrid (ORH), and regional climate models (RCMs) are evaluated against station data obtained from the respective national weather services and international databases. We did so by performing a comparison in three ways: point to pixel, point to area grid cell average, and stations' average to area grid cell average over 21 regions of East Africa: 17 in Ethiopia, 2 in Kenya, and 2 in Tanzania. We found that the latter method provides better correlation and significantly reduces biases and errors. The correlations were analysed at daily, dekadal (10 days), and monthly resolution for rainfall and maximum and minimum temperature (T_{max} and T_{min}) covering the period of 1983–2005. At a daily timescale, CHIRPS, followed by ARC2 and CHIRP, is the best performing rainfall product compared to ORH, individual RCMs (I-RCM), and RCMs' mean (RCMs). CHIRPS captures the daily rainfall characteristics well, such as average daily rainfall, amount of wet periods, and total rainfall. Compared to CHIRPS, ARC2 showed higher underestimation of the total (−30 %) and daily (−14 %) rainfall. CHIRP, on the other hand, showed higher underestimation of the average daily rainfall (−53 %) and duration of dry periods (−29 %). Overall, the evaluation revealed that in terms of multiple statistical measures used on daily, dekadal, and monthly timescales, CHIRPS, CHIRP, and ARC2 are the best performing rainfall products, while ORH, I-RCM, and RCMs are the worst performing products.

For T_{max} and T_{min}, ORH was identified as the most suitable product compared to I-RCM and RCMs. Our results indicate that CHIRPS (rainfall) and ORH (T_{max} and T_{min}), with higher spatial resolution, should be the preferential data sources to be used for climate change and hydrological studies in areas of East Africa where station data are not accessible.

1 Introduction

In Sub-Saharan Africa (SSA) about 80 % of people living in poverty will continue to depend on the agriculture sector as their major income sources under continuing global change (Dixon et al., 2001; IFPRI, 2009). Unlike in other regions of the world, agricultural activities in SSA are marked by low production, mainly due to poor natural resource management, rainfall amount and variability, economy, and technologies. According to IFPRI (2009), reducing poverty in SSA is becoming more challenging due to rapid population growth and associated decline in the quality and availability of environmental resources (e.g. water and soil). Additionally, food security and livelihoods of people are threatened by the direct impacts of change in climate such as the increasing frequency of extreme events and weather variability impacts on the production and productivity of agricultural lands (Malo et al., 2012). In general, the impact of climate change in Africa ranges from social and economic to health, water, and food security, which is a threat to the lives of Africans (Urama and Ozor, 2010; Gan et al., 2016).

These challenges outlined hold in particular for the eastern parts of SSA, including Ethiopia, Kenya, and Tanzania. The population (> 80 %) mainly depends on agriculture for their livelihood in this region and agriculture-based income contributes 40 % to the country's gross domestic product (GDP) (FAO, 2014). Observed changes in extreme climate events such as recurring droughts and floods have a tremendous impact on the socio-economy of the region (Gebrechorkos et al., 2018). Devastating droughts in SSA linked to the high variability (seasonal and inter-annual) of rainfall (Sheffield et al., 2013) are projected to increase in frequency (IPCC, 2007, 2014; Niang et al., 2014). In addition to the projected impact, the region is already facing significant food security issues and natural-resource-based clashes (UNEP, 2011; World Bank, 2012).

The impacts of future climate change in East Africa vary from region to region. In order to understand the impacts of future climate at the regional and local scale, ground station data with high spatial and temporal resolution are crucial. Regions with poor ground observations are highly vulnerable to climate threats (Wilby and Yu, 2013), which holds particularly for developing countries. In Africa, high-quality climate data from meteorological field stations are scarce (Dinku et al., 2013) and inconsistencies exist between other data products, largely due to a limited number of ground stations, merging and interpolation methods (Huffman et al., 2009; Nikulin et al., 2012; Sylla et al., 2013), limited time resolution, and limited documentation quality. In addition, climate data with high temporal and spatial resolution, even if collected by the national meteorological agencies, are often not available due to data sharing policies. With advancements of technologies and research activities, a number of climate data products from different sources (remote sensing, climate model, and reanalysis) have been produced over the last decades that can fill the data gap particularly for drought-prone regions (Gan et al., 2016) and can be used for hydrological and climate change studies.

Several satellite-based rainfall estimates have been developed over the last decades (Sapiano and Arkin, 2009; Zambrano-Bigiarini et al., 2017). In Africa, a list of rainfall and temperature products are available that can be used for climate change studies, such as the African Rainfall Climatology version 2.0 (ARC2) from the Climate Prediction Center (CPC) of the National Oceanic and Atmospheric Administration (NOAA) with a spatial resolution of 0.1° (Novella et al., 2013) and Climate Hazards Group InfraRed Precipitation (CHIRP), and CHIRP with Station data (CHIRPS) from the Climate Hazard Group (CHG) with a spatial resolution of 0.05° (Funk et al., 2015). In addition, the Multi-Source Weighted-Ensemble Precipitation (MSWEP) (Beck et al., 2017), Tropical Applications of Meteorology using Satellite and ground-based observations (TAMSAT) (Tarnavsky et al., 2014), TAMSAT African Rainfall Climatology And Time series (TARCAT) (Maidment et al., 2014), and data from the Enhancing National Climate Services (ENACTS) initiative (Dinku et al., 2014) are available at varying spatial and temporal resolutions and for longer periods.

As another source of climate information, climate model-derived data are suitable tools for assessing climate variability and change. Globally, reanalysis-based climate products, such as the Observational Reanalysis Hybrid (Sheffield et al., 2006), Modern-Era Retrospective analysis for Research and Applications version 2 (MERRA-2) (Gelaro et al., 2017), and Climate Forecast System Reanalysis (CFSR) (Saha et al., 2010), are widely used for climate and hydrological studies. Moreover, dynamically downscaled data from global climate models (GCMs) are widely used in regional- and local-scale climate studies. Regional climate models (RCMs) produced from dynamically downscaled GCMs provide spatial resolutions that suit end users (Sun et al., 2006). However, downscaling of climate information from GCMs to assess the impact of climate change on environmental resources at regional or smaller scale has only recently been performed, e.g. as dynamical downscaling within the CORDEX community (CORDEX-Africa; see e.g. Abiodun et al., 2016). In Africa (CORDEX-Africa domain) the spatial resolution of RCMs is available at about 0.44° ($\sim$ 50 km) and at varying temporal resolutions. In East Africa, a number of studies have been done with the applications of RCMs for climate studies (Anyah and Semazzi, 2006, 2007; Diro et al., 2011; Endris et al., 2013; Segele et al., 2009).

Before being used as input to different climate or hydrological models, climate data products need to be evaluated against field-based meteorological stations. For studying climate change and climate extremes, data with high accuracy and from long periods (> 30 years) are required. In addition, current hydrological (e.g. Soil and Water Assessment Tool, SWAT; Neitsch et al., 2002) and climate models (e.g. Statistical DownScaling Model, SDSM; Wilby and Dawson, 2004) require daily time series of rainfall and temperature covering long periods. Considering these requirements, concerning lengths of time series and temporal resolution on the one hand and the limited availability of station data on the other hand, it is not surprising that comprehensive evaluations of climate data products, particularly on a daily timescale, are not available for East Africa to the best of our knowledge. However, a few studies are available based on monthly gridded data (e.g. Cattani et al., 2016; Kimani et al., 2017), for limited time periods. Moreover, Kimani et al. (2017) only considered CHIRPS, while a more comprehensive evaluation and comparison of different data sources would be highly desirable. Based on the data requirements of impact models, the climate data products to be included in such an evaluation should be selected or excluded based on high spatial and temporal (i.e. daily) resolution, quality (missing values), and temporal coverage (length of time series), while also taking the results from previous studies (e.g. Cattani et al., 2016; Bayissa et al., 2017; Kimani et al., 2017) in to account.

Therefore, this study aims at comparing and evaluating the available climate data products for Ethiopia, Kenya, and Tanzania at the highest possible spatial and temporal (i.e. daily, for reasons of comparability extended also to dekadal and monthly) resolution against station data using the most widely applied and accepted statistical and graphical evaluation methods. Results of our study will help overcome the data scarcity in the study area, in terms of spatial coverage and temporal resolution gaps of daily, dekadal, and monthly climate data products that can be used for hydrological and climate change and impact studies at a watershed or regional scale. In addition, the data sets can be used for local and regional climate projections using climate models, such as the Statistical DownScaling Model (SDSM) (Wilby and Dawson, 2004).

2 Study area and data

2.1 Study region

The study focuses on the evaluation of daily, dekadal, and monthly climate data sources for regions of East Africa, particularly Ethiopia, Kenya, and Tanzania (Fig. 1). The region is divided by the Great Rift Valley and is topographically one of the most diverse and complex parts of Africa, characterized by multiple rainfall regimes. Generally, the rainfall cycle (climatological annual cycle) in East Africa is linked to the position changes of the intertropical convergence zone (ITCZ) (Endris et al., 2013). Variability in the rainfall patterns in this region is partly induced by local factors such as the heterogeneity of the land surface and complex topography and their interaction with global climate forcing systems. Countries of the region face similar weather and climate variabilities (spatial and temporal variabilities) and increasing temperature and decreasing precipitation trends (Pricope et al., 2013). In addition, all East African countries face similar issues, such as frequent droughts, floods, poverty, and a lack of clean and adequate water supply. The conditions could worsen in the near future due to climate change; therefore, sustainable adaptation and mitigation strategies are required, which rely on advanced climate and hydrological models and the respective data inputs.

2.2 Data sets

The reference data sets used for the evaluation of multiple data products in this study are based on daily rainfall, maximum temperature (T_{max}), and minimum temperature (T_{min}) derived from 332 rain gauges and synoptic stations. Station data for Ethiopia were provided by the National Meteorological Agency (NMA) of Ethiopia for the period 1954–2016. The daily data provided by NMA were carefully and extensively checked for their quality and some missing data were filled in from hard copies. For Kenya and Tanzania, the global summary of the day available at the National Climate Data Center (NCDC) (https://www.ncdc.noaa.gov/, last access: 14 March 2017) is used. For evaluation, based on the criteria outlined above, we considered satellite-based rainfall estimates, Observational-Reanalysis Hybrid (ORH), and a historical period of RCMs driven by GCMs to be compared against field-based meteorological stations. The three satellite-based rainfall estimates fulfilling all criteria are the African Rainfall Climatology Version 2.0 (ARC2) (Novella et al., 2013), the Climate Hazards Group InfraRed Precipitation (CHIRP) and CHIRP with Station data version 2 (CHIRPS) (Funk et al., 2015). Not included was, for example, TAMSAT, which is available at higher spatial and temporal resolution and for a longer time period but contains considerable data gaps (Maidment et al., 2017, https://icdc.cen.uni-hamburg.de/1/daten/atmosphere/tamsat-rainfall-africa/, last access: 7 January 2017) during the evaluation period. ENACTS and TARCAT are available only on dekadal (10 days) timescales. In addition, MERRA-2 and CFS-R are not included in this study due to their coarse spatial resolution compared to the other reanalysis products (i.e. ORH).

ARC2 is the second version of the ARC and is compatible with the algorithm of the Rainfall Estimation Version 2 (RFE 2.0) (Novella et al., 2013). The product is a composite of 3-hourly geostationary infrared data, which makes it different from RFE, centred over Africa provided by the European Organization for the Exploitation of Meteorological Satellites (EUMETSAT) and quality-controlled daily rainfall records acquired from the Global Telecommunication System (GTS) gauges. ARC2 is consistent with the historical data sets of the Climate Prediction Center Merged Analysis of Precipitation (CMAP) (Xie and Arkin, 1997) and Global Precipitation Climatology Project (GPCP) (Novella et al., 2013). The data set is updated regularly and it is available at a spatial resolution of 0.1° covering the period from 1983 to present. ARC2 is available at the International Research Institute climate data library (IRI/LDEO, 2016).

CHIRPS is a semi-global rainfall product designed for drought monitoring and global environmental changes (Funk et al., 2015). The product provides daily, pentadal, dekadal, and monthly data at a 0.05° spatial resolution available at the Climate Hazards Group (CHG; ftp://ftp.chg.ucsb.edu/pub/org/chg/products) and the International Research Institute climate data library (IRI/LDE, 2016). CHIRPS combines a 0.05° resolution of satellite images and data from ground stations to form a gridded rainfall time series. Station data (see also below) are used to produce a preliminary 2-day rainfall product by blending data from sparsely located GTS

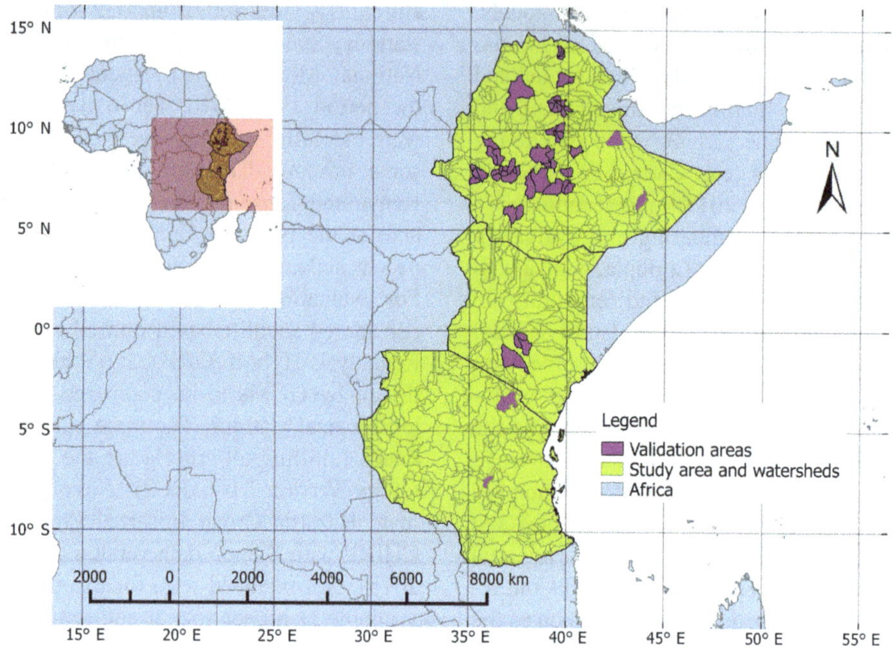

Figure 1. Map of Africa and study regions (Ethiopia, Kenya, and Tanzania) with data validation areas (EthioShed1–17, KenShed1 and 2, and TanzShed1&2). The basins are retrieved from the WaterBase global data portal (http://www.waterbase.org/, last access: 18 January 2017).

gauges with rainfall estimates retrieved from the cold cloud duration (CCD) at every pentad. In addition, the final product is developed by blending the best available monthly and pentadal station data with the monthly and pentadal rainfall estimates retrieved from the CCD, respectively, to produce a gridded rainfall product (Funk et al., 2015). The development process of CHIRPS includes the 0.05° monthly precipitation climatology (CHPclim), satellite-only Climate Hazards Group InfraRed Precipitation (CHIRP), and station blending techniques. CHIRP is available at the Climate Hazards Group (CHG, 2017). The second version of CHIRPS, which is updated regularly, provides an improved daily rainfall time series (1981–present) with a spatial resolution of 0.05° ranging from 50° S to 50° N (and all longitudes) (Funk et al., 2015). The development process of CHIRPS and its application in drought monitoring in Africa (e.g. Ethiopia) is explained in detail by Funk et al. (2015). CHIRPS is not only used for drought monitoring, but also for other global environmental applications (Zambrano-Bigiarini et al., 2017), water resource management, and climate dynamics (Ceccherini et al., 2015; Deblauwe et al., 2016). ORH is a global (Sheffield et al., 2006) and regional (Northern/West/East Africa) (Chaney et al., 2014) 3-hourly, daily, and monthly meteorological data set covering the period between 1901 and 2012. ORH is developed by a spatial downscaling of the NCEP–NCAR reanalysis (Kalnay et al., 1996) up to a spatial resolution of 0.1° using a bilinear interpolation. ORH merges multiple data products such as the NASA Langley Surface Radiation Budget (SRB), the monthly temperature data from the University of East Anglia Climate Research Unit (CRU), Tropical Rainfall Measuring Mission Multi-satellite Precipitation Analysis (TMPA) (Huffman et al., 2007), and other observational-based rainfall products (Chaney et al., 2014). The spatial downscaling of ORH is done with the inclusion of changes in elevation and it is evaluated against ground stations (global summary of the day) available at the US National Climatic Data Center (NCDC). ORH is corrected for temporal inhomogeneity and biases, and random errors are omitted through assimilation with quality-controlled and gap-filled ground station data available at NCDC (https://www.ncdc.noaa.gov/, last access: 28 March 2017) as a global summary of the day (Chaney et al., 2014). These data are freely available from the Terrestrial Hydrology Research Group, Princeton University (http://hydrology.princeton.edu, last access: 12 May 2016; Terrestrial Hydrology Research Group Princeton University, 2016). Even though ORH is not updated regularly, it has been widely used in climate and hydrological studies (e.g. Troy et al., 2011; Wang et al., 2011; Demaria et al., 2012; Sheffield et al., 2014)

Compared to the other rainfall products, monthly ground station data from Ethiopia, Kenya, and Tanzania are included in CHIRPS. Evaluating CHIRPS based on ground station data might thus raise concerns about the independence of data. However, not all stations used in this study are included in CHIRPS and in general the stations are not consistently used in the development process of CHIRPS. In addition, the station data used in CHIRPS are mainly a monthly total from a limited number of stations. For example, in Ethiopia, where all station data originate from NMA, during January 1983–February 1983 and August 2005–

December 2005, the monthly stations used in CHIRPS declined from 139 to 132 and from 175 to 169, respectively. In 2015, the number of stations included in CHIRPS even declined to below 10. Moreover, in Kenya and Tanzania, during the period of January 1983–December 2005 the number of stations used in CHIRPS declined from 142 to 62 and from 171 to 55, respectively (ftp://chg-ftpout.geog.ucsb.edu/pub/org/chg/products/CHIRPS-2.0/diagnostics/). Besides the difference in temporal resolution (monthly vs. daily) and the number of stations between station data included in CHIRPS and the validation data set, the latter deviated from the former since we used original data provided by NMA (Ethiopia) which were quality-controlled and extended by adding data from hard copies. Overall, while not fully independent, the relation between CHIRPS and the validation data set should be weak, besides the fact that there is no other (fully independent) validation data set available.

Historical data (control model runs) of the CORDEX RCMs are also used as a potential source for rainfall, T_{max}, and T_{min} data. RCMs are climate models with a higher spatial resolution compared to GCMs. The driving data of RCMs are derived from GCMs or reanalysis data and can include greenhouse gases (GHGs) and aerosol forcing. Compared to GCMs, RCMs consider local factors such as complex topography and land cover inhomogeneity in a physically based manner (IPCC, 2007). In Africa, dynamical downscaling was performed in a large effort within the CORDEX community (CORDEX-Africa). Within CORDEX-Africa the continent's climate was dynamically modelled by an international consortium, providing a spatial resolution of about 50 km. According to the IPCC report (2007), RCMs can be used for a wide range of applications such as climate change studies. Following the recommendation of Endris et al. (2015), the historical data derived from two CORDEX RCMs, RCA (Samuelsson et al., 2011), and COSMO-CLM or CCLM (Baldauf et al., 2011), driven by HadGEM2-ES (MOHC, United Kingdom), MPI-ESM-LR (MPI, Germany), and GFDL-ESM2M (NOAA/GFDL, United States) are used. Rainfall, T_{max}, and T_{min} products of both RCMs are retrieved from the Earth System Grid Federation (ESGF, 2016) data portal.

3 Methodology

3.1 Selection of validation areas and ground stations

The evaluation of multiple daily, dekadal (10 days), and monthly rainfall, T_{max}, and T_{min} products were conducted for selected basins of Ethiopia (EthioShed1–EthioShed17), Kenya (KenShed1 and KenShed2), and Tanzania (TanzShed1 and TanzShed2) (Fig. 1). The polygons in Fig. 1 are river basins retrieved from the global river basins available in the WaterBase data portal hosted by the United Nations University (UNU-INWEH: http://www.waterbase.org/, last access: 18 January 2017). In most regions of Africa not only are the density and availability of field-based meteorological stations limited, but their accessibility is very restricted for many reasons. For this study, it was only possible to get daily station data from the National Meteorological Agency (NMA) of Ethiopia with a reasonable spatial and temporal coverage. Therefore, the selection of validation areas is based on the availability, quality, and density of field-based meteorological stations during the period of 1983–2005. It was almost impossible to find multiple stations in one satellite grid cell. For Kenya and Tanzania, therefore, stations with more than 10 years (> 50 % of the study period) were included for evaluation (Table 1).

The quality of selected stations was checked and extremely high rainfall records during dry seasons, such as daily rainfall of > 480 mm preceding and following dry days, were excluded. Finally, a total of 132 stations were found suitable for comparison, 2 to 12 stations located in the validation areas. In addition to these stations in the validation areas, 78 stations, randomly distributed over the region, are used to compare on an individual basis with the rainfall and temperature products. Compared to Kenya and Tanzania, the quality, continuity, and spatial and temporal coverage of stations were better in Ethiopia and only stations with missing values of less than 20 % were considered. The availability of multiple stations in a validation area helps to check the quality of individual stations by using methods such as the double mass curve (Vernimmen et al., 2012) and allows for replacement of missing values of one station from a nearby station.

3.2 Comparing ground data with satellite, observational reanalysis, and climate model-based data

The most commonly used method to compare ground observations with other data products such as satellite-based rainfall estimates and climate model outputs is point (station) to pixel comparison. When comparing daily rainfall, particularly in very complex topography, on a point to pixel basis it can be challenging to acquire reasonable agreement. Therefore, in this study we used point to pixel, point to area grid cell average, and stations' average to area grid cell average comparison to evaluate the accuracy of each product. The area grid cell average is the average number of pixels covering the basin or the validation area. Similarly, the station average is used to indicate the average value of the stations inside the validation area. Therefore, during the comparison process, the individual or the stations' average is compared to the area grid cell average of the product. The most commonly used statistical methods such as the Pearson correlation coefficient (CC), bias, relative bias (R_{bias}), mean absolute error (MAE), root mean square error (RMSE), and index of agreement (IA) (Cohen Liechti et al., 2012; Daren Harmel and Smith, 2007; Moazami et al., 2013) are used. CC (Eq. 1) is

Table 1. General characteristics of selected validation areas and meteorological stations covering the time period 1983–2005.

Validation areas/basins	Basin area (km^2)	Average area elevation (m)	Number of stations	Average station elevation (m)	Average annual rainfall (mm)	Average T_{max}/T_{min}
EthioShed1	8980	1516	9	1881	1758	26.3/13.6
EthioShed2	12 828	2279	12	2009	968.6	25.9/11.4
EthioShed3	15 123	2192	9	2104	1202.6	25.3/11.7
EthioShed4	8323	2180	7	1954	994.42	31.8/16.5
EthioShed5	5625	1720	10	1800	1039.1	26.4/13.4
EthioShed6	11 204	2830	8	2510	1168.7	22.1/8.0
EthioShed7	12 445	1830	8	1973	1524.53	25.7/12.4
EthioShed8	6522	1930	5	2022	1628.35	26.0/14.0
EthioShed9	4666	1526	4	1738	578.4	28.0/14.4
EthioShed10	5986	2520	8	2580	1133.1	21.2/9.3
EthioShed11	11 496	1256	7	1468	945	27.4/15.2
EthioShed12	3868	520	2	400	343.8	34.1/22.3
EthioShed13	4934	1301	4	2413	588	26.2/13.1
EthioShed14	2835	1360	4	1239	706	31.8/16.5
EthioShed15	1121	2307	4	2183	495	24.3/11.1
EthioShed16	3012	2102	5	2148	1110	26.0/11.8
EthioShed17	9909	1998	12	2056	2075	23.8/10.2
KenShed1	11 712	1980	4	1024	1156.1	25/13.5
KenShed2	7861	2328	3	1602	1418.6	24/13.2
TanzShed1	8092	1244	3	1137	1137.8	28.7/17.5
TanzShed2	2154	1097	3	1428	1136.2	28.2/17.8

applied to evaluate the agreement of individual products (P) to station data (O). A value of CC close to 1 shows a perfect positive fit between the products and station data.

$$\mathrm{CC} = \frac{\sum_{i=1}^{N}\left(P_i - \overline{P}\right)\cdot\left(O_i - \overline{O}\right)}{\sqrt{\sum_{i=1}^{N}\left(P_i - \overline{P}\right)^2}\cdot\sqrt{\sum_{i=1}^{N}(O_i - \overline{O})^2}} \quad (1)$$

The average differences and systematic bias of each product are given as bias (Eq. 2) and R_{bias} (Eq. 3). Bias can be positive (overestimation) or negative (underestimation) according to the accuracy of each product.

$$\mathrm{Bias} = \frac{\sum(P_i - O_i)}{N} \quad (2)$$

$$R_{\mathrm{bias}} = \frac{\sum_{i=1}^{N}(P_i - O_i)}{\sum_{i=1}^{N}O_i} \times 100 \quad (3)$$

The MAE and RMSE (Eqs. 4 and 5) are well known and accepted indicators of goodness of fit, which shows the differences between ground observations and model or other product outputs (Legates and McCabe, 1999).

$$\mathrm{MAE} = \frac{\sum_{i=1}^{N}|O_i - P_i|}{N} \quad (4)$$

$$\mathrm{RMSE} = \sqrt{\frac{\sum_{i=1}^{N}(O_i - P_i)^2}{N}} \quad (5)$$

The IA (Willmott, 1981) is another widely used indicator of goodness of fit between observed and model output. IA (Eq. 6) describes how much of the model or product output (rainfall, T_{max}, and T_{min} products) are error-free compared to the ground observations.

$$\mathrm{IA} = \frac{\sum(P_i - O_i)^2}{\sum(|P - \overline{O}| + |O - \overline{O}|)} \quad (6)$$

In addition to the above statistical methods, the Taylor diagram (Taylor, 2001) is used to summarize the statistical relationship between ground station data and the products for rainfall, T_{max}, and T_{min}. In this diagram, the relationships between the two fields are explained by the correlation coefficient (R), centred mean square (rms) difference (E'), and standard deviation (σ). The diagram is useful for evaluating the accuracy of multiple data sources or model output against a reference or observational data (IPCC, 2001). A single point on the diagram displays three statistical values (R, E', and σ) and their relationship is given by Eq. (7).

$$E'^2 = \sigma_{\mathrm{f}}^2 + \sigma_{\mathrm{r}}^2 - 2\sigma_{\mathrm{f}}\sigma_{\mathrm{r}}R, \quad (7)$$

where σ_{f}^2 and σ_{r}^2 are the variance of the model and observation fields and R is the correlation coefficient between the two fields (Eq. 8).

$$R = \frac{\frac{1}{N}\sum_{n=1}^{N}(f_n - \overline{f})(r_n - \overline{r})}{\sigma_{\mathrm{f}}\sigma_{\mathrm{r}}} \quad (8)$$

In the diagram, the distance from the reference point (observed data) is given as the centred rms difference of the two fields (Eq. 9). A model with no error would show a perfect correlation to the observations.

$$E'^2 = \frac{1}{N}\sum_{n=1}^{N}[(f_n - \overline{f}) - (r_n - \overline{r})]^2, \quad (9)$$

where f is the test (e.g. model or satellite) field and r is the reference (observed) field, whereas σ_f and σ_r are the standard deviations of the model and reference fields (Eqs. 10a and b).

$$\sigma_f = \sqrt{\frac{1}{N}\sum_{n=1}^{N}(f_n - \overline{f})^2} \quad (10)$$

$$\sigma_r = \sqrt{\frac{1}{N}\sum_{n=1}^{N}(r_n - \overline{r})^2} \quad (11)$$

Additionally, rainfall characteristics such as the number of wet days, duration and amount of wet periods, duration of dry periods, and daily and total rainfall are used to evaluate the accuracy of individual rainfall products by comparing to the observed data. Rainfall characteristics are widely used indicators in rainfall modelling (Wilby and Dawson, 2007; Jebari et al., 2012) and include the number of wet days (days yr^{-1}), which is the count of days with rainfall per year; duration (days) of wet and dry periods, indicating the average number of consecutive wet and dry days during the study period; and the amount of wet periods (mm), indicating the amount of rainfall observed during the identified wet period.

4 Results

4.1 Validation of satellite, observational reanalysis, and climate model-based products

The correlation of each rainfall product with station data is summarized in Fig. 2 for the 21 validation areas, and details (scatter plots) are provided in the Supplement (Fig. S1). The results show that in most of the validation areas, CHIRPS, followed by CHIRP and ARC2, is more strongly correlated with station data compared to ORH and the individual RCMs (I-RCM). In addition to the lower correlation, ORH and I-RCM showed large biases in all the validation areas (Fig. S1).

Based on the results in Fig. 2 and the scatter plots provided in the Supplement (Fig. S1), CHIRPS and CHIRP are the most accurate rainfall products, with higher correlation and lower biases, and ARC2 and ORH are the second best products. I-RCM and RCMs' mean (RCMs) correlate weakly in most of the validation areas. In addition, I-RCM (not shown in Fig. 2 but provided in the Supplement Table S1) and RCMs show a strong over- and underestimation of monthly rainfall compared to the other products. In EthioShed1, for example, CHIRPS and CHIRP are shown to be the most accurate products, while ARC2 and ORH showed higher dispersion above and below the regression line (see Fig. S1). Similarly, in EthioShed4 both CHIRP and CHIRPS have an equal R^2, but in terms of biases (points below and above the regression line) CHIRPS performed better. The observed biases in CHIRP and higher correlations in CHIRPS in all the validation areas highlight the role of the station-satellite data blending techniques. Compared to other validation areas, the agreement of products in EthioShed16 is comparably weak, and CHIRPS and CHIRP showed the higher correlation (0.7) compared to ARC2, ORH, and RCMs.

As for the daily, dekadal, and monthly resolution, the comparison is performed in three ways: point to pixel, point to area grid cell average, and stations' average to area grid cell average using the methods described in Sect. 3.2. An explanatory example is given in Table 2, using stations of EthioShde1, displaying the difference in comparing products through point (station) to pixel, point to area grid cell average, and stations' average to area grid cell average. The agreement of each product with station data on a daily timescale and in the point to pixel comparison is weak, with significantly higher biases and errors. For rainfall, in general, the latter method, stations' average to area grid cell average, provides better correlation, higher index of agreement, and lower biases and errors. Compared to the point to pixel method, the stations' average to area grid cell average improves the correlations of ARC2, CHIRP, and CHIRPS by 81.3 %, 65.7 %, and 8 %, respectively. In addition to the correlation, the method reduces the RMSE by more than 66 %. Compared to ARC2 and CHIRP (Table 2), CHIRPS gives a significantly higher correlation and IA and lower biases and RMSE. During area averaging, extremely high rainfall events obtained for a location from the various data products are levelled off by averaging, and this makes the product more representative of the area. In most of the rainfall products, there are occasionally higher daily rainfall values recorded and the averaging removes these extremes, which are much higher than the observed station data in the area. Compared to the point to pixel method, the second method, the point to area grid cell average, provides a reasonable correlation.

For T_{max} and T_{min}, only ORH, I-RCM, and RCMs are compared with station data. For 21 validation areas, ORH proved to be the most accurate product for both T_{max} (Fig. 4) and T_{min} (Fig. 5). In comparison to I-RCM and RCMs, ORH showed a significantly higher correlation and lower biases and errors in most of the validation areas. In 7 of the 21 validation areas, RCMs showed a higher correlation in T_{max} than ORH and I-RCM. However, for T_{min}, ORH in 20 of the 21 validation areas showed a higher correlation. In general, I-RCM and RCMs showed higher RMSE and biases in most of the validation areas compared to ORH. Next to ORH and compared to I-RCM, RCMs appeared to be the best data source, particularly for T_{max}. RCMs showed a relatively higher correlation and lower biases and errors compared to I-RCM in most of the validation areas.

The agreement of each product increases with decreasing temporal resolution, from daily to dekadal and monthly resolutions. Including the historical data of each individual RCM, all RCMs, and ORH, the overall comparison using some of the statistical methods is summarized in Figs. 3, 4, and 5

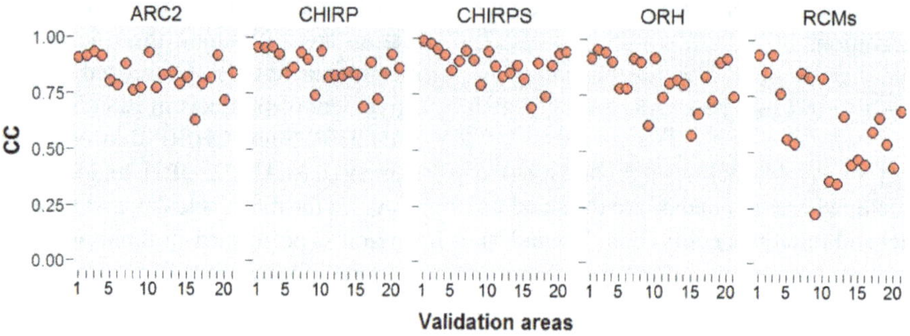

Figure 2. The correlation coefficient (CC) values between the observed monthly rainfall and data from ARC2, CHIRP, CHIRPS, ORH, and RCMs for the 21 validation areas (1–17 for EthioShed1–17, 18–19 for KenShed1 and 2, and 20–21 for TanShed1–2; see Table 1) covering the period of 1983–2005.

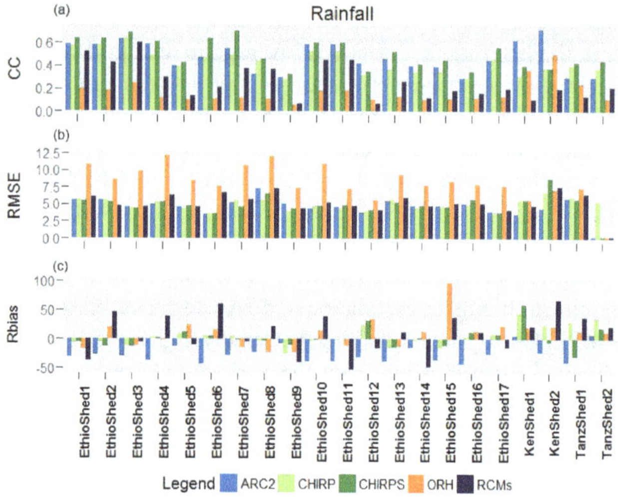

Figure 3. Statistical evaluation of daily rainfall retrieved from ARC2, CHIRP, CHIRPS, ORH, and RCMs against ground observations: CC **(a)**, RMSE **(b)**, and R_{bias} **(c)** during the period of 1983–2005 for 21 validation areas of East Africa (see Table 1).

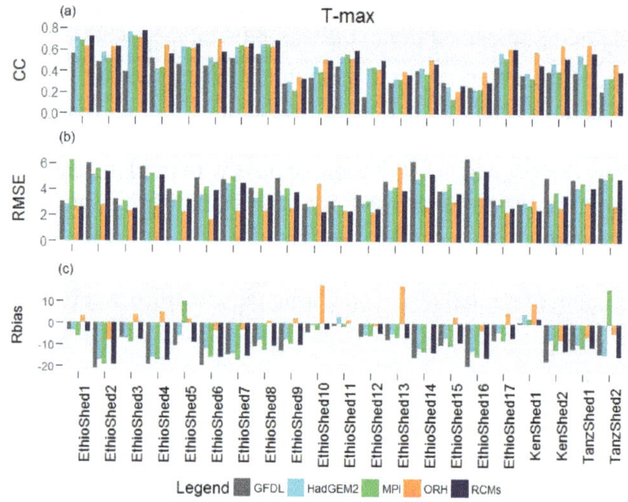

Figure 4. Statistical evaluation of daily T_{max} retrieved from GFDL, HadGEM2, MPI, ORH, and RCMs against ground observations: CC **(a)**, RMSE **(b)**, and R_{bias} **(c)** over the period of 1983–2005 for 21 validation areas of East Africa.

for rainfall, T_{max}, and T_{min}, respectively. The evaluation of each rainfall product (ARC2, CHIRP, CHIRPS, ORH, and RCMs) showed a different degree of agreement with station data (Fig. 3). The performance of the individual RCMs (I-RCM) for all the validation areas is provided in the Supplement (Table S1). At a daily timescale, CHIRPS, followed by ARC2 and CHIRP, proved to be the most accurate rainfall product compared to ORH, I-RCM, and RCMs in all the validation areas. In general, out of the 21 validation areas, CHIRPS, ARC2, and CHIRP showed a higher correlation in 17, 3, and 1 validation areas, respectively. In addition to the higher correlation, CHIRPS, CHIRP, and ARC showed lower RMSE than ORH, I-RCM, and RCMs. Similarly, CHIRPS and CHIRP showed lower biases than observed in ARC2, ORH, I-RCM, and RCMs in most of the validation areas.

Moreover, the observed daily rainfall characteristics are well represented by CHIRPS and ARC2 in most of the validation areas (Table S2). On average, over the 21 validation areas, CHIRPS captures the number of wet days (−0.17 % deviation), average duration of wet (−13.4 % deviation) and dry periods (−17.6 % deviation), total rainfall (−4.5 % deviation), average amount of wet periods (−17 % deviation), and average daily rainfall (−7.7 % deviation) well (Table 3). Next to CHIRPS, ARC2 showed higher agreement in producing the average duration of wet (−20 % deviation) and dry periods (11.3 % deviation) and average amount of wet periods (−38 % deviation). CHIRP, on the other hand, showed a higher agreement in the total amount of rainfall with a 2.75 % deviation, which is higher than CHIRPS, ARC2, ORH, I-RCM, and RCMs. In contrast, ARC2 and GFDL showed higher underestimation (−34.7 % deviation) and overestimation (27.2 % deviation), respectively, of the total amount of rainfall compared to the other products. In addition, ARC2 showed a higher underestimation of the number of wet days

Table 2. An example of the statistics used to compare ground rainfall data with satellite products (e.g. ARC2, CHIRP, and CHIRPS) in EthioShed1. The three modes of comparison are compared based on a range of statistical variables (Sect. 3.2). The point (station) to area grid cell average is computed by comparing individual stations to the area grid cell average of each product. Best fit of the last three rows is indicated in bold; best fits of the nine stations are also highlighted.

Station	ARC2						CHIRP						CHIRPS					
	CC	Bias	R_{bias}	MAE	RMSE	IA	CC	Bias	R_{bias}	MAE	RMSE	IA	CC	Bias	R_{bias}	MAE	RMSE	IA
Anger	0.32	−0.55	−14.7	4.64	10.48	0.52	0.37	0.45	11.9	4.87	9.58	0.56	0.40	−0.15	−4.1	4.40	9.40	0.6
Arijo	0.29	−1.25	−37.6	5.02	10.14	0.53	0.23	0.12	3.4	5.94	10.29	0.49	0.37	−0.42	−12.5	5.08	9.33	0.61
Bedele	0.33	−1.40	−28.7	5.00	10.67	0.55	0.34	−0.45	−9.3	5.21	9.32	0.55	0.41	−0.54	−11.1	4.96	9.30	0.62
Dedesa	0.29	−0.71	−18.0	4.77	10.53	0.51	0.28	0.13	3.4	5.16	10.49	0.50	0.34	−0.23	−5.8	4.81	9.92	0.55
Gimbi	0.32	−1.03	−23.9	5.10	10.78	0.55	0.39	0.12	2.8	5.17	9.85	0.59	0.42	−0.20	−4.5	4.91	9.76	0.64
Nekemt	0.44	−1.20	−23.4	4.71	10.69	0.64	0.38	0.02	0.3	5.79	10.96	0.59	0.41	−0.75	−14.6	5.30	10.44	0.62
Alge	0.32	−1.08	−27.6	5.13	10.2	0.55	0.36	−0.45	−11.7	4.99	9.12	0.56	0.37	−0.36	−9.4	5.36	10.01	0.6
Ayira	0.3	−1.02	−26.1	5.18	10.68	0.53	0.40	−0.18	−4.80	4.95	8.98	0.60	0.36	−0.37	−9.6	5.41	10.07	0.6
Hurma	0.31	−1.01	−25.7	5.20	10.44	0.54	0.38	−0.56	−14.5	4.59	8.86	0.54	0.37	−0.60	−15.8	5.15	9.59	0.6
Average of point–pixel	0.32	−1.03	−25.1	**4.97**	10.51	0.55	0.35	−0.09	**−2.06**	5.19	**9.72**	0.55	**0.38**	**−0.40**	−9.7	5.04	9.76	**0.6**
Average of point–area grid cell average	0.37	−1.1	−26.8	**4.82**	9.59	0.53	0.37	−0.21	**−5.5**	10.18	9.51	0.54	**0.40**	**−0.46**	−11.5	4.86	**9.38**	**0.58**
Stations' average–area grid cell average	0.58	−1.3	−27.3	5.66	3.22	0.74	0.58	−0.27	**−5.7**	5.6	3.29	0.75	**0.64**	**−0.59**	−12.0	**5.38**	**3.08**	**0.79**

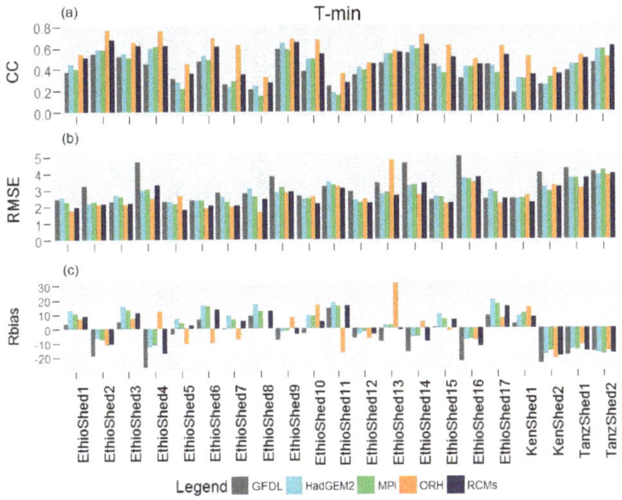

Figure 5. Statistical evaluation of daily T_{min} retrieved from GFDL, HadGEM2, MPI, ORH, and RCMs against ground observations: CC **(a)**, RMSE **(b)**, and R_{bias} **(c)** over the period of 1983–2005 for 21 validation areas of East Africa (see Table 1).

(−15.1 % deviation) and average daily rainfall (−23.8 % deviation) compared to CHIRPS and ORH. CHIRP, on the other hand, showed higher overestimation of the number of wet days and duration and amount of wet periods (> 59.7 % deviation) and underestimation of the duration of dry periods and average daily rainfall (−62 % deviation) compared to the other products. Moreover, RCMs, next to CHIRP, showed higher overestimation of the number of wet days and duration and amount of wet periods (> 44.9 % deviation) and total rainfall amount (11.4 % deviation) and underestimation of the average duration of dry periods and daily rainfall by about 41 %. In general, the observed rainfall characteristics are captured well by CHIRPS, with a percentage difference from the observations of −0.17 % to −17.6 % for the number of wet days and duration of dry periods, respectively, compared to CHIRP, ARC2, ORH, I-RCM, and RCMs (Table 3).

The average daily rainfall of the study region retrieved from CHIRPS is displayed in Fig. 6 for the study period 1983–2005. Using CHIRPS as reference rainfall data, the absolute difference is presented in Fig. 6 for ARC2, CHIRP, ORH, and individual RCMs (HadGEM2, GFDL, and MPI). The average daily rainfall from the individual RCMs (I-RCM) shows large discrepancies from CHIRPS compared to CHIRP, ARC2 and ORH. The comparison at a daily timescale, particularly of rainfall, is challenging and more emphasis is given to this evaluation compared to dekadal and monthly resolutions. RCMs (RCA4 and CCLM) driven by HadGEM2-ES (HadGEM2), MPI-ESM-LR (MPI), and GFDL-ESM2M (GFDL) are used in this study. For RCMs driven by all GCMs, the average is used. The daily rainfall, T_{max}, and T_{min} maps of GFDL display the result of a single RCM (RCA4) driven by GFDL-ESM2M for the period of 1983–2005. Higher and lower average daily rainfall values are displayed by GFDL and ORH, respectively (Fig. 6). However, all the products showed a similar tendency in capturing the daily rainfall distribution: higher in the western and lower in the eastern part of the region. In addition, the average daily T_{max} and T_{min} (Fig. 7) of the region show relatively higher disagreement between ORH and I-RCM, which is given as an absolute difference from ORH. Even though I-RCM shows high deviation from ORH, it showed higher agreement in Ethiopia, Kenya, and Tanzania for T_{max} and T_{min}.

Table 3. Summary of daily rainfall characteristics retrieved from multiple rainfall products and averaged over the validation areas of Ethiopia, Kenya, and Tanzania. Values in brackets give the deviation from the observed value (%). The value which comes closest to the observed value is highlighted in bold.

Rainfall characteristics	Obs.	ARC2	CHIRP	CHIRPS	ORH	HadGEM2	MPI	GFDL	RCMs
Number of wet days (days yr^{-1})	189.58	162.98 (−15.1)	351.06 (59.7)	**189.26** (−0.17)	192.14 (1.34)	205.08 (7.85)	243.55 (24.92)	210.42 (10.42)	299.36 (44.9)
Average duration of wet periods (days)	5.86	4.78 (−20)	167.96 (186)	**5.13** (−13.4)	3.02 (−64)	11.70 (66.5)	12.17 (69.9)	9.37 (46)	21.36 (113.9)
Total amount of precipitation (mm yr^{-1})	953.63	671.62 (−34.7)	**980.24** (2.75)	912.0 (−4.5)	1027.02 (7.41)	841.73 (−12.5)	1055.7 (10.2)	1253.38 (27.2)	1068.6 (11.4)
Average amount of wet periods (mm)	30.20	20.56 (−38)	498.43 (177)	**25.46** (−17)	15.64 (−63.5)	50.12 (69.6)	55.45 (59)	59.64 (65.6)	78.88 (89.3)
Average duration of dry periods (days)	5.37	6.01 (11.3)	1.53 (−111.3)	4.5 (−17.6)	2.55 (−71.31)	6.91 (25.04)	**5.67** (5.4)	6.55 (19.8)	3.55 (−41)
Average daily precipitation (mm day^{-1})	5.28	4.16 (−23.8)	2.78 (−62)	4.88 (−7.7)	**5.4** (2.3)	3.88 (−31.5)	4.19 (−22.8)	5.69 (7.6)	3.48 (−41)

4.2 Validation of satellite, observational reanalysis, and climate model-based products at dekadal and monthly resolutions

To understand the role of higher spatial resolution in improving the agreement with station data, a similar statistical evaluation was performed using the coarse resolution of CHIRPS (0.25°). Compared to the coarse resolution of CHIRPS, the daily improved version (0.05°) used in this study showed an increased correlation of up to 3.2 % in all the validation areas. In line with the daily evaluation, the comparison was extended to dekadal and monthly resolutions for rainfall, T_{max}, and T_{min} using the same statistical methods. For this analysis the observed daily ground observations and data from ARC2, CHIRP, CHIRPS, ORH, I-RCM, and RCMs were aggregated to dekadal and monthly resolutions. With decreasing temporal resolution (daily to monthly), the agreement of each product showed a marked improvement in all the validation areas. In addition to the increase in correlation, biases (bias and R_{bias}) and errors (MAE and RMSE) in rainfall are decreased at dekadal and monthly resolutions.

At dekadal and monthly resolution, the agreement of all rainfall products with station data increased compared to daily resolutions and the results for eight validation areas of Ethiopia, Kenya, and Tanzania are given in Fig. 8. The same plots – with similar results – for another 13 areas are provided in the Supplement (Fig. S2). Similar to the daily evaluation, CHIRPS appeared to be the most accurate rainfall product both at dekadal and monthly resolutions in most of the validation areas compared to the other products. In addition to the higher correlation of CHIRPS with station data at monthly and dekadal timescales, the centred mean square (rms) difference and the standard deviation are close to the observations in most of the validation areas. Following CHIRPS, CHIRP appeared to be the second best data source for dekadal and monthly rainfall and in three validation areas (EthioShed3, 15, and 16) showed a slightly higher correlation than CHIRPS. In two validation areas (KenShed1 and 2), ARC2 showed a slightly higher correlation than CHIRP and CHIRPS. However, in KenShed2, ARC2 showed a higher deviation from the observed value compared to CHIRP and CHIRPS. CHIRPS has, for example, almost identical standard deviation to the station data in all the validation ar-

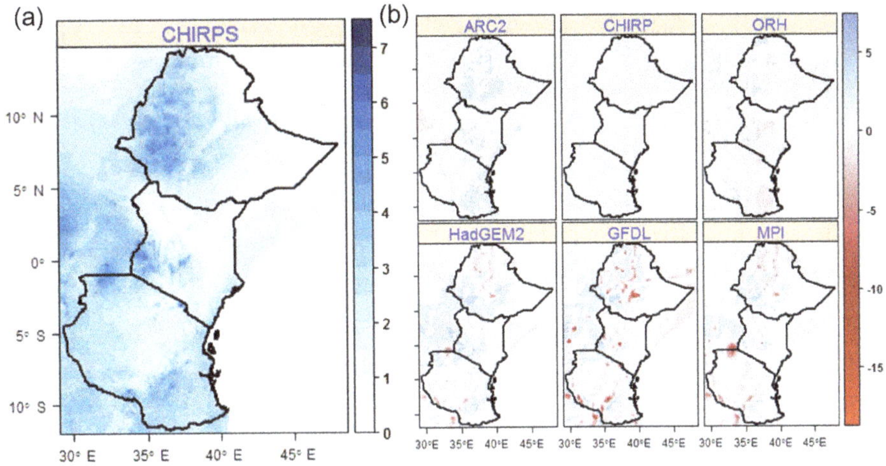

Figure 6. Average daily rainfall (mm day^{-1}) map of East Africa retrieved from CHIRPS **(a)** for the study period 1983–2005. For rainfall data from ARC2, CHIRP, ORH, and RCMs **(b)**, the absolute difference (mm day^{-1}) from CHIRPS is displayed. All the maps are given in a 0.05° spatial resolution.

eas except in areas with a lower number of ground stations (EthioShed12–15 and TanzShed1). Overall, CHIRPS, CHIRP, and ARC2 were found to be the best performing rainfall products, while ORH, I-RCM, and RCMs are the worst performing products.

Moreover, for T_{max} and T_{min}, the correlation of ORH, I-RCM, and RCMs increased from daily to dekadal and monthly resolutions. The agreement of each product with station data, for eight validation areas of Ethiopia, Kenya, and Tanzania, is given in Figs. 9 and 10 for T_{max} and T_{min}, respectively. The same plots – with similar results – for another 13 areas are provided in the Supplement (Figs. S3 and S4 for T_{max} and T_{min}, respectively). Compared to I-RCM and RCMs, the correlation between ORH and station data is higher in most of the validation areas. In addition, ORH showed lower centred mean square (rms) difference and biases (bias and R_{bias}). In addition, compared to the I-RCM and RCMs, the standard deviation of ORH is close to the respective observations in most of the validation areas. Compared to I-RCM, the standard deviation and centred mean square (rms) difference of RCMs are lower in most of the validation areas.

5 Discussion

Detection of rainfall characteristics by satellite observations or climate model simulations' output (GCMs and RCMs) is very challenging as compared to temperature. This is especially evident in East Africa, where the topography is complex and characterized by multiple rainfall regimes. In particular, it is difficult to estimate rainfall with satellite imageries in the mountainous region of East Africa (Cattani et al., 2016) because these products inevitably do not represent the regional rainfall patterns and complexity of the region's topography (Romilly and Gebremichael, 2011). Here, for an improved understanding of the climatic conditions of this complex region and their impact on environmental resources, daily rainfall, T_{max}, and T_{min} products from high-resolution satellite imageries, observational reanalysis, and climate models outputs are compared against ground observations. Such an evaluation was not available until now for the considered region. Therefore, an in-depth evaluation was performed, more specifically, on a daily timescale, of the satellite-based rainfall products (ARC2, CHIRPS, and CHIRP), ORH, and RCMs (CCLM and RCA) driven by three GCMs. ARC2, CHIRP, and CHIRPS are rainfall products, whereas ORH and RCMs provide rainfall, T_{max}, and T_{min}.

From the comparison (using point to pixel, point to area grid cell average, and stations' average to area grid cell average methods), the stations' average to area grid cell average showed the best correlation and least biases and errors in all the validation areas. A study by Duan et al. (2016) in the Adige basin (Italy) found that comparing rainfall products such as CHIRPS on a watershed scale showed a marked improvement in overall agreement compared to the point to pixel method on daily and monthly timescales. Comparing the coarse resolution of satellite products and of RCMs using the point to pixel method cannot be expected to result in a high agreement with station data. Ground stations provide point data measured over continuous time periods, whereas satellite products provide area averages based on discontinuous (rain) estimates. Field-based stations (as point measurements) cannot be considered as reference data for the evaluation of area-based rainfall estimates (Cohen Liechti et al., 2012; Wang and Wolff, 2010), if not compared at a monthly or annual timescale. This is similar to our finding that the point to pixel comparison for all products inside and outside the validation areas show weak statistical relations with

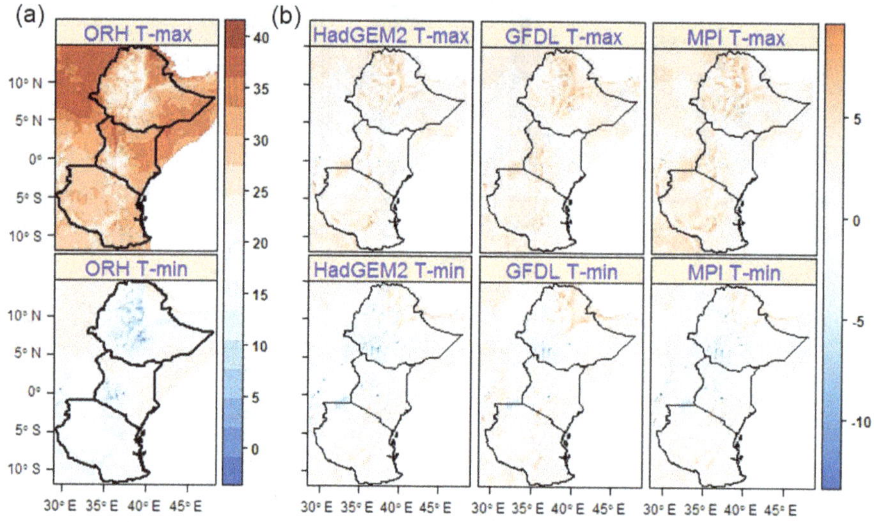

Figure 7. Map of average daily T_{max} and T_{min} (°C) for East Africa generated from ORH **(a)** for the study period 1983–2005. For temperature data from HadGEM2, GFDL, and MPI **(b)**, the absolute difference (°C) from ORH is displayed. All the maps are given in a 0.1° spatial resolution.

ground stations (see e.g. Table 2). The correspondence of all products at a daily timescale and in all the validation areas was found to be comparably weak, and the findings are in agreement with earlier studies (Cohen Liechti et al., 2012; Dembélé and Zwart, 2016).

At a daily timescale, CHIRPS, followed by ARC2 and CHIRP, showed higher correlation and lower errors and biases in all the validation areas compared to ORH, I-RCM, and RCMs. In addition, CHIRPS captured the daily rainfall characteristics well, while ARC2 showed higher underestimation of the average daily and total rainfall. The agreement of all the rainfall products increases from daily to dekadal and monthly timescales (Fig. 8), and this is consistent with other studies (Cohen Liechti et al., 2012; Dembélé and Zwart, 2016; Kimani et al., 2017). Generally, CHIRPS, with a high spatial resolution, followed by CHIRP and ARC2, was the best performing rainfall product in terms of correlation, biases, and errors and in characterizing regional rainfall characteristics. By contrast, ORH, I-RCM, and RCMs appeared to be less precise rainfall products at all timescales and in all validation areas. When looking at the performance of different data products in the selected validation areas (Fig. S1), dispersion is comparably higher in areas with a lower number of ground stations. An additional confounding factor could be the very complex topography of the region. This might explain why products with coarser spatial resolution (ORH, I-RCM, and RCMs) showed higher dispersion compared to products with higher spatial resolution (CHIRPS, CHIRP, and ARC2).

The daily rainfall data (global summary of the day) available at the National Climate Data Center (NCDC) needed to be controlled for quality before application. In East Africa, particularly Ethiopia, the data available at NCDC are very poor and only few stations are available. Therefore, products developed based on the global summary of the day such as ORH cannot be expected to provide as accurate results, particularly for the most complex climate variable, rainfall, as CHIRPS and ARC2. CHIRPS incorporates monthly station data obtained from different regional meteorological organizations, e.g. from Ethiopia, Kenya, and Tanzania. In all the validation areas one to seven stations were included in the development of CHIRPS in different months during 1981–2005. In EthioShed1 (Table 2), for example, six of the nine stations we considered in this study are included in CHIRPS. The inclusion of monthly station data can be assumed to improve CHIRPS' performance compared to other rainfall products. This particular feature of CHIRPS (compared to CHIRP and other data products) is somewhat problematic for our analysis, since the correlated data are not fully independent. However, since only monthly data from a limited number of stations were included in CHIRPS, the dependency is rather weak and indirect. In fact, the improved performance of CHIRPS was even shown in areas where station data are not included (e.g. Arijo, Bedele, and Hurma stations in EthioShed1) and on a daily timescale.

Even though ORH was one of the worst performing rainfall products, it appeared to be the most accurate data source for T_{max} and T_{min} at daily, dekadal, and monthly resolutions compared to I-RCM and RCMs. Nikulin et al. (2012) presented a detailed comparison of daily gridded observations with multiple RCMs, including RCA and CCLM, and they found large discrepancies over the whole region of Africa. However, in this region, RCMs appeared to be the second best data source for both T_{max} and T_{min}, and I-RCM is less precise, with slightly higher biases and errors. In this region, other studies (Endris et al., 2013; Kim et al., 2014) concluded

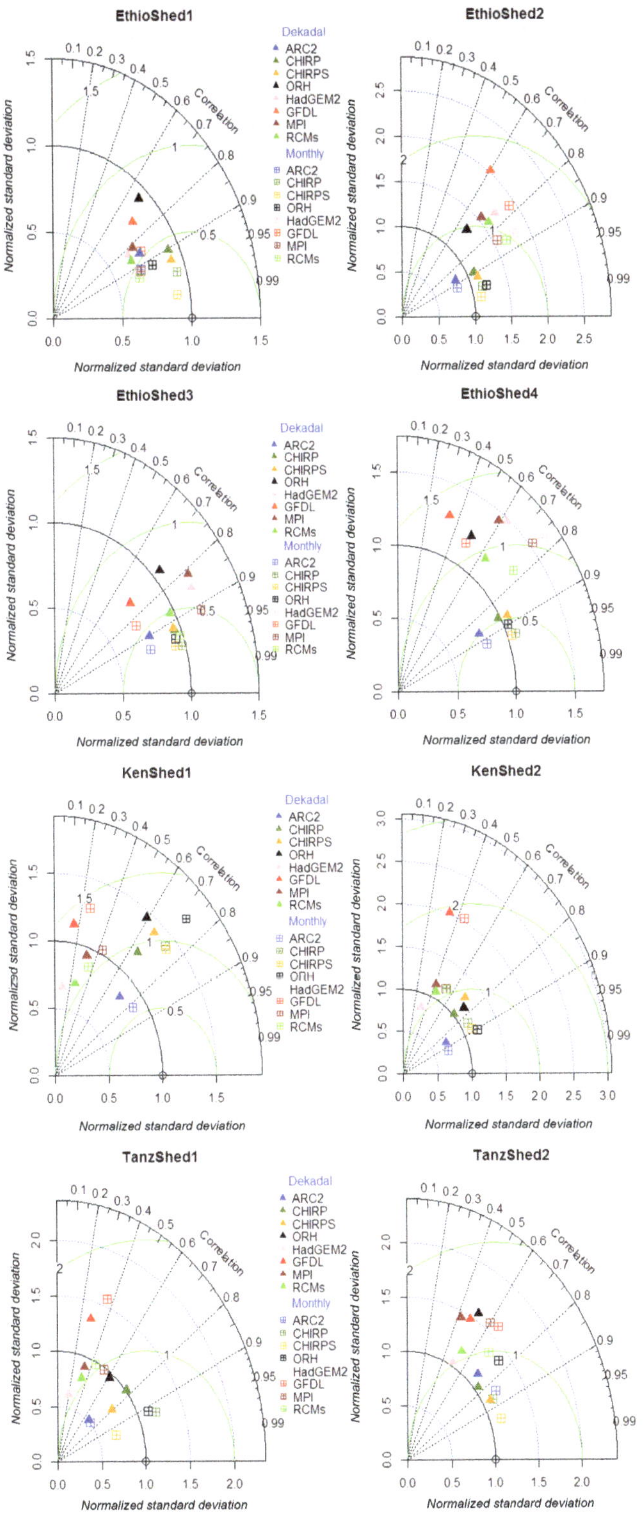

Figure 8. Taylor diagram displaying the agreement between ground observations and synthesized dekadal and monthly rainfall over eight validation areas of Ethiopia, Kenya, and Tanzania covering the period of 1983–2005.

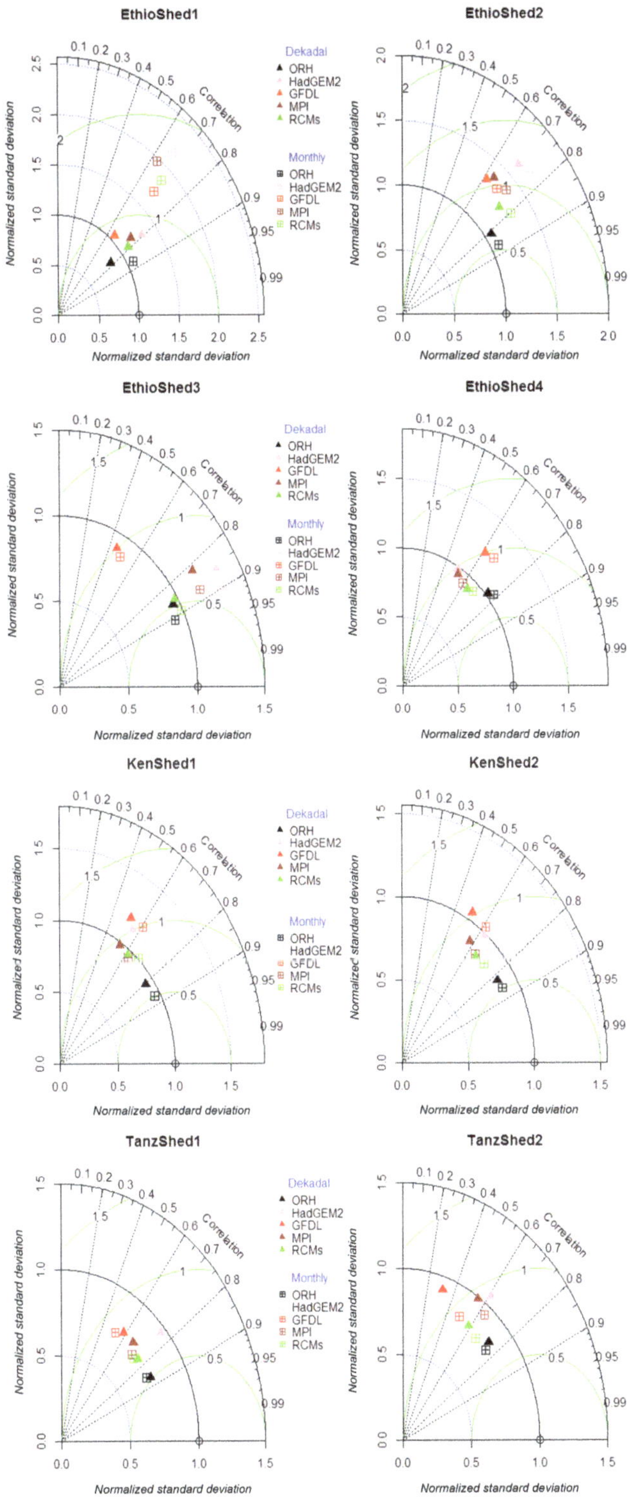

Figure 9. Taylor diagram displaying the agreement between ground observations and synthesized dekadal and monthly T_{max} over eight validation areas of Ethiopia, Kenya, and Tanzania covering the period of 1983–2005.

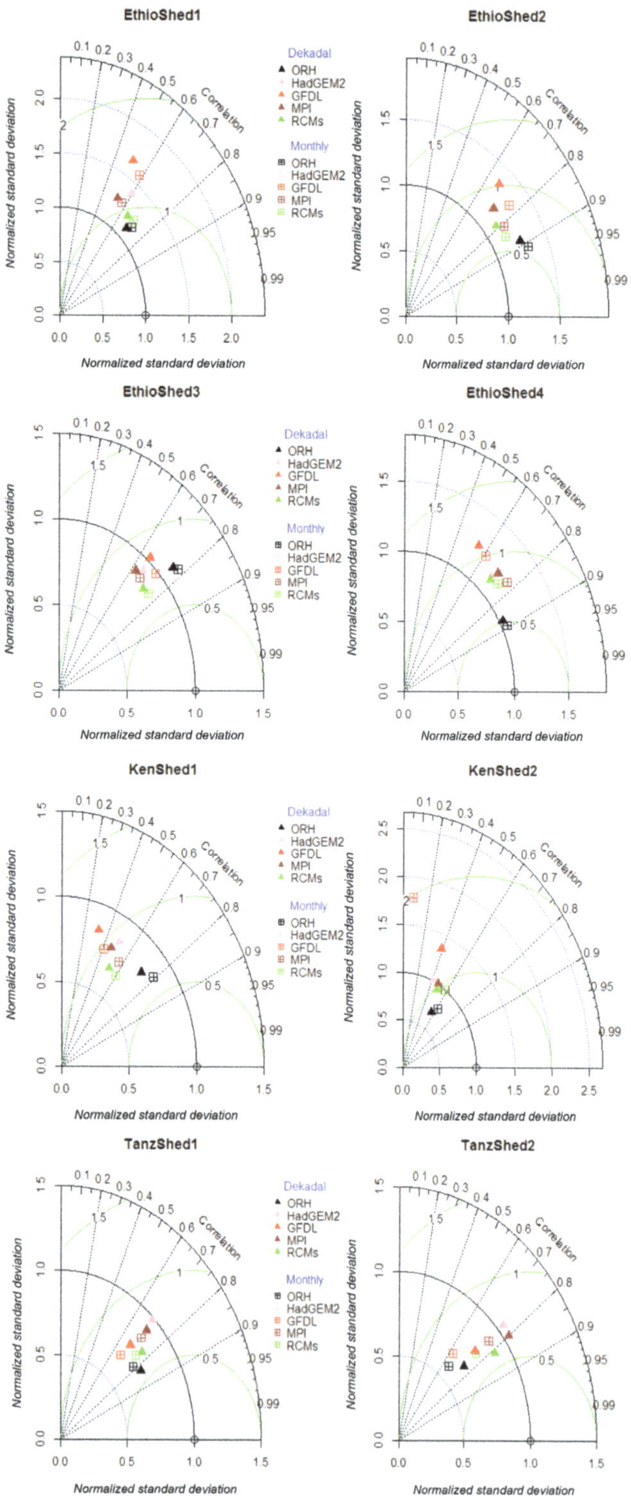

Figure 10. Taylor diagram displaying the agreement between ground observations and synthesized dekadal and monthly T_{min} over eight validation areas of Ethiopia, Kenya, and Tanzania covering the period of 1983–2005.

that the multi-model or ensemble mean of CORDEX RCMs provides reasonable results compared to individual RCMs (I-RCM). The systematic bias of I-RCM and RCMs is higher in most of the validation areas compared to the other products, particularly for rainfall, that can be improved by applying different bias correction techniques such as empirical quantile mapping (Lafon et al., 2013; Maraun, 2013; Teng et al., 2015) before application to different hydrological and climate models (e.g. SDSM). In general, in topographically complex regions such as East Africa, RCMs require further improvements in terms of spatial resolution and accuracy by adding more local information to the modelling process, particularly for precipitation.

6 Summary and conclusion

The evaluation of rainfall, T_{max}, and T_{min} from different sources against station data was performed for large parts of East Africa (Ethiopia, Kenya, and Tanzania) using three methods: point to pixel, point to area grid cell average, and stations' average to area grid cell averages. Compared to the other two methods, the last-mentioned method (stations' average to area grid cell average) provides a better correlation and index of agreement (IA) and lower errors (MAE and RMSE) and biases (bias and R_{bias}). Using this method, individual rainfall, T_{max}, and T_{min} products were compared at daily, dekadal (10 days), and monthly resolutions. At a daily timescale, CHIRPS, ARC2, and CHIRP provide a better agreement with station data compared to ORH, I-RCM, and RCMs. Compared to CHIRPS and CHIRP, ARC2, ORH, I-RCM, and RCMs showed higher biases and errors in most of the validation areas. Overall, the performance of CHIRPS is higher than the other rainfall products in capturing the daily rainfall characteristics, such as the number of wet days, duration of wet and dry periods, total and daily rainfall, and amount of wet periods. ARC2 better captures duration of wet and dry periods, but showed higher underestimation of the total and daily rainfall and number of wet days compared to CHIRPS and CHIRP. I-RCM and RCMs, on the other hand, showed higher overestimation of the number of wet days, duration and amount of wet periods, and total rainfall and underestimation of the average duration of dry periods and daily rainfall.

ORH, conversely, appeared to be one of the worst performing rainfall products for the study region, but the most accurate product, compared to I-RCM and RCMs, for T_{max} and T_{min} at a daily timescale in most of the validation areas. The evaluation of the above products at dekadal and monthly timescales showed that CHIRPS with high spatial resolution (0.05°) has a higher correlation and lower errors and biases than the other rainfall products. As the temporal resolution gets coarser (e.g. monthly), the correlation between ground observations and the above products significantly increases. In addition, biases (bias and R_{bias}) and errors (MAE and

RMSE) significantly decreased. Similar to that of rainfall, the comparison at dekadal and monthly resolution showed an improved correlation and lower errors and biases for both T_{max} and T_{min}. Compared to I-RCM and RCMs, ORH with higher spatial resolution was found to be more accurate at dekadal and monthly resolutions. Next to ORH, RCMs showed a better performance than I-RCM, with lower biases and errors.

In general, CHIRPS for rainfall and ORH for T_{max} and T_{min} performed best in the considered regions of Ethiopia, Kenya, and Tanzania. Further studies need to confirm whether this finding holds for other regions as well, and our approach may represent a blueprint of how to address this question. Since CHIRPS and ORH are available with higher spatial and temporal resolution and for longer periods, these data sources can be used for long-term climate studies (trend, variability, and extreme indices) and input for climate or hydrological models. Considering the typical need for daily data for model input, it remains to be investigated whether poor daily data with a limited bias and similar variance are an acceptable replacement of missing station data when used for impact model studies. In addition, the products can be used to check the plausibility of available ground stations or substitute ground observations in regions of Ethiopia, Kenya, and Tanzania where ground station data are not available or accessible.

Data availability. In this study, publicly available daily rainfall and maximum and minimum temperature data are retrieved from multiple climate data sources, as provided below. The rainfall products used in this study are the Africa Rainfall Climatology version 2.0 (ARC2), available at the International Research Institute climate data library (https://iridl.ldeo.columbia.edu/SOURCES/.NOAA/.NCEP/.CPC/.FEWS/.Africa/.DAILY/.ARC2/; IRI/LDEO, 2016), Climate Hazards Group InfraRed Precipitation (CHIRP), available at the Climate Hazards Group (ftp://ftp.chg.ucsb.edu/pub/org/chg/products/CHIRP/daily/; CHG, 2017), and Climate Hazards Group InfraRed Precipitation with Station (CHIRPS), available at the International Research Institute climate data library (https://iridl.ldeo.columbia.edu/SOURCES/.UCSB/.CHIRPS/; IRI/LDE, 2016). In addition, daily rainfall and maximum and minimum temperature data are retrieved from the Observational-Reanalysis Hybrid (ORH), available at the Terrestrial Hydrology Research Group, Princeton University (http://hydrology.princeton.edu/data.php; Terrestrial Hydrology Research Group Princeton University, 2016), and the regional climate models (RCMs) of the CORDEX-Africa domain, available at the Earth System Grid Federation (https://esgf-index1.ceda.ac.uk/projects/esgf-ceda/; ESGF, 2016).

Author contributions. SHG, SH, and CB designed the research, and SHG performed the data analysis and wrote the paper, with substantial contributions from SH and CB.

Competing interests. The authors declare that they have no conflict of interest.

Acknowledgements. We would like to thank the National Meteorological Agency (NMA) of Ethiopia for providing adequate data for the study and the World Meteorological Organization (WMO) representative for Eastern and Southern Africa for their kind support in data collection. We would also like to thank the Graduate Academy of Technische Universität Dresden (TU Dresden) for its financial support during the study period.

Edited by: Shraddhanand Shukla

References

Abiodun, B. J., Abba Omar, S., Lennard, C., and Jack, C.: Using regional climate models to simulate extreme rainfall events in the Western Cape, South Africa: Simulating Extreme Rainfall Events in Western Cape, Int. J. Climatol., 36, 689–705, https://doi.org/10.1002/joc.4376, 2016.

Anyah, R. O. and Semazzi, F. H. M.: Climate variability over the Greater Horn of Africa based on NCAR AGCM ensemble, Theor. Appl. Climatol., 86, 39–62, https://doi.org/10.1007/s00704-005-0203-7, 2006.

Anyah, R. O. and Semazzi, F. H. M.: Variability of East African rainfall based on multiyear Regcm3 simulations, Int. J. Climatol., 27, 357–371, https://doi.org/10.1002/joc.1401, 2007.

Baldauf, M., Seifert, A., Förstner, J., Majewski, D., Raschendorfer, M., and Reinhardt, T.: Operational Convective-Scale Numerical Weather Prediction with the COSMO Model: Description and Sensitivities, Mon. Weather Rev., 139, 3887–3905, https://doi.org/10.1175/MWR-D-10-05013.1, 2011.

Bayissa, Y., Tadesse, T., Demisse, G., and Shiferaw, A.: Evaluation of Satellite-Based Rainfall Estimates and Application to Monitor Meteorological Drought for the Upper Blue Nile Basin, Ethiopia, Remote Sens., 9, 669, https://doi.org/10.3390/rs9070669, 2017.

Beck, H. E., van Dijk, A. I. J. M., Levizzani, V., Schellekens, J., Miralles, D. G., Martens, B., and de Roo, A.: MSWEP: 3-hourly 0.25° global gridded precipitation (1979–2015) by merging gauge, satellite, and reanalysis data, Hydrol. Earth Syst. Sci., 21, 589–615, https://doi.org/10.5194/hess-21-589-2017, 2017.

Cattani, E., Merino, A., and Levizzani, V.: Evaluation of Monthly Satellite-Derived Precipitation Products over East Africa, J. Hydrometeorol., 17, 2555–2573, https://doi.org/10.1175/JHM-D-15-0042.1, 2016.

Ceccherini, G., Ameztoy, I., Hernández, C., and Moreno, C.: High-Resolution Precipitation Datasets in South America and West Africa based on Satellite-Derived Rainfall, Enhanced Vegetation Index and Digital Elevation Model, Remote Sens., 7, 6454–6488, https://doi.org/10.3390/rs70506454, 2015.

Chaney, N. W., Sheffield, J., Villarini, G., and Wood, E. F.: Development of a High-Resolution Gridded Daily Meteorological Dataset over Sub-Saharan Africa: Spatial Analysis of Trends in Climate Extremes, J. Climate, 27, 5815–5835, https://doi.org/10.1175/JCLI-D-13-00423.1, 2014.

Climate Hazards Group (CHG): Climate Hazards Group InfraRed Precipitation (CHIRP): available at ftp://ftp.chg.ucsb.edu/pub/org/chg/products/CHIRP/daily/, last access: 24 July 2017.

Cohen Liechti, T., Matos, J. P., Boillat, J.-L., and Schleiss, A. J.: Comparison and evaluation of satellite derived precipitation products for hydrological modeling of the Zambezi River Basin, Hydrol. Earth Syst. Sci., 16, 489–500, https://doi.org/10.5194/hess-16-489-2012, 2012.

Daren Harmel, R. and Smith, P. K.: Consideration of measurement uncertainty in the evaluation of goodness-of-fit in hydrologic and water quality modeling, J. Hydrol., 337, 326–336, https://doi.org/10.1016/j.jhydrol.2007.01.043, 2007.

Deblauwe, V., Droissart, V., Bose, R., Sonké, B., Blach-Overgaard, A., Svenning, J.-C., Wieringa, J. J., Ramesh, B. R., Stévart, T., and Couvreur, T. L. P.: Remotely sensed temperature and precipitation data improve species distribution modelling in the tropics: Remotely sensed climate data for tropical species distribution models, Global Ecol. Biogeogr., 25, 443–454, https://doi.org/10.1111/geb.12426, 2016.

Demaria, E. M. C., Maurer, E. P., Sheffield, J., Bustos, E., Poblete, D., Vicuña, S., and Meza, F.: Using a Gridded Global Dataset to Characterize Regional Hydroclimate in Central Chile, J. Hydrometeorol., 14, 251–265, https://doi.org/10.1175/JHM-D-12-047.1, 2012.

Dembélé, M. and Zwart, S. J.: Evaluation and comparison of satellite-based rainfall products in Burkina Faso, West Africa, Int. J. Remote Sens., 37, 3995–4014, https://doi.org/10.1080/01431161.2016.1207258, 2016.

Dinku, T., Kinfe, H., Ross, M., Elena, T., and Stephen, C.: Combined use of satellite estimates and rain gauge observations to generate high-quality historical rainfall time series over Ethiopia, Int. J. Climatol., 34, 2489–2504, https://doi.org/10.1002/joc.3855, 2013.

Dinku, T., Block, P., Sharoff, J., Hailemariam, K., Osgood, D., Corral, J. del, Cousin, R., and Thomson, M. C.: Bridging critical gaps in climate services and applications in Africa, Earth Perspect., 1, 15, https://doi.org/10.1186/2194-6434-1-15, 2014.

Diro, G. T., Grimes, D. I. F., and Black, E.: Teleconnections between Ethiopian summer rainfall and sea surface temperature: part I – observation and modelling, Clim. Dynam., 37, 103–119, https://doi.org/10.1007/s00382-010-0837-8, 2011.

Dixon, J., Gulliver, A., and Gibbon, D.: Farming systems and poverty, Food and Agricultural Organization of the United Nations and World Bank, Rome and Washington, DC, available at: http://www.fao.org/docrep/003/y1860e/y1860e00.htm (last access: 3 August 2015), 2001.

Duan, Z., Liu, J., Tuo, Y., Chiogna, G., and Disse, M.: Evaluation of eight high spatial resolution gridded precipitation products in Adige Basin (Italy) at multiple temporal and spatial scales, Sci. Total Environ., 573, 1536–1553, https://doi.org/10.1016/j.scitotenv.2016.08.213, 2016.

Earth System Grid Federation (ESGF): Regional Climate Models (RCMs), available at: https://esgf-index1.ceda.ac.uk/projects/esgf-ceda/, last access: 16 November 2016.

Endris, H. S., Omondi, P., Jain, S., Lennard, C., Hewitson, B., Chang'a, L., Awange, J. L., Dosio, A., Ketiem, P., Nikulin, G., Panitz, H.-J., Büchner, M., Stordal, F., and Tazalika, L.: Assessment of the Performance of CORDEX Regional Climate Models in Simulating East African Rainfall, J. Climate, 26, 8453–8475, https://doi.org/10.1175/JCLI-D-12-00708.1, 2013.

Endris, H. S., Lennard, C., Hewitson, B., Dosio, A., Nikulin, G., and Panitz, H.-J.: Teleconnection responses in multi-GCM driven CORDEX RCMs over Eastern Africa, Clim. Dynam., 46, 2821–2846, https://doi.org/10.1007/s00382-015-2734-7, 2015.

FAO: Adapting to climate change through land and water management in Eastern Africa, Food and Agricultural Organization of the United Nations and World Bank, Rome, available at: http://www.fao.org/3/a-i3781e.pdf (last access: 3 August 2015), 2014.

Funk, C., Peterson, P., Landsfeld, M., Pedreros, D., Verdin, J., Shukla, S., Husak, G., Rowland, J., Harrison, L., Hoell, A., and Michaelsen, J.: The climate hazards infrared precipitation with stations – a new environmental record for monitoring extremes, Sci. Data, 2, 150066, https://doi.org/10.1038/sdata.2015.66, 2015.

Gan, T. Y., Ito, M., Hülsmann, S., Qin, X., Lu, X. X., Liong, S. Y., Rutschman, P., Disse, M., and Koivusalo, H.: Possible climate change/variability and human impacts, vulnerability of drought-prone regions, water resources and capacity building for Africa, Hydrolog. Sci. J., 61, 1209–1226, https://doi.org/10.1080/02626667.2015.1057143, 2016.

Gebrechorkos, S. H., Hülsmann, S., and Bernhofer, C.: Changes in temperature and precipitation extremes in Ethiopia, Kenya, and Tanzania, Int. J. Climatol., 1–13, https://doi.org/10.1002/joc.5777, 2018.

Gelaro, R., McCarty, W., Suárez, M. J., Todling, R., Molod, A., Takacs, L., Randles, C. A., Darmenov, A., Bosilovich, M. G., Reichle, R., Wargan, K., Coy, L., Cullather, R., Draper, C., Akella, S., Buchard, V., Conaty, A., da Silva, A. M., Gu, W., Kim, G.-K., Koster, R., Lucchesi, R., Merkova, D., Nielsen, J. E., Partyka, G., Pawson, S., Putman, W., Rienecker, M., Schubert, S. D., Sienkiewicz, M., and Zhao, B.: The Modern-Era Retrospective Analysis for Research and Applications, Version 2 (MERRA-2), J. Climate, 30, 5419–5454, https://doi.org/10.1175/JCLI-D-16-0758.1, 2017.

Huffman, G. J., Bolvin, D. T., Nelkin, E. J., Wolff, D. B., Adler, R. F., Gu, G., Hong, Y., Bowman, K. P., and Stocker, E. F.: The TRMM Multisatellite Precipitation Analysis (TMPA): Quasi-Global, Multiyear, Combined-Sensor Precipitation Estimates at Fine Scales, J. Hydrometeorol., 8, 38–55, https://doi.org/10.1175/JHM560.1, 2007.

Huffman, G. J., Adler, R. F., Bolvin, D. T., and Gu, G.: Improving the global precipitation record: GPCP Version 2.1, Geophys. Res. Lett., 36, L17808, https://doi.org/10.1029/2009GL040000, 2009.

IFPRI: Economywide Impacts of Climate Change on Agriculture in Sub-Saharan Africa – climatechange-agriculture.pdf, available at: http://www.indiaenvironmentportal.org.in/files/climatechange-agriculture.pdf (last access: 3 August 2015), 2009.

IPCC: IPCC Third Assessment Report: Climate Change 2001 (TAR), Geneva Switzerland, available at: http://www.ipcc.ch/ipccreports/tar/wg1/ (last access: 3 August 2015), 2001.

IPCC: Climate Change 2007: The Physical Science Basis: Contribution of Working Group I to the Fourth Assessment Report of the of the IPCC, edited by: Solomon, S. et al., CAmbridge University Press, available at: https://www.ipcc.ch/publications_and_data/ar4/wg1/en/contents.html (last access: 30 November 2016), 2007.

IPCC: AR5 IPCC Whats in it for Africa, available at: https://cdkn.org/resource/highlights-africa-ar5/ (last access: 4 January 2017), 2014.

International Research Institute climate data library (IRI/LDE): Climate Hazards Group InfraRed Precipitation with Station data (CHIRPS), available at: https://iridl.ldeo.columbia.edu/SOURCES/.UCSB/.CHIRPS/, last access: 12 May 2016.

International Research Institute climate data library (IRI/LDEO): Africa Rainfall Climatology version 2 (ARC2), available at: https://iridl.ldeo.columbia.edu/SOURCES/.NOAA/.NCEP/.CPC/.FEWS/.Africa/.DAILY/.ARC2/, last access: 31 October 2016.

Jebari, S., Berndtsson, R., Olsson, J., and Bahri, A.: Soil erosion estimation based on rainfall disaggregation, J. Hydrol., 436–437, 102–110, https://doi.org/10.1016/j.jhydrol.2012.03.001, 2012.

Kalnay, E., Kanamitsu, M., Kistler, R., Collins, W., Deaven, D., Gandin, L., Iredell, M., Saha, S., White, G., Woollen, J., Zhu, Y., Leetmaa, A., Reynolds, R., Chelliah, M., Ebisuzaki, W., Higgins, W., Janowiak, J., Mo, K. C., Ropelewski, C., Wang, J., Jenne, R., and Joseph, D.: The NCEP/NCAR 40-Year Reanalysis Project, B. Am. Meteorol. Soc., 77, 437–471, https://doi.org/10.1175/1520-0477(1996)077<0437:TNYRP>2.0.CO;2, 1996.

Kimani, M. W., Hoedjes, J. C. B., and Su, Z.: An Assessment of Satellite-Derived Rainfall Products Relative to Ground Observations over East Africa, Remote Sens., 9, 430, https://doi.org/10.3390/rs9050430, 2017.

Kim, J., Waliser, D. E., Mattmann, C. A., Goodale, C. E., Hart, A. F., Zimdars, P. A., Crichton, D. J., Jones, C., Nikulin, G., Hewitson, B., Jack, C., Lennard, C., and Favre, A.: Evaluation of the CORDEX-Africa multi-RCM hindcast: systematic model errors, Clim. Dynam., 42, 1189–1202, https://doi.org/10.1007/s00382-013-1751-7, 2014.

Lafon, T., Dadson, S., Buys, G., and Prudhomme, C.: Bias correction of daily precipitation simulated by a regional climate model: a comparison of methods, Int. J. Climatol., 33, 1367–1381, https://doi.org/10.1002/joc.3518, 2013.

Legates, D. R. and McCabe, G. J.: Evaluating the use of "goodness-of-fit" Measures in hydrologic and hydroclimatic model validation, Water Resour. Res., 35, 233–241, https://doi.org/10.1029/1998WR900018, 1999.

Maidment, R. I., Grimes, D., Allan, R. P., Tarnavsky, E., Stringer, M., Hewison, T., Roebeling, R., and Black, E.: The 30 year TAMSAT African Rainfall Climatology And Time series (TARCAT) data set, J. Geophys. Res.-Atmos., 119, JD021927, https://doi.org/10.1002/2014JD021927, 2014.

Maidment, R. I., Grimes, D., Black, E., Tarnavsky, E., Young, M., Greatrex, H., Allan, R. P., Stein, T., Nkonde, E., Senkunda, S., and Alcántara, E. M. U.: A new, long-term daily satellite-based rainfall dataset for operational monitoring in Africa, Sci. Data, 4, 170063, https://doi.org/10.1038/sdata.2017.63, 2017.

Malo, M., Jember, G., and Woodfine, A.: Strenghtening Capacity for Climate Change Adaptation in the Agriculture Sector in Ethiopia, Proceedings from National Workshop, Food and Agricultural Organization of the United Nations and World Bank, Nazreth, Ethiopia, available at: http://www.fao.org/docrep/014/i2155e/i2155e00.pdf (last access: 4 August 2015), 2012.

Maraun, D.: Bias Correction, Quantile Mapping, and Downscaling: Revisiting the Inflation Issue, J. Climate, 26, 2137–2143, https://doi.org/10.1175/JCLI-D-12-00821.1, 2013.

Moazami, S., Golian, S., Kavianpour, M. R., and Hong, Y.: Comparison of PERSIANN and V7 TRMM Multi-satellite Precipitation Analysis (TMPA) products with rain gauge data over Iran, Int. J. Remote Sens., 34, 8156–8171, https://doi.org/10.1080/01431161.2013.833360, 2013.

Neitsch, S., Arnold, J., Kiniry, J., Williams, J., and King, K.: Soil and Water Assement Tool Theoretical Documentation, available at: https://swat.tamu.edu/media/1290/swat2000theory.pdf (last access: 23 May 2018), 2002.

Niang, I., Ruppel, O. C., Abdrabo, M. A., Essel, A., Lennard, C., Padgham, J., and Urquhart, P.: Africa. In: Climate Change 2014: Impacts, Adaptation, and Vulnerability. Part B: Regional Aspects, Contribution of Working Group II to the Fifth Assessment Report of the Intergovernmental Panel on Climate Change, edited by: Barros, V. R., Field, C. B., Dokken, D. J., Mastrandrea, M. D., Mach, K. J., Bilir, T. E., Chatterjee, M., Ebi, K. L., Estrada, Y. O., Genova, R. C., Girma, B., Kissel, E. S., Levy, A. N., MacCracken, S., Mastrandrea, P. R., and White, L. L., Cambridge University Press, Cambridge, United Kingdom and New York, NY, USA, 1199–1265, available at: http://www.ipcc.ch/pdf/assessment-report/ar5/wg2/WGIIAR5-Chap22_FINAL.pdf (last access: 4 January 2017), 2014.

Nikulin, G., Jones, C., Giorgi, F., Asrar, G., Buechner, M., Cerezo-Mota, R., Christensen, O. B., Deque, M., Fernandez, J., Haensler, A., van Meijgaard, E., Samuelsson, P., Sylla, M. B., and Sushama, L.: Precipitation Climatology in an Ensemble of CORDEX-Africa Regional Climate Simulations, J. Climate, 25, 6057–6078, https://doi.org/10.1175/JCLI-D-11-00375.1, 2012.

Novella, N. S., Thiaw, W. M., Novella, N. S., and Thiaw, W. M.: African Rainfall Climatology Version 2 for Famine Early Warning Systems, Httpdxdoiorg101175JAMC–11-02381, available at: http://journals.ametsoc.org/doi/abs/10.1175/JAMC-D-11-0238.1 (last access: 30 November 2016), 2013.

Pricope, N. G., Husak, G., Lopez-Carr, D., Funk, C., and Michaelsen, J.: The climate-population nexus in the East African Horn: Emerging degradation trends in rangeland and pastoral livelihood zones, Global Environ. Chang., 23, 1525–1541, https://doi.org/10.1016/j.gloenvcha.2013.10.002, 2013.

Romilly, T. G. and Gebremichael, M.: Evaluation of satellite rainfall estimates over Ethiopian river basins, Hydrol. Earth Syst. Sci., 15, 1505–1514, https://doi.org/10.5194/hess-15-1505-2011, 2011.

Saha, S., Moorthi, S., Pan, H.-L., Wu, X., Wang, J., Nadiga, S., Tripp, P., Kistler, R., Woollen, J., Behringer, D., Liu, H., Stokes, D., Grumbine, R., Gayno, G., Wang, J., Hou, Y.-T., Chuang, H., Juang, H.-M. H., Sela, J., Iredell, M., Treadon, R., Kleist, D., Van Delst, P., Keyser, D., Derber, J., Ek, M., Meng, J., Wei, H., Yang, R., Lord, S., van den Dool, H., Kumar, A., Wang, W., Long, C., Chelliah, M., Xue, Y., Huang, B., Schemm, J.-K., Ebisuzaki, W., Lin, R., Xie, P., Chen, M., Zhou, S., Higgins, W., Zou, C.-Z., Liu, Q., Chen, Y., Han, Y., Cucurull, L., Reynolds, R. W., Rutledge, G., and Goldberg, M.: The NCEP Climate Forecast System Reanalysis, B. Am. Meteorol. Soc., 91, 1015–1058, https://doi.org/10.1175/2010BAMS3001.1, 2010.

Samuelsson, P., Jones, C. G., WilléN, U., Ullerstig, A., Gollvik, S., Hansson, U., Jansson, C., KjellströM, E., Nikulin, G., and Wyser, K.: The Rossby Centre Regional Climate model RCA3: model description and performance: THE ROSSBY CENTRE REGIONAL CLIMATE MODEL RCA3, Tellus A, 63, 4–23, https://doi.org/10.1111/j.1600-0870.2010.00478.x, 2011.

Sapiano, M. R. P. and Arkin, P. A.: An Intercomparison and Validation of High-Resolution Satellite Precipitation Estimates with 3-Hourly Gauge Data, J. Hydrometeorol., 10, 149–166, https://doi.org/10.1175/2008JHM1052.1, 2009.

Segele, Z. T., Leslie, L. M., and Lamb, P. J.: Evaluation and adaptation of a regional climate model for the Horn of Africa: rainfall climatology and interannual variability, Int. J. Climatol., 29, 47–65, https://doi.org/10.1002/joc.1681, 2009.

Sheffield, J., Goteti, G., Wood, E. F., Sheffield, J., Goteti, G., and Wood, E. F.: Development of a 50-Year High-Resolution Global Dataset of Meteorological Forcings for Land Surface Modeling, https://doi.org/10.1175/JCLI3790.1, 2006.

Sheffield, J., Wood, E. F., Chaney, N., Guan, K., Sadri, S., Yuan, X., Olang, L., Amani, A., Ali, A., Demuth, S., and Ogallo, L.: A Drought Monitoring and Forecasting System for Sub-Sahara African Water Resources and Food Security, B. Am. Meteorol. Soc., 95, 861–882, https://doi.org/10.1175/BAMS-D-12-00124.1, 2013.

Sheffield, J., Wood, E. F., Chaney, N., Guan, K., Sadri, S., Yuan, X., Olang, L., Amani, A., Ali, A., Demuth, S., and Ogallo, L.: A Drought Monitoring and Forecasting System for Sub-Sahara African Water Resources and Food Security, B. Am. Meteorol. Soc., 95, 861–882, https://doi.org/10.1175/BAMS-D-12-00124.1, 2014.

Sun, L., Li, H., Zebiak, S. E., Moncunill, D. F., Filho, F. D. A. D. S., and Moura, A. D.: An Operational Dynamical Downscaling Prediction System for Nordeste Brazil and the 2002–04 Real-Time Forecast Evaluation, J. Climate, 19, 1990–2007, https://doi.org/10.1175/JCLI3715.1, 2006.

Sylla, M. B., Giorgi, F., Coppola, E., and Mariotti, L.: Uncertainties in daily rainfall over Africa: assessment of gridded observation products and evaluation of a regional climate model simulation, Int. J. Climatol., 33, 1805–1817, https://doi.org/10.1002/joc.3551, 2013.

Tarnavsky, E., Grimes, D., Maidment, R., Black, E., Allan, R. P., Stringer, M., Chadwick, R., and Kayitakire, F.: Extension of the TAMSAT Satellite-Based Rainfall Monitoring over Africa and from 1983 to Present, J. Appl. Meteorol. Clim., 53, 2805–2822, https://doi.org/10.1175/JAMC-D-14-0016.1, 2014.

Taylor, K. E.: Summarizing multiple aspects of model performance in a single diagram, J. Geophys. Res.-Atmos., 106, 7183–7192, https://doi.org/10.1029/2000JD900719, 2001.

Teng, J., Potter, N. J., Chiew, F. H. S., Zhang, L., Wang, B., Vaze, J., and Evans, J. P.: How does bias correction of regional climate model precipitation affect modelled runoff?, Hydrol. Earth Syst. Sci., 19, 711–728, https://doi.org/10.5194/hess-19-711-2015, 2015.

Terrestrial Hydrology Research Group Princeton University: Observational-Reanalysis hybrid (ORH), available at: http://hydrology.princeton.edu/data.php, last access: 12 May 2016.

Troy, T. J., Sheffield, J., and Wood, E. F.: Estimation of the Terrestrial Water Budget over Northern Eurasia through the Use of Multiple Data Sources, J. Climate, 24, 3272–3293, https://doi.org/10.1175/2011JCLI3936.1, 2011.

UNEP: The Democratic Republic of the Congo Post-Conflict Environmental Assessment United Nations Environment Programme Synthesis for Policy Makers, available at: http://postconflict.unep.ch/publications/UNEP_DRC_PCEA_EN.pdf (last access: 30 November 2016), 2011.

Urama, K. and Ozor, N.: Impacts of climate change on water resources in Africa?: the Role of Adaptation, African Technology Policy Studies Network (ATPS), available at: https://www.researchgate.net/publication/267218899_Impacts_of_climate_change_on_water_resources_in_Africa_the_role_of_adaptation (last access: 5 August 2015), 2010.

Vernimmen, R. R. E., Hooijer, A., Mamenun, Aldrian, E., and van Dijk, A. I. J. M.: Evaluation and bias correction of satellite rainfall data for drought monitoring in Indonesia, Hydrol. Earth Syst. Sci., 16, 133–146, https://doi.org/10.5194/hess-16-133-2012, 2012.

Wang, A., Lettenmaier, D. P., and Sheffield, J.: Soil Moisture Drought in China, 1950–2006, J. Climate, 24, 3257–3271, https://doi.org/10.1175/2011JCLI3733.1, 2011.

Wang, J. and Wolff, D. B.: Evaluation of TRMM Ground-Validation Radar-Rain Errors Using Rain Gauge Measurements, J. Appl. Meteorol. Clim., 49, 310–324, https://doi.org/10.1175/2009JAMC2264.1, 2010.

Wilby, R. L. and Dawson, C. W.: sdsm – a decision support tool for the assessment of regional climate change impacts, Environ. Modell. Softw., 17, 145–157, https://doi.org/10.1016/S1364-8152(01)00060-3, 2004.

Wilby, R. L. and Dawson, C. W.: SDSM 4.2 – A decision support tool for the assessment of regional climate change impacts, United Kingdom, available at: https://sdsm.org.uk/SDSMManual.pdf (last access: 11 January 2017), 2007.

Wilby, R. L. and Yu, D.: Rainfall and temperature estimation for a data sparse region, Hydrol. Earth Syst. Sci., 17, 3937–3955, https://doi.org/10.5194/hess-17-3937-2013, 2013.

Willmott, C. J.: On the Validation of Models, Phys. Geogr., 2, 184–194, 1981.

World Bank: Doing business in The East African Community, IFC/World Bank Rep., 116 pp., available at: http://www.tzdpg.or.tz/fileadmin/_migrated/content_uploads/DB12-EAC_01.pdf (last access: 30 November 2016), 2012.

Xie, P. and Arkin, P. A.: Global Precipitation: A 17-Year Monthly Analysis Based on Gauge Observations, Satellite Estimates, and Numerical Model Outputs, B. Am. Meteorol. Soc., 78, 2539–2558, https://doi.org/10.1175/1520-0477(1997)078<2539:GPAYMA>2.0.CO;2, 1997.

Zambrano-Bigiarini, M., Nauditt, A., Birkel, C., Verbist, K., and Ribbe, L.: Temporal and spatial evaluation of satellite-based rainfall estimates across the complex topographical and climatic gradients of Chile, Hydrol. Earth Syst. Sci., 21, 1295–1320, https://doi.org/10.5194/hess-21-1295-2017, 2017.

3

Are we using the right fuel to drive hydrological models? A climate impact study in the Upper Blue Nile

Stefan Liersch[1], Julia Tecklenburg[1], Henning Rust[2], Andreas Dobler[2], Madlen Fischer[2], Tim Kruschke[3], Hagen Koch[1], and Fred Fokko Hattermann[1]

[1]Potsdam Institute for Climate Impact Research (PIK), Telegraphenberg A31, 14473 Potsdam, Germany
[2]Free University of Berlin (FUB), Institute of Meteorology, Carl-Heinrich-Becker-Weg 6–10, 12165 Berlin, Germany
[3]GEOMAR Helmholtz Centre for Ocean Research Kiel, Wischhofstr. 1–3, 24148 Kiel, Germany

Correspondence: Stefan Liersch (liersch@pik-potsdam.de)

Abstract. Climate simulations are the fuel to drive hydrological models that are used to assess the impacts of climate change and variability on hydrological parameters, such as river discharges, soil moisture, and evapotranspiration. Unlike with cars, where we know which fuel the engine requires, we never know in advance what unexpected side effects might be caused by the fuel we feed our models with. Sometimes we increase the fuel's octane number (bias correction) to achieve better performance and find out that the model behaves differently but not always as was expected or desired. This study investigates the impacts of projected climate change on the hydrology of the Upper Blue Nile catchment using two model ensembles consisting of five global CMIP5 Earth system models and 10 regional climate models (CORDEX Africa). WATCH forcing data were used to calibrate an eco-hydrological model and to bias-correct both model ensembles using slightly differing approaches. On the one hand it was found that the bias correction methods considerably improved the performance of average rainfall characteristics in the reference period (1970–1999) in most of the cases. This also holds true for non-extreme discharge conditions between Q_{20} and Q_{80}. On the other hand, bias-corrected simulations tend to overemphasize magnitudes of projected change signals and extremes. A general weakness of both uncorrected and bias-corrected simulations is the rather poor representation of high and low flows and their extremes, which were often deteriorated by bias correction. This inaccuracy is a crucial deficiency for regional impact studies dealing with water management issues and it is therefore important to analyse model performance and characteristics and the effect of bias correction, and eventually to exclude some climate models from the ensemble. However, the multi-model means of all ensembles project increasing average annual discharges in the Upper Blue Nile catchment and a shift in seasonal patterns, with decreasing discharges in June and July and increasing discharges from August to November.

1 Introduction

Ethiopia is a country where about 80 % of the population is engaged in the agricultural sector (Dile et al., 2013; Deressa et al., 2011), the main source of income for rural communities (Bryan et al., 2009). Around 90 % of the country's grain is produced by smallholder farms. Subsistence and rain-fed farming systems dominate and, with few exceptions, irrigation is not practised[1]. Consequently, agricultural and livestock production, people's livelihoods, and food security depend strongly on weather conditions, mainly on rainfall patterns such as amounts and timing. Hence, a large share of Ethiopia's population is very vulnerable to weather conditions and in particular to its inter-annual variability (Busby et al., 2014; Megersa et al., 2014; Headey et al., 2014; Zaitchik et al., 2012; Simane et al., 2012).

The Ethiopian highlands, where the Blue Nile rises, are considered to be the "water tower" of East Africa. The Blue Nile, for instance, contributes about 55–65 % of the flow of the Nile at the confluence with the White Nile (King, 2013;

[1]http://www.fao.org/wairdocs/ilri/x5548e/x5548e0k.htm

Sutcliffe and Parks, 1999). The river is therefore the most important water resource, not only for Ethiopia but also for the downstream riparian countries of Sudan and Egypt. Water politics in the Nile basin have a long history and are a central geopolitical feature in this region (Gebreluel, 2014; Ibrahim, 2012). With growing populations, industrialization, and climate change and its variability, the situation is becoming more and more tense (Gebreluel, 2014). Knowledge about availability of future water resources in this region and therefore studies providing insights into climate change and variability, and their impacts on hydrology, are of utmost importance.

A review of future hydrological and climate studies in the River Nile basin is provided by Di Baldassarre et al. (2011) and a review on hydrological extremes in the Upper Blue Nile catchment (UBN) by Taye et al. (2015). Recent studies on climate change and variability in the UBN or its tributaries served different purposes. The studies by Mengistu et al. (2014), Taye and Willems (2012), Conway and Schipper (2011), and Conway and Hulme (1993) investigated for instance trends of past climate change using observed and/or generated climate data. Diro et al. (2009) analysed the quality of rainfall data using two numerical weather prediction models. Another category of studies investigates the performance and projected trends of climate models (e.g. Conway and Schipper, 2011; Diro et al., 2011).

Studies performed to assess impacts of climate change in the UBN can be categorized into (i) studies applying simple approaches, assuming for instance a fixed percentage of decrease or increase of a climatic variable or discharge (Jeuland and Whittington, 2014); (ii) studies using a single climate model (e.g. McCartney and Menker Girma, 2012; Soliman et al., 2009; Abdo et al., 2009); and (iii) studies analysing complex climate model ensembles (e.g. Teklesadik et al., 2017; Liersch et al., 2017; Aich et al., 2014; Mengistu and Sorteberg, 2012; Setegn et al., 2011; Beyene et al., 2010; Elshamy et al., 2009; Kim et al., 2008).

As a matter of fact, climatic variables such as air temperature, precipitation, and radiation simulated by global and regional climate models usually have a bias in the historical (reference) period (e.g. Addor and Seibert, 2014; Berg et al., 2012; Gudmundsson et al., 2012; Hagemann et al., 2011). Moreover, they often fail to adequately represent spatio-temporal dynamics at the regional scale. In climate studies, the absolute or relative changes between historical and projection periods are analysed and reported in the following manner: model X projects a temperature increase of 2.5 K in 2021–2050 and an increase of 8 % of rainfall relative to its reference period. Here, it does not matter whether model X was too cold/warm or too dry/wet during the reference period. Only the rate of change matters, which might be reasonable in this context. Moreover, in climate change studies it is common practice nowadays to analyse the entire available model ensemble and to calculate the multi-model mean, which is superior to any one individual climate model (Pierce et al., 2009). Unfortunately, a daily multi-model mean climate time series does not serve as reasonable input for impact models operating at the daily time step. Therefore, the application of climate model ensembles is always recommended for hydrological studies (Teutschbein and Seibert, 2010) and is considered nowadays as state of the art.

Quantitative and application-oriented impact studies require a certain accuracy of input data as well as adequate representation of the relevant processes by the models used. Small biases already present in temperature or precipitation may lead to considerable biases in impact models (Maraun et al., 2010). Therefore, various bias correction approaches were developed, particularly for hydrological applications (Piani et al., 2010; Dosio and Paruolo, 2011). The expectation of using bias-corrected input data is that they are quantitatively more precise than their uncorrected counterparts.

The authors of studies using complex model ensembles in the UBN, cited above, applied different approaches to generate climate input time series for hydrological modelling. Elshamy et al. (2009) used a distribution mapping approach to simultaneously downscale and bias-correct 17 CMIP3[2] GCMs (SRES A1B) and applied the corrected climate data to run the Nile Forecasting System in the UBN. The delta-change method was used by Mengistu and Sorteberg (2012) and Kim et al. (2008) to generate time series of temperature and precipitation used as input for hydrological modelling. Mengistu and Sorteberg (2012) used 19 GCMs of the CMIP3 model ensemble (SRES scenarios A2, A1B, and B1) to generate climate inputs for the SWAT model and Kim et al. (2008) used six GCMs (SRES A2) to run a monthly water balance model. Setegn et al. (2011) applied a downscaling approach for daily temperature and precipitation data to 15 CMIP3 GCMs (SRES scenarios A2, A1B, and B1) using a cumulative frequency distribution approach. They used the climate data to run the SWAT model in the Lake Tana basin. Beyene et al. (2010) performed a quantile mapping approach to bias-correct 11 CMIP3 GCMs (SRES A2 and B1) to run the VIC hydrological model for the entire Nile basin. Recently, Teklesadik et al. (2017) published a study comparing climate change impacts, particularly on actual evapotranspiration, using six hydrological models driven by the same four CMIP5 GCMs used in the study at hand. Liersch et al. (2017) used a climate model ensemble to analyse the impacts of the Grand Ethiopian Renaissance Dam on downstream discharges under current and future climate conditions based on the 10 "best" global and regional climate models identified in this study.

The study at hand falls into the same category using the most recent global and regional climate projections released for the IPCC 5th Assessment Report (IPCC, 2013). Uncorrected and bias-corrected climate simulations of five CMIP5[3]

[2] http://cmip-pcmdi.llnl.gov/cmip3_overview.html?submenuheader=1

[3] http://cmip-pcmdi.llnl.gov/cmip5/

Earth system models (ESMs) and 10 uncorrected and bias-corrected regional climate models (RCMs) from CORDEX Africa[4] were used to run the Soil and Water Integrated Model (SWIM). The climate scenarios used by both model ensembles are the Representative Concentration Pathways (RCPs) RCP 4.5 and RCP 8.5 (van Vuuren et al., 2011; Meinshausen et al., 2011). Hence, we analyse 60 discharge simulations (two RCPs and 15 uncorrected and 15 bias-corrected climate model runs) for the reference period 1970–1999 and two future periods 2030–2059 and 2070–2099.

The first objective of this study is to assess climate change and its impacts on the availability of future water resources in the UBN defined at gauge El Diem (Sudan border). The second objective is to discuss the implications of using different model ensembles to project future discharges by comparing the results of the whole range of uncorrected and bias-corrected ESMs and RCM ensembles. Eventually an ensemble is assembled including only those members fulfilling certain performance criteria. These criteria are used to characterize the suitability of simulations for different purposes, such as for qualitative or quantitative studies. A qualitative impact study may have lower demands on the quality of climate simulations than a study investigating hydrological extremes or water management strategies. In the latter case, the requirements in terms of quantitative accuracy are much higher. The following questions were central to our investigations.

a. What are the likely impacts of climate change on future discharges in the UBN?

b. Is there an agreement on the signal of climate change impacts in the 21st century using different climate model ensembles?

c. To what extent can bias correction alter the magnitudes of change signals in hydrological simulations in the study area?

d. In how far can we trust simulations that require a strong correction?

2 Study area

The entire Blue Nile River basin covers an area of about 296 000 km^2. The study area considered here is the Upper Blue Nile catchment (UBN) defined at gauge El Diem at the border between Ethiopia and Sudan that covers an area of 172 000 km^2. Elshamy et al. (2009) estimates a catchment area of 185 000 km^2 and Mengistu and Sorteberg (2012) an area of 174 000 km^2 for the UBN. These discrepancies are certainly based on different digital elevation models and GIS algorithms used to delineate the catchment area and thus may add to the uncertainties of such studies, which are not easily quantifiable. In Fig. 1, the UBN is encircled by a red line. In addition, it shows the 576 subbasins that were delineated for the hydrological modelling exercise, the three gauging stations used to calibrate the hydrological model, and the coordinates of the climate data grid. The source of the Blue Nile River is Lake Tana in the Ethiopian highlands and the catchment is located in the north-western part of Ethiopia (Taye and Willems, 2012). It drains a major part of the western highlands (Sutcliffe and Parks, 1999) that is predominantly governed by a unimodal rainfall regime depending on the movement of the intertropical convergence zone (ITCZ). The inter-annual variability of annual rainfall amounts in the Ethiopian highlands is high (Zaitchik et al., 2012) and ranges between 800 and 2200 mm, and the elevation of the UBN varies from 4000 to 500 m.a.s.l. (Taye and Willems, 2012). The river has a length of almost 1000 km from the Lake Tana outlet to the Sudan border.

3 Methods

3.1 Data

Freely available WATCH Forcing Data (WFD) (Weedon et al., 2011) based on ERA-40 (Uppala et al., 2005) reanalysis and climate observations were used to bias-correct five ESMs and 10 RCM runs and to calibrate and validate the hydrological model SWIM (Soil and Water Integrated Model), developed by Krysanova et al. (2005). Although the quality of WFD varies in space (Rust et al., 2015), this gridded product with a spatial resolution of 0.5° was used as input because observed climate data were not available for this study. The SRTM digital elevation model (Jarvis et al., 2008) was used to delineate the 576 subbasins and to derive some terrain-specific parameters. Required soil parameters were derived from the Digital Soil Map of the World (FAO et al., 2009) and land use cover data were reclassified from Global Land Cover (GLC2000) (Bartholomé and Belward, 2005). Observed monthly discharge data for model calibration and validation were provided by the Global Runoff Data Centre (GRDC[5]).

3.2 Hydrological model

The Soil and Water Integrated Model (SWIM), developed by Krysanova et al. (2005), is a semi-distributed, process-based eco-hydrological model that operates at the daily time step. It was developed on the basis of the MATSALU (Krysanova et al., 1989) and SWAT (Arnold et al., 1993) models and is continuously being further developed and adapted to new or specific requirements (Krysanova et al., 2015). Hydrological response units (HRUs), considered as areas with similar hydrological characteristics, are the smallest model units where all hydrological, nutrient, and vegetation processes

[4]http://start.org/cordex-africa/about/

[5]http://www.bafg.de/GRDC/EN/Home/homepage_node.html

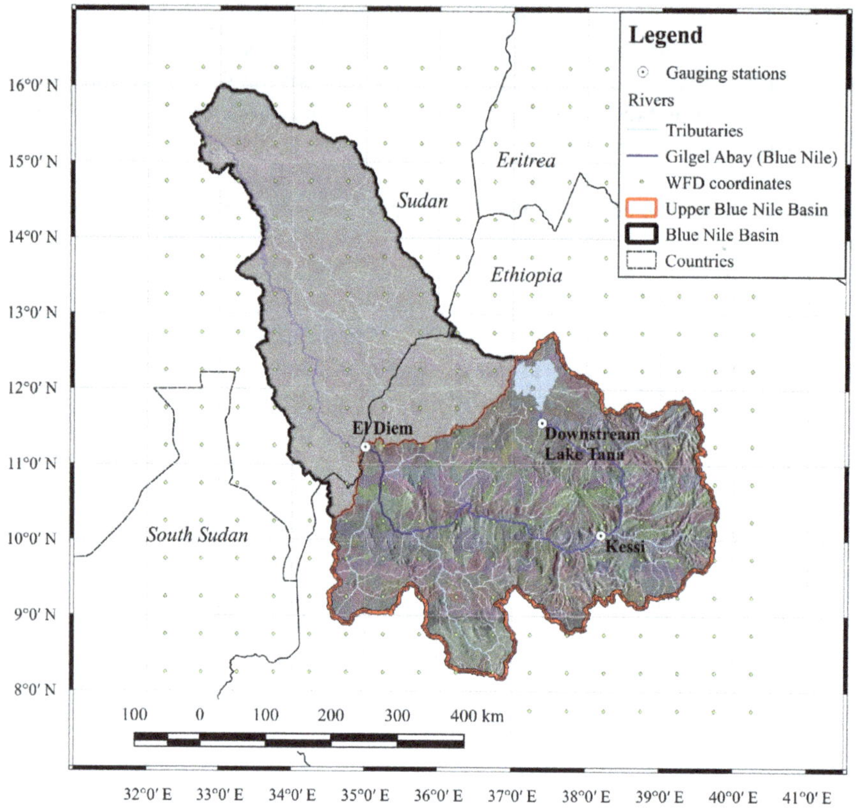

Figure 1. Map of the Blue Nile River basin. The Upper Blue Nile (UBN) catchment (172 000 km^2) is enclosed by the red line. The three gauges used for model calibration and validation are represented by white circles.

are calculated. There is no lateral interaction between HRUs but area-weighted daily fluxes are calculated and aggregated at the subbasin scale and routed through the river network. SWIM distinguishes three flow components: surface runoff, subsurface runoff, and contributions of the shallow groundwater aquifer. Actual evapotranspiration is determined by simulated soil evaporation and transpiration from the vegetation cover. Water percolating from the shallow groundwater aquifer into the deep groundwater aquifer is lost from the system but is considered in the water balance.

A reservoir module, developed by Koch et al. (2013), was incorporated in SWIM and parameterized to better account for Lake Tana's storage effects and to consider the impact of the weir at the lake's outlet in future simulations that was constructed in the year 1996.

Radiation data required by SWIM as essential climate input were not available in all RCM runs. To maintain consistency and comparability in hydrological simulations, daily radiation data were computed after Hargreaves and Samani (1985) from daily minimum and maximum air temperature and the latitude of the respective subbasin. The simulated radiation data were calibrated to fit average annual observed radiation data of about 1800 kWh m^{-2}.

3.3 Climate models

The ESM ensemble used in this study consists of the following five CMIP5 models: GFDL-ESM2M, HadGEM2-ES, IPSL-CM5A-LR, MIROC-ESM-CHEM, and NorESM1-M. Projections of these five ESMs were linearly downscaled and bias-corrected by Hempel et al. (2013) in the frame of the Inter-Sectoral Impact Model Intercomparison Project (ISIMIP)[6] (Warszawski et al., 2014). The uncorrected ESM simulations were interpolated to the WFD 0.5° grid.

Table S1 in the Supplement provides an overview of the RCM runs organized by the CORDEX Africa initiative[7]. The ensemble consists of four RCMs driven by different ESMs. The RCM SMHI-RCA4 was driven by seven ESMs, CanRCM4 by CanESM2, and the RCMs KNMI-RACMO22T and DMI-HIRHAM4 by EC-EARTH. The 10 RCM runs were bias-corrected by the authors of this paper. Table S2 shows the model IDs of all 15 climate models used in some figures and tables.

[6]https://www.pik-potsdam.de/research/climate-impacts-and-vulnerabilities/research/rd2-cross-cutting-activities/isi-mip

[7]http://start.org/cordex-africa/about/

3.4 Climate scenarios

For both the global and regional climate model ensembles, the two scenarios RCP 4.5 and RCP 8.5 were used because they represent a broad range of uncertainties with regard to possible future pathways and related climate projections. According to van Vuuren et al. (2011) and Meinshausen et al. (2011), RCP 4.5 represents the medium stabilization scenario (stabilization without overshoot pathway leading to $+4.5\,\mathrm{W\,m^{-2}}$ radiative forcing (relative to pre-industrial forcing) and $\sim 650\,\mathrm{ppm}\,CO_2$ equiv. by 2100) and RCP 8.5 the highest emission scenario (rising radiative forcing pathway leading to $+8.5\,\mathrm{W\,m^{-2}}$ and $\sim 1370\,\mathrm{ppm}\,CO_2$ eq. by 2100), assuming no stabilization in global greenhouse gas emissions.

3.5 Bias correction

Despite regional downscaling to finer resolution, RCM simulations often show considerable biases when compared to observed data (Addor and Seibert, 2014; Christensen et al., 2008). A review of bias correction methods (linear scaling, local intensity scaling, power transformation, and distribution or quantile mapping) is provided by Teutschbein and Seibert (2012). The authors conclude that the distribution or quantile mapping method achieves the best performance for most of the selected criteria. Although quantile mapping is a successful method to improve the representation of daily rainfall characteristics, it fails to correct multi-day and inter-annual variables, such as mean maximum 4-day precipitation, mean minimum 14-day precipitation, and inter-annual variability (Addor and Seibert, 2014). The drawback that all approaches have in common is that they are based on the stationarity assumption, which presumes that future physical processes in the atmosphere are comparable to the period used to correct the simulations. Bias correction of climate simulation data is nowadays a widely used practice in hydrological impact modelling, but it should be treated with caution. As Maraun et al. (2010) point out, the origins of the bias in climate simulations (mathematical formulations in climate models) are not solved by the post-processing and may disrupt internal physical coherence between weather variables. Hence, the correction is usually based on wrong reasons (Addor and Seibert, 2014). Alternatives to bias correction are so-called delta-change methods. Sophisticated approaches of this method are described by Anandhi et al. (2011), Bosshard et al. (2011), and Chiew et al. (2009).

3.5.1 Bias correction of ESMs

Bias-corrected data of five CMIP5 ESMs were available and provided by ISIMIP. In a first step ESM data were linearly interpolated to the WFD 0.5° grid, implementing the standard Gregorian calendar. Temperature data were corrected using a trend-preserving additive approach where monthly mean values were adjusted for a systematic bias by adding a grid-point-specific and month-specific constant offset. Therefore, the absolute projected temperature changes of the ESMs are not changed. The daily variability of ESM temperatures was adjusted to reproduce WFD variability by adding a monthly correction factor on temperature anomalies.

Precipitation data were corrected using a multiplicative approach where monthly mean precipitation was multiplied by a grid-point-specific and month-specific constant correction factor. Relative changes projected by the ESMs are thereby preserved. A known problem of this method is that extraordinarily high values of daily precipitation can occur in the bias-corrected simulation if very high simulated daily precipitation data are multiplied by high correction factors. Therefore, the correction factor was limited to a value of 10. Remaining extremely high daily precipitation values were truncated to 400 mm. After the method introduced by Piani et al. (2010), daily precipitation variability and the frequency of dry days were corrected by applying a transfer function to fit the normalized simulated time series of wet months to the normalized WFD time series. A more detailed description of the bias correction procedure applied to the five CMIP5 ESMs used in this study is provided by Hempel et al. (2013).

3.5.2 Bias correction of RCMs

Precipitation biases in most CORDEX RCMs show a high seasonality for grid boxes within the evaluation domain of the UBN. This limits a bias correction based on seasonal or annual means. However, as some of these grid boxes do show almost no precipitation events for single months, a harmonic-based bias correction method analogous to the one applied to temperature is not feasible for precipitation. Furthermore, this results in a large uncertainty in the estimation of the corresponding monthly biases. Thus, based on the recommendation from Dobler and Ahrens (2008), a bias correction is only applied on months and grid boxes with more than 100 rainy days (rainfall above $1\,\mathrm{mm\,day^{-1}}$) within the calibration period (1951–2001).

The method applied is based on a local rainy day intensity scaling, correcting the frequency of rainy days and the mean precipitation on rainy days to fit the observed values in a specific calibration period (Schmidli et al., 2006). Details on the implementation and an evaluation are given in Dobler and Ahrens (2008). The method has been successfully applied before as a downscaling and bias-correcting method for precipitation in alpine regions (Dobler and Ahrens, 2008; Dobler et al., 2011).

The underlying idea is the assumption of a smooth seasonal cycle for the variables simulated by the RCM and the observational reference (WFD). These cycles are modelled with a series of harmonic functions using vector generalized linear models (Yee, 2015), and the difference in cycles between an RCM reference simulation and the observational product is used for bias correction of the RCM projection.

The seasonality in the location parameter of a quantity (i.e. the expectation value in the case of a Gaussian distribution) can be modelled as

$$\mu(t) = \mu_0 + \sum_{k=1}^{K} \mu_k \sin(k\,\omega t) + \sum_{l=1}^{L} \mu_l \cos(l\,\omega t), \qquad (1)$$

with $\omega = \frac{2\pi}{365.25}$, $t = 1, \ldots, 366$ being the time variable running over all possible days of the year; K and L are the orders of the harmonic function expansion for μ. A scale parameter σ can be modelled analogously in this framework. The result is a climatological distribution, i.e. a description of the probability distribution throughout the year.

Selection of orders K and L is based on a 10-fold cross validation using the Continuous Rank Probability Score (CRPS, Wilks, 2011) as the cost function. The difference in parameters between the RCM reference and the observational product (WFD) is subtracted from the parameters of the RCM projections for bias correction. Quantile mapping (e.g. Vrac and Friederichs, 2015) now maps the values from the uncorrected to the corrected climatological distribution.

Particular care needs to be taken when correcting minimum and maximum temperature to avoid inconsistencies such as $T_{\max} < T_{\min}$. Here, a variable transformation ensures physical consistency:

$$T_1 = \log(T_{\max} - T) \qquad (2)$$

$$T_2 = \log(T - T_{\min}). \qquad (3)$$

After bias-correcting T_1 and T_2, corrected values for $T_{\max}$ and $T_{\min}$ can be obtained by back-transforming the variables.

3.6 Evaluating the suitability of climate simulations

Evaluating the suitability of climate simulations for regional impact studies is a process that includes seemingly objective components (e.g. analysing performance criteria) and subjective components (choosing criteria and setting their thresholds). Data visualization and interpretation by the user might be considered as a mixture of both objectivity and subjectivity. The choice of periods used as reference and future projection does also influence the results. The former is often predetermined by data availability or conventions and the latter usually by the client. Moreover, there are uncertainties with regard to the quality of the dataset used as the comparison baseline, mostly observed and/or generated climate data.

Evaluation of climate model performance is complicated by the fact that climate simulations cannot be compared to the reference dataset on a real-time daily, monthly, or annual basis, as is common practice with discharge simulations in hydrological modelling. Climate simulations are not supposed to reproduce or predict the weather for a certain day, month, or year. Hence, only statistical parameters, summarized over a period of usually 30 years (e.g. the annual cycle represented by average daily or monthly time series), or the mean, quantile values, and standard deviation of the entire daily time series can be used as a basis for comparison.

In the first step of climate model evaluation, daily and monthly precipitation characteristics of uncorrected (UC) and bias-corrected (BC) climate simulations were compared to WFD characteristics (reference climate). In a second step, SWIM was employed to simulate daily discharge using all climate simulations for reference and future periods. Since the main purpose of this study is to assess climate change impacts on hydrology, using hydrological performance indicators to evaluate climate simulations is a straightforward method. A similar approach was used by Elshamy et al. (2013) who used a GLUE-like methodology to exclude and weigh climate model performance. Another benefit of this approach is that a spatially semi-distributed hydrological model does not only account for temporal but also for spatial patterns of climate inputs. Therefore, the annual cycle represented by daily ($n = 365$) discharge simulations (sim), averaged over the 30-year reference period, was compared against the baseline simulation using WFD (ref). The performance criteria applied to these time series are the coefficient of determination (R^2), PBIAS, standard deviation (SD), and the normalized SD of discrepancies (SD_D) or the centred root mean square errors.

The characteristics of daily discharges were analysed using flow duration curves (FDCs), where every single discharge value is related to the percentage of time it is equalled or exceeded (Smakhtin, 2000). FDCs summarize discharge variability of a time series and display the complete range from low flows to flood events. In order to analyse and visualize average, low, and high flow characteristics, 17 percentile values ($Q_{0.01}$–$Q_{99.99}$) were used to compute FDCs based on the entire daily discharge time series of the 30-year reference period. This method was applied to assess whether model performance is suitable to study non-extreme discharge conditions (NED) and/or high and low flow situations as well as their extremes.

$$\mathrm{PBIAS} = \frac{\sum_{i=1}^{n} (\mathrm{sim}_i - \mathrm{ref}_i) \cdot 100}{\sum_{i=1}^{n} (\mathrm{ref}_i)} \qquad (4)$$

$$\mathrm{SD_D} = \frac{\mathrm{SD}(\mathrm{sim}_i - \mathrm{ref}_i)}{\mathrm{SD_{ref}}} \qquad (5)$$

In addition to the criteria used to evaluate model performance in the reference period, it is also important to consider model behaviour in future periods. In fact, unexpected behaviour in projection periods was observed in several simulations, particularly in some BC simulations. The hypothesis is that the stronger the necessity of bias correction, the higher the risk that the BC simulation will show unexpected behaviour in future periods. Therefore, another criterion was introduced that indicates the rate of change of PBIAS between the future and the reference period. Note that the definition of threshold values is somewhat subjective and was influenced by the simulation results of the model ensemble. However, if the thresholds had been set more critically, almost no climate model would have passed the evaluation process successfully. The model selection process and the definition of criteria thresholds are described in the following section.

3.7 Model selection

Beside analysing the impact of climate projections on future discharges using the whole UC and BC ESM and RCM ensembles, a climate model ensemble was assembled containing only those models that fulfil the criteria and their thresholds defined below. In order to become a member of the selected ensemble, a model must basically achieve all the following three criteria.

- *Seasonality.* The annual cycle based on average daily discharge simulations must achieve $R^2 \geq 0.85$. Models with $R^2 < 0.85$ are assumed to represent discharge seasonality only poorly.
- *Volumetric deviation.* Average daily discharge simulations must achieve a PBIAS $\leq \pm 30\,\%$.
- *Non-extreme discharges (NED).* NED represent discharge conditions between FDC percentile values between $> Q_{10}$ and $< Q_{90}$ ($Q_{20}, Q_{30}, \ldots, Q_{80}$). Percentiles in this range should not deviate more than $\pm 30\,\%$ from WFD discharge simulation.

Models meeting these three criteria are assumed to be suitable for a qualitative impact assessment and are indicated in the column "pre" (preselection) in Table 1. In addition, the columns HF (high flows, FDC percentiles Q_{10}, Q_5, Q_1, $Q_{0.1}$, $Q_{0.01}$) and LF (low flows, FDC percentiles Q_{90}, Q_{95}, Q_{99}, $Q_{99.9}$, $Q_{99.99}$) indicate further whether a particular model adequately represents extreme discharge conditions and might be used for specific investigations. Again, the FDC values in the respective range should not exceed the threshold of $\pm 30\,\%$.

After simulating discharges using all climate scenarios it was found that several simulations project enormous increase in annual river discharge already in the period 2030–2059. This was particularly the case in simulations where bias correction resulted in stupendous increase of extreme daily rainfall and therefore extraordinary high peak discharges. Hence, another criterion was defined representing the rate of change. Simulations where average annual discharges changed by more than $\pm 30\,\%$ in the period 2030–2059 (RCP 8.5) relative to the reference period were omitted from the selected ensemble, even if the first three criteria were achieved. This criterion is represented in Table 1 in the column "Change", which reveals that both UC and BC models either always achieve or do not achieve this criterion.

4 Results

4.1 Model calibration and validation

The eco-hydrological model SWIM was calibrated to three discharge gauges in the UBN: (1) downstream Lake Tana, (2) Kessie, and (3) El Diem. Due to limited data availability, the model was calibrated to the monthly time step using a semi-automated approach. The calibration (1981–1986) and validation (1987–1992) periods for gauge El Diem were on the one hand chosen according to data availability and on the other hand to cover periods of wet and dry years. Data availability for the gauges Lake Tana and Kessie was limited to the years 1969–1975 and 1976–1979, respectively. The gauges were successively calibrated where a parameter sensitivity analysis was performed in a first step to assess reasonable parameter ranges as boundary conditions for the automatic calibration algorithm PEST (Model-Independent Parameter Estimation & Uncertainty Analysis software)[8]. The objective functions to measure model performance are the Nash–Sutcliffe efficiency (NSE) (Nash and Sutcliffe, 1970) and PBIAS, where NSE was the primary criterion.

Figure 2 shows the results of monthly and average monthly discharges at gauge El Diem for calibration (left panel) and validation (right panel). According to Moriasi et al. (2007), NSE values of 0.92 (calibration) and 0.90 (validation) are considered to be very good for the monthly time step. The same classification is achieved for the volumetric errors in both periods. The percent bias (PBIAS) between simulated and observed data is $-6.7\,\%$ (calibration) and $-14.4\,\%$ (validation). SWIM simulates peak discharges adequately in most years with few exceptions of rather large underestimation in the years 1983, 1987, and 1988. One explanation for this is the lack of accuracy of WFD inputs and/or observed discharge in some years. The simulated amount of water percolating into the deep aquifer is about 7 % on average. Without this recharge component, it was not possible to achieve good simulations during the dry period.

Figure S1a and b in the Supplement show the calibration results for the gauges downstream Lake Tana and Kessie. The available GRDC discharge time series for both gauges are rather short and in the case of Tana, the data of the years 1973–1975 are not reliable. Compared to the discharge data given in Dile et al. (2013) and Setegn et al. (2011), maximum discharges are usually around 200–250 $m^3\,s^{-1}$, as is the case in the years 1969–1972 (Fig. S1a). Monthly WFD precipitation volumes do not explain the high discharges observed in the last 3 years. Hence, only the first 4 years were used for calibration, where an NSE of 0.67 and a PBIAS of 23.1 % were achieved. Monthly discharges at gauge Kessie in the four years where GRDC data were available are underestimated by $-18.8\,\%$ and achieved an NSE of 0.92. According

[8]http://www.pesthomepage.org/

Table 1. Selection of uncorrected (UC) and bias-corrected (BC) Earth system models (ESMs) and regional climate models (RCMs).

	R^2	PBIAS	FDC ±30 %			Change	Selection	
Climate model	> 0.85	±30 %	NED	HF	LF	±30 %	pre	final
UC GFDL	×	×	–	×	–	×	–	–
BC GFDL	×	×	×	–	–	×	×	×
UC HadGEM	×	×	×	–	–	×	×	–
BC HadGEM	×	×	×	×	–	×	×	×
UC IPSL	×	–	–	–	–	–	–	–
BC IPSL	×	×	×	–	–	–	×	–
UC MIROC	×	×	×	~	–	×	×	–
BC MIROC	×	×	×	–	–	×	×	×
UC NorESM	–	×	~	~	–	×	–	–
BC NorESM	×	×	×	–	–	×	×	×
UC CanESM2/RCM4	×	×	–	×	–	–	–	–
BC CanESM2/RCM4	×	×	×	–	–	–	×	–
UC CanESM2/RCA4	×	–	–	–	–	×	–	–
BC CanESM2/RCA4	×	×	–	–	–	×	–	–
UC CNRM-CM5/RCA4	×	–	–	–	–	–	–	–
BC CNRM-CM5/RCA4	×	×	×	–	–	–	×	–
UC GFDL/RCA4	×	×	×	×	–	×	×	×
BC GFDL/RCA4	×	×	×	–	–	×	×	–
UC EC-EARTH/Hirham	×	–	–	~	–	×	–	–
BC EC-EARTH/Hirham	×	×	×	–	–	×	×	×
UC EC-EARTH/RACMO	×	–	–	×	–	×	–	–
BC EC-EARTH/RACMO	×	×	×	–	–	×	×	×
UC EC-EARTH/RCA4	×	–	–	–	–	×	–	–
BC EC-EARTH/RCA4	×	×	×	–	–	×	×	×
UC MIROC5/RCA4	×	×	–	×	–	–	–	–
BC MIROC5/RCA4	×	×	~	–	–	–	×	–
UC MPI-M-ESM-LR/RCA4	×	–	–	–	–	×	–	–
BC MPI-M-ESM-LR/RCA4	×	×	×	–	–	×	×	×
UC NorESM1/RCA4	×	×	~	×	×	×	×	×
BC NorESM1/RCA4	×	×	–	–	–	×	–	–

"×" is the criterion achieved; "~" is the criterion almost achieved; "–" is the criterion not achieved. "HF" refers to the high flows ($\leq Q_{10}$); "LF" refers to the low flows ($\geq Q_{90}$). "Change ±30" is the volumetric change between the reference period and RCP 8.5 in 2030–2059. The abbreviation "pre" refers to the preselection; "final" refers to the models selected in the final ensemble.

to Moriasi et al. (2007) the results for the two gauges can be classified to be between good and very good.

4.2 Model performance

4.2.1 Performance of daily and monthly precipitation

Monthly medians and average annual precipitation sums of UC ESM and RCM simulations deviate sometimes strongly from WFD (see Figs. S2, S3, and S4 in the Supplement). The underlying data for the box plots are monthly precipitation sums of the 30-year reference period averaged over the UBN catchment area. Bias correction improved the performance of both indicators considerably in both model ensembles. Deviations of average annual precipitation of all BC ESMs are lower than ±2 %. The results for the BC RCM ensemble are more diverse. Five RCMs deviate $\leq \pm 2$ %, three RCMs $\leq \pm 5$ %, and two RCMs $\leq \pm 7$ %.

Despite the improvement of monthly medians and average annual precipitation sums, bias correction increased the range of monthly precipitation sums critically in several models in both ensembles. This phenomenon can be observed particularly if the deviation of monthly medians between UC simulation and WFD is rather large (e.g. IPSL from May to October, MIROC in July, NorESM in July and August). The effect of increasing variability of monthly precipitation sums is even higher with the method used to bias-correct RCMs and is true for all RCMs (Figs. S3 and S4). The extreme outliers in many models generated by both correction methods are also noticeable.

Not all UC models do adequately represent the unimodal rainfall regime in the UBN. UC NorESM shows for instance a distinct bimodal regime, which is also visible but less pro-

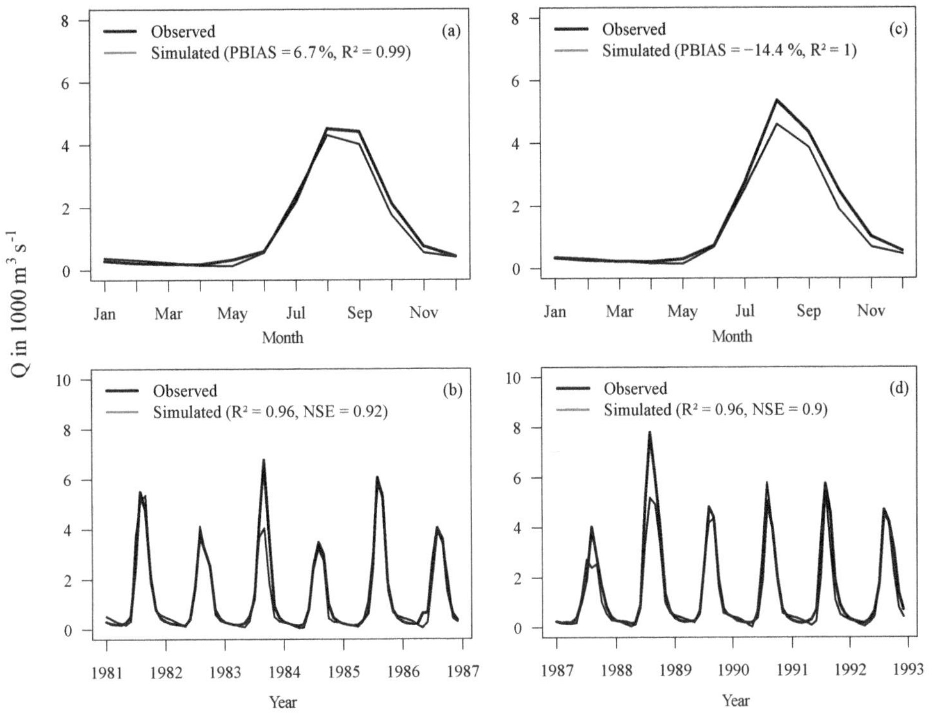

Figure 2. Simulated discharges for calibration **(a, b)** and validation **(c, d)** periods at gauge El Diem (Sudan border) using WATCH Forcing Data (WFD). The annual cycle is shown in the top row and average monthly discharges in the bottom row.

nounced in GFDL and MIROC (Fig. S2) and only weakly visible in MIROC/RCA4 (Fig. S4). Although bias correction eliminated this deficiency, it is questionable at what cost. The physical basis was certainly disrupted by the correction method applied.

Tables S3 and S4 in the Supplement show the following statistical parameters of daily precipitation averaged over the catchment: average number of days with precipitation > 1 mm per annum (*n*Days > 1 mm), average daily precipitation (ave), maximum daily precipitation (max), standard deviation (SD), average precipitation in July, August, and September (ave JAS), and the standard deviation of daily precipitation in July, August, and September (SD JAS). Where Table S3 shows absolute values, Table S4 shows the differences to WFD precipitation (sim-WFD). The two SD parameters were computed by division, SD_{sim}/SD_{WFD}. The Tables show for instance that maximum daily precipitation is underestimated by all UC models except MIROC. Bias correction resulted in overestimation in 13 out of 15 models. All BC RCMs overestimate maximum daily precipitation, many of them significantly; yet the differences in average daily precipitation of BC simulations are, with exceptions, usually rather small. Large deviations in maximum daily precipitation and in the number of rainy days at the same time, while achieving only small differences in average daily precipitation, indicate that the distribution of daily rainfall can differ sometimes strongly among simulations. It is also noticeable that the SD of daily precipitation of all UC models is lower than the WFD SD. Almost all BC simulations show higher SD than the UC simulations, where all ESM SD values are still lower than WFD SD and all RCM SD values are greater than or equal to WFD SD.

4.2.2 Performance of average daily discharge using UC and BC climate input

Bias correction improved the performance of averaged daily discharge simulations ($n = 365$) considerably for all members of the ESM ensemble and for most members of the RCM ensemble. Figures 3 and 4 show the simulated hydrographs in the reference period comparing UC and BC simulations with WFD using R^2 and PBIAS to indicate discharge performance of the annual cycle.

All UC discharge simulations using ESM climate input, except the one based on GFDL, underestimate average annual discharges, which is indicated by negative PBIAS values (Fig. 3). IPSL shows the largest deviations, with a PBIAS of −84 %. All other models deviate less than 30 % from WFD discharges. R^2 values indicate that seasonal discharge patterns are more or less adequately represented by all models, except NorESM, which simulates a bimodal regime with a small peak in June and a high peak in October instead of one single major peak between August and September. Peak discharges simulated with GFDL and MIROC climate input occur approximately 4 weeks later than the peak simulated with WFD. Discharges simulated with HadGEM achieve an R^2 of

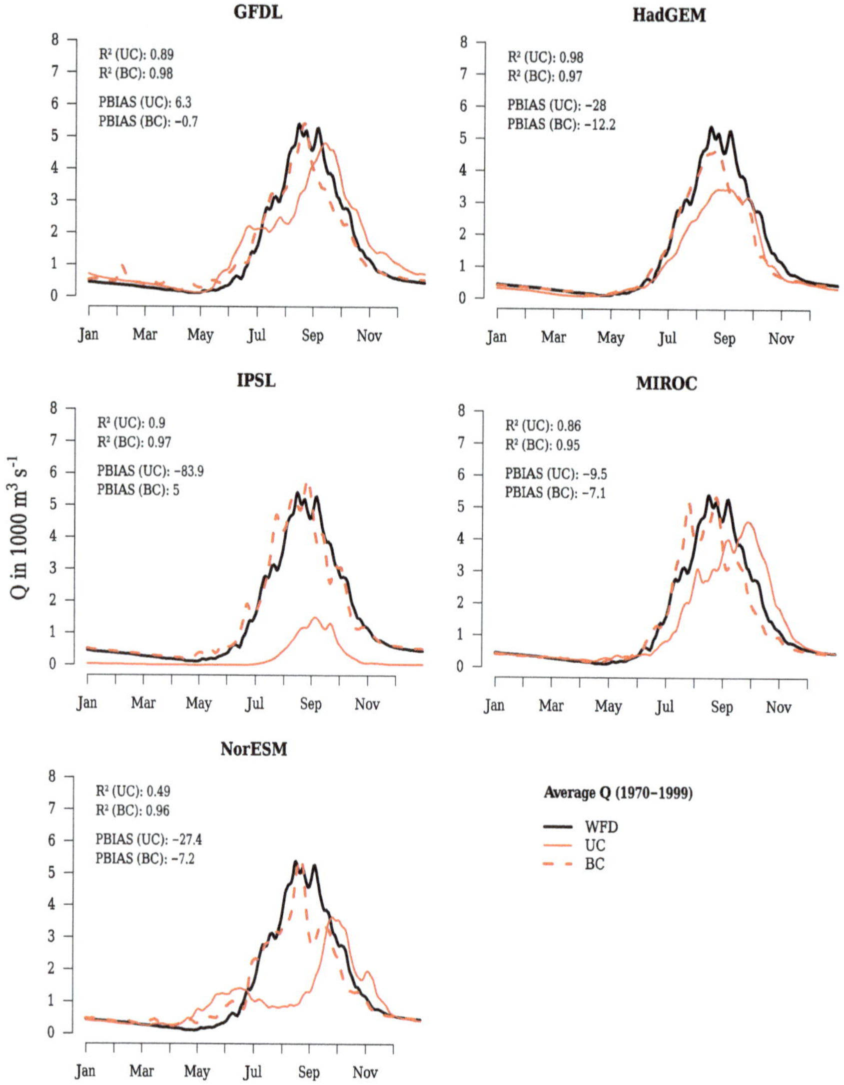

Figure 3. Annual cycle of average daily uncorrected (UC) and bias-corrected (BC) simulated discharges at gauge El Diem using Earth system model input and WATCH Forcing Data (WFD) in the reference period (1970–1999).

0.98 but are too low during the high flow season. Another example is the UC IPSL model, which achieves an R^2 of 0.9, although it underestimates discharge by $-84\,\%$. Hence, high R^2 values can be misleading if they are not combined with a volumetric criterion such as PBIAS.

In contrast to ESMs, the majority of discharge simulations based on UC RCMs overestimate average annual discharges in the reference period (Fig. 4). The deviations of six UC RCMs are larger than 30 %. However, seasonal discharge patterns are generally better represented using UC RCM climate input than UC ESM input. The lowest UC RCM R^2 value is 0.93 compared to an R^2 of 0.49 by NorESM of the UC ESM ensemble. Hence, bias correction improved R^2 values only slightly for 50 % of RCMs. In 60 % of the cases, the volumetric deviation (PBIAS) of BC RCMs is significantly lower than in the corresponding UC models. Based on these two indicators, the performance of BC RCM simulations is generally better than UC RCMs. However, there is a strong tendency of peak flow overestimation in six out of ten BC RCMs, which is not captured by R^2 and PBIAS. Therefore, a visual assessment of hydrographs is important as well as an analysis of daily discharge characteristics using FDCs (see following section).

Taylor diagrams (Taylor, 2001) are another method to visualize model performance showing three performance indicators (R^2, normalized SD, and SD_D) in a single plot (see Fig. 5). They facilitate the visual assessment of model performance where outliers can be easily identified. A model with similar statistical characteristics to the reference dataset would be represented by a point at 1.0 on the x-scale and 0.0 on the y-scale. However, interpretation of normalized values is difficult in terms of numerical thresholds, though

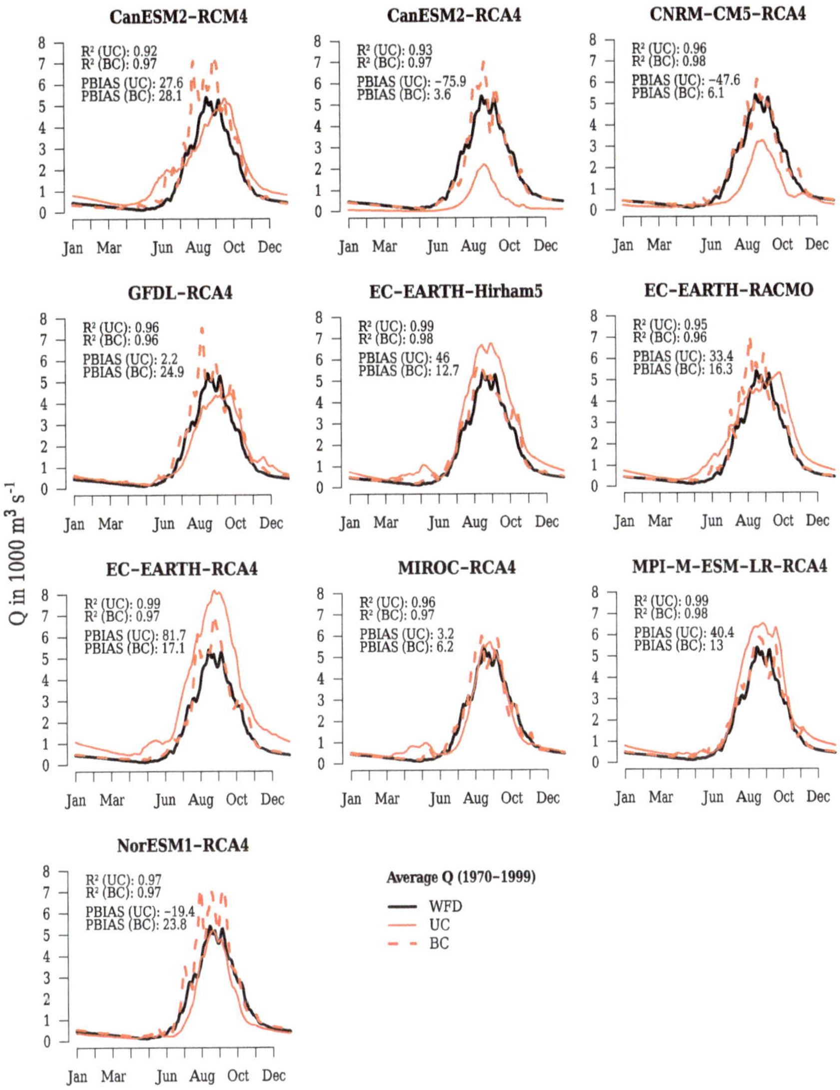

Figure 4. Annual cycle of average daily uncorrected (UC) and bias-corrected (BC) simulated discharges at gauge El Diem using regional climate model input and WATCH Forcing Data (WFD) in the reference period (1970–1999).

Fig. 5a identifies UC IPSL and UC NorESM clearly as outliers. IPSL is, for instance, an outlier because it shows deficiencies at representing SD (0.25 where 1.0 would be ideal) and SD_D (0.79 where 0.0 would be ideal). UC NorESM performs poorly in terms of all indicators. After bias correction all ESMs show rather good performance (see Fig. 5b). Except BC IPSL, all models have lower SD than WFD.

The charac-teristics of RCMs are different. Half of the UC RCMs' SDs (Fig. 5c) deviate more than ±0.25 from standardized WFD but perform much better in terms of R^2.

Interestingly, after bias correction (Fig. 5d), all models show a higher SD than WFD, which is consistent with higher SD of daily rainfall as described in the previous section.

4.2.3 Flow duration curves

FDCs are employed here to analyse and characterize strengths and weaknesses of daily discharge simulations with regard to NED conditions, high flows, low flows, and their extremes. Figure S5 in the Supplement shows FDCs of all ensembles, where the black line represents simulations using WFD. At least one obvious outlier can be clearly identified in both UC ensembles (IPSL and CanESM2-RCA4). Apart from the outliers, NED characteristics are slightly better represented by the UC ESM ensemble (Fig. S5a) than by the UC RCM ensemble (Fig. S5c). Most of the UC RCMs tend to overestimate NED and low flows. At a first glance, the biases were significantly reduced by the correction methods (Fig. S5b and d), especially for NED. However, compared to UC simulations, the correction led to higher biases in the

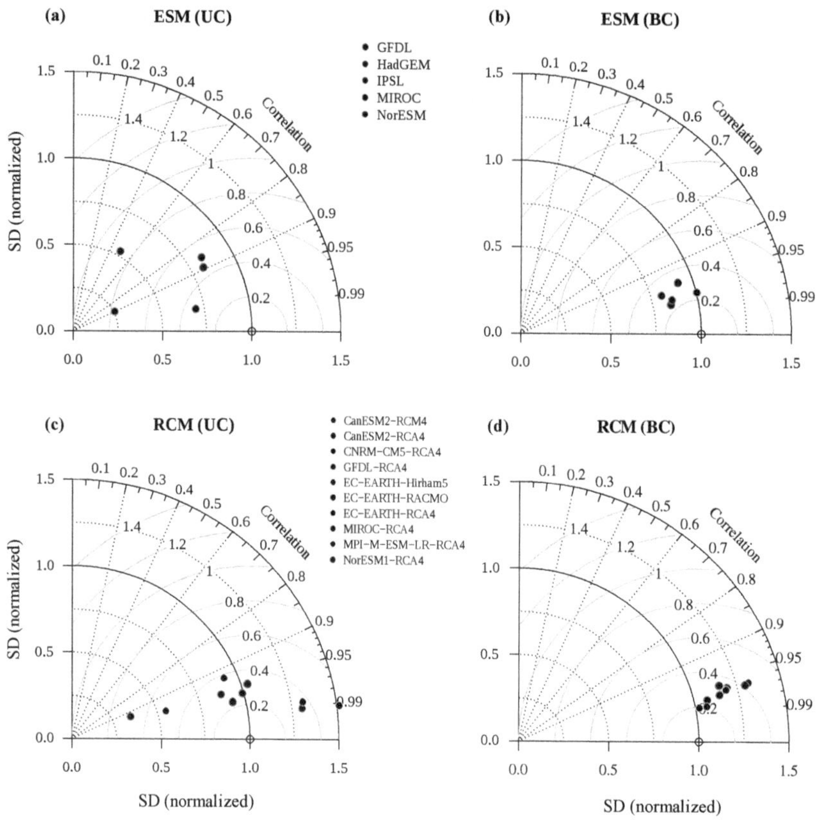

Figure 5. Taylor diagram of average daily discharges at gauge El Diem in the reference period (1970–1999). It shows R^2, standard deviation (SD) normalized by SD_{ref}, and normalized SD_D of discrepancies for Earth system model (ESM) input in the top row and regional climate model (RCM) input in the bottom row.

high and low flow segments and especially in their extreme values. Note that a logarithmic y-scale is used where large deviations in the extreme high flow section appear rather small on this plot although they are in fact extremely high.

Figure 6 overcomes this problem by showing relative deviations of FDCs between discharge time series simulated with climate model inputs and the baseline using WFD. The values corresponding to Fig. 6 are provided by Tables S5–S8 in the Supplement. Assuming that deviations in the range of ±30 % are tolerable, there is not a single UC model (Fig. 6a and c) which fulfils these requirements for all percentile values. However, the UC ESMs' MIROC and HadGEM (Fig. 6a) show acceptable deviations (±30 %) in NED conditions, but there is not a single UC RCM representing NED conditions in the given range (Fig. 6c). The best UC RCM result was achieved with NorESM1-RCA4. Figure 6b and d show that bias correction was successful in correcting the biases of NED for all ESMs and seven out of ten RCMs. The correction method applied to ESMs leads to different patterns in the high and low flow sections compared to the method used to bias-correct RCMs.

Between Q_1 and Q_{10} (high flows), the BC ESMs tend to underestimate values (but in the given range of acceptable deviations), whereas BC RCMs overestimate flows corresponding to these percentiles. There is not a single BC RCM that represents Q_1 conditions in the given range of ±30 %. The smallest overestimation for Q_1 is 52.4 %. All BC RCMs strongly overestimate extreme high flows $Q_{0.1}$ and $Q_{0.01}$. The highest $Q_{0.01}$ overestimation is 656.9 % and the lowest 100.4 % (Table S8). The BC ESMs perform better in the extreme high flow segments. However, only GFDL and HadGEM simulate $Q_{0.1}$ values in the acceptable range and only HadGEM for $Q_{0.01}$ (Table S6).

In the low flow section (between Q_{90} and Q_{99}) there is no BC ESM that performs adequately for all percentile values. Except HadGEM that overestimates low flows, the other models tend to underestimate values. Extreme low flows ($Q_{99.9}$ and $Q_{99.99}$) are only represented by GFDL within the acceptable range. The BC RCMs all underestimate low flows, where four models are within the acceptable range of deviations for Q_{95}; there is only one model within this range for Q_{99} (CanESM2-RCM4). Extreme low flow conditions ($Q_{99.9}$ and $Q_{99.99}$) are only represented adequately by EC-EARTH-RCA4; the other RCMs severely underestimate extreme low flows.

To summarize the evaluation of model performance based on FDCs, it can be stated that bias correction improved the performance of simulated NED significantly. However, with

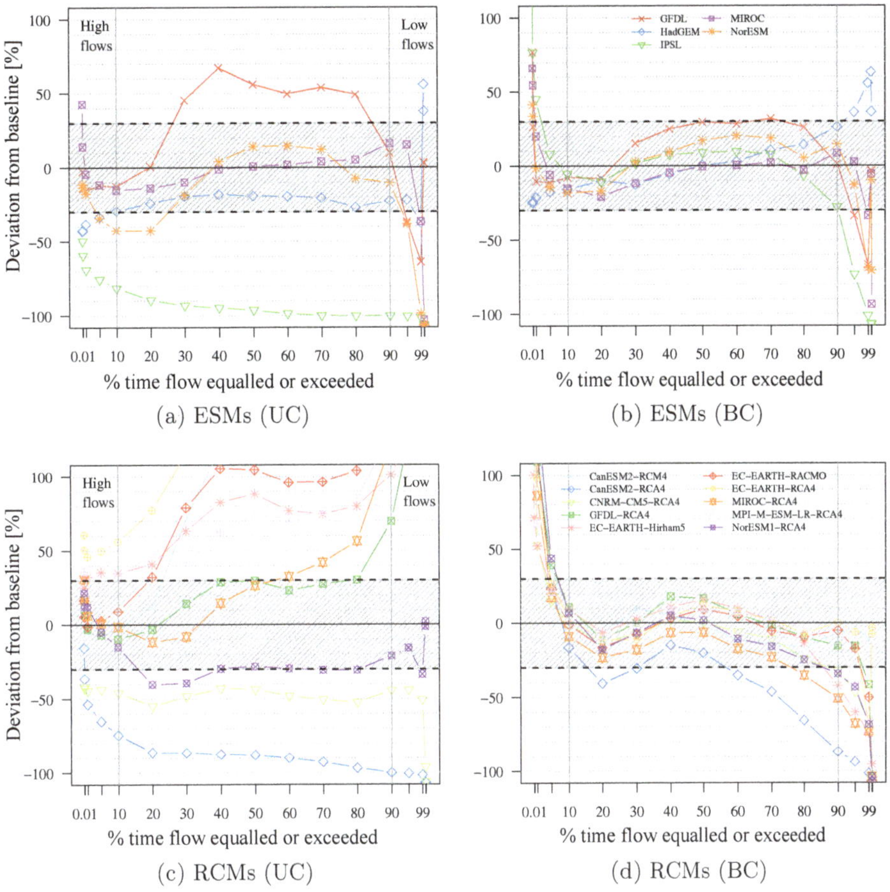

Figure 6. Relative deviations of FDCs from baseline discharge simulation at gauge El Diem using WATCH Forcing Data (WFD) in the reference period (1970–1999). Simulations based on uncorrected (UC) and bias-corrected (BC) Earth system model (ESM) input in the top row and regional climate model (RCM) input in the bottom row.

a few exceptions, both bias correction methods did not improve the performance of high and low flows. This is particularly true for extreme values, which are strongly exaggerated in most cases.

4.3 Temperature, precipitation, and evapotranspiration projections

Figures 7, 8, and 9 show precipitation, temperature, and actual evapotranspiration projections of the selected model ensemble (Sect. 4.5) for the 21st century for RCP 4.5 and RCP 8.5 as anomalies to the reference period in the UBN. They indicate the total range of change and the 5-year moving average (MA5) for both scenarios. The precipitation MA5 does not show a distinct trend of change over the century, but average annual precipitation is projected to be up to 100 mm ($\sim$7 %) higher than in the reference period. The increase is only marginally higher in RCP 8.5 than in RCP 4.5. In Fig. S6 it is shown that a maximum of only three out of 15 UC climate models project decreasing average annual precipitation. The multi-model mean of the CMIP5 ESM ensemble projects showed increasing annual precipitation of 5 % in 2030–2059 and 6 % in 2070–2099 under RCP 4.5 and 8.4 % in 2030–2059 and 15.6 % in 2070–2099 under RCP 8.5. Figure S7 shows where the five ESMs used in this study are situated within the entire CMIP5 ensemble. It is noticeable that only three out of 26 ESMs show declining precipitation trends under RCP 8.5.

Projected surface air temperatures show a clearly increasing trend over the 21st century in both RCPs. Compared to the reference period, the multi-model mean of the selected ensemble projects an increase of 1.7 K (1.5 to 1.9 K) in RCP 4.5 and 2.2 K (1.9 to 3.5 K) in RCP 8.5 in 2050. At the end of the century average temperatures climb up to 2.5 K (1.9 to 4.1 K) under RCP 4.5 and 4.9 K (3.0 to 6.5 K) under RCP 8.5. The multi-model mean of the CMIP5 ESM ensemble projects showed increasing average annual temperatures of 1.6 K in 2030–2059 and 2.3 K in 2070–2099 under RCP 4.5 and 1.7 K in 2030–2059 and 3.9 K in 2070–2099 under RCP 8.5.

Although surface air temperature already increases until 2050 in both scenarios by up to 2.2 K, actual evapotranspiration remains rather stable on the level of the reference pe-

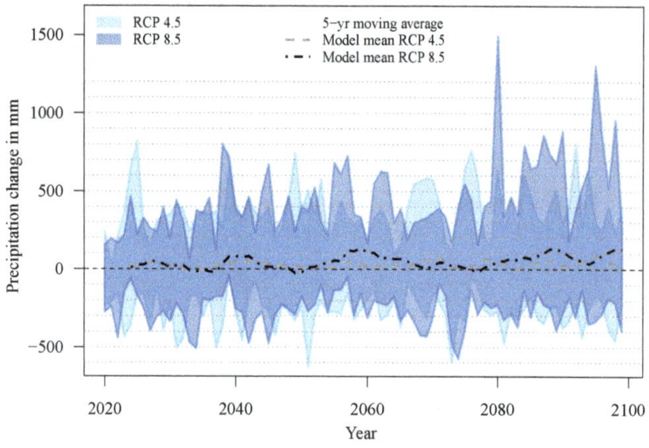

Figure 7. Anomalies of annual precipitation amounts relative to the reference period (1970–1999). Range of selected model ensemble.

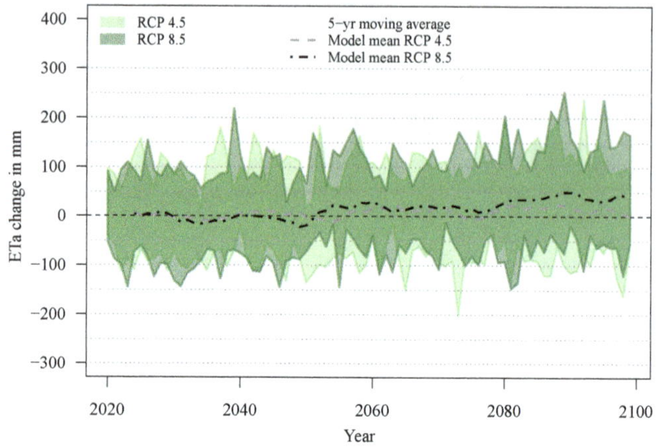

Figure 9. Anomalies of annual actual evapotranspiration (ETa) amounts relative to the reference period (1970–1999). Range of selected model ensemble.

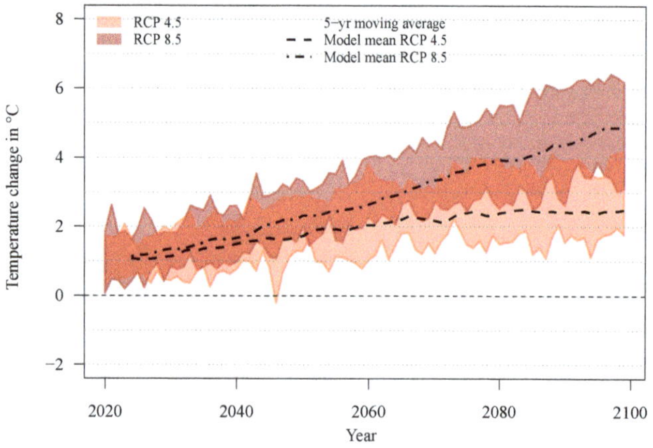

Figure 8. Anomalies of average annual mean air temperature relative to the reference period (1970–1999). Range of selected model ensemble.

riod. Only in the second half of the 21st century do the projected values increase by up to 50 mm per annum. Hence, it can be concluded that actual evapotranspiration is already at its maximum and can only increase if water availability increases too, as is the case after 2050.

4.4 Impact of bias correction on discharge projections

Figures 10 and 11 show projected discharge changes of each single model under RCP 8.5 in the period 2030–2059. The changes are relative to the models' reference period. The figures allow the changes between the reference and the future period of UC and BC models to be investigated, as well as the differences of projected changes between UC and BC simulations. The indicators R^2 and PBIAS are not used to measure the performance, but they indicate the magnitude of change between the reference and the projection period.

The IPSL model shows the largest deviations between the future and the reference period (Fig. 10) for both UC and BC simulations. The UC IPSL model projects an increase of 95.4 % in average annual discharge. A visual assessment supports the previously made assumptions that the IPSL model does not provide adequate climate simulations in the study area. This is true for both UC and BC climate simulations. Aich et al. (2014) applied the same five BC ESMs in four large African river basins and found that also in the Niger basin (comparable climate zone to the Blue Nile River) one of the five models projects extreme and unexplainable changes although it performed adequately in the historical period. In the case of the Niger River basin, it was the MIROC model that behaved awkwardly in the projection period, whereas the IPSL behaved normally in the range of the other models.

The HadGEM model is the only model where bias correction changed the sign of the discharge signal. The simulation with UC climate input projects a decrease of average annual discharges of −2.9 % and the BC simulation an increase of +2.2 %. The results of the NorESM1 model are interesting. The UC model simulates a bimodal rainfall and runoff system with a dry period during the rainy season in July to September. Although the model was forced by bias correction into a completely different system, by pushing the dry season into a rainy season, the projections do not seem anywhere near as disrupted as the IPSL simulation. Hence, the NorESM1 results do not support the assumption that strong bias correction necessarily results in unexpected behaviour in future periods. Looking at the change of average peak magnitudes between UC and BC ESM simulations in the reference and the future period, the change signals are in a similar order, except for simulations based on IPSL. They are also in the order of average peaks simulated with WFD input; compare with Fig. 3.

Figure 11 shows that maximal discharge peaks simulated with RCM climate input are often much higher than average peaks simulated with WFD ($\sim 6000\,\mathrm{m^3\,s^{-1}}$). Where

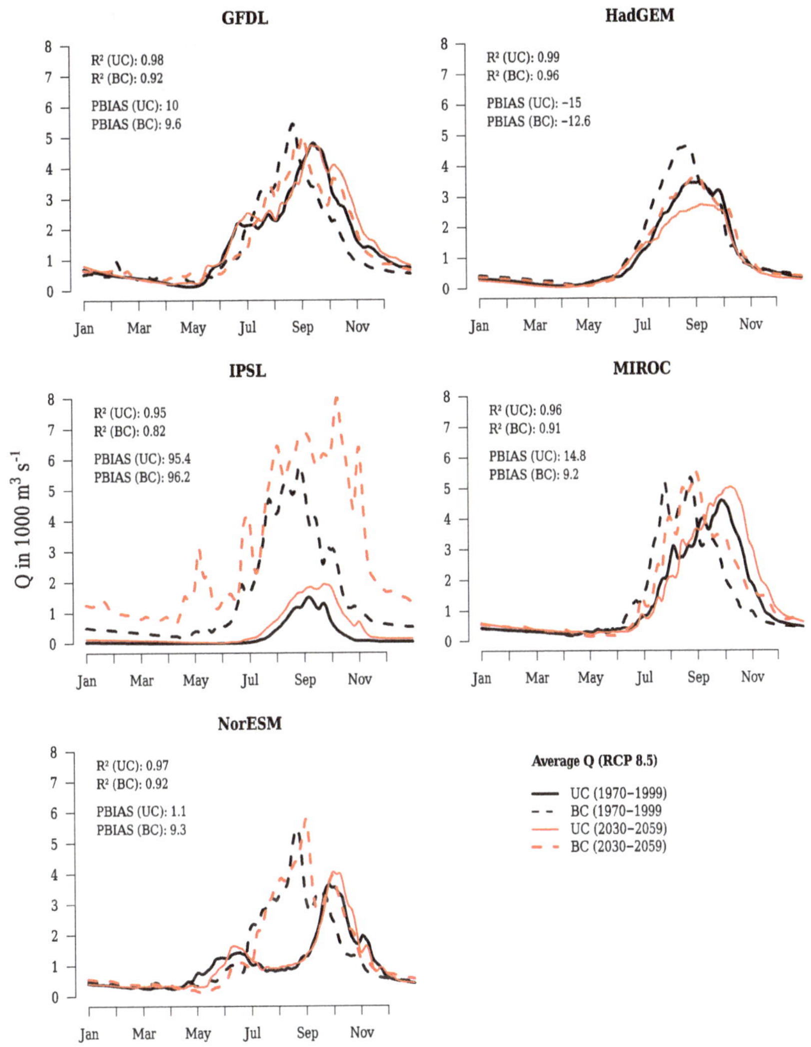

Figure 10. Changes of average daily discharges at gauge El Diem based on uncorrected (UC) and bias-corrected (BC) Earth system model (ESM) input in the period (2030–2059) under RCP 8.5 relative to the models' reference period (1970–1999). R^2 and PBIAS values are computed to show the differences between the projection period and the reference period.

only two UC RCMs simulate higher peaks in the reference period (EC-EARTH-Hirham5 and EC-EARTH-RCA4), five BC RCMs simulate peaks higher than 7000 m^3 s^{-1}. Looking at projected peaks in the period 2030–2059 (RCP 8.5) shows that nine out of ten BC RCM-driven and five UC RCM simulations simulate peaks that are higher than 7000 m^3 s^{-1}. The projected changes of peak discharge magnitudes between UC and BC RCMs are significantly higher in BC simulations in 50 % of the models. This is not surprising because bias correction of RCMs already led to significant overestimation of high flows in the reference period, as was discussed in Sect. 4.2.3. This behaviour is exaggerated in future periods.

4.5 Selected model ensemble

Table 1 summarizes the performance criteria for all UC and BC simulations using R^2, PBIAS, deviations from FDC values, and the change rate. The seasonality criterion $R^2 > 0.85$ was achieved by all simulations except the one based on UC NorESM. Seven out of 30 simulations failed to represent the volumetric deviation criterion PBIAS ± 30 %. Concerning the FDC criteria, 12 simulations passed the NED test, seven simulations the high flow criterion, and only one simulation the low flow criterion. The column "pre" (preselection) shows whether a model fulfilled the criteria in the first three columns. These models might be chosen for a qualitative impact assessment. However, four models that passed the preselection criteria were omitted from the selected model ensemble because they project very high changes in average annual discharges (column "Change"). Sometimes both the UC and BC simulations were judged to be suitable. In order not to put too much weight on the results of one model, only the better simulation (UC or BC) was selected for the final

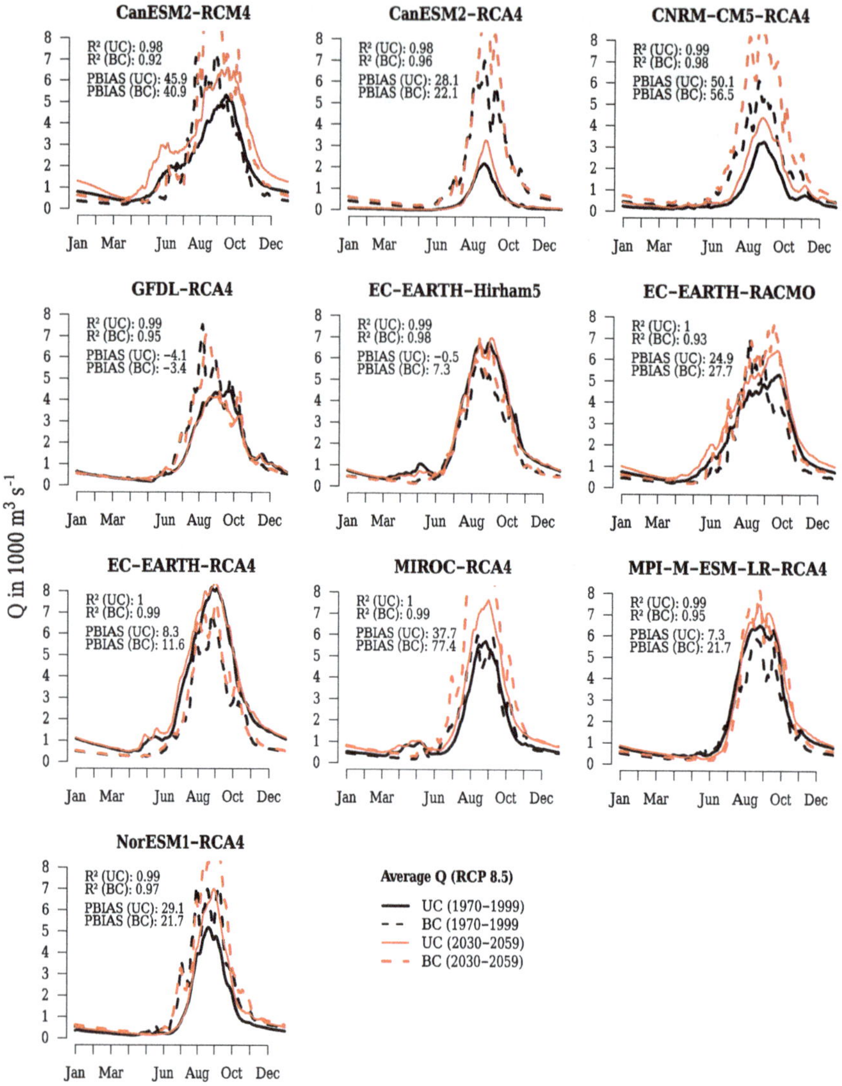

Figure 11. Changes of average daily discharges at gauge El Diem based on uncorrected (UC) and bias-corrected (BC) regional climate model (RCM) input in the period (2030–2059) under RCP 8.5 relative to the models' reference period (1970–1999). R^2 and PBIAS values are computed to show the differences between the projection period and the reference period.

model ensemble and is denoted in the column "final". The latter column indicates that 10 out of 30 simulations passed all performance criteria and thus become members of the selected model ensemble. This ensemble consists of four BC ESMs, four BC RCMs, and two UC RCMs.

4.6 Climate impacts on discharges

In this section, the similarities and differences of projected climate change impacts on Blue Nile discharges at gauge El Diem are discussed. The two UC and BC ESM and RCM ensembles and the selected model ensemble are considered (see Table 1, column "final"). In Figs. 12 and 13 and S8–S11, each model simulation is represented by a semi-transparent polygon, where blueish colours indicate an increase and reddish colours a decrease in monthly discharges. The more saturated the colour, the more models project the same rate of change. The figures show monthly changes relative to average annual discharges in the reference period. This method was chosen in order to avoid overemphasizing large relative changes in dry periods which are not significant compared to annual discharges.

Table 2 shows the total range of changes in average annual discharges projected by the multi-model means of UC and BC ESMs and RCMs and the selected model ensembles. In the near future (2030–2059) in both RCPs, the range of UC models is between 7.4 and 19 %, the range of BC models between 11.3 and 27.7 %, and the range of the selected ensemble between 5.8 and 11.3 %. In the far future (2070–2099) considering both RCPs, the range of UC models is between 7.5 and 21.6 %, the range of BC models between 20.3

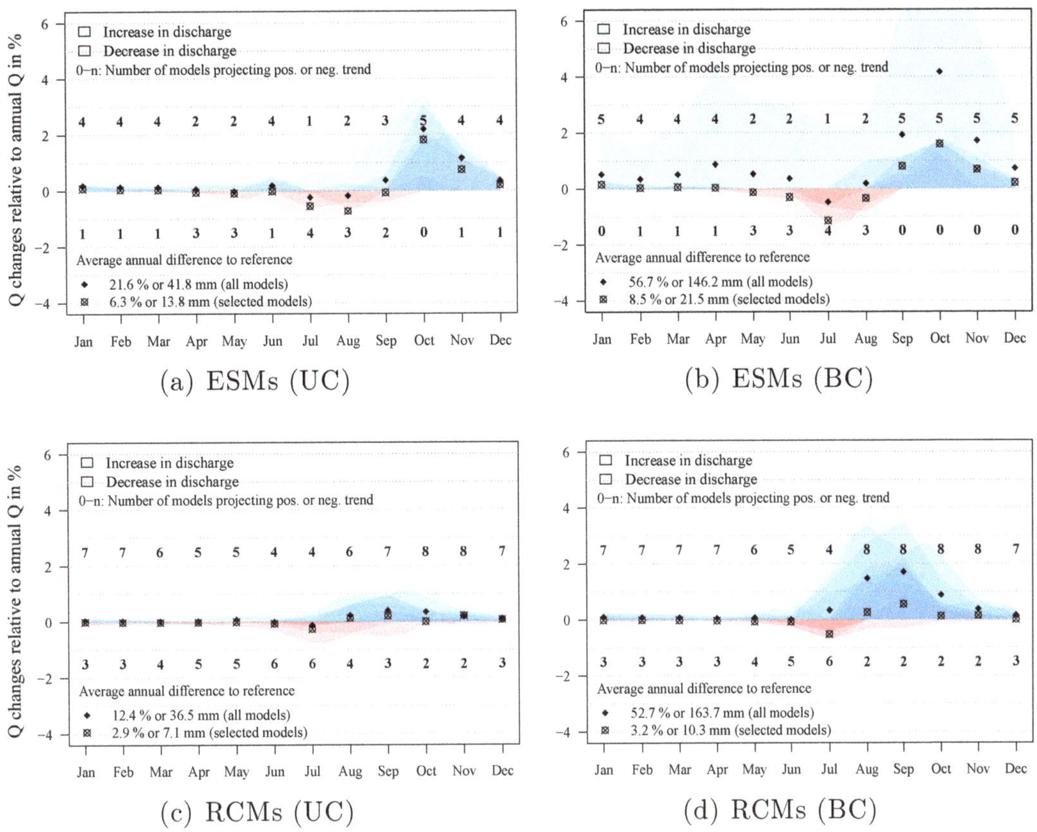

Figure 12. Monthly discharge changes of uncorrected (UC) and bias-corrected (BC) Earth system model (ESM) and regional climate model (RCM) simulations in % under RCP 8.5 (2070–2099). Monthly changes are relative to average annual discharge in the reference period (1970–1999) at gauge El Diem.

Table 2. Projected changes in average annual discharges relative to 1970–1999 in %.

	RCP 4.5		RCP 8.5	
Model ensemble	2030–2059	2070–2099	2030–2059	2070–2099
UC ESMs	7.4	7.5	8.2	21.6
UC RCMs	18.5	14.2	19.0	12.4
BC ESMs	11.3	20.3	24.5	56.7
BC RCMs	23.5	22.3	27.7	52.7
Selected	5.8	8.4	11.3	13.2

and 56.7 %, and the range of the selected ensemble between 8.4 and 13.2 %. The following conclusions summarize the projected changes of average annual discharges more specifically.

- All ensembles in all RCPs and future periods have in common that they all project an increase of average annual discharges. An exception is the selected model ensemble of the UC ESMs under RCP 4.5 (2030–2059), which projects a decrease of −0.4 % (Fig. S8a).

- The multi-model means of both UC and BC RCM ensembles (all models) usually project a higher increase of average annual discharges than the ESM ensembles, except under RCP 8.5 (2070–2099); see Figs. S9d and S11d.

- The multi-model means of BC simulations (both RCPs and periods) always project higher increases in average annual discharges than the UC multi-model means.

- The magnitude of change signals projected by selected models in the respective ensemble is always lower than the magnitude of the whole ensemble. This is mainly caused by the fact that models projecting changes of > ±30 %, between the reference period and 2030–2059

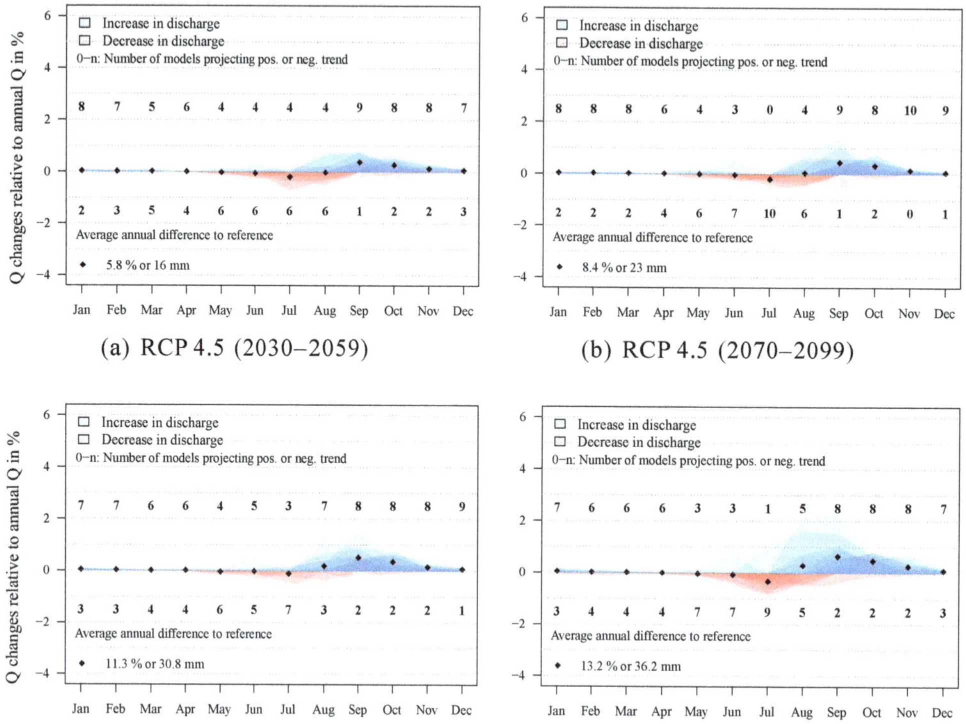

(a) RCP 4.5 (2030–2059) (b) RCP 4.5 (2070–2099)

(c) RCP 8.5 (2030–2059) (d) RCP 8.5 (2070–2099)

Figure 13. Monthly discharge changes of the selected model ensemble (10 models) relative to average annual discharge in the reference period (1970–1999) at gauge El Diem.

under RCP 8.5, were omitted from the ensemble of selected models.

- A noticeable difference between the UC RCM and ESM ensembles is that projected average annual discharges in the far future are lower (RCMs) and higher (ESMs) than in the near future.

There are also general findings concerning changes in seasonality.

- There is a trend of decreasing discharges at the end of the dry season projected by all ensembles in both RCPs and periods. The period indicating a drying trend projected by the ESM ensemble tends to be longer and starts a bit earlier (June/July to August) than the trend projected by RCMs (only July).

- There is a trend of increasing discharges during the rainy season projected by all ensembles in both RCPs and periods. The period indicating higher discharges starts earlier in the RCM ensembles (August to November) than in the ESM ensembles (September to November).

- Both ensembles agree that there is almost no change projected in the dry period between December and May.

5 Discussion and conclusions

Are we using the right fuel to drive hydrological models? What are the likely impacts of climate change on future discharges in the UBN and is there a strong agreement of projected trends? How far does bias correction influence the results and can we trust models that require strong correction? These questions, posed in the introduction, are discussed in the following.

The majority ($\geq 80\,\%$) of the 15 climate models used in this study agree that average annual discharges in the UBN are likely to increase in future. The models project a trend towards decreasing discharges at the end of the dry period (June and July) and an increase during the rainy season (August to November). Due to the use of different climate model ensembles, downscaling approaches, study areas within the UBN, and periods of analysis, a direct comparison with other studies is difficult but clearly reveals that the selection of climate models predominantly influences the results and conclusions made. Setegn et al. (2011) found for instance that the CMIP3 GCMs they used to investigate climate impacts on discharges in the Lake Tana catchment (Blue Nile headwaters) project decreasing trends, but they also state that "...it seems that, by chance, the nine GCMs used in this study are those that show a precipitation decrease...". On the other hand, Dile et al. (2013) conclude that discharges may increase by up to 135 % in the same region. Taking the, sometimes contradicting, results of recent studies into account (Teklesadik et al., 2017; Dile et al., 2013; Mengistu and Sorteberg, 2012; McCartney and Menker Girma, 2012; Setegn et al., 2011; Conway and Schipper, 2011; Diro et al., 2011; Elshamy et al., 2009), one can conclude that climate impacts in the UBN are uncertain but there is a bias towards a wetter future. The findings of this study, using the most recent global and regional climate models as well as precipitation projections of the entire CMIP5 ensemble, underline the latter statement.

Apart from discussing whether the future in the UBN will become generally wetter or drier, decisions with regard to the adaptation of land and water management to changing climatic conditions requires not only information on qualitative but also accurate seasonal quantitative changes. The value of using uncorrected climate simulations to answer those questions is, due to the lack of spatio-temporal accuracy and the lack of statistically representative observed weather characteristics, usually rather limited. Bias correction of climate simulations is an attempt to overcome at least some of these deficiencies.

The reference dataset used to bias-correct climate models and to calibrate and validate the hydrological model is another source of uncertainty. WFD were used in this study because bias correction on ESMs, provided by ISIMIP, was performed on the basis of this dataset. Moreover, WFD provide a sound basis as climate input, particularly in data-scarce regions, as was shown in various studies (Vetter et al., 2015; Aich et al., 2014; Liersch et al., 2013). The use of a different reference dataset would certainly require different calibration parameter settings and correction factors but would probably not impact the change signals. The most important issue in this connection is the consistency in using the same reference for calibration, validation, and bias correction.

As was shown in this study, monthly medians and average annual precipitation amounts of UC ESM and RCM simulations deviate sometimes strongly from reference climate. Although bias correction improved the performance of average climate conditions, the range of monthly precipitation amounts increased critically in several models, producing some extreme outliers in both ensembles. This phenomenon was particularly observed in simulations where deviations of monthly medians between UC simulations and WFD were rather large in the reference period. Average daily precipitation and the number of rainy days were considerably improved by bias correction, but 13 out of 15 BC models overestimate daily precipitation maxima, and many of them significantly. Hence, the bias correction methods applied to ESMs and RCMs in this study could be considered to be only partly successful. While achieving significant improvement in terms of average daily, monthly, and annual precipitation characteristics, increasing variability of precipitation amounts, and therefore under- and overestimation of extremes, was the result in many simulations.

This phenomenon is problematic for impact studies and the application of hydrological models, particularly if changes of extreme values are the subject of investigation. Large overestimation of precipitation on some days or in some months, for instance, which are balanced by dry months in the long term, can lead to large amounts of excess water that may be simulated almost entirely as surface runoff by the hydrological model. Therefore, it is reasonable to use hydrological performance indicators to evaluate the suitability of climate simulations, particularly for quantitative impact studies, and to create a subset of models for the impact assessment. Another way to deal with low performance in the simulation of extremes in impact studies is to analyse changes in return periods of extreme events (Hattermann et al., 2016).

Due to the fact that discharge simulations, based on climate simulations, cannot be compared to observed discharges on a real-time daily, monthly, or annual basis, the methods to evaluate discharge performance are limited. In this study, the annual cycle (daily time series averaged over the simulation period) was characterized by R^2 and PBIAS, where R^2 was a measure of seasonality and PBIAS a measure of volumetric deviations. Flow duration curves (FDCs) were used to characterize the distribution of average flow conditions, high and low flows, as well as their extremes, by using the whole time series of daily discharge simulations. Unsurprisingly, discharge simulations show similar deficiencies to precipitation simulations. Using bias-corrected climate simulations improved the performance of non-extreme discharges (NED) significantly but, with few exceptions, the performance of high and low flows did not improve; in fact, it worsened in most of the simulations. Many BC discharge simulations tend to exaggerate high (overestimation) and low flows (underestimation). Comparing peak discharges using UC and BC climate input, for instance, showed a tremendous increase in some BC simulations, although average monthly precipitation patterns of BC models achieved a much better fit than their UC counterparts. Moreover, the multi-model means of BC simulations (both RCPs and periods) always project higher increases in average annual discharges than the UC multi-model means. However, a hydrological impact study in the Danube River basin showed in turn that relative changes in average monthly discharges projected using UC and BC climate models are overall comparable (Stagl and Hattermann, 2015).

Knowing these limitations, one should carefully consider the model's suitability and the purpose it is being used for. An impact study focusing on relative changes of future water availability may have lower requirements in terms of model accuracy than a study with the aim of investigating future extremes, such as floods and droughts or a study addressing land and water management issues including irrigation and/or reservoir operations. Whenever complex water management is involved, bias correction is often unavoidable because the simulation of reservoir and irrigation operations requires rather accurate hydrological input. However, to simply trust in climate input only because it was bias-corrected would be naive. Therefore, the question of model selection is valid. Why should one use or trust models to assess changes in seasonal patterns, for instance, that have not represented those patterns in the past or use a model to investigate future flood risk that completely fails to represent rainfall extremes? Again, bias correction may help to overcome some quality issues but it was also found in this study that improving climate simulations in the reference period does not guarantee higher quality or reliability in simulating future periods. On the contrary, the greater the necessity to correct a particular model, the higher the risk that BC simulations will show unexpected behaviour in future periods, where exceptions confirm the rule. Examples confirming this assumption are the following models: IPSL, CanESM2-RCA4, CNRM-CM5-RCA4, and MIROC-RCA4. However, the NorESM1 model is an exception here, because the BC simulation does not show extreme changes in future periods although strong bias correction was necessary in some months to force the model from a bimodal into a unimodal rainfall regime. It should be emphasized that the analysis of climate model performance in this study is only valid for the region of the UBN. It does not imply that a model which performed poorly in this study area will generally perform poorly in other regions, too.

The authors of this study conclude that a purpose-driven selection of a climate model subset is a reasonable approach, particularly in a regional context. To identify models that perform to a good level, the selection process should include an analysis of climate inputs, seasonal discharge patterns, volumetric deviations, and daily dynamics (FDCs), and an assessment of the magnitude of projected future changes. It is also worth mentioning that the thresholds defined to evaluate model performance have a subjective component and are based on statistical parameters, graphical data interpretation, and modelling expertise. If the thresholds had been set more critically in this study, almost no climate model would have passed the evaluation process successfully. The rather weak thresholds were a compromise and reveal the fact that the performance of many climate models is still far beyond being adequate for applied quantitative impact studies. This statement includes bias-corrected simulations and implies that the ability of bias correction can, depending on the approach, be rather limited and thus does not necessarily improve the reliability per se. In another river basin with different characteristics, e.g. with a nival regime or a bimodal rainfall regime, the performance criteria and their thresholds may have been defined differently. Hence, the model selection method can be applied to other river basins but it is always necessary to consider region-specific characteristics that may require the introduction of new criteria adapted to the situation at hand. However, model selection for regional impact studies is only a reasonable, justifiable, and recommended approach if the uncertainties of the selected ensemble are communicated within the context of the whole model ensemble.

This study demonstrated that neither the trend-preserving method applied to the five ESMs nor the harmonic-based method used to bias-correct the 10 RCMs was able to generate fully satisfactory climate inputs for a regional hydrological impact study with high demands in terms of quantitative accuracy. Hence, further research is required to improve regional climate simulations and/or to investigate alternative correction methods or approaches to make climate simulations meaningful for application-oriented regional studies available. Currently, the most promising solutions seem to be sophisticated delta-change methods, as suggested by Anandhi et al. (2011), Bosshard et al. (2011), and Chiew et al. (2009).

Competing interests. The authors declare that they have no conflict of interest.

Acknowledgements. This research was funded by the German Federal Foreign Office and supported by the Ethiopian Environmental Protection Authority and the German Embassy in Addis Ababa. In addition we thank the two anonymous reviewers for their constructive comments which helped us to improve the quality of the manuscript.

Edited by: Ralf Merz

References

Abdo, K. S., Fiseha, B. M., Rientjes, T. H. M., Gieske, A. S. M., and Haile, A. T.: Assessment of climate change impacts on the hydrology of Gilgel Abay catchment in Lake Tana basin, Ethiopia, Hydrol. Process., 23, 3661–3669, https://doi.org/10.1002/hyp.7363, 2009.

Addor, N. and Seibert, J.: Bias correction for hydrological impact studies – beyond the daily perspective, Hydrol. Process., 28, 4823–4828, https://doi.org/10.1002/hyp.10238, 2014.

Aich, V., Liersch, S., Vetter, T., Huang, S., Tecklenburg, J., Hoffmann, P., Koch, H., Fournet, S., Krysanova, V., Müller, E. N., and Hattermann, F. F.: Comparing impacts of climate change on streamflow in four large African river basins, Hydrol. Earth Syst. Sci., 18, 1305–1321, https://doi.org/10.5194/hess-18-1305-2014, 2014.

Anandhi, A., Frei, A., Pierson, D. C., Schneiderman, E. M., Zion, M. S., Lounsbury, D., and Matonse, A. H.: Examination of change factor methodologies for climate change impact assessment, Water Resour. Res., 47, https://doi.org/10.1029/2010WR009104, 2011.

Arnold, J., Allen, P., and Bernhardt, G.: A comprehensive surface groundwater flow model, J. Hydrol., 142, 47–69, 1993.

Bartholomé, E. and Belward, A.: GLC2000: a new approach to global land cover mapping from Earth observation data, Int. J. Remote Sens., 26, 1959–1977, https://doi.org/10.1080/01431160412331291297, 2005.

Berg, P., Feldmann, H., and Panitz, H.-J.: Bias correction of high resolution regional climate model data, J. Hydrol., 448–449, 80–92, https://doi.org/10.1016/j.jhydrol.2012.04.026, 2012.

Beyene, T., Lettenmaier, D., and Kabat, P.: Hydrologic impacts of climate change on the Nile River Basin: implications of the 2007 IPCC scenarios, Climatic Change, 100, 433–461, https://doi.org/10.1007/s10584-009-9693-0, 2010.

Bosshard, T., Kotlarski, S., Ewen, T., and Schär, C.: Spectral representation of the annual cycle in the climate change signal, Hydrol. Earth Syst. Sci., 15, 2777–2788, https://doi.org/10.5194/hess-15-2777-2011, 2011.

Bryan, E., Deressa, T. T., Gbetibouo, G. A., and Ringler, C.: Adaptation to climate change in Ethiopia and South Africa: options and constraints, Environ. Sci. Policy, 12, 413–426, https://doi.org/10.1016/j.envsci.2008.11.002, 2009.

Busby, J., Cook, K., Vizy, E., Smith, T., and Bekalo, M.: Identifying hot spots of security vulnerability associated with climate change in Africa, Climatic Change, 124, 717–731, https://doi.org/10.1007/s10584-014-1142-z, 2014.

Chiew, F. H. S., Teng, J., Vaze, J., Post, D. A., Perraud, J. M., Kirono, D. G. C., and Viney, N. R.: Estimating climate change impact on runoff across southeast Australia: Method, results, and implications of the modeling method, Water Resour. Res., 45, W10414, https://doi.org/10.1029/2008WR007338, 2009.

Christensen, J. H., Boberg, F., Christensen, O. B., and Lucas-Picher, P.: On the need for bias correction of regional climate change projections of temperature and precipitation, Geophys. Res. Lett., 35, L20709, https://doi.org/10.1029/2008GL035694, 2008.

Conway, D. and Hulme, M.: Recent fluctuations in precipitation and runoff over the Nile sub-basins and their impact on main Nile discharge, Climatic Change, 25, 127–151, https://doi.org/10.1007/BF01661202, 1993.

Conway, D. and Schipper, E. L. F.: Adaptation to climate change in Africa: Challenges and opportunities identified from Ethiopia, Global Environ. Change, 21, 227–237, https://doi.org/10.1016/j.gloenvcha.2010.07.013, 2011.

Deressa, T. T., Hassan, R. M., and Ringler, C.: Perception of and adaptation to climate change by farmers in the Nile basin of Ethiopia, J. Agr. Sci., 149, 23–31, https://doi.org/10.1017/S0021859610000687, 2011.

Di Baldassarre, G., Elshamy, M., van Griensven, A., Soliman, E., Kigobe, M., Ndomba, P., Mutemi, J., Mutua, F., Moges, S., Xuan, Y., Solomatine, D., and Uhlenbrook, S.: Future hydrology and climate in the River Nile basin: a review, Hydrolog. Sci. J., 56, 199–211, https://doi.org/10.1080/02626667.2011.557378, 2011.

Dile, Y. T., Berndtsson, R., and Setegn, S. G.: Hydrological Response to Climate Change for Gilgel Abay River, in the Lake Tana Basin – Upper Blue Nile Basin of Ethiopia, PLOS ONE, 8, https://doi.org/10.1371/journal.pone.0079296, 2013.

Diro, G. T., Grimes, D. I. F., Black, E., O'Neill, A., and Pardo-Iguzquiza, E.: Evaluation of reanalysis rainfall estimates over Ethiopia, Int. J. Climatol., 29, 67–78, https://doi.org/10.1002/joc.1699, 2009.

Diro, G. T., Toniazzo, T., and Shaffrey, L.: Ethiopian Rainfall in Climate Models, in: African Climate and Climate Change, edited by: Williams, C. J. R. and Kniveton, D. R., Vol. 43 of Advances in Global Change Research, 51–69, Springer Netherlands, https://doi.org/10.1007/978-90-481-3842-5_3, 2011.

Dobler, A. and Ahrens, B.: Precipitation by a regional climate model and bias correction in Europe and South Asia, Meteorol. Z., 17, 499–509, 2008.

Dobler, A., Yaoming, M., Sharma, N., Kienberger, S., and Ahrens, B.: Regional climate projections in two alpine river basins: Upper Danube and Upper Brahmaputra, Adv. Sci. Res., 7, 11–20, https://doi.org/10.5194/asr-7-11-2011, 2011.

Dosio, A. and Paruolo, P.: Bias correction of the ENSEMBLES high-resolution climate change projections for use by impact models: Evaluation on the present climate, J. Geophys. Res.-Atmos., 116, D16106, https://doi.org/10.1029/2011JD015934, 2011.

Elshamy, M., di Baldassarre, G., and van Griensven, A.: Characterizing Climate Model Uncertainty Using an Informal Bayesian Framework: Application to the River Nile, J. Hydrol. Eng. ASCE, 18, 582–589, https://doi.org/10.1061/(ASCE)HE.1943-5584.0000656, 2013.

Elshamy, M. E., Seierstad, I. A., and Sorteberg, A.: Impacts of climate change on Blue Nile flows using bias-corrected GCM scenarios, Hydrol. Earth Syst. Sci., 13, 551–565, https://doi.org/10.5194/hess-13-551-2009, 2009.

FAO, IIASA, ISRIC, ISSCAS, and JRC: Harmonized World Soil Database (version 1.1), FAO, Rome, Italy and IIASA, Laxenburg, Austria, 2009.

Gebreluel, G.: Ethiopia's Grand Renaissance Dam: Ending Africa's Oldest Geopolitical Rivalry?, Wash. Quart., 37, 25–37, https://doi.org/10.1080/0163660X.2014.926207, 2014.

Gudmundsson, L., Bremnes, J. B., Haugen, J. E., and Engen-Skaugen, T.: Technical Note: Downscaling RCM precipitation to the station scale using statistical transformations – a comparison of methods, Hydrol. Earth Syst. Sci., 16, 3383–3390, https://doi.org/10.5194/hess-16-3383-2012, 2012.

Hagemann, S., Chen, C., Haerter, J. O., Heinke, J., Gerten, D., and Piani, C.: Impact of a Statistical Bias Correction on the Projected Hydrological Changes Obtained from Three GCMs and Two Hydrology Models, J. Hydrometeorol., 12, 556–578, https://doi.org/10.1175/2011JHM1336.1, 2011.

Hargreaves, G. and Samani, Z.: Reference crop evapotranspiration from temperature, T. ASAE, 11, 96–99, 1985.

Hattermann, F. F., Huang, S., Burghoff, O., Hoffmann, P., and Kundzewicz, Z. W.: Brief Communication: An update of the article "Modelling flood damages under climate change conditions – a case study for Germany", Nat. Hazards Earth Syst. Sci., 16, 1617–1622, https://doi.org/10.5194/nhess-16-1617-2016, 2016.

Headey, D., Taffesse, A. S., and You, L.: Diversification and Development in Pastoralist Ethiopia, World Dev., 56, 200–213, https://doi.org/10.1016/j.worlddev.2013.10.015, 2014.

Hempel, S., Frieler, K., Warszawski, L., Schewe, J., and Piontek, F.: A trend-preserving bias correction – the ISI-MIP approach, Earth Syst. Dynam., 4, 219–236, https://doi.org/10.5194/esd-4-219-2013, 2013.

Ibrahim, A.: The Nile Basin Cooperative Framework Agreement: The Beginning of the End of Egyptian Hydro-Political Hegemony, Missouri Environmental Law and Policy Review, 18, 284–312, available at: https://scholarship.law.missouri.edu/cgi/viewcontent.cgi?article=1395&context=jesl (last access: 24 January 2018), 2012.

IPCC: Climate Change 2013. The Physical Science Basis. Working Group I Contribution to the Fifth Assessment Report of the Intergovernmental Panel on Climate Change, Tech. rep., IPCC, available at: http://www.ipcc.ch/report/ar5/wg1/ (last access: 14 April 2016), 2013.

Jarvis, A., Reuter, H., Nelson, A., and Guevara, E.: Hole-filled seamless SRTM data V4, International Centre for Tropical Agriculture (CIAT), available at: http://srtm.csi.cgiar.org (last access: 9 February 2016), 2008.

Jeuland, M. and Whittington, D.: Water resources planning under climate change: Assessing the robustness of real options for the Blue Nile, Water Resour. Res., 50, 2086–2107, https://doi.org/10.1002/2013WR013705, 2014.

Kim, U., Kaluarachchi, J. J., and Smakhtin, V. U.: Climate Change Impacts on Hydrology and Water Resources of the Upper Blue Nile River Basin, Ethiopia, Research Report 126, IWMI, 2008.

King, A.: An Assessment of Reservoir Filling Policies under a Changing Climate for Ethiopias Grand Renaissance Dam, PhD thesis, Drexel University, 2013.

Koch, H., Liersch, S., and Hattermann, F.: Integrating water resources management in eco-hydrological modelling, Water Sci. Technol., 67, 1525–1533, https://doi.org/10.2166/wst.2013.022, 2013.

Krysanova, V., Meiner, A., Roosaare, J., and Vasilyev, A.: Simulation modelling of the coastal waters pollution from agricultural watershed, Ecol. Model., 49, 7–29, 1989.

Krysanova, V., Hattermann, F., and Wechsung, F.: Development of the ecohydrological model SWIM for regional impact studies and vulnerability assessment, Hydrol. Process., 19, 763–783, https://doi.org/10.1002/hyp.5619, 2005.

Krysanova, V., Hattermann, F., Huang, S., Hesse, C., Vetter, T., Liersch, S., Koch, H., and Kundzewicz, Z. W.: Modelling climate and land use change impacts with SWIM: lessons learnt from multiple applications, Hydrolog. Sci. J., 60, 606–635, https://doi.org/10.1080/02626667.2014.925560, 2015.

Liersch, S.: Discharge simulations for the Blue Nile at gauge El Diem based on uncorrected and bias-corrected GCM and RCM inputs, Potsdam Institute for Climate Impact Research, Dataset, https://doi.org/10.4121/uuid:05b9f40f-583d-479b-a79e-f961f72436db, 2018.

Liersch, S., Cools, J., Kone, B., Koch, H., Diallo, M., Aich, V., Fournet, S., and Hattermann, F.: Vulnerability of food production in the Inner Niger Delta to water resources management under climate variability and change, Environ. Sci. Policy, 34, 18–33, https://doi.org/10.1016/j.envsci.2012.10.014, 2013.

Liersch, S., Koch, H., and Hattermann, F. F.: Management Scenarios of the Grand Ethiopian Renaissance Dam and Their Impacts under Recent and Future Climates, Water, 9, 728, https://doi.org/10.3390/w9100728, 2017.

Liersch, S., Rust, H., Dobler, A., Kruschke, T., and Fischer, M.: Bias-corrected CORDEX precipitation, min/mean/max temperature for Ethiopia, RCP 4.5 and RCP 8.5, GFZ Data Services, https://doi.org/10.5880/PIK.2018.009, 2018.

Maraun, D., Wetterhall, F., Ireson, A. M., Chandler, R. E., Kendon, E. J., Widmann, M., Brienen, S., Rust, H. W., Sauter, T., Themeßl, M., Venema, V. K. C., Chun, K. P., Goodess, C. M., Jones, R. G., Onof, C., Vrac, M., and Thiele-Eich, I.: Precipitation downscaling under climate change: Recent developments to bridge the gap between dynamical models and the end user, Rev. Geophys., 48, RG3003, https://doi.org/10.1029/2009RG000314, 2010.

McCartney, M. P. and Menker Girma, M.: Evaluating the downstream implications of planned water resource development in the Ethiopian portion of the Blue Nile River, Water Int., 37, 362–379, https://doi.org/10.1080/02508060.2012.706384, 2012.

Megersa, B., Markemann, A., Angassa, A., Ogutu, J. O., Piepho, H.-P., and Zarate, A. V.: Impacts of climate change and variability on cattle production in southern Ethiopia: Perceptions and empirical evidence, Agr. Syst., 130, 23–34, https://doi.org/10.1016/j.agsy.2014.06.002, 2014.

Meinshausen, M., Smith, S., Calvin, K., Daniel, J., Kainuma, M., Lamarque, J.-F., Matsumoto, K., Montzka, S., Raper, S., Riahi, K., Thomson, A., Velders, G., and Vuuren, D.: The RCP greenhouse gas concentrations and their extensions from 1765 to 2300, Climatic Change, 109, 213–241, https://doi.org/10.1007/s10584-011-0156-z, 2011.

Mengistu, D., Bewket, W., and Lal, R.: Recent spatiotemporal temperature and rainfall variability and trends over the Upper Blue Nile River Basin, Ethiopia, Int. J. Climatol., 34, 2278–2292, https://doi.org/10.1002/joc.3837, 2014.

Mengistu, D. T. and Sorteberg, A.: Sensitivity of SWAT simulated streamflow to climatic changes within the Eastern Nile River basin, Hydrol. Earth Syst. Sci., 16, 391–407, https://doi.org/10.5194/hess-16-391-2012, 2012.

Moriasi, D. N., Arnold, J. G., Van Liew, M. W., Bingner, R. L., Harmel, R. D., and Veith, T. L.: Model evaluation guidelines for systematic quantification of accuracy in watershed simulations, T. ASABE, 50, 885–900, 2007.

Nash, J. and Sutcliffe, J.: River flow forecasting through conceptual models, Part 1 – a discussion of principles, J. Hydrol., 10, 282–290, https://doi.org/10.1016/0022-1694(70)90255-6, 1970.

Piani, C., Weedon, G., Best, M., Gomes, S., Viterbo, P., Hagemann, S., and Haerter, J.: Statistical bias correction of global simulated daily precipitation and temperature for the application of hydrological models, J. Hydrol., 395, 199–215, https://doi.org/10.1016/j.jhydrol.2010.10.024, 2010.

Pierce, D. W., Barnett, T. P., Santer, B. D., and Gleckler, P. J.: Selecting global climate models for regional climate change studies, P. Natl. Acad. Sci. USA, 106, 8441–8446, https://doi.org/10.1073/pnas.0900094106, 2009.

Rust, H. W., Kruschke, T., Dobler, A., Fischer, M., and Ulbrich, U.: Discontinuous daily Temperatures in the WATCH forcing data setes, J. Hydrometeorol., 16, 465–472, https://doi.org/10.1175/JHM-D-14-0123.1, 2015.

Schmidli, J., Frei, C., and Vidale, P. L.: Downscaling from GCM precipitation: A benchmark for dynamical and statistical downscaling methods, Int. J. Climatol., 26, 679–689, https://doi.org/10.1002/joc.1287, 2006.

Setegn, S. G., Rayner, D., Melesse, A. M., Dargahi, B., and Srinivasan, R.: Impact of climate change on the hydroclimatology of Lake Tana Basin, Ethiopia, Water Resour. Res., 47, W04511, https://doi.org/10.1029/2010WR009248, 2011.

Simane, B., Zaitchik, B. F., and Mesfin, D.: Building Climate Resilience in the Blue Nile/Abay Highlands: A Framework for Action, Int. J. Environ. Res. Pu., 9, 610–631, https://doi.org/10.3390/ijerph9020610, 2012.

Smakhtin, V.: Estimating daily flow duration curves from monthly streamflow data, Water SA, 26, 13–18, 2000.

Soliman, E. S., Sayed, M. A. A., and Jeuland, M.: Impact Assessment of Future Climate Change for the Blue Nile Basin, Using a RCM Nested in a GCM, Nile Basin Water Engineering Scientific Magazine, 2, 15–30, 2009.

Stagl, J. C. and Hattermann, F. F.: Impacts of Climate Change on the Hydrological Regime of the Danube River and Its Tributaries Using an Ensemble of Climate Scenarios, Water, 7, 6139–6172, https://doi.org/10.3390/w7116139, 2015.

Sutcliffe, J. and Parks, Y.: The Hydrology of the Nile, no. 5 in Special Publication, IAHS, Institute of Hydrology, Wallingford, UK, 1999.

Taye, M. T. and Willems, P.: Temporal variability of hydroclimatic extremes in the Blue Nile basin, Water. Resour. Res., 48, W03513, https://doi.org/10.1029/2011WR011466, 2012.

Taye, M. T., Willems, P., and Block, P.: Implications of climate change on hydrological extremes in the Blue Nile basin: A review, Journal of Hydrology: Regional Studies, 4, 280–293, https://doi.org/10.1016/j.ejrh.2015.07.001, 2015.

Taylor, K. E.: Summarizing multiple aspects of model performance in a single diagram, J. Geophys. Res.-Atmos., 106, 7183–7192, https://doi.org/10.1029/2000JD900719, 2001.

Teklesadik, A. D., Alemayehu, T., van Griensven, A., Kumar, R., Liersch, S., Eisner, S., Tecklenburg, J., Ewunte, S., and Wang, X.: Inter-model comparison of hydrological impacts of climate change on the Upper Blue Nile basin using ensemble of hydrological models and global climate models, Climatic Change, 141, 517–532, https://doi.org/10.1007/s10584-017-1913-4, 2017.

Teutschbein, C. and Seibert, J.: Regional Climate Models for Hydrological Impact Studies at the Catchment Scale: A Review of Recent Modeling Strategies, Geography Compass, 4, 834–860, https://doi.org/10.1111/j.1749-8198.2010.00357.x, 2010.

Teutschbein, C. and Seibert, J.: Bias correction of regional climate model simulations for hydrological climate-change impact studies: Review and evaluation of different methods, J. Hydrol., 456–457, 12–29, https://doi.org/10.1016/j.jhydrol.2012.05.052, 2012.

Uppala, S. M., Kållberg, P. W., Simmons, A. J., Andrae, U., Da Costa Bechtold, V., Fiorino, M., Gibson, J. K., Haseler, J., Hernandez, A., Kelly, G. A., Li, X., Onogi, K., Saarinen, S., Sokka, N., Allan, R. P., Andersson, E., Arpe, K., Balmaseda, M. A., Beljaars, A. C. M., Van De Berg, L., Bidlot, J., Bormann, N., Caires, S., Chevallier, F., Dethof, A., Dragosavac, M., Fisher, M., Fuentes, M., Hagemann, S., Hólm, E., Hoskins, B. J., Isaksen, L., Janssen, P. A. E. M., Jenne, R., Mcnally, A. P., Mahfouf, J.-F., Morcrette, J.-J., Rayner, N. A., Saunders, R. W., Simon, P., Sterl, A., Trenberth, K. E., Untch, A., Vasiljevic, D., Viterbo, P., and Woollen, J.: The ERA-40 re-analysis, Q. J. Roy. Meteor. Soc., 131, 2961–3012, https://doi.org/10.1256/qj.04.176, 2005.

van Vuuren, D., Edmonds, J., Kainuma, M., Riahi, K., Thomson, A., Hibbard, K., Hurtt, G., Kram, T., Krey, V., Lamarque, J.-F., Masui, T., Meinshausen, M., Nakicenovic, N., Smith, S., and Rose, S.: The representative concentration pathways: an overview, Climatic Change, 109, 5–31, https://doi.org/10.1007/s10584-011-0148-z, 2011.

Vetter, T., Huang, S., Aich, V., Yang, T., Wang, X., Krysanova, V., and Hattermann, F.: Multi-model climate impact assessment and intercomparison for three large-scale river basins on three continents, Earth Syst. Dynam., 6, 17–43, https://doi.org/10.5194/esd-6-17-2015, 2015.

Vrac, M. and Friederichs, P.: Multivariate–intervariable, spatial, and temporal–bias correction, J. Climate, 28, 218–237, 2015.

Warszawski, L., Frieler, K., Huber, V., Piontek, F., Serdeczny, O., and Schewe, J.: The Inter-Sectoral Impact Model Intercomparison Project (ISI–MIP): Project framework, P. Natl. Acad. Sci. USA, 111, 3228–3232, https://doi.org/10.1073/pnas.1312330110, 2014.

Weedon, G. P., Gomes, S., Viterbo, P., Shuttleworth, W. J., Blyth, E., Österle, H., Adam, J. C., Bellouin, N., Boucher, O., and Best, M.: Creation of the WATCH Forcing Data and its use to assess global and regional reference crop evaporation over land during the twentieth century, J. Hydrometeorol., 12, 823–848, https://doi.org/10.1175/2011JHM1369.1, 2011.

Wilks, D. S.: Statistical methods in the atmospheric sciences, Academic Press, San Diego, CA, 3rd Edn., 2011.

Yee, T. W.: Vector Generalized Linear and Additive Models: With an Implementation in R, Springer, New York, 2015.

Zaitchik, B. F., Simane, B., Habib, S., Anderson, M. C., Ozdogan, M., and Foltz, J. D.: Building Climate Resilience in the Blue Nile/Abay Highlands: A Role for Earth System Sciences, Int. J. Environ. Res. Pu., 9, 435–461, https://doi.org/10.3390/ijerph9020435, 2012.

4

Evaluation of the ability of the Weather Research and Forecasting model to reproduce a sub-daily extreme rainfall event in Beijing, China using different domain configurations and spin-up times

Qi Chu[1,2,3], **Zongxue Xu**[1,2], **Yiheng Chen**[3], **and Dawei Han**[3]

[1]College of Water Sciences, Beijing Normal University, Beijing, 100875, China
[2]Beijing Key Laboratory of Urban Hydrological Cycle and Sponge City, Beijing, 100875, China
[3]Department of Civil Engineering, University of Bristol, Bristol, BS8 1TR, UK

Correspondence: Zongxue Xu (zongxuexu@vip.sina.com)

Abstract. The rainfall outputs from the latest convection-scale Weather Research and Forecasting (WRF) model are shown to provide an effective means of extending prediction lead times in flood forecasting. In this study, the performance of the WRF model in simulating a regional sub-daily extreme rainfall event centred over Beijing, China is evaluated at high temporal (sub-daily) and spatial (convective-resolving) scales using different domain configurations and spin-up times. Seven objective verification metrics that are calculated against the gridded ground observations and the ERA-Interim reanalysis are analysed jointly using subjective verification methods to identify the likely best WRF configurations. The rainfall simulations are found to be highly sensitive to the choice of domain size and spin-up time at the convective scale. A model run covering northern China with a 1 : 5 : 5 horizontal downscaling ratio (1.62 km), 57 vertical layers (less than 0.5 km), and a 60 h spin-up time exhibits the best performance in terms of the accuracy of rainfall intensity and the spatial correlation coefficient (R'). A comparison of the optimal run and the initial run performed using the most common settings reveals clear improvements in the verification metrics. Specifically, R' increases from 0.226 to 0.67, the relative error of the maximum precipitation at a point rises from −56 to −11.7 %, and the root mean squared error decreases by 33.65 %. In summary, re-evaluation of the domain configuration options and spin-up times used in WRF is crucial for improving the accuracy and reliability of rainfall outputs used in applications related to regional sub-daily heavy rainfall (SDHR).

1 Introduction

The possibility that sub-daily heavy rainfall (SDHR) will increase with climate change is of significant societal concern. SDHR-driven flash floods (FFs) are among the most destructive natural hazards that threaten many urban areas in northern and central China and many other parts of the world. In these regions, SDHR is mainly triggered by regional mesoscale circulation systems (MCSs) and occurs with increased intensity and frequency in warm seasons (Yu et al., 2007; Chen et al., 2012). Records from the Emergency Events Database (EM-DAT) indicate that the damages and losses caused by FF events in China have increased significantly over the past several decades. The risks are expected to continue to grow, given the increase in the magnitude of SDHR predicted by most general circulation models (Chen et al., 2012; Willems et al., 2012; Westra et al., 2014). The accelerating pace of urbanization also contributes to the increase in risk; urbanization has already changed the hydrologic characteristics of the land surface considerably, resulting in higher peak flows and shorter flow concentration times (Xu and Zhao, 2016; Gao et al., 2017). In such cases, very short-term (< 6 h) rainfall predictions are not sufficient to provide adequate warning and mobilize emergency response activities. Recently developed statistically based rainfall generation methods and remote sensing data have been shown to enable the extension of the lead time to 24 h (Yu et al., 2016). However, this lead time is still insufficient to provide effective flood mitigation for medium or large urban areas with very short hydrologic response times (Shih et al., 2014; Li et al., 2017). Therefore, numerical weather prediction (NWP), which represents a means of forecasting heavy rainfall with lead times exceeding 24 h, has come into wide use in flood-related studies and applications (Cuo et al., 2011).

Precipitation uncertainty accounts for a large proportion of the uncertainty in flood forecasts. Hence, given the large uncertainties in NWP, its use in flood forecasting has long been questioned (Castelli, 1995; Bartholmes and Todini, 2005). Its usefulness was not realized until the end of the 20th century; substantial improvements in the predictive skill of NWP were made that resulted from the increases in computational power and storage capacity, which enable parallel processing of high-resolution forcing data and the resolution of convective-scale physical processes (Done et al., 2004; Clark et al., 2016). The NWP models developed during and after this period can perform regional and convective-scale modelling and display good performance in simulating heavy rainfall. Experimental studies have shown that NWP models of this kind, such as the WRF model (Skamarock et al., 2008), tend to capture greater numbers of small-scale processes and the triggers of convective storms (Klemp, 2006; Prein et al., 2015). Increasing numbers of meteorological operational centres and research groups are adopting these new NWP models to carry out simulations of heavy rainfall events or real-time forecasting. The resolutions of the rainfall products have improved from tens of kilometres to less than a kilometre, and the lead times have increased from less than a day to more than a week (WMO, 2013). Meanwhile, case studies have been carried out using regional convective-resolving models to evaluate the local rainfall predictions generated by sophisticated regional nesting techniques or the global smooth grid transition approach on unstructured grids (Swinbank and James Purser, 2006; Hong and Lee, 2009; Soares et al., 2012; Sikder and Hossain, 2016; Heinzeller et al., 2016). The results of these studies demonstrate that, over relatively short periods of time, regional modelling is often superior to large-scale modelling because it better resolves surface heterogeneities, topography, and small-scale features in air flow such as growing instabilities (Miguez-Macho et al., 2004; Yu et al., 2010; Prein et al., 2015; Brömmel et al., 2018).

Despite the great potential of NWP models to predict heavy rainfall, a number of uncertainties remain that must be considered. The errors induced by the initial and boundary conditions represent one source of these uncertainties; others stem from cognitive errors and the scale effect in the solution of physical models, both of which may be exacerbated by the chaotic nature of NWP. In regional simulations, these uncertainties are expected to be further magnified by downscaling or the use of mesh transition procedures, so re-evaluation and calibration of the related model configurations are commonly required (Warner, 2011; Vrac et al., 2012; Liu et al., 2012). As an example, running the WRF model at convective scales means that convective processes are more likely to be resolved by explicit physical schemes than when sub-grid parameterizations are used, which may incorporate new structural uncertainties related to the model physics (Done et al., 2004; Ruiz et al., 2010; Crétat et al., 2012). In addition to model physics, several other aspects of model configuration, such as the domain size, the spatial resolution, and the spin-up time, may also have a substantial impact on the uncertainty in rainfall forecasts through their effects on the initial and boundary conditions (Aligo et al., 2009; Fierro et al., 2009; Cuo et al., 2011). However, these aspects of model configuration have received less attention in regional case studies because of their insignificant effects on rainfall forecasts in coarse-resolution and long-term model simulations when compared to the physics of the WRF model. Generally, these model configuration aspects are left at the common settings recommended by the official website of the WRF model and by some experimental regional heavy rainfall studies.

Precipitation is one of the most sensitive variables to NWP model uncertainties. In this study, a re-evaluation of WRF is performed to explore whether the recommended configuration of WRF represents the best choice in reproducing a regional SDHR event that happened in Beijing. The WRF model is assessed here because of its superior scalability and computational efficiency; these traits are valued in interdisciplinary studies (Klemp, 2006; Foley et al., 2012; Coen et al., 2013; Yucel et al., 2015). As the latest NWP community model, WRF incorporates up-to-date developments in physics, numerical methods, and data assimilation and is thus widely used in theoretical studies and practical applications (Powers et al., 2017). The selected regional SDHR event occurred on 21 July 2012 and was centred over Beijing, China. Beijing is among the most vulnerable cities to SDHR-induced floods in central China (Yu et al., 2007). The precipitation in this area is caused mainly by monsoon weather systems and enhanced by local orographic effects, and 60–80 % of the total annual precipitation occurs during just a few SDHR events (Xu and Chu, 2015). The SDHR event that occurred on 21 July 2012 caused the most disastrous urban flood in Beijing since 1950. The national operational NWP system failed to predict this event, which resulted in 79 deaths and more than USD 1.6 billion in damage (Brömmel et al., 2018; Wang et al., 2013; Zhou et al., 2014). Thus, several convective-scale studies have been carried out to re-evaluate the optimal combination of the physics options used in the WRF model, such as Di et al. (2015) and Wang et al. (2015). These studies represent the background information that stimulates this research.

The second question we attempt to explore is to what extent rainfall simulations could be improved through the use of the likely best set of settings if the recommended model configurations are not the best choices. The aspects of the model configuration that are evaluated in this study are the domain size, vertical resolution, horizontal resolution, and spin-up time. These options have been found to have substantial impacts on daily-scale extreme rainfall outputs (Leduc and Laprise, 2009; Aligo et al., 2009; Goswami et al., 2012).

A comparative test with four scenarios is designed. Each scenario evaluates one model configuration option to ensure that the simulated disparities can be attributed solely to a single factor each time. In addition, the test is conceived as a progressive process: the optimal setting identified in each scenario will be adopted as the primary choice for the next scenario to help quantify the overall improvement in the accuracy of rainfall outputs. The "ground truth" datasets are gridded observations obtained from Beijing Normal University and the China Meteorological Administration. A coarser-resolution reanalysis called ERA-Interim (Dee et al., 2011) is also employed in identifying departures of the WRF simulations from the driving weather fields as the model setup is varied. Seven objective verification metrics that reflect different features of the model performance are adopted and considered jointly as part of a subjective verification process because no single verification approach has been shown to provide comprehensive information about the quality of rainfall simulations (Sikder and Hossain, 2016). Most of the metrics adopted here are those used to assess the performance of WRF over daily or longer time periods (Liu et al., 2012; Tian et al., 2017). In this research, these metrics are calculated on an hourly basis and averaged over different sub-daily time spans to evaluate the performance of the WRF model using different configurations from a sub-daily and convective-scale perspective.

2 Numerical model used to forecast heavy rainfall

The advanced WRF (ARW-WRF) model, version 3.7.1, is used as the dynamical downscaling tool. ARW-WRF is a compressible non-hydrostatic and convection-permitting regional NWP model that employs the conservative form of the dynamic Euler equations. As the latest regional NWP community system, WRF is composed of two dynamic cores, a data assimilation system and a platform that facilitates parallel computation and function portability. Observations, model output or assimilated reanalysis output can be used to initialize WRF. In terms of discretization, WRF uses a third-order Runge–Kutta method for temporal separation and an Arakawa C-grid staggering scheme for spatial discretization. The model is capable of conducting either one-way or two-way nested runs for regional downscaling. A detailed introduction to the physics and numerical properties of ARW-WRF can be found in Skamarock et al. (2008). Given its emphasis on efficiency, portability, and updates to reflect the state of the art, WRF has been employed in settings ranging from research to applications and has been incorporated into various operational systems, such as the Hurricane-WRF system for hurricane forecasting and the WRF-Hydro system for hydrologic prediction.

In WRF, the domain size implicitly determines the large-scale dynamics and terrain effects, whereas the vertical and horizontal grid spacings determine the smallest resolvable scale (Goswami et al., 2012). Together, these domain configuration options affect the spectrum of the resolved scales and the nature of scale interactions in the model dynamics (Leduc and Laprise, 2009). Thus, they are responsible for the generation and distribution of precipitation. In regional simulations, small domain sizes are commonly preferred for computational efficiency. Seth and Rojas (2003) demonstrated that simulations with small domain sizes are more likely to benefit from the lateral boundary conditions (LBCs) by dampening the feedback from local perturbations on the large-scale general circulation. However, insufficiently large domains have been shown to prevent the full development of small-scale features over areas of interest. To solve this issue, the official website of WRF provides general guidance (Warner, 2011). This guidance recommends that the ranges of domains should include the major features of the leading MCSs and local surface perturbations, and more than five grid points should exist between adjacent nested domains to allow for sufficient relaxation.

As for grid spacing, it appears plausible that WRF model runs performed with relatively small grid spacings would provide more accurate outputs because such runs would resolve more small-scale phenomena of interest that are not present in the LBCs. This statement is generally accepted as true when a relatively coarse-resolution run (> 10 km horizontally or > 1 km vertically) is compared with a relatively finely resolved run at the convective scale (1–5 km horizontally or < 1 km vertically) in representing a convective storm. However, this conclusion is controversial when the comparison is conducted among convective-scale model runs. Taking the horizontal resolution as an example, although there is evidence to show that WRF runs performed at relatively high resolution capture more convective-scale features, the accuracy of rainfall outputs either shows considerable or no statistical improvement (Roberts and Lean, 2008; Kain et al., 2008; Schwartz et al., 2009). In one study, Fierro et al. (2009) suggested that some features detected in convective-scale runs with too small horizontal grid spacings tend to weaken the kinetic structures that favour torrential rainfall. A similar conclusion was drawn by Aligo et al. (2009) in evaluating the impact of the vertical grid spacing on simulations of summer rainfall performed using WRF. Thus, horizontal and vertical grid spacings of approximately 4 and 1 km, respectively, have been employed as a reasonable compromise between accuracy and computational efficiency in several regional studies.

In regional modelling, a spin-up period is often required to balance the inconsistencies between the results simulated by the model physics and the initial and boundary conditions provided by the forcing data (Luna et al., 2013). The proper spin-up time depends on the time needed for initialization, which can be affected by the size of the domain and the local boundary perturbations (Warner et al., 1997; Kleczek et al., 2014). Moreover, the presence of chaotic behaviour, which causes reductions in the predictive skill of models over time, imposes an upper bound on the spin-up time. Therefore, in

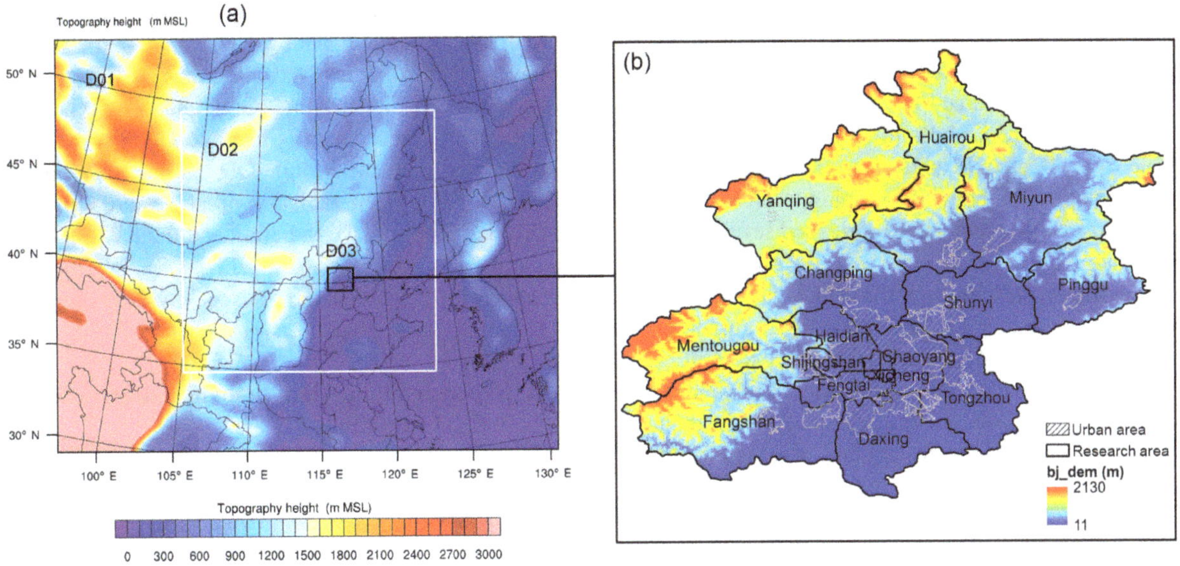

Figure 1. Relative location of the study area. **(a)** Shows the three nested domains adopted in most of the experiments, of which domain three (D03) covers the entire Beijing area; **(b)** depicts the geographic features of the Beijing area.

cases where short forecast lead times are expected, e.g. real-time rainfall forecasting, the spin-up time is mainly determined by the domain size and the regional initial and boundary conditions. However, in cases where long forecast lead times are needed, e.g. warnings of extreme rainfall, the effects of chaotic behaviour should be relatively evident. In practice, this issue is commonly addressed by regularly updating the lateral boundary information derived from the latest forecasts or analyses to maintain consistency between the regional model solutions and the atmospheric forcing conditions. In such cases, the best-fit performance may occur for model runs with long spin-up times. Based on most previous studies, a spin-up time of 12 h is recommended to obtain an initial state; however, this spin-up time is often regarded as the suitable choice in many regional case studies without further verification.

3 Studied event and experimental design

As mentioned above, one aim of this study is to re-evaluate whether the recommended WRF domain configuration options and spin-up time represent the optimal model configuration for reproducing a regional SDHR event when evaluated at a sub-daily timescale. Here, the SDHR event that occurred on 21 July 2012 and was centred on Beijing, China is selected as a case study. The reasons why this event is selected, the synoptic and physical features that drove this event, and the model physics adopted in this study are presented before the entire procedure of the experimental design is introduced.

3.1 Study event selection and WRF physical schemes

Beijing is selected as the study area because it is one of the most vulnerable cities to SDHR-induced FF hazards in China. Beijing is located in central China. It has an area of 16 411 km^2, and its weather is mainly affected by the semi-humid warm continental monsoon climate. The flows of air that favour local precipitation are cold, dry flows of air from high-latitude areas to the north and hot, wet flows of air from the ocean to the south. The interactions between these two flows of air lead to clear divergence in the temporal distribution of rainfall amount; 60–80 % of the annual precipitation occurs during just a few heavy rainfall events during the warm season. Of all of the heavy rainfall events, the intensity and frequency of SDHR events have been shown to display the greatest increasing tendencies over the past several decades. Meanwhile, Beijing, as the capital of China, has experienced a significant expansion of its urban area and rapid increases in its population and economic development. The negative effects of this expansion, such as losses of natural water bodies, increases in land cover with low permeability, and increases in urban drainage pipe networks, have led to continuous decreases in the hydrologic response time. In addition, most of the population lives in the southwestern plain area. This region is downstream of mountainous areas with steep terrain that varies in elevation from 60 to 2300 m (Fig. 1). All of these factors contribute to the continuing increase in the exposure of this city to the high risks of flooding and waterlogging caused by SDHR events (Xu and Chu, 2015).

The case study examines the largest heavy rainfall event that has occurred in Beijing in the past 65 years. The rainfall event lasted for 16 h (from 02:00 to 18:00 UTC) on 21 July 2012, and the highest hourly rainfall intensity ($100\,mm\,h^{-1}$) was experienced in the southwestern part of the plain area. The associated FF hazard led to 79 deaths and damages totalling USD 1.6 billion, and more than 1.6 million people were affected. In addition to Beijing, the adjacent provinces, including Hubei and Liaoning, were all significantly affected by this event and experienced severe FF hazards. The synoptic features that triggered the rainfall were an eastward-moving vortex in the middle to high troposphere, a northward-moving zone of subtropical high pressure, and sharp vertical wind shear (Sun et al., 2013). The rainfall event as a whole can be divided into two phases. From 02:00 to 14:00 UTC, the convective rain was dominated and enhanced by the orographic effect. The frontal rain was then followed by the arrival of a cold front moving from the northwest until 18:00 UTC (Guo et al., 2015). The rainfall intensity in the second phase was relatively low compared to that in the first phase, due to the lack of strong kinetic forcing to maintain the occurrence of precipitation.

The ERA-Interim reanalysis and 30-second static geographical data are employed to initialize the surface and meteorological fields of the WRF. ERA-Interim is produced by an integrated forecasting system (IFS) used by the European Centre for Medium-Range Weather Forecasts (ECMWF). The IFS is an Earth system model that incorporates a data assimilation system and an atmospheric model that is fully coupled with land-surface and oceanic processes. The atmospheric model provides output every 30 min at a spectral resolution of T255 (approximately 81 km over Beijing). This output is then employed as prior information and combined with available observations twice a day to produce the reanalysis output using the four-dimensional variation (4D-Var) assimilation system. The final reanalysis product, ERA-Interim, is a global gridded dataset that is available at a spectral resolution of T255 and at both the 60 levels used in the model and 38 interpolated pressure levels for all dates beginning on 1 January 1979 (Berrisford et al., 2009; Dee et al., 2011). Here, the ERA-Interim pressure-level data are selected as the initial forcing. One reason is that, as is necessary, the vertical grid spacing between the adjacent pressure layers is less than 1 km in the free troposphere, where the convective processes mainly occurred during the Beijing SDHR event. In addition, the NWP models used by the China Meteorological Administration mainly employ 31 vertical levels in regional forecasting (WMO, 2013).

As shown in Fig. 2, ERA-Interim captures the vortex and the subtropical high pressure well that occurred at the beginning of the rainfall event. In addition, the patterns of the leading MCSs and the primary synoptic features shown in this figure also correspond well to those described in previous studies (Zhou et al., 2014). The setup of the model physics is based mainly on the results of sensitive, high-resolution studies on the physics of the WRF model in simulating the same event (Wang et al., 2015; Di et al., 2015). The "resolved rain" is driven by the single-moment 6-class microphysics scheme (Hong and Lim, 2006), whereas the "convective rain" is resolved using the Grell–Dévényi cumulus parameterization scheme (Grell and Dévényi, 2002). The Noah land-surface model (Chen and Dudhia, 2001) is used and coupled with the Monin–Obukhov surface layer model (Ek et al., 2003). The radiation processes are represented by the RRTMG shortwave radiation and the RRTMG longwave radiation schemes (Iacono et al., 2008). For the planetary boundary layer scheme, the Yonsei University method (Hong et al., 2006) is adopted.

3.2 Experimental design: domain configuration options and spin-up time

The comparative test is designed as a progressive process to help quantify the overall improvement in the performance of WRF after re-evaluating the WRF experiments performed using different domain configuration options and spin-up times. The test is classified into four successive scenarios. The first three scenarios investigate the domain configuration options, including the domain size, vertical resolution, and horizontal resolution; the fourth scenario concerns the spin-up time. During the entire procedure, the optimum configuration identified in each scenario is then adopted as the primary choice for the corresponding configuration in the following scenario. The initial datasets and the model physics are the same for all of the domains throughout the entire comparative procedure. Because the area of interest is located in the middle latitudes, the Lambert conformal projection is employed in all of the experiments, which is centred on the same latitude (42.25° N) and longitude (114.0° E). Moreover, sigma vertical coordinates with a top level of 50 hPa are used in all of the experiments.

Initially, the WRF domain configuration options and the spin-up time are set to the recommended values described in Sect. 2. Three levels of two-way nested domains are adopted so that the horizontal resolution in the smallest domain is sufficiently high to explicitly resolve convective-scale processes (Fig. 1). An odd downscaling ratio (1 : 3 : 3) is selected to reduce the initial error introduced by interpolating the initial fields to the assigned Arakawa grid. For the same reason, the boundaries of each domain are set along specific grid lines of the ERA-Interim dataset. Of the three nested domains, the outermost domain (D01) has the largest horizontal grid spacing of 40.5 km over north-central China, where the main perturbed synoptic features occur. The innermost domain (D03) has the smallest horizontal grid spacing of nearly 4.5 km over the area of interest, Beijing. The second domain (D02) is the child of D01 and the parent of D03 and has a horizontal grid spacing of 13.5 km. The distance between D01 and D02 is similar to that between D02 and D03, both of which exceed five grid points. The grid numbers of D01, D02, and D03 are 40×40, 72×72, and 90×90, respectively. The eta values utilized in the initial run are set based on the pressure values

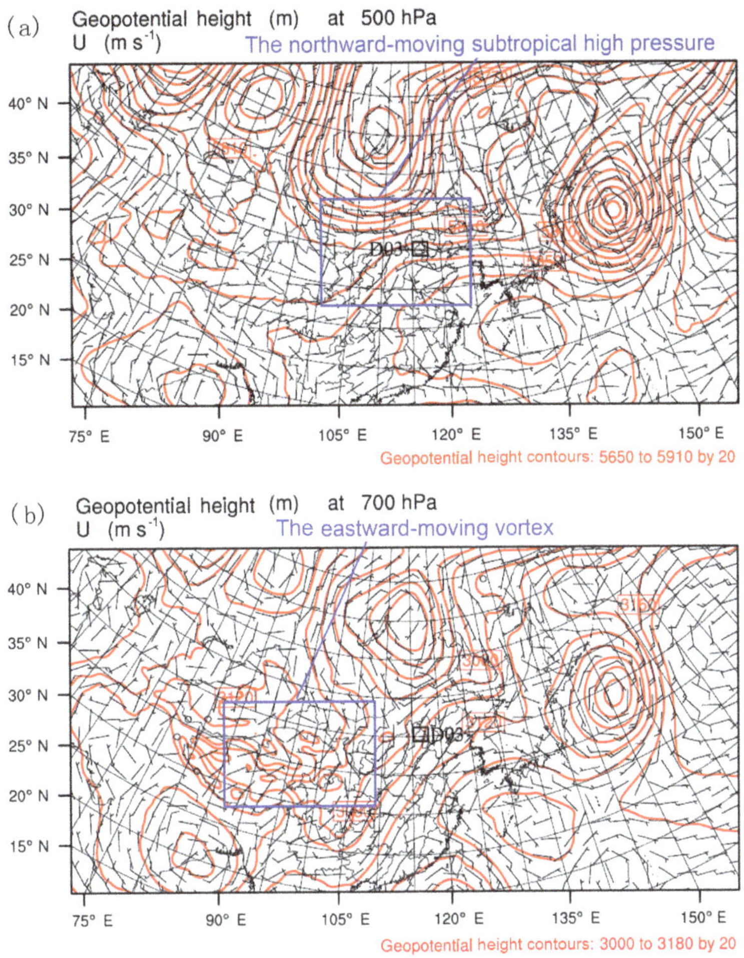

Figure 2. Initial wind field and geopotential height field at 00:00 UTC on 20 July 2012 over the Northern Hemisphere obtained from the ERA Interim reanalysis. **(a)** The fields at 500 hPa; **(b)** the fields at 700 hPa.

at the 29 vertical layers of the ERA-Interim pressure-level data. A spin-up time of 12 hours (12 h) is selected; the outputs are saved every 3 h in D03 and every hour in D02. The LBCs are updated every 6 h using ERA-Interim.

As shown in Table 1, the first experiment (C0) adopts the model configuration options mentioned above. To determine whether the domain configuration options and the spin-up time used in C0 are the likely best set, four scenarios are designed. The first scenario (S1) focuses on evaluating the effect of the WRF domain size. For computational efficiency, the MCS systems that drive the local synoptic features are not completely contained within the outermost domain of C0, the information of which is compensated by the updated LBCs from ERA-Interim. Two comparative experiments, C1 and C2, are devised to verify that the domain size assigned to C0 is large enough to enable the full development of small-scale features. Of the three experiments, C2 has the largest outermost domain size, which incorporates the leading MCS systems over the entire Northern Hemisphere. The intermediate domain, which is centred between the outermost and innermost domains, is then adopted as the outermost domain of C1. The purpose of scenario two (S2) is to evaluate whether the use of a higher vertical resolution in a WRF model run results in better performance. In this scenario, the starting experiment is the optimal experiment identified in S1 (OS1), forced by the ERA-Interim pressure-level data with 29 vertical levels. This starting experiment is then followed by two experiments, C3 and C4, which incorporate 1 and 2 times more vertical levels than OS1 (57 and 85 vertical levels), respectively. In the Beijing SDHR event, the pressure-level data meet the requirement of a grid spacing of less than 1 km in the troposphere; however, this condition is not necessarily satisfied in other regions. Thus, an experiment forced by the ERA-Interim model-level data with 38 vertical levels (C5) is also designed for comparison. The three experiments (OS2, C6, and C7) in scenario three (S3) differ in terms of their horizontal resolutions and nesting ratios, with increased nesting ratio of 1 : 3 : 3 (4.5 km grid spacing in D03), 1 : 5 : 5 (1.62 km in D03), and 1 : 7 : 7 (0.826 km in D03). The last scenario (S4) is designed to identify a reasonable optimal

Table 1. Categories of experiments with different domain sizes, vertical resolutions, horizontal resolutions and spin-up times.

Scenario	Experiment number	Domain size	Vertical levels	Horizontal resolution (nesting ratio)	Spin-up time
Domain size (S1)	Case 0 (C0)	D01 40 × 40; D02 72 × 72; D03 90 × 90	29 (pressure level)	D01 40.5 km; D02 13.5 km; D03 4.5 km 1 : 3 : 3	12 h
	Case 1 (C1)	D01 80 × 64; D02 120 × 120	as C0	as C0	as C0
	Case 2 (C2)	D01 160 × 128; D02 240 × 192	as C0	as C0	as C0
Vertical	Optimal case in S1 (OS1)	as OS1	29	as C0	as C0
resolution	Case 3 (C3)	as OS1	57	as C0	as C0
(S2)	Case 4 (C4)	as OS1	85	as C0	as C0
	Case 5 (C5)	as OS1	38 (model level)	as C0	as C0
Horizontal resolution	Optimal case in S2 (OS2)	as OS1	as OS2	1 : 3 : 3	as C0
(S3)	Case 6 (C6)	as OS1	as OS2	D01 40.5 km; D02 8.1 km; D03 1.62 km 1 : 5 : 5	as C0
	Case 7 (C7)	as OS1	as OS2	D01 40.5 km; D02 5.785 km; D03 0.826 km 1 : 7 : 7	as C0
Spin-up time	Optimal case in S3 (OS3)	as OS1	as OS2	as OS3	12 h
(S4)	Case 8 (C8)	as OS1	as OS2	as OS3	0 h
	Case 9–Case 19 (C9–C19)	as OS1	as OS2	as OS3	24–144 h per 12 h

model run with the maximum spin-up time after minimizing the uncertainties introduced by inappropriate domain configuration options. It contains 1 starting experiment (OS3) and 12 comparative experiments (C8–C19). Except for C8, which includes no spin-up time, the remaining experiments (C9–C19) include spin-up times that increase from 24 to 144 h by every 12 h.

4 Verification schemes

Both objective and subjective verification methods are applied to the innermost domain (D03) at a sub-daily scale. D03 is selected because it covers the area of interest, Beijing, and the convective processes in this domain can be explicitly resolved in all of the experiments. The rainfall data used for comparison in D03 are 3-hourly 0.05° data that were produced by fusing rain gauge observations and the CMORPH data (Huang et al., 2013). The ERA-Interim reanalysis is utilized as well to monitor the possible departures of the model simulations from the driving fields. Because the sub-daily rainfall is not available from the reanalysis, the atmospheric precipitable water vapour (PW), which determines the possible maximum precipitation, is instead compared with the model outputs every 6 h. In addition, the model outputs that cover a larger domain (D02) are compared with an hourly 0.1° gridded dataset obtained from the China Meteorological Administration (http://data.cma.cn/data/cdcdetail/dataCode/SEVP_CLI_CHN_MERGE_CMP_PRE_HOUR_GRID_0.10.html, last access: 17 June 2018). The comparison over domain two is used only as an auxiliary method for subjective verification, based on the assumption that an experiment with good performance in the inner domain should also capture the large-scale features in the outer domain, as the appropriate representation of these large-scale features will result in more accurate boundary conditions.

Seven error metrics that describe different features of precipitation are selected for use as objective verification metrics. Five are rainfall-related and compared by bilinear interpolation of the output of the simulations to the grid of the ground truth data. The accumulated areal rainfall is assessed using the relative error of the total precipitation (RE_{TP}). The percentage of correct rainfall hits is measured using the probability of detection (POD) with a threshold of 0.1 mm. The root mean squared error (RMSE) represents the amount of continuous error in the predicted precipitation. Detailed illustrations of these three metrics can be found in Liu et al. (2012) and Tian et al. (2017). The other two rainfall-

Table 2. Correlations between the original and rescaled objective verification metrics.

Original metrics	Representative meaning	Rescaled metrics	Threshold value
POD	probability of detection	$\mathrm{POD}' = \mathrm{POD}$	n/a
RMSE	root mean squared error	$\mathrm{RMSE}' = 1 - \mathrm{RMSE}/\mathrm{RMSE}_{\mathrm{max}}$	+62.5 max
R	Pearson correlation coefficients	$R' = R$	n/a
WRMSE	RMSE of the precipitable water	$\mathrm{WRMSE}' = 1 - \mathrm{WRMSE}/\mathrm{WRMSE}_{\mathrm{max}}$	+8.3 max
WR	R of the precipitable water	$\mathrm{WR}' = \mathrm{WR}$	n/a
$\mathrm{RE}_{\mathrm{PMAX}}$	relative error of the maximum precipitation	$\mathrm{PMAX}' = \mathrm{RE}_{\mathrm{PMAX}} + 1$	n/a
$\mathrm{RE}_{\mathrm{TP}}$	relative error of the total precipitation	$\mathrm{TP}' = \mathrm{RE}_{\mathrm{TP}} + 1$	n/a

n/a = not applicable.

related metrics are the relative error of the maximum grid precipitation ($\mathrm{RE}_{\mathrm{PMAX}}$) and the Pearson correlation coefficient (R), which describe the spatial association between the simulations and the ground truth data (Eqs. 1 and 2). The two metrics selected for the verification of PW (PW-related metrics) are the root mean squared error (WRMSE) and the Pearson correlation coefficient (WR). For comparison, the PW fields of the reanalysis are remapped to the grids of the model outputs using the WRF Preprocessing System (WPS). In this study, all of the metrics are calculated between the simulations and the reference data on the same grid at each time step (3 h in D03). The values of these metrics are then averaged over four different time periods (6, 12, 18, and 24 h) counted from 00:00 UTC on 21 July 2012. Different time periods are selected with the purpose of determining whether the performance of WRF differs when the evaluation is conducted using different durations.

$$\mathrm{RE} = \frac{1}{N}\sum_{i=1}^{N}\left[\frac{f-r}{r} \times 100\,\%\right], \tag{1}$$

$$R = \frac{1}{N}\sum_{i=1}^{N}\left(\frac{\sum_{j=1}^{M}\left(f_j-\overline{f}\right)\left(r_j-\overline{r}\right)}{\sqrt{\sum_{j=1}^{M}\left(f_j-\overline{f}\right)^2\sum_{j=1}^{M}\left(r_j-\overline{r}\right)^2}}\right). \tag{2}$$

Here, R is the empirical spatial correlation coefficient; M is the total number of grid points within the evaluated domain of the starting experiment; f_j is the value of the jth grid point in the tested field at time step i; r_j is the value of the reference field; N is the total number of time steps, depending on the time period considered; and RE is the relative error. For the maximum precipitation, f is the tested value of the maximum gridded precipitation over the area of interest, and r is the reference value of the maximum gridded precipitation over the same area.

To facilitate evaluation, the metrics are further adjusted to ensure that the ideal value of all of the metrics is 1. In this study, RMSE and WRMSE are first divided by a rescaling factor to fall into the range of 0–1 and then subtracted from 1 to provide an indication of good performance. The rescaled metrics, RMSE′ and WRMSE′, have the value 1 representing the lowest accumulated error (highest accuracy). The factor used for rescaling is determined by the largest values of each error metric in all of the experiments and is kept at the same value for all of the evaluated time periods (Sikder and Hossain, 2016). $\mathrm{RE}_{\mathrm{PMAX}}$ and $\mathrm{RE}_{\mathrm{TP}}$ are added by 1 to have the ideal value of 1. The rescaled metrics are PMAX′ and TP′, respectively. The other metrics are not rescaled because they already have ideal values of 1, but they are assigned a new set of symbols to distinguish them from the original metrics used before rescaling. For example, POD is replaced with POD′, and R is replaced with R'. Table 2 shows the correlations between the original metrics and the rescaled metrics. Given that the metrics describe different features of the rainfall simulations, the values of these metrics are checked and considered together in subjective verification to determine the likely best set of domain configuration options and to search for the longest reasonable spin-up time.

5 Results and analyses

In each scenario, the metrics are compared among the experiments that consider different durations and cover the same domain (D03). The results are presented in four sub-graphs; each sub-graph shows the values of the metrics calculated for individual evaluated time periods. The spatial distribution of rainfall is also presented over domain two (D02) when evident discrepancies are noted in the results obtained for the inner domain (D03) and the outer domain (D02). Table 1 shows the categories of the scenarios and the model configurations adopted in each experiment. In the following section, the domain size scenario (S1) is evaluated first, followed by the vertical resolution scenario (S2) and the horizontal resolution scenario (S3).

5.1 Results of the domain size scenario

Figure 3 shows the spatial values of the verification metrics for the WRF domain size experiments. The performance of

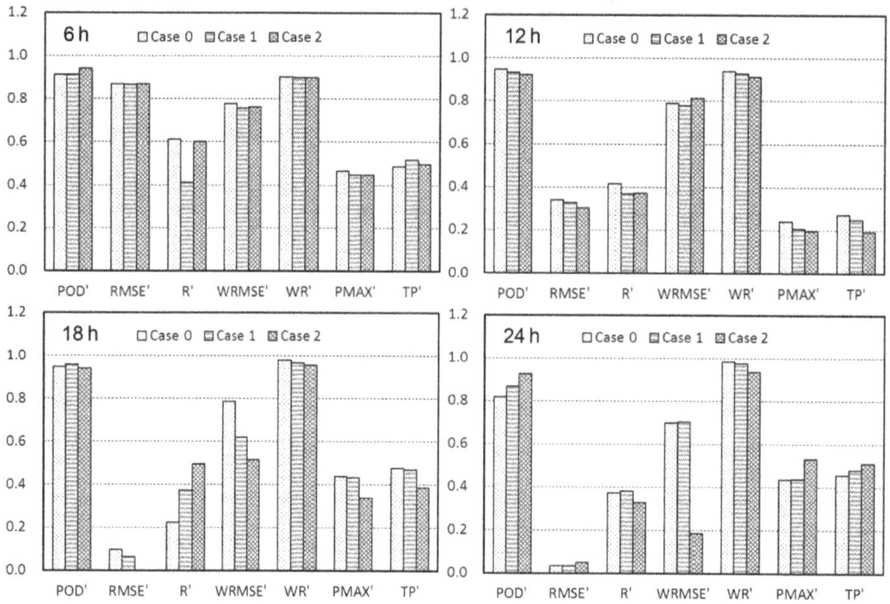

Figure 3. Spatial values of the verification metrics for the WRF domain size experiments, calculated over different temporal durations and over domain three. Case 0 (C0) incorporates the smallest domain, which covers north-central China; Case 1 (C1) incorporates a domain of intermediate size that covers northern China and part of Mongolia; and Case 2 (C2) incorporates the largest domain, which covers the Northern Hemisphere. The metrics are calculated over time periods of 6, 12, 18, and 24 h that begin at 12:00 UTC on 21 July 2012.

the experiments clearly worsens as the evaluated temporal duration increases from 6 to 24 h. The most evident deteriorations are detected in the point-to-point accuracy of the rainfall; the reversed root mean squared error (RMSE′) decreases by 0.8, which represents a 6-fold increase in the cumulative spatial error. The spatial association between the simulations and the gridded observations also declines; the Pearson correlation coefficient (R') decreases by 0.3 on average. Although a slight increase is observed in the percentage of correct hits (POD′) during the first 18 h, this increase is followed by a rapid decrease of nearly 14 % during the last stage of the rainfall event. The relative bias in the accumulated areal rainfall (TP′) indicates that the total rainfall amount is underestimated throughout the entire evaluated temporal period. The maximum gridded precipitation (PMAX′) is also underestimated; the largest negative bias occurs during the heavy convective rainfall stage. For PW, a slight decrease is found in the reversed accumulated error (WRMSE′), whereas an increase of 5–9 % is detected in the spatial correlation coefficient (WR′). Such variations may be attributable to the role of the updated boundary conditions in adjusting the local model solutions to approach the large-scale atmospheric circulation conditions.

Comparison of the four sub-graphs shows that the values of the metrics do not point to a single perfect experiment in a given period, and their ranked predictive skills determined using a given metric differ when evaluated over different time periods. During the early stage of the rainfall event (6 h), C0 yields better performance than C1 and C2 in terms of RMSE′, R', and PMAX′; it simultaneously displays the lowest value of POD′ and the largest bias in estimating the total precipitation. Although the superiority of C0 is more evident in the second period, a sharp deterioration is then observed in capturing the point-to-point accuracy of precipitation for the 18 h duration, where the lowest R' is obtained. Meanwhile, C1, which employs a domain of moderate size, displays greater skill than C0 in capturing the correct hits and the spatial pattern of the simulated rainfall. C2 employs the largest domain. Although it shows the best fit to the rainfall observations on the daily scale (24 h), it displays the worst performance over the three shorter time periods. For the PW fields, the highest similarity with the ERA-Interim reanalysis is found for C0, whereas the lowest similarity is found for C2. These results demonstrate indirectly that small domains are more likely to be influenced by updated boundary conditions.

In this scenario, if the experiments are merely evaluated in D03, the conclusion that C0 displays the best performance during most of the evaluated time periods may be reached. However, when evaluated in D02, clear differences between C0 and the ground truth in both the spatial characteristics of the rainfall and the magnitude of the maximum precipitation are detected. Figure 4 shows the spatial distribution of the accumulated 6 h precipitation over the domain two area of C0. Note that the speed of movement of the belt of heavy rain simulated in C0 is a few kilometres per hour faster than those in C1 and C2, leading to an early end of the heavy rainfall event. This difference may explain why the modelling skill of C0 declines significantly as the end of the rainfall event

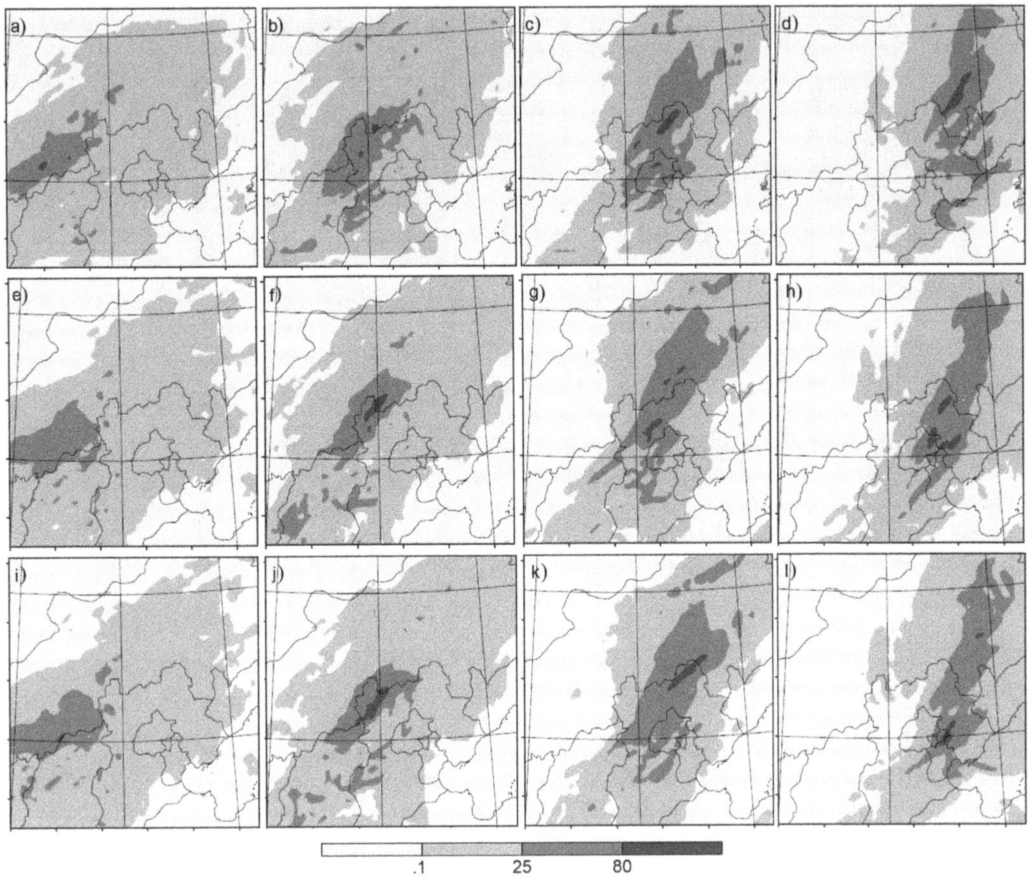

Figure 4. Spatial distribution of 6-h accumulated precipitation for the domain size experiments over the domain two area of C0 during the Beijing heavy rainfall event beginning at 12:00 UTC on 21 July 2012. **(a)** Accumulated precipitation (AP) in C0 during the first 6 h period (00:00–06:00 UTC); **(b)** AP in C0 during the second 6 h period (06:00–12:00 UTC); **(c)** AP in C0 during the third 6 h period (12:00–18:00 UTC); **(d)** AP in C0 during the fourth 6 h period (18:00–00:00 UTC); **(e)** AP in C1 during the first 6 h period; **(f)** AP in C1 during the second 6 h period; **(g)** AP in C1 during the third 6 h period; **(h)** AP in C1 during the fourth 6 h period; **(i)** AP in C2 during the first 6 h period; **(j)** AP in C2 during the second 6 h period; **(k)** AP in C2 during the third 6 h period; and **(l)** AP in C2 during the fourth 6 h period.

approaches. The belt of heavy rain in C0 displays an orientation that is shifted nearly 10 degrees northward from those simulated in C1 and C2 during the first 6 h, and the storm centre in C0 displays the smallest range; it is nearly half of the area in C2. The results indicate that the domain size of C0 is not broad enough to allow the model physics to fully develop the small-scale features that favour heavy rainfall. The spatial characteristics of precipitation are relatively similar in the other two experiments, but C1 outperforms C2 in both the rainfall-related and the PW-related features over domain two.

It may be that C2 does not yield better performance than C1 because of its inefficient use of boundary conditions to ad-just the false perturbations generated by the local model run. Therefore, C1 is verified as reasonable from both statistical and physical perspectives and is chosen as the optimal experiment in the domain size scenario (OS1).

5.2 Results of the vertical resolution scenario

Based on the analysed results, C1 is selected as the starting experiment in the vertical resolution scenario. As mentioned above, C1 is forced with the ERA-Interim pressure-level data with 29 vertical levels. C3 and C4 are forced with the same pressure-level data with 57 and 85 vertical levels, respectively, whereas C5 is forced with the model-level data with 38 vertical levels. As shown in Fig. 5, a decline in model performance is also obtained for all of the vertical resolution experiments as the evaluated time period increases in length. Moreover, the largest deterioration in RMSE′ is also observed; it decreases by 0.82 on average. The values of TP′ and PMAX′ derived from the simulations are slightly higher than those predicted in S1 but are still less than those calculated for the actual precipitation over the entire rainfall event. POD′ displays an evident decrease during the end stage of the rainfall event, and its magnitude decreases 50 % less relative

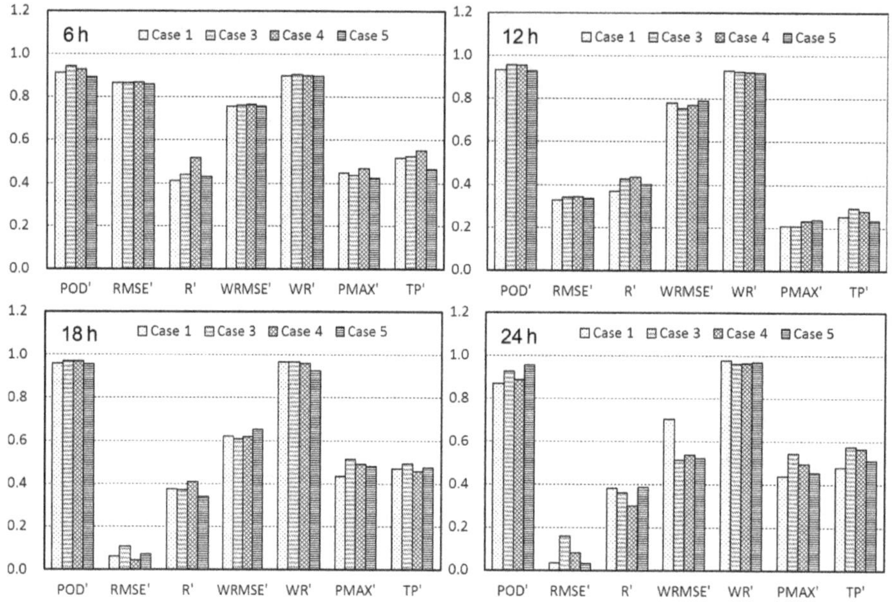

Figure 5. As in Fig. 3, but for the experiments in scenario two with different vertical resolutions. Case 1 is forced by the ERA-Interim pressure-level data with 29 vertical levels; cases 3 and 4 are forced by the same data but include double and triple the number of vertical levels, respectively; Case 5 is forced by the ERA-Interim model-level data with 38 vertical levels.

to that shown in C1. The most obvious difference from the domain size scenario is that the values of R' calculated between the simulations and the ground truth vary slightly and remain almost the same between the different time periods. In addition, the performance of the vertical resolution experiments seems to be less sensitive to the boundary conditions because they result in relatively small variations in WRMSE$'$ and WR$'$.

Unlike the apparent discrepancies noted in the metrics obtained for the domain size experiments, the differences in the rainfall-related metrics among the experiments with different numbers of vertical levels are not evident, especially during the less rainy period (6 h) and the period when convective rainfall dominates (12 h). During the first 12 h, C4 displays better agreement with the gridded observations than the other three experiments in terms of the accuracy and spatial correlation of the rainfall amount. However, over the longer time periods, C3 displays the greatest skill, according to most of the verification metrics. Comparing C3 and C1 shows that increases in the vertical resolution may increase WRF's ability to explicitly resolve small-scale physical processes and improve the accuracy of the amount and distribution of the simulated rainfall. Comparing C3 and C4 shows that, although C4 includes further refinement of the vertical resolution, the C4 performance is worse than that of C3 when the evaluated time period increases from 6 h to more than 12 h. This result may occur because progressive reductions in the vertical grid spacing magnify the propagation of surface perturbations through the vertical grid columns, potentially weakening the kinetic energy that favours precipitation. Examining the values of WRMSE$'$ and WR$'$ shows that the differences between the simulations and the reanalysis are more distinct in C3 and C4 than in C1. This discrepancy may occur due to the exaggeration of the initial errors introduced by the interpolation process and the incorporation of false surface perturbations introduced by the limited accuracy and resolution of the initial forcing data. C5 shows either better or worse performance than C1 in each period but produces less accurate rainfall simulations than C3 over most of the evaluated durations. As such, C3 is identified as yielding the best performance in the vertical resolution scenario.

5.3 Results of the horizontal resolution scenario

Based on the results obtained for scenario S2, C3 is selected as the starting experiment in the horizontal resolution scenario. The modelling skill of the S3 experiments shows similar temporal trends as that of the S2 experiments (Figs. 5 and 6). However, the sensitivity of the metrics to the variation in the horizontal resolution is more evident than that with different vertical resolutions. Over most of the evaluated time periods, C6, which has a grid spacing of 1.62 km, displays better performance than C3 and C7 having grid spacings of 4.5 and 0.826 km, respectively. Comparison of C3 and C6 shows that C6 tends to produce more accurate spatial patterns of rainfall throughout the heavy rainfall event in Beijing. Higher values of PMAX$'$ and TP$'$ are also detected in C6 when compared to C3. This result stems in part from the explicit resolution of the convective processes by the WRF microphysics scheme, which may explain why the PMAX$'$ of C7 is higher than C6 over most of the tested durations. Note that the modelling skill of C7 deteriorates

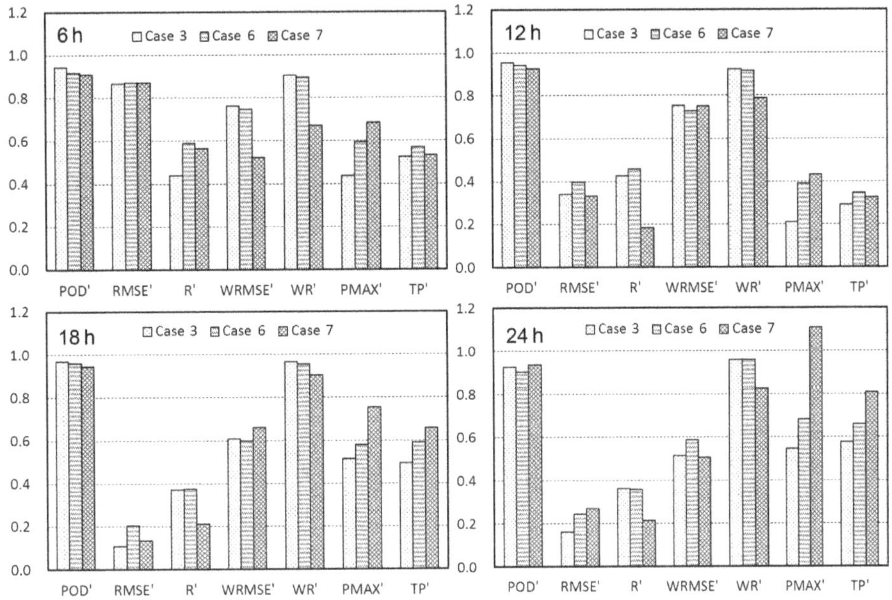

Figure 6. As in Fig. 3, but for the experiments in scenario three with different horizontal resolutions. Case 3 has an initial downscaling ratio of 1 : 3 : 3 with horizontal grid spacing of 40.5, 13.5, and 4.5 km, whereas cases 6 and 7 have the same large horizontal grid spacing with nesting ratios of 1 : 5 : 5 and 1 : 7 : 7, respectively. The innermost grid spacing is 1.62 km in Case 6 and 0.826 km in Case 7.

rapidly after the heavy rain begins (12 h); the lowest POD′ and R' values of the three experiments are obtained for this simulation and time period. Analysis of the WRMSE′ values suggests that simulation C7 displays significant departures from the coarser-scale PW fields that are used to force the model. Thus, model simulations with excessively high horizontal resolutions may also display poor performance. Theoretically, this deterioration may be attributed to the accumulated errors introduced by the imperfect model physics or biases in the initial and boundary conditions, which can be exaggerated by the chaotic nature of NWP systems. According to the above analysis, C6 yields the best agreement with the ground truth data among the horizontal resolution experiments.

5.4 Searching for the likely ideal spin-up time

To limit the effects of the chaotic nature of NWP on the model simulations and extend the lead time, the scenario in which the spin-up time used in WRF is varied is placed at the end of the experimental design, after the possible errors introduced by inappropriate domain configuration options have been reduced. In S4, C6 is adopted as the starting experiment (OS3). Unlike the previous scenarios, the ranks of the spin-up time experiments, as sorted by the metrics, are nearly the same across the different time periods. Hence, Fig. 7 presents only the modelling skill of the spin-up time experiments over the time period of 18 h. The model performance of WRF in simulating heavy rainfall clearly varies with the spin-up time. For most of the metrics, an obvious diurnal tendency is found from 0 to 60 h, followed by a short-term

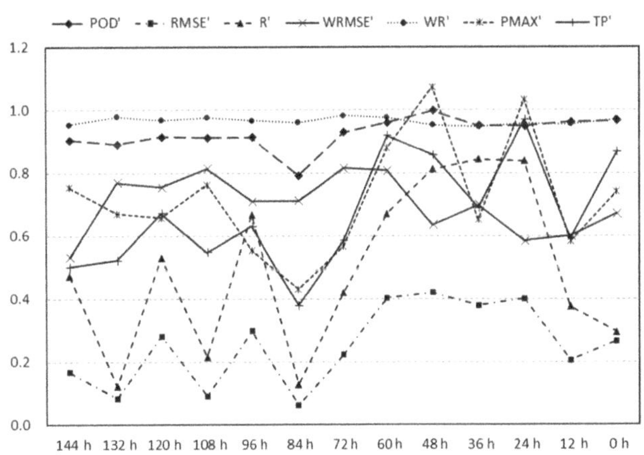

Figure 7. Spatial values of the verification metrics for the WRF spin-up experiments, calculated over 18 h periods and over domain three. Case 6 employs an initial spin-up time of 12 h; Case 8 employs a spin-up time of 0 h; and from Case 9 to Case 19, the spin-up time is increased from 24 to 144 h by every 12 h.

decrease until 72 h; random fluctuations occur after 72 h. Before 72 h, the variations in the rainfall and PW metrics are almost consistent; thus, the good fits of the simulations produced by the model runs with longer spin-up times are also physically reasonable within this period. The discrepancies among these experiments may be due to differences in the initial conditions (e.g. the water vapour amounts and the times of day when the simulations begin).

From TP′, it is found that all of the spin-up time experiments underestimate the total rainfall amount during the

Table 3. Comparison of the values of the error metrics in the initial experiment and the optimum experiments identified for each scenario.

Experiment number	POD′	RMSE′	R'	WRMSE′	WR′	PMAX′	TP′
Case 0 (C0)	0.950	0.098	0.226	0.789	0.980	0.440	0.478
Case 1 (C1)	0.960	0.064	0.376	0.622	0.967	0.436	0.471
Case 3 (C3)	0.969	0.110	0.373	0.610	0.967	0.515	0.496
Case 6 (C6)	0.963	0.205	0.375	0.600	0.956	0.582	0.592
Case 12 (C12)	0.959	0.402	0.670	0.807	0.977	0.883	0.920

heavy rainfall event. Of all of the rainfall-related metrics, POD′ is found to display the least sensitivity to the spin-up time; however, it displays similar variations over time as PMAX′, R', and RMSE′ before 72 h, with the highest values shown in the experiment with a spin-up time of 48 h (C11). Positive biases are detected in PMAX′ in C9 (which is run 24 h ahead) and C11, in which the largest positive biases are detected in the simulated amount of water vapour across the analysed periods and earlier (during the initialization period). This result may occur because the atmospheric water vapour content determines the maximum possible rainfall amount. C12, which includes a spin-up time of 60 h, is ranked third in terms of PMAX′, whereas it displays better performance than C9 and C11 in terms of TP′, WR′, and WRMSE′. As seen in Fig. 8, C9, C11 and C12 also rank in the top three, based on the values of the rainfall-related metrics calculated over domain two. However, larger departures from the forcing PW fields are seen in C9 and C11 than in C12. The difference is that C12 shows the best agreement with the ground truth data in terms of both the rainfall- and PW-related fields. Overall, C12 is regarded as the experiment that best reproduces the Beijing SDHR event with the optimal set of domain configuration options and the longest spin-up time.

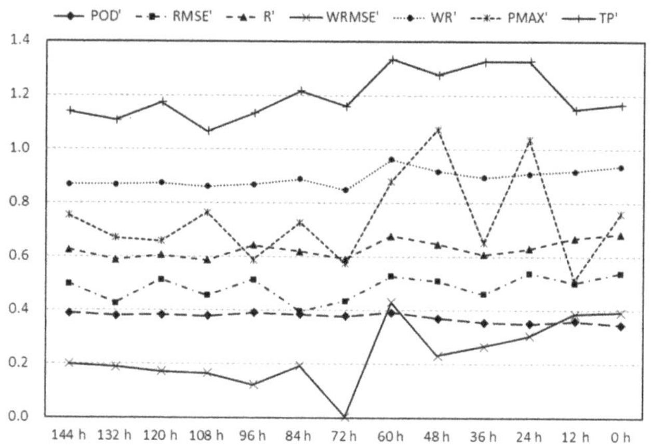

Figure 8. As in Fig. 7, but the metrics are calculated over 18 h periods and over domain two in Case 6.

6 Discussion

The results reveal that the initial experiment with the most commonly employed WRF domain settings does not yield the best performance in reproducing the temporal and spatial characteristics of SDHR on the convective scale. In S1, the assigned domain size of C0 is not sufficiently broad to allow the model physics to fully develop local small-scale features, resulting in obvious reductions in modelling skill as the evaluated time duration increases from 12 to 24 h. Further refinement of the grid spacing of C0 in S2 and S3 is shown to enable more explicit resolution of convective processes, leading to more accurate rainfall simulations. The comparison made in S4 suggests that the proper spin-up time is determined by both the time needed for model initialization and the accuracy of the initial conditions fed into the model run. Moreover, experiments with too large domains, too high spatial resolutions, or too long spin-up times also yield poor performance in rainfall simulations. Therefore, the reasonableness of these WRF settings should be checked before the model is used in regional NWP systems for flood forecasting or as a reference for the design of flood mitigation strategies.

In addition to exploring whether the recommended WRF domain configuration options and spin-up time are optimal for application in SDHR-prone urban areas, the performance of the model is quantified, and its total improvement is evaluated by comparing the values of the verification metrics yielded by the experiments. Table 3 compares the values of the verification metrics obtained for the optimal experiments in each scenario with the values obtained for the initial experiment. Here, the 18 h time duration is selected for evaluation because it covers most of the heavy rainfall event, and the metrics calculated over this period display a greater range and thus greater ability in identifying the simulation with the best performance. One exception is the domain size scenario, in which C0 presents the most obvious reduction in performance during the last stage of the rainfall event (24 h). Therefore, the improvement in C1 relative to C0 is mainly represented by R' and POD′ across D03 over the 18 h time period. The improvement produced by refining the vertical resolution is indicated by all of the rainfall-related metrics but is accompanied by a decrease in WRMSE′ that stems in part from the reduction in kinetic energy, which promotes rainfall. C6 yields higher values of POD′, RMSE′, R', and PMAX′ when compared with C3, indicating that appropriate increases in the horizontal resolution can increase the accuracy of rainfall simulations. The largest differences in the metrics between C6 and C12 occurs for PMAX′, which may relate to the different initial weather conditions at the different starting times of the model runs.

Overall, although the magnitudes of the increases in the rainfall metrics differ, they all reflect an increase in model skill after the re-evaluation process has been conducted. Specifically, R' increases from 0.226 in C0 to 0.67 in C12, RMSE′ increases from 0.098 to 0.402, and PMAX′ increases from 0.44 to 0.883. As the complete assessment is based on objective verification metrics and checked by subjective verification methods, it can be concluded that the domain configuration options and the spin-up time have significant effects on regional simulations of SDHR. Therefore, re-evaluating the values of those settings used in high-resolution regional studies is certainly worthwhile, and the accuracy of predictions of heavy rain clearly benefit from these analyses. For the evaluated metrics, evaluations based on a single type of metric or a single time period may clearly result in partially accurate conclusions. The use of datasets from multiple sources in verification can help increase the comprehensiveness of the analyses, such as the use of WRMSE′ and WR′ in this study. The use of different time periods helps to determine the optimal configurations with higher physical rationality, such as the selection of the proper domain size. In addition, the verification results may also depend on the fields and temporal–spatial scales of interest. To further understand the effects of WRF model configuration options on regional simulations of sub-daily heavy rainfall, more objective verification metrics for SDHR should be developed, and more case studies of SDHR events are also needed. Given that the uncertainties in the regional NWP studies result mainly from the inaccurate boundary conditions associated with grid nesting techniques, methods that can serve as alternate schemes to reduce these uncertainties are also worth studying. One example includes the mesh transitions approach used on irregular grids. In addition, more accurate simulations are expected when the model is driven with forcing data with higher temporal or spatial resolutions than those of the ERA-Interim reanalysis because the uncertainties and errors introduced by the input data could be further reduced.

7 Conclusions

In this study, a comparative test is designed to evaluate the effects of WRF domain configuration options and the spin-up time on simulations of the precipitation during the SDHR event that occurred on 21 July 2012 in Beijing, China. Three nested domains are established: D01 is the largest, has the coarsest resolution, and covers the leading synoptic features; and D03 is the smallest and covers the area of interest, Beijing. The initial conditions of the three domains are provided by the ERA-Interim reanalysis and the 30-second static geographical datasets. For the LBCs, D01 is forced by the ERA-Interim reanalysis, whereas D02 is forced by D01, and D03 is forced by D02. The reference ground truth data used for verification is 3-hourly 0.05 gridded rainfall observations and the coarser-scale ERA-Interim reanalysis. Five rainfall-related error metrics and two PW-related indices that monitor the departure of the model simulations from the driving fields are calculated at the convective-resolving scale over different sub-daily time spans. These metrics are then checked and considered together as part of a subjective verification process that is intended to pinpoint the likely best combination of the domain configuration options and spin-up time and to help quantify the possible improvements in the model performance of WRF in reproducing severe SDHR events after carrying out the entire re-evaluation process.

Precipitation simulations are sensitive to changes in domain size, vertical resolution, horizontal resolution, and spin-up time. Of all of the configurations, the most obvious variations are found when adjusting the domain size and the spin-up time. This analysis shows that domains that cover only the area of interest may be insufficiently broad to permit full development of small-scale features, resulting in poor performance in capturing the spatial pattern of heavy rainfall, especially in the early stages of rainfall events. Despite the dominant role of chaotic processes, it is still possible that model runs with longer spin-up times may result in better rainfall simulations, given favourable initial weather conditions. The effects of the vertical and horizontal resolutions are smaller, but the accuracy of the rainfall amount and the correct hits exhibit evident increases in runs with slightly higher spatial resolutions. A comparison of C12, which uses the evaluated optimum configurations, and C0, which uses the recommended settings, shows that the metrics clearly increase. Specifically, R' increases from 0.226 to 0.67, RE_{PMAX} rises from -56 to $-11.7\,\%$, and RMSE decreases by 33.65 %. Thus, substantial benefits may result from re-evaluating the WRF domain configuration options and spin-up times used in regional studies of SDHR.

Given the intensification of SDHR and the increased risks posed by SDHR-induced hazards, the demands of the operational flood management community for more accurate rainfall predictions with longer lead times, especially over highly affected areas with very short hydrologic response times, are increasing. One method that has now been proven to be effective is to dynamically downscale freely available global NWP products to areas of interest using high-resolution regional NWP models (e.g. WRF). Therefore, the uncertainties associated with the downscaling process, such as errors in boundary conditions and the issues associated with grid nesting, should be carefully evaluated to ensure that the rainfall simulations produced are both statistically accurate and physically reasonable before they are employed in flood forecasting systems. This study illustrates the importance of re-evaluating the domain configuration options and spin-up times used in WRF for improving regional rainfall simulations. Comparisons of the metrics indicate that evaluations based on just one category of metrics or values of metrics calculated over only one time period (e.g. 24 h) do not result in comprehensive comparisons and may lead to partially accurate conclusions. The use of PW fields calculated against reanalysis output is verified to be helpful in determining the optimal set of model configurations when analyses of rainfall-related metrics do not yield uniform conclusions. In addition, evaluations conducted over larger-scale domains are demonstrated to be useful in establishing the reasonableness of the evaluated results. Overall, the evaluation process is partly subjective. To simplify the assessment process, verification methods that can replace this subjective verification procedure should be developed. More regional case studies are also needed to further investigate the effects of configuration options in simulations of regional SDHR and to explore methods of reducing the uncertainties in regional NWP modelling associated with the scale-variation procedures. In addition, the use of more accurate forcing data with higher temporal and spatial resolutions is also expected to reduce the errors in the initial and boundary conditions and could thus be helpful in further improving the accuracy of rainfall simulations and extending the lead times of forecasts.

Competing interests. The authors declare that they have no conflict of interest.

Acknowledgements. This study is supported by the key research projects "Sponge city construction and urban flooding/waterlogging disaster in the sub-centre of Beijing City" (Z171100002217080), Beijing Municipal Science and Technology Commission, and "Urban storm flooding/waterlogging disasters under changing environment" (2017YFC1502701), national key research and development plan, Ministry of Science and Technology, China. Support is also received from the the Resilient Economy and Society by Integrated SysTems modelling (RESIST), the Newton Fund via Natural Environment Research Council (NERC) and the Economic and Social Research Council (ESRC; NE/N012143/1), and the National Natural Science Foundation of China (no. 4151101234). The China Scholarship Council supports the first author for her academic visit to the University of Bristol, UK.

Edited by: Uwe Ehret

References

Aligo, E. A., Gallus Jr., W. A., and Segal, M.: On the impact of WRF model vertical grid resolution on Midwest summer rainfall forecasts, Weather Forecast., 24, 575–594, 2009.

Bartholmes and Todini: Coupling meteorological and hydrological models for flood forecasting, Hydrol. Earth Syst. Sci., 9, 333–346, https://doi.org/10.5194/hess-9-333-2005, 2005.

Berrisford, P., Dee, D. P., Fielding, K., Fuentes, M., Kallberg, P., Kobayashi, S., and Uppala, S. M.: The ERA-Interim Archive, ERA Report Series, 1, 1–16, 2009.

Brömmel, D., Frings, W., and Wylie, B.: Technical Report Juqueen Extreme Scaling Workshop 2015, Tech. rep., Jülich, Germany, available at: http://hdl.handle.net/2128/8435, last access: 17 June 2018.

Castelli, F.: Atmosphere modeling and hydrologic-prediction uncertainty, U.S. – Italy Research Workshop on the Hydrometeorology, impacts and management of extreme floods, Perugia, 1995.

Chen, F. and Dudhia, J.: Coupling an advanced land surface-hydrology model with the Penn State-NCAR MM5 modeling system. Part I: Model implementation and sensitivity, Mon. Weather Rev., 129, 569–585, 2001.

Chen, H., Sun, J., Chen, X., and Zhou, W.: CGCM projections of heavy rainfall events in China, Int. J. Climatol., 32, 441–450, 2012.

Clark, P., Roberts, N., Lean, H., Ballard, S. P., and Charlton-Perez, C.: Convection-permitting models: a step-change in rainfall forecasting, Meteorol. Appl., 23, 165–181, 2016.

Coen, J. L., Cameron, M., Michalakes, J., Patton, E. G., Riggan, P. J., and Yedinak, K. M.: WRF-Fire: coupled weather–wildland fire modeling with the weather research and forecasting model, J. Appl. Meteorol. Clim., 52, 16–38, 2013.

Crétat, J., Pohl, B., Richard, Y., and Drobinski, P.: Uncertainties in simulating regional climate of Southern Africa: sensitivity to physical parameterizations using WRF, Clim. Dynam., 38, 613–634, 2012.

Cuo, L., Pagano, T. C., and Wang, Q. J.: A review of quantitative precipitation forecasts and their use in short-to medium-range streamflow forecasting, J. Hydrometeorol., 12, 713–728, 2011.

Dee, D. P., Uppala, S. M., Simmons, A. J., Berrisford, P., Poli, P., Kobayashi, S., Andrae, U., Balmaseda, M. A., Balsamo, G., Bauer, P., Bechtold, P., Beljaars, A. C. M., van de Berg, L., Bidlot J., Bormann, N., Delsol, C., Dragani, R., Fuentes, M., Geer, A. J., Haimberger, L., Healy, S. B., Hersbach, H., Hólm, E. V., Isaksen, L., Kållberg, P., Köhler, M., Matricardi, M., McNally, A. P., Monge-Sanz, B. M., Morcrette, J.-J., Park, B.-K., Peubey, C., de Rosnay, P., Tavolato, C., Thépaut, J.-N., and Vitart, F.: The ERA-Interim reanalysis: configuration and performance of the data assimilation system, Q. J. Roy. Meteorol. Soc., 137, 553–597, https://doi.org/10.1002/qj.828, 2011.

Di, Z. H., Duan, Q. Y., Gong, W., Wang, C., Gan, Y. J., Quan, J. P., Li, J. D., Miao, C. Y., Ye, A. Z., and Tong, C.: Assessing WRF model parameter sensitivity: A case study with five-day summer precipitation forecasting in the Greater Beijing Area, Geophys. Res. Lett., 42, 579–587, 2015.

Done, J., Davis, C. A., and Weisman, M.: The next generation of NWP: Explicit forecasts of convection using the Weather Research and Forecasting (WRF) model, Atmos. Sci. Lett., 5, 110–117, 2004.

Ek, M. B., Mitchell, K. E., Lin, Y., Rogers, E., Grunmann, P., Koren, V., Gayno, G., and Tarpley, J. D.: Implementation of Noah land surface model advances in the National Centers for Environmental Prediction operational mesoscale Eta model, J. Geophys. Res.-Atmos., 108, 8851, https://doi.org/10.1029/2002JD003296, 2003.

Fierro, A. O., Rogers, R. F., Marks, F. D., and Nolan, D. S.: The impact of horizontal grid spacing on the microphysical and kinematic structures of strong tropical cyclones simulated with the WRF-ARW model, Mon. Weather Rev., 137, 3717–3743, 2009.

Foley, A. M., Leahy, P. G., Marvuglia, A., and McKeogh, E. J.: Current methods and advances in forecasting of wind power generation, Renewable Energy, 37, 1–8, 2012.

Gao, Y., Yuan, Y., Wang, H., Schmidt, A. R., Wang, K., and Ye, L.: Examining the effects of urban agglomeration polders on flood events in Qinhuai River basin, China with HEC-HMS model, Water Sci. Technol., 75, 2130–2138, 2017.

Goswami, P., Shivappa, H., and Goud, S.: Comparative analysis of the role of domain size, horizontal resolution and initial conditions in the simulation of tropical heavy rainfall events, Meteorol. Appl., 19, 170–178, 2012.

Grell, G. A. and Dévényi, D.: A generalized approach to parameterizing convection combining ensemble and data assimilation techniques, Geophys. Res. Lett., 29, 1693, https://doi.org/10.1029/2002GL015311, 2002.

Guo, C., Xiao, H., Yang, H., and Tang, Q.: Observation and modeling analyses of the macro-and microphysical characteristics of a heavy rain storm in Beijing, Atmos. Res., 156, 125–141, 2015.

Heinzeller, D., Duda, M. G., and Kunstmann, H.: Towards convection-resolving, global atmospheric simulations with the Model for Prediction Across Scales (MPAS) v3.1: an extreme scaling experiment, Geosci. Model Dev., 9, 77–110, https://doi.org/10.5194/gmd-9-77-2016, 2016.

Hong, S. Y. and Lee, J. W.: Assessment of the WRF model in reproducing a flash-flood heavy rainfall event over Korea, Atmos. Res., 93, 818–831, 2009.

Hong, S. Y. and Lim, J. O. J.: The WRF single-moment 6-class microphysics scheme (WSM6), J. Korean Meteorol. Soc., 42, 129–151, 2006.

Hong, S. Y., Noh, Y., and Dudhia, J.: A new vertical diffusion package with an explicit treatment of entrainment processes, Mon. Weather Rev., 134, 2318–2341, 2006.

Huang, C., Zheng, X., Tait, A., Dai, Y., Yang, C., Chen, Z., Li, T., and Wang, Z.: On using smoothing spline and residual correction to fuse rain gauge observations and remote sensing data, J. Hydrol., 508, 410–417, 2013.

Iacono, M. J., Delamere, J. S., Mlawer, E. J., Shephard, M. W., Clough, S. A., and Collins, W. D.: Radiative forcing by long-lived greenhouse gases: calculations with the AER radiative transfer models, J. Geophys. Res.-Atmos., 113, D13103, https://doi.org/10.1029/2008JD009944, 2008.

Kain, J. S., Weiss, S. J., Bright, D. R., Baldwin, M. E., Levit, J. J., Carbin, G. W., Schwartz, C. S., Weisman, M. L., Droegemeeier, K. K., Weber, D. B., and Thomas, K. W.: Some practical considerations regarding horizontal resolution in the first generation of operational convection-allowing NWP, Weather Forecast., 23, 931–952, 2008.

Kleczek, M. A., Steeneveld, G. J., and Holtslag, A. A.: Evaluation of the weather research and forecasting mesoscale model for GABLS3: impact of boundary-layer schemes, boundary conditions and spin-up, Bound.-Lay. Meteorol., 152, 213–243, 2014.

Klemp, J. B.: Advances in the WRF model for convection-resolving forecasting, Adv. Geosci., 7, 25–29, https://doi.org/10.5194/adgeo-7-25-2006, 2006.

Leduc, M. and Laprise, R.: Regional climate model sensitivity to domain size, Clim. Dynam., 32, 833–854, 2009.

Li, J., Chen, Y., Wang, H., Qin, J., Li, J., and Chiao, S.: Extending flood forecasting lead time in a large watershed by coupling WRF QPF with a distributed hydrological model, Hydrol. Earth Syst. Sci., 21, 1279–1294, https://doi.org/10.5194/hess-21-1279-2017, 2017.

Liu, J., Bray, M., and Han, D.: Sensitivity of the Weather Research and Forecasting (WRF) model to downscaling ratios and storm types in rainfall simulation, Hydrol. Process., 26, 3012–3031, 2012.

Luna, T., Castanheira, M., and Rocha, A.: Assessment of WRF-ARW forecasts using warm initializations, available at: http://climetua.fis.ua.pt/publicacoes/APMG_extended_abstract_2013_Luna_et_al.pdf (last access: 17 June 2018), 2013.

Miguez-Macho, G., Stenchikov, G. L., and Robock, A.: Spectral nudging to eliminate the effects of domain position and geometry in regional climate model simulations, J. Geophys. Res.-Atmos., 109, D13104, https://doi.org/10.1029/2003JD004495, 2004.

Powers, J. G., Klemp, J. B., Skamarock, W. C., Davis, C. A., Dudhia, J., Gill, D. O., Coen, J. L., and Grell, G. A.: The Weather Research and Forecasting Model: Overview, System Efforts, and Future Directions, B. Am. Meteorol. Soc., 98, 1717–1737, https://doi.org/10.1175/BAMS-D-15-00308.1, 2017.

Prein, A. F., Langhans, W., Fosser, G., Ferrone, A., Ban, N., Goergen, K., Keller, M., Tölle, M., Gitjahr, O., Feser, F., Brisson, E., Kollet, S., Schmidli, J., van Lipzig, N. P. M., and Leung, R.: A review on regional convection-permitting climate modeling: Demonstrations, prospects, and challenges, Rev. Geophys., 53, 323–361, 2015.

Roberts, N. M. and Lean, H. W.: Scale-selective verification of rainfall accumulations from high-resolution forecasts of convective events, Mon. Weather Rev., 136, 78–97, 2008.

Ruiz, J. J., Saulo, C., and Nogués-Paegle, J.: WRF model sensitivity to choice of parameterization over South America: validation against surface variables, Mon. Weather Rev., 138, 3342–3355, 2010.

Schwartz, C. S., Kain, J. S., Weiss, S. J., Xue, M., Bright, D. R., Kong, F. Y., Thomas, K. W., Levit, J. J., and Coniglio, M. C.: Next-day convection-allowing WRF model guidance: A second look at 2-km versus 4-km grid spacing, Mon. Weather Rev., 137, 3351–3372, 2009.

Seth, A. and Rojas, M.: Simulation and sensitivity in a nested modeling system for South America. Part I: Reanalyses boundary forcing, J. Climate, 16, 2437–2453, 2003.

Shih, D. S., Chen, C. H., and Yeh, G. T.: Improving our understanding of flood forecasting using earlier hydro-meteorological intelligence, J. Hydrol., 512, 470–481, 2014.

Sikder, S. and Hossain, F.: Assessment of the weather research and forecasting model generalized parameterization schemes for advancement of precipitation forecasting in monsoon-driven river basins, J. Adv. Model. Earth Syst., 8, 1210–1228, 2016.

Skamarock, W. C., Klemp, J. B., Dudhia, J., Gill, D. O., Barker, D. M., Duda, M. G., Huang, X. Y., Wang, W., and Powers, J. G.: A Description of the Advanced Research WRF Version 3, NCAR Note NCAR/TN-475, NCAR, Boulder, Colorado, USA, 2008.

Soares, P. M., Cardoso, R. M., Miranda, P. M., de Medeiros, J., Belo-Pereira, M., and Espirito-Santo, F.: WRF high resolution dynamical downscaling of ERA-Interim for Portugal, Clim. Dynam., 39, 2497–2522, 2012.

Sun, M. S., Yang, L. Q., Yin, Q., Niu, Z. Y., and Gao, L. M.: Analysis of the cause of a torrential rain occurring in Beijing on 21 July 2012, Torrent. Rain Disast., 32, 218–223, 2013.

Swinbank, R. and James Purser, R.: Fibonacci grids: A novel approach to global modeling, Q. J. Roy. Meteorol. Soc., 132, 1769–1793, 2006.

Tian, J., Liu, J., Yan, D., Li, C., and Yu, F.: Numerical rainfall simulation with different spatial and temporal evenness by using a WRF multiphysics ensemble, Nat. Hazards Earth Syst. Sci., 17, 563–579, https://doi.org/10.5194/nhess-17-563-2017, 2017.

Vrac, M., Drobinski, P., Merlo, A., Herrmann, M., Lavaysse, C., Li, L., and Somot, S.: Dynamical and statistical downscaling of the French Mediterranean climate: uncertainty assessment, Nat. Hazards Earth Syst. Sci., 12, 2769–2784, https://doi.org/10.5194/nhess-12-2769-2012, 2012.

Wang, K., Wang, L., Wei, Y. M., and Ye, M.: Beijing storm of July 21, 2012: observations and reflections, Nat. Hazards, 67, 969–974, 2013.

Wang, S. L., Kang, H. W., Gu, X. Q., and Ni, Y. Q.: Numerical Simulation of Mesoscale Convective System in the Warm Sector of Beijing '7.21' Severe Rainstorm, Meteorol. Mon., 41, 544–553, 2015.

Warner, T. T.: Quality assurance in atmospheric modeling, B. Am. Meteorol. Soc., 92, 1601–1610, 2011.

Warner, T. T., Peterson, R. A., and Treadon, R. E.: A tutorial on lateral boundary conditions as a basic and potentially serious limitation to regional numerical weather prediction, B. Am. Meteorol. Soc., 78, 2599–2617, 1997.

Westra, S., Fowler, H. J., Evans, J. P., Alexander, L. V., Berg, P., Johnson, F., Kendon, E. J., Lenderink, G., and Roberts, N. M.: Future changes to the intensity and frequency of short-duration extreme rainfall, Rev. Geophys., 52, 522–555, 2014.

Willems, P., Arnbjerg-Nielsen, K., Olsson, J., and Nguyen, V. T. V.: Climate change impact assessment on urban rainfall extremes and urban drainage: methods and shortcomings, Atmos. Res., 103, 106–118, 2012.

WMO: Anticipated advances in numerical weather prediction, and the growing technology gap in weather forecast, available at: https://www.wmo.int/pages/prog/www/swfdp/Meetings/documents/Advances_NWP.pdf (last access: 17 June 2018), 2013.

Xu, Z. X. and Chu, Q.: Climatological features and trends of extreme precipitation during 1979–2012 in Beijing, China, P. Int. Assoc. Hydrolog. Sci., 369, 97–102, 2015.

Xu, Z. X. and Zhao, G.: Impact of urbanization on rainfall-runoff processes: case study in the Liangshui River Basin in Beijing, China, P. Int. Assoc. Hydrolog. Sci., 373, 7–12, 2016.

Yu, E.-T., Wang, H.-J., and Sun, J.-Q.: A quick report on a dynamical downscaling simulation over China using the nested model, Atmos. Ocean. Sc. Lett., 3, 325–329, 2010.

Yu, R., Xu, Y., Zhou, T., and Li, J.: Relation between rainfall duration and diurnal variation in the warm season precipitation over central eastern China, Geophys. Res. Lett., 34, L13703, https://doi.org/10.1029/2007GL030315, 2007.

Yu, W., Nakakita, E., Kim, S., and Yamaguchi, K.: Impact Assessment of Uncertainty Propagation of Ensemble NWP Rainfall to Flood Forecasting with Catchment Scale, Adv. Meteorol., 4, 1–17, https://doi.org/10.1155/2016/1384302, 2016.

Yucel, I., Onen, A., Yilmaz, K. K., and Gochis, D. J.: Calibration and evaluation of a flood forecasting system: Utility of numerical weather prediction model, data assimilation and satellite-based rainfall, J. Hydrol., 523, 49–66, 2015.

Zhou, Y. S., Liu, L., Zhu, K. F., and Li, J. T.: Simulation and evolution characteristics of mesoscale systems occurring in Beijing on 21 July 2012, Chinese J. Atmos. Sci., 38, 885–896, 2014.

5

Recent changes to the hydrological cycle of an Arctic basin at the tundra–taiga transition

Sebastian A. Krogh and John W. Pomeroy

Centre for Hydrology, University of Saskatchewan, 121 Research Dr., Saskatoon, SK S7N 1K2, Canada

Correspondence: Sebastian A. Krogh (seba.krogh@usask.ca)

Abstract. The impact of transient changes in climate and vegetation on the hydrology of small Arctic headwater basins has not been investigated before, particularly in the tundra–taiga transition region. This study uses weather and land cover observations and a hydrological model suitable for cold regions to investigate historical changes in modelled hydrological processes driving the streamflow response of a small Arctic basin at the treeline. The physical processes found in this environment and explicit changes in vegetation extent and density were simulated and validated against observations of streamflow discharge, snow water equivalent and active layer thickness. Mean air temperature and all-wave irradiance have increased by 3.7 °C and 8.4 W m^{-2}, respectively, while precipitation has decreased 48 mm (10 %) since 1960. Two modelling scenarios were created to separate the effects of changing climate and vegetation on hydrological processes. Results show that over 1960–2016 most hydrological changes were driven by climate changes, such as decreasing snowfall, evapotranspiration, deepening active layer thickness, earlier snow cover depletion and diminishing annual sublimation and soil moisture. However, changing vegetation has a significant impact on decreasing blowing snow redistribution and sublimation, counteracting the impact of decreasing precipitation on streamflow, demonstrating the importance of including transient changes in vegetation in long-term hydrological studies. Streamflow dropped by 38 mm as a response to the 48 mm decrease in precipitation, suggesting a small degree of hydrological resiliency.

These results represent the first detailed estimate of hydrological changes occurring in small Arctic basins, and can be used as a reference to inform other studies of Arctic climate change impacts.

1 Introduction

Rapid warming in the Arctic (Hansen et al., 2010; Przybylak et al., 2010; Wanishsakpong et al., 2016) has produced significant environmental changes (Hinzman et al., 2005), such as decreasing snow cover duration (Brown et al., 2010) and permafrost thaw (Liljedahl et al., 2016). A reduced snow cover period can result in smaller and slower snowmelt, larger evapotranspiration losses and reduced sublimation losses from headwater basins in cold regions (Pomeroy et al., 2015; Rasouli et al., 2015). Permafrost thaw can impact regional and local hydrology by increasing surface and subsurface connectivity and baseflow (Connon et al., 2014; Liljedahl et al., 2016; Walvoord and Kurylyk, 2016). Increases in vegetation cover and density have been observed and are especially pronounced near the tundra–taiga ecozone transition (Lantz et al., 2013; Myers-Smith et al., 2011; Sturm et al., 2001; Tape et al., 2006; Xu et al., 2013); however, the impact on the hydrology of these transition Arctic basins is poorly understood. These environmental changes will likely continue in the future, representing challenges for water resources managers and engineers throughout the Arctic.

Precipitation trends over the Arctic are highly uncertain due to a sparse monitoring network (Serreze et al., 2003) and difficulties in measuring snowfall in windswept environments (Goodison et al., 1998; Pan et al., 2016). Nevertheless, positive and negative trends have been calculated for the largest Arctic river basins (Walsh, 2005, Table 6.12) and throughout the Arctic (Whitfield et al., 2004). Over northern Canada, an overall increasing trend in annual precipitation has been observed (DeBeer et al., 2016; Vincent et al., 2015); however, there is great spatial variability and uncertainty due to the low-density observational network (Milewska and Hogg, 2001). Mean annual temperatures in northwestern

Canada have increased more than anywhere else in Canada by roughly 3–3.5 °C between 1948 and 2012 (Vincent et al., 2015); moreover, mean winter temperatures show the largest increase of up to 6.5 °C (DeBeer et al., 2016).

Arctic vegetation has changed in response to warmer temperatures (Hinzman et al., 2005; Martin et al., 2017; Myers-Smith and Hik, 2018). The tundra–taiga treeline in Alaska, USA, has advanced from 80 to 100 m in the last 200 years (Suarez et al., 1999). Payette and Filion (1985) studied white spruce (picea glauca) expansion into northern Quebec, Canada, and found that the treeline has not changed substantially over the past centuries; however, below the treeline, its density has increased. On the other hand, both shrub coverage and density have increased in the Arctic. Lantz et al. (2013) reported that between 1972 and 2004, shrub density and cover have increased substantially in the upland tundra east of the Mackenzie River delta of northwestern Canada. Similar results were found by Tape et al. (2006) in northern Alaska and pan-Arctic. Overall, these previous studies observed that the Arctic treeline has not undergone a substantial change over the last century, but that shrub expansion is ubiquitous near the Arctic treeline in North America. Wildfires can rapidly modify vegetation cover and are important to nutrient cycling, biodiversity and the control of pathogens and pests (Bond and Keeley, 2005). Warmer temperatures and longer dry seasons are increasing vulnerability to wildfire (Romero-Lankao et al., 2014), resulting in increased frequency and duration of wildfires since the mid-1980s (Westerling et al., 2006; Williamson et al., 2009). Changes in vegetation are important, as they have been shown to control snow redistribution (Ellis et al., 2013; Essery and Pomeroy, 2004; Ménard et al., 2014; Pomeroy and Brun, 2001) and energy fluxes (Ménard et al., 2012; Pomeroy et al., 2006; Sturm et al., 2000).

Many studies have looked at observed changes in large northward-flowing river basins. There is an increase in annual discharge from large river basins to the Arctic Ocean (McClelland et al., 2006; Overeem and Syvitski, 2010; Peterson et al., 2002; Rood et al., 2017; Yang et al., 2002), a decrease in river ice thickness (Peterson et al., 2002) and earlier river/lake ice break-up dates (Janowicz, 2010; Prowse et al., 2011). However, most of these large river basins have their headwaters and primary zones of runoff generation well below the Arctic Circle, and therefore are not necessarily representative of changes in the Arctic hydrological cycle. As limited observations are available in the Arctic, model outputs have also been used to investigate change. Increasing trends were found in simulated monthly evapotranspiration and streamflow for the Mackenzie River basin, Canada (Yip et al., 2012), and in simulated Arctic soil temperature and active layer thickness (Oelke et al., 2004), whereas decreasing trends were found in simulated Arctic snow accumulation and snow cover duration (Liston and Hiemstra, 2011). López-Moreno et al. (2016) analyzed simulated historical trends in the snow processes of a small basin above the Arctic treeline in Svalbard, using a physically based hydrological model that accounted for blowing snow redistribution and energy balance snowmelt. They found that simulated snow accumulation, snow-covered season and days with snowfall decreased significantly, driven by a significant increase in air temperature. No study has looked at changes in Arctic hydrological processes from headwater basins that originate near the Arctic treeline, nor has the relationship between changes in hydrological processes due to climate and vegetation change been investigated.

Using hydrological models to simulate the hydrological cycle can compensate for an inability to observe it due to ungauged basins (Pomeroy et al., 2013b) and decline in the coverage of Arctic monitoring networks (Laudon et al., 2017). Previous studies acknowledged the need for robust hydrological models suitable for cold regions to simulate Arctic hydrology (Quinton and Carey, 2008; Woo et al., 2008), particularly due to the complex interaction between subsurface and surface mass and energy fluxes (Kane et al., 1991; Krogh et al., 2017; Zhang et al., 2000). Physical processes that must be accounted for include snow accumulation and melt (Marsh et al., 2010), snow interception and sublimation from forest canopies (Hedstrom and Pomeroy, 1998; Pomeroy et al., 1998; Schmidt and Gluns, 1991), blowing snow sublimation and redistribution (Pomeroy et al., 1997; Schmidt, 1982), evapotranspiration (Wessel and Rouse, 1994), infiltration into frozen and unfrozen soils (Gray et al., 2001; Kane, 1980; Kane and Stein, 1983), water flow through snowpack (Colbeck, 1972; Marsh and Woo, 1984a, b), ground freeze and thaw (Juminikis, 1977), surface and subsurface flow (Quinton and Gray, 2001; Quinton and Marsh, 1999), and groundwater (Cederstrom et al., 1953) and streamflow routing (Woo and Sauriol, 1980). The Cold Regions Hydrological Model (CRHM) platform was used to create the Arctic Hydrology Model (AHM) configuration (CRHM-AHM) by Krogh et al. (2017). This spatially distributed and physically based model includes the key hydrological processes found at the Arctic treeline, such as blowing snow, snow and rain interception, sublimation, snowmelt, flow through snow, infiltration to frozen and unfrozen soils, evapotranspiration, runoff as overland flow and subsurface flow through organic terrain, frozen ground dynamics including active layer thaw and groundwater flow and streamflow routing. CRHM-AHM was shown to properly represent the winter and summer hydrology of this environment with minimal calibration of some uncertain routing and storage parameters (Krogh et al., 2017). A great advantage of this model is its flexibility and potential to be adapted for the simulation of other Arctic basins.

The aim of this study is to understand, diagnose and quantify the long-term hydrological changes of a small Arctic treeline basin, including transient changes in vegetations and climate over a multidecadal period, using the CRHM-AHM model. The study addresses the following research questions: what hydrological changes are caused by individual transient

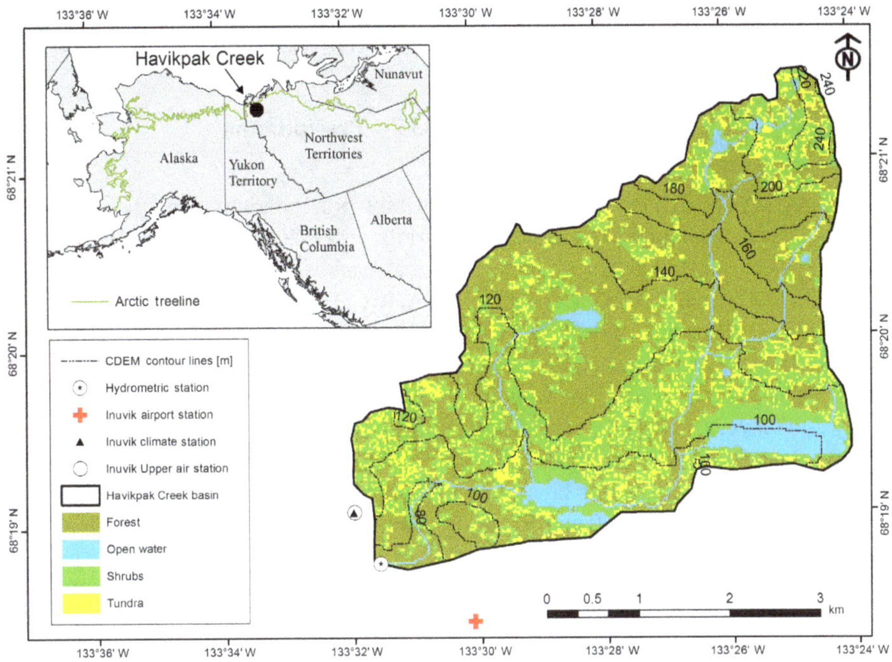

Figure 1. Havikpak Creek basin including elevation contour lines (based on the Canadian Digital Elevation Model – 20 m), the location of weather and hydrometric stations and the 1992 landcover map based on Krogh et al. (2017). Inset plot shows the location of the study site within North America and the approximate location of the Arctic treeline.

changes in climate and vegetation? What are the coupled hydrological impacts of changes in climate and vegetation? Does transient vegetation change enhance or dampen climate change? Does Arctic hydrology show resiliency to the impacts of climate change? To address these questions, the study compares three observation-driven hydrological modelling scenarios: (1) observed climate change and constant vegetation cover, (2) observed changes in vegetation with climate held constant and (3) observed changes in climate and vegetation.

2 Study site

Havikpak Creek (HPC), with an area of 16.4 km^2, is located east of Inuvik, Northwest Territories, Canada (Fig. 1), near the tundra–taiga transition. HPC is in the continuous permafrost region, with an elevation rising from 60 m a.s.l. in the southwest to 240 m a.s.l. in the northeast. This basin was selected as it has a history of process-based hydrological studies, which provides a good understanding of dominating hydrological processes, has long-term meteorological records and has been part of important international initiatives, such as the Mackenzie GEWEX study (MAGS). HPC is also within the domain of the NASA Arctic-Boreal Vulnerability Experiment (ABoVE; https://above.nasa.gov/, last access: 12 July 2018), which aims to better understand the vulnerability and resiliency of Arctic boreal ecosystems, and therefore, its great relevance.

Estimates of mean annual temperature and precipitation between 1981 and 2010 at Inuvik, using observations at the Meteorological Service of Canada weather station (Climate ID: 2202570) by Environment and Climate Change Canada (ECCC), are −8.2 °C and 240.6 mm, respectively (http://climate.weather.gc.ca/climate_normals/index_e.html, last access: 12 July 2018). However, Krogh et al. (2017) showed that the corrected mean annual precipitation between 1980 and 2009 at Inuvik, based on the Adjusted and Homogenized Canadian Climate Dataset (AHCCD; Mekis and Vincent, 2011) and additional local weather stations, is 327 mm. Differences between precipitation estimates published by ECCC and Krogh et al. (2017) are due to Krogh's use of the AHCCD dataset with its corrections of snowfall wind undercatch and trace events. Such large adjustments to corrected precipitation are not uncommon at high latitudes in Canada and can influence trend detection (Pomeroy and Goodison, 1997). Using corrected data, 59 % of the mean annual precipitation is snowfall; however, peak monthly precipitation occurs as rainfall in August (∼ 45 mm; Krogh et al., 2017). Snow accumulation typically starts in mid-September, with peak accumulation at the end of April or beginning of May, and snowmelt lasts until early to mid-June (Krogh et al., 2017). The streamflow regime of Havikpak Creek is measured by the Water Survey of Canada and is characterized by a rapid increase due to snowmelt in May and June, during which the annual peak streamflow occurs (1–4 m^3 s^{-1}), followed by decreasing streamflow interrupted by sporadic summer peaks due to intense rainfall

(Krogh et al., 2017). No streamflow was observed during winter.

In 1992, HPC was predominantly covered by black spruce (*Picea mariana*) forest (50.0 %) followed by alder shrubs (31.7 %), short grass, moss and lichen tundra (11.6 %) and open water (6.7 %) (Krogh et al., 2017). However, as shrubs colonize the tundra (Lantz et al., 2013) these percentages have changed. No changes in forest cover have been reported, though investigations into this are understood to be underway. A slight "greening" of the region has been detected through Normalized Difference Vegetation Index (NDVI) analysis of Landsat satellite imagery, but is not attributed to specific vegetation changes (Ju and Masek, 2016). Soils in HPC are characterized by a top layer (roughly 10 cm) of decomposed and highly porous organic matter (upper peat), followed by a highly decomposed and denser organic layer underneath (lower peat), estimated to be between 20 and 50 cm thick on top of a mineral soil layer (Krogh et al., 2017). No soil changes have been reported. For a detailed description regarding HPC climate, land cover, soils, weather and hydrometric stations, the reader is referred to Krogh et al. (2017).

3 Data

Reconstructed weather time series used in this study are based on a combination of observations, adjusted and homogenized time series from the AHCCD dataset (station ID: 2202578; Mekis and Vincent, 2011), ERA-40 (Uppala et al., 2005) and ERA-Interim (Dee et al., 2011) reanalysis over the period 1960–2016. Reanalysis data have been used in the past to complement meteorological observations for hydrological studies (e.g. Krogh et al., 2015). Six-hourly timestep variables were used to drive CRHM-AHM (see Sect. 4.1): precipitation, air temperature, relative humidity, wind speed and incoming short- and long-wave radiation (Fig. 2). Data used for model validation consist of observed daily streamflow (Sect. 5.2). A reconstructed vegetation cover map, topographic information and a site visit informed the spatial model configuration.

3.1 Temperature

Daily minimum, maximum and mean temperatures from the AHCCD dataset are available from 1957 to 2014. Hourly temperature is available from Inuvik Airport from 1980 to 2016 and from Inuvik Climate from 2003 to 2016. To generate a uniform time series of hourly temperature, the following steps were followed: (1) minimum and maximum from the AHCCD dataset (1957–2014) were used to generate hourly temperature by fitting a sinusoidal function, as presented by Chow and Levermore (2007; Eq. 6); (2) hourly temperature measured by the Inuvik Airport station (1980–2016) was used to correct hourly temperature from the AHCCD dataset (1960–1980) through a linear regression model ($R^2 = 0.97$); and (3) Inuvik Airport hourly data were used for the period 1980–2016.

3.2 Precipitation

Daily precipitation from the AHCCD dataset for the period 1960–2006 is available; however, after 1994, several gaps were found. Precipitation measurements from the AHCCD at Inuvik were all made by observers and are considered reliable. After 1994, automatic systems were sometimes used to improve the corrections from snow ruler measurements (Mekis and Vincent, 2011). For measurements from 1994 to 2007, a combination of AHCCD and the local ECCC automatic weather stations Inuvik Climate, Inuvik Upper Air and Inuvik Airport (Fig. 1) was used. From 2007 onward, the Inuvik Climate station (automatic) was the only station recording precipitation. The automatic station snowfall data were corrected for wind undercatch using the expression presented by Smith (2008) for the Alter-shielded Geonor solid precipitation gauge. A specific snowfall correction had to be applied between October and March for the water years 2010 to 2012, as winter precipitation from the Inuvik Climate precipitation gauge was not found to be credible. Observed snow accumulation (snow water equivalent, SWE) in sheltered sites and observed streamflow suggest that snowfall measured during these years was grossly underestimated. The ratios between SWE measured at the end of winter (1 April) and cumulative snowfall in 2011–2013 were 2.6, 1.7 and 2.5, respectively (after wind undercatch corrections); the ratios associated with the other years with both SWE and streamflow data (2003 to 2015) show values around 1. A solution to this problem was proposed and implemented by Pomeroy et al. (1997) at a nearby location (Trail Valley Creek) and consists of estimating "true" winter snowfall from late season snow surveys in a small glade within a forest. Pomeroy et al. (1997) argued that the wind- and sun-sheltered and cold conditions of the site ensured that the snow on the ground in the glade was not redistributed, sublimated or melted, and was therefore equal to the cumulative snowfall. SWE measurements used in this study have the same conditions as those found by Pomeroy et al. (1997) (i.e. sheltered site with mild winds and cold environment), and therefore, their approach was used to estimate true snowfall.

To disaggregate daily into hourly precipitation, the same procedure used in Krogh et al. (2017) was followed. This employs the microcanonical cascade model presented by Güntner et al. (2001). This disaggregation technique assumes that the probability distribution function of the weights factors, defined as the ratio between a lower and upper disaggregation level (e.g. 12 and 24 h), from the different disaggregation levels (e.g. 3, 6, 12 and 24 h) is constant, and it was obtained aggregating hourly precipitation records. The reader is referred to Güntner et al. (2001) and Krogh et al. (2017) for

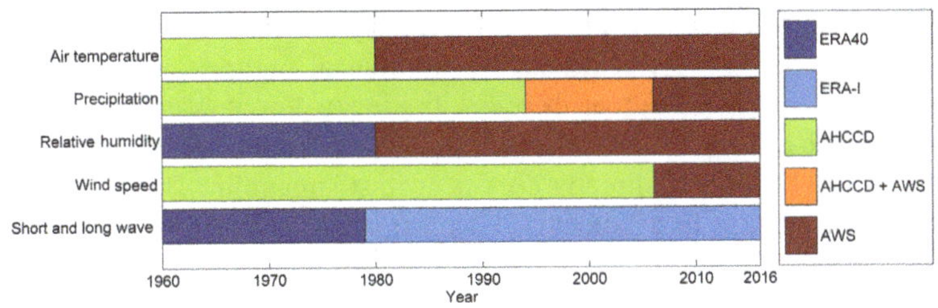

Figure 2. Data source for each of the weather variables during the period 1960–2016. AWS: automatic weather station. AHCCD: Adjusted and Homogenized Canadian Climate Data. ERA-I: ERA-Interim.

further details of this methodology and the particular application to the Inuvik precipitation dataset, respectively.

3.3 Relative humidity

Relative humidity was calculated using 6-hourly air temperature and dew point temperature from ERA-40 for the period 1960–2002, using the expression from Lawrence (2005). A linear interpolation was then used to calculate hourly values. ERA-40 values from 1960 to 1980 were corrected using a linear relationship for the period 1980–2002 between hourly ERA-40 and measured relative humidity at Inuvik Airport ($R^2 = 0.7$). Finally, hourly corrected values from ERA-40 were used from 1960 to 1980 and observed values from 1980 to 2016. Relative humidity was not permitted to exceed 100 % in this estimation.

3.4 Wind speed

Hourly 10 m height wind speed from the AHCCD dataset and Inuvik Airport station for the period 1960–2006 and 2006–2016 were used, respectively.

3.5 Short- and long-wave irradiance

Short- and long-wave irradiance were not measured and so were obtained from the ERA-40 (1960–2002) and ERA-Interim (1979–2016) atmospheric model reanalyses at 3 h time steps. A linear interpolation was used to obtain hourly values for each dataset. The ERA-Interim is a more advanced reanalysis, and has shown small biases in Arctic environments (Lindsay et al., 2014), so it was used as true incoming radiation and ERA-40 outputs were corrected to match the ERA-Interim. The overlapping period between ERA-40 and ERA-Interim is 1979–2002 (23 years); this period was used to bias-correct ERA-40 over 1960–1979 using the quantile mapping technique. Quantile mapping is a statistical approach used in hydrometeorological studies to bias-correct weather-variable time series from atmospheric models against measurements (e.g. Boé et al., 2007); it corrects each quantile by matching the empirical cumulative distribution functions. The irradiance time series created uses the bias-corrected ERA-40 for 1960–1979 and ERA-I for 1979–2016.

3.6 Streamflow

Daily streamflow discharge at HPC was observed and estimated at the hydrometric station (ID: 10LC017) by the ECCC Water Survey of Canada (WSC). This station is downstream from the Havikpak Creek crossing with the Dempster Highway and its drainage area defines the basin for modelling purposes. Discharge estimates for this station start in 1995 and are available to 2015; however, the year 2005 is not available. Measuring small stream discharge in the Arctic is challenging and problems or uncertainties associated with the estimates are acknowledged in the metadata provided by the ECCC through the Environment Canada Data Explorer. The main issues in the hydrometric record are due to the presence of ice and snow in the cross section during snowmelt including peak streamflows, as ice and snow cause substantial variability in rating curves and make streamflow and water stage measurements quite difficult.

3.7 Vegetation cover and shrub density

The vegetation cover map and shrub density used in this study are based on the map and values presented by Krogh et al. (2017) from 1992, and the changing shrub cover and density rates presented by Lantz et al. (2013) for a larger region that includes Havikpak Creek. Lantz et al. (2013) showed that between 1972 and 2004 (32 years) shrub cover increased by 15 % (±3.6) and shrub density increased by 68 % (±24.1) on average. These average rates were recalculated to an annual basis, resulting in rates of 0.47 % yr^{-1} and 2.13 % yr^{-1} for shrub cover and density increases, respectively. To reconstruct times series of vegetation cover and shrub density, the average rates presented by Lantz et al. (2013) were used to linearly extrapolate forwards and backwards from the values used in Krogh et al. (2017), creating a time series of vegetation cover and shrub density from 1960 to 2016. As shrubs colonize the tundra, any increase in shrub cover is compensated by a decrease in the tundra cover, maintaining a constant drainage area. It is unclear when shrub expansion in the Arctic began (Tape et al., 2006), mostly because satellite images started to be available in the 1970s, limiting our understanding of vegetation changes to the 1970s onward.

The HPC forest was held constant in this study, as there are no published studies quantifying forest cover or density change in the region. However, we acknowledge that there are ongoing investigations about changes in forest structure in the region. Greening of the NDVI is not directly attributable to forest change and could be due to documented shrubification. There are no recorded wildfires in Havikpak Creek during the study period as it is close to the airport and so fire suppression by local authorities is very effective.

4 Methodology

4.1 Hydrological modelling

The Cold Regions Hydrological Modelling platform (CRHM) is a process-based and spatially distributed hydrological modelling system with a flexible modular structure that allows the selection of different hydrological processes from an extensive library to create a customized hydrological model. Most of the modules available in the CRHM have a strong physical basis, with particular emphasis on, but not restricted to, cold region processes. The CRHM Arctic Hydrology Model configuration (CRHM-AHM) developed and verified by Krogh et al. (2017) includes the following hydrological processes: forest canopy interception, sublimation and evaporation, snow melt and snow accumulation, evapotranspiration, blowing snow redistribution and sublimation, ground freeze and thaw, water flow through snowpack and organic terrain, infiltration into frozen and unfrozen soils, soil moisture storage and flow, surface water flow and streamflow routing. The model was run over the period from October 1960 to October 2016 on an hourly basis. A 4-year spin-up period was used by repeating the years 1960–1963.

CRHM uses hydrological response units (HRUs; Flügel, 1995) as the spatial unit of discretization for application of the continuity equation to compute mass and energy fluxes. In the CRHM-AHM, the HPC basin was discretized into 11 HRUs initially classified by land cover: tundra, sparse shrubs, close shrubs, taiga, forest, wetland and open water. To include the different near-surface wind regimes observed by Pomeroy and Marsh (1996) over the basin, the tundra and sparse shrub HRUs were each split into an upper and lower HRU to reflect stronger wind regimes in the hilly, higher elevation, upper basin. To simulate the long-lasting snow drifts found in steep gullies and around small lakes, a gully/drift HRU was created following the criteria from Pomeroy and Marsh (1996). The physiographic characteristics of the HRUs used in the CRHM-AHM applied in HPC are as in Krogh et al. (2017, Table 2).

The parameterization of the CRHM-AHM followed the deduction–induction–abduction approach (DIA; Pomeroy et al., 2013a) by first using field information (e.g. slope and vegetation cover), parameters from previous studies in Havikpak Creek and other research basins with similar hydrological regimes and physical processes, and then calibrating a few subsurface and surface hydraulic and storage parameters, for which there was poor understanding, against streamflow. The CRHM-AHM represents the snow, permafrost and streamflow regimes of Havikpak Creek well when compared to observations (Krogh et al., 2017).

4.2 Modelling scenarios

Three modelling scenarios representing only historical climate change (ΔC), only historical vegetation change (ΔV) and both historical climate and vegetation change (ΔCV) were developed to examine the hydrological impacts of changes in HPC since 1960 and are described below.

4.2.1 Model scenario 1 (ΔC): changing climate and constant vegetation

This scenario uses the reconstructed climate time series presented in Sect. 3 for the period 1960–2016 with a constant vegetation cover and density representative of the year 1988, which is the average vegetation cover of the modelling period.

4.2.2 Model scenario 2 (ΔV): constant climate and changing vegetation

This scenario uses a "normal" water year in terms of precipitation and temperature to generate the stationary climate. The mean annual (October to September) precipitation and temperature for the period October 1960 to October 2015 is 332 mm and −8.2 °C, respectively. To select a normal water year, the residual between mean annual precipitation and air temperature for the entire period (1960–2016) was calculated to select the water year with the minimum combined residual. This was the water year 1962–1963 as the mean annual precipitation and temperature were 327 mm and −8.0 °C, respectively. Seasonal representability was also investigated by looking at the standard deviation of the absolute difference between mean monthly values and the 1962–1963 water year monthly values, resulting in 10 mm and 1.1 °C for precipitation and temperature, respectively, suggesting that 1962–1963 is a good representation of the monthly variations. Given the importance of snowmelt to streamflow in the Arctic, winter precipitation (October to April) was compared; for 1962–1963 it was 194 mm, and on average over the period it was 166 mm, suggesting that this normal year is somewhat snowier than average.

This scenario includes transient changes in vegetation using the vegetation cover and density time series as described in Sect. 3.7. The increase in shrub cover was proportionally applied to the upper and lower sparse shrub HRUs, whereas the area of the wetland and gully/drift HRUs was kept constant as their delineation does not depend on the shrub-covered area, but on wetland and topographic criteria (Krogh

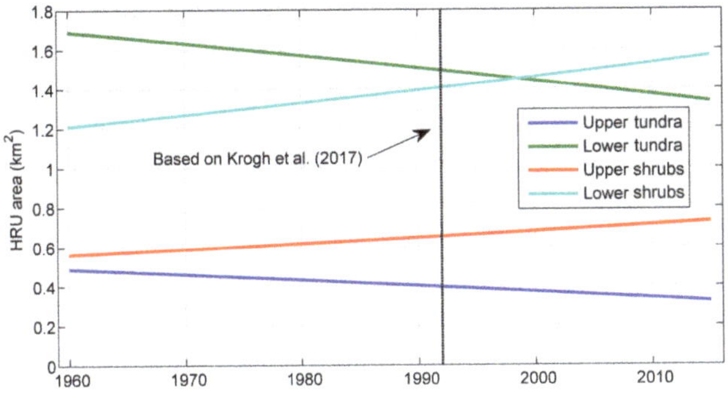

Figure 3. Annual changes in the tundra and shrub HRUs area used in the CRHM-AHM model.

et al., 2017). To implement this transient change, the model was run annually and the shrub cover and density parameters were incremented every 1 November. Figure 3 presents the change in area for the sparse shrubs and tundra HRUs during the modelling period, and the year 1992, which is the vegetation cover used by Krogh et al. (2017).

4.2.3 Model scenario 3 (ΔCV): changing climate and vegetation

This scenario includes changing climate and transient vegetation as presented for the scenarios ΔC and ΔV, and represents the hydrology of Havikpak Creek as realistically as possible.

4.2.4 Transferring initial conditions

In ΔV and ΔCV, the CRHM-AHM was run annually to permit the updating of vegetation parameters at the end of the hydrological year; therefore, final conditions from one year needed to be transferred to the next, and updated with the change in the HRU area. To transfer the initial condition of a given state variable "S" (e.g. volumetric soil moisture or snow water equivalent) from the year (t) and HRU1 (S_1^t) to the next year and HRU2 (S_2^{t+1}), the following relationship can be obtained through mass conservation, assuming that area is transferred from HRU1 (tundra) to HRU2 (sparse shrub):

$$S_2^{t+1} = \frac{A_2^t \cdot S_2^t + \left(A_1^{t+1} - A_1^t\right) \cdot S_1^t}{A_2^{t+1}} \quad (1)$$

$$S_1^{t+1} = S_1^t. \quad (2)$$

Equations (1) and (2) were used to pass on soil moisture, soil recharge and snow water equivalent state variables from year to year as HRU areas changed.

4.3 Trend and change point analysis

The non-parametric Mann–Kendall test (Kendall, 1975; Mann, 1945) was used to perform trend analysis on simulated hydrological variables and observed weather data using a significance threshold of $p \leq 0.05$. The Mann–Kendall test has been extensively used to analyze linear trends in hydrological datasets (e.g. Burn and Hag Elnur, 2002; Hamed, 2008; Yip et al., 2012), proving better results than other methods (Hess et al., 2001). As recommended by Hamed and Rao (1998) time series autocorrelation was removed before performing the Mann–Kendall test to eliminate the detection of false trends. The trend of slopes was calculated using Sen (1968) based on Kendall's rank correlation τ. Variables presented as a percentage of annual precipitation (i.e. rainfall and snowfall ratios) were log-transformed ($y = \log(x/(1-x))$ first. Single change points in the time series were detected using the R-Package "changepoint" version 2.2.2 (Killick et al., 2016) based upon Hinkley (1970). These two techniques (Mann–Kendall and change point analysis) were used together as they complement each other and can be used to look at changes in different ways. For example, the detection of significant trends using the Mann–Kendall test depends on the arbitrary significance threshold, whereas the change point analysis assumes that the time series is normally distributed. Although both techniques have their own limitations they are both equally legitimate, resulting in potentially two different results, such as a time series with no statistically significant trend but a detectable mean change point.

4.4 Teleconnections

To determine the influence of climatological teleconnections on hydrometeorological conditions in HPC, basin-scale mass fluxes were correlated to five climatic indexes representing large-scale circulation features over 1960–2016: (1) Arctic Oscillation (AO; Thompson and Wallace, 1998), (2) North Atlantic Oscillation (NAO; Hurrell et al., 2001), (3) North Pacific Index (NPI; Trenberth and Hurrell, 1994), (4) Southern Oscillation Index (SAO; Rasmusson and Carpenter, 1982) and (5) Pacific Decadal Oscillation (PDO; Mantua and Hare, 2002). These climatic indexes have been used to investigate teleconnections in Arctic and subarctic environments (Bonsal et al., 2006; Déry and Wood, 2004; Serreze et al., 2002). Teleconnections analysis was restricted to ΔCV, as this scenario fully represents the observed changes in HPC.

5 Results

5.1 Meteorological trends

Figure 4 shows point changes and trends in seasonal and water year (October to September) precipitation for the period October 1960–October 2016. Seasons were defined based on local hydrology: winter is from October to April when

Table 1. Changes in precipitation and temperature for the period 1960–2016 and statistically significant trends at the $p \leq 0.05$ significance threshold using the Mann–Kendall test. Changes in precipitation as a percentage with respect to 1960 are also presented.

Period	Precipitation (mm)	Minimum air temperature (°C)	Mean air temperature (°C)	Maximum air temperature (°C)
Winter	–	8.0	5.2	–
Spring	–15.1 (27 %)	-	2.7	–
Summer	–	–	0.8	1.8
Fall	–	–	1.6	–
Annual	–	8.0	3.7	1.8

the snowpack forms and redistributes, spring is May when most snowmelt occurs, summer is from June to August and is a season of rainfall, soil thaw and minimal snowmelt and fall is September when the active layer of the grounds starts to refreeze and precipitation shifts to snowfall. No trends were found for seasonal or annual precipitation, except spring, which had a significant and decreasing trend of -2.7 mm decade^{-1}. Conversely to the trend analysis, the change point analysis shows changes in most seasons and annually. Winter, spring and summer precipitation decreases from 187 to 160 mm, 25 to 13 mm and 146 to 108 mm, respectively, whereas fall precipitation increases from 16 to 34 mm. Annual precipitation decreases from 369 to 321 mm (48 mm) in the water year 1972. Analysis of the number of days with precipitation above the thresholds 1, 2, 5, 10 and 25 mm day^{-1} showed a decreasing trend for events greater than 1, 2 and 5 mm day^{-1} with a slope of -3.8, -1.7 and -0.7 days decade^{-1}, respectively. There are no changes in measurement methods associated with these changes.

Figure 5 shows seasonal and annual changes points and trends for minimum, maximum and mean daily air temperature. Increasing trends for mean air temperature were found annually and in every season, with the largest positive trend of 0.9 °C decade^{-1} in winter. Maximum air temperatures increased significantly annually and in summer, at 0.3 °C decade^{-1} in both cases. Winter, spring and fall maximum air temperatures did not show significant trends. Minimum air temperatures increased rapidly annually and in winter, at 1.4 °C decade^{-1} in both cases. Spring, summer and fall minimum annual temperatures did not show significant trends. Change point analysis showed that these trends are reflected by an increase in mean annual temperature during the water year 1992, from -9.1 to -7.1 °C. Seasonally, the change point analysis shows warming in all seasons but in summer and fall for minimum and mean temperature, respectively. Table 1 presents the changes in temperature for the period 1960–2016 for variables with statistically significant trends. The 8 °C increase in annual and winter minimum temperatures and 3.2 °C increase in annual (5.2 °C winter) mean

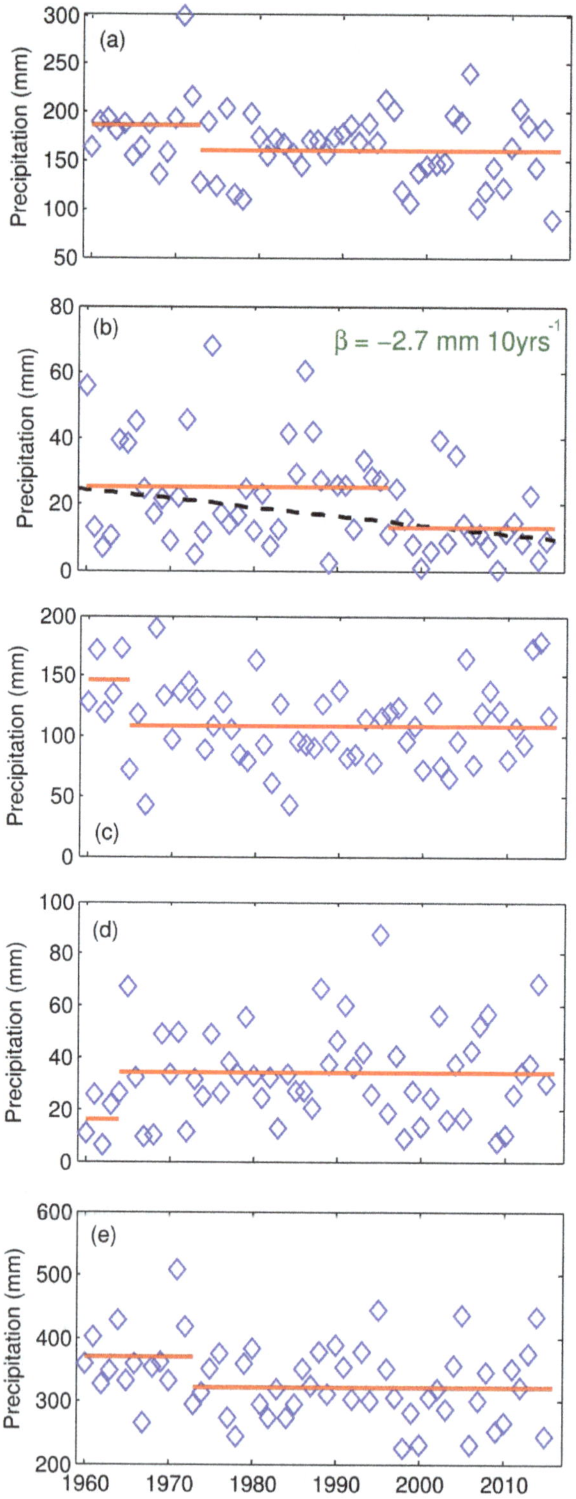

Figure 4. Observed seasonal and annual precipitation for each water year (October–September) from 1960 to 2015. **(a)** Winter (October–April), **(b)** spring (May), **(c)** summer (June–August), **(d)** fall (September) and **(e)** annual. The slope (β) is shown in mm decade^{-1} for statistically significant trends at the $p \leq 0.05$ significance threshold. The solid red line shows the annual change point.

temperatures over 56 years are remarkable and amongst the highest recorded on Earth.

Table 2 presents the statistically significant trends for the other meteorological forcing variables used by CRHM-AHM at seasonal and annual scales. Mean annual short-wave irradiance has been decreasing by $-1.4\,\mathrm{W\,m^{-2}\,decade^{-1}}$ driven by decreases in spring and summer, whilst mean annual long-wave irradiance has been increasing by $2.9\,\mathrm{W\,m^{-2}\,decade^{-1}}$, with greater increases in summer and fall than in winter and spring. Mean annual all-wave irradiance (short- and long-wave irradiance) has been increasing by $1.5\,\mathrm{W\,m^{-2}\,decade^{-1}}$; however, summer all-wave irradiance has been decreasing by $-2.9\,\mathrm{W\,m^{-2}\,decade^{-1}}$. Mean annual wind speed did not change and relative humidity has been increasing by $0.8\,\%\,\mathrm{decade^{-1}}$. Table 3 shows the change point analysis for the atmospheric variables forcing CRHM-AHM. Mean annual short- and long-wave irradiance have change points in the water year 1969, from 112 to 104 and 230 to $242\,\mathrm{W\,m^{-2}}$, whereas all-wave irradiance has a change point in 1997, from 344 to $348\,\mathrm{W\,m^{-2}}$. Mean annual relative humidity has a change point in the water 2013, from 69 to 75 %. No change point was found for mean annual wind speed. Three wind speed thresholds representing non-blowing snow ($2\,\mathrm{m\,s^{-1}}$), light-drifting ($6\,\mathrm{m\,s^{-1}}$) and strong blizzards ($12\,\mathrm{m\,s^{-1}}$) were analyzed. Significant decreases in the hours of events larger than 2 and $6\,\mathrm{m\,s^{-1}}$ were found at -71 and -23 events $\mathrm{decade^{-1}}$. The number of hourly events with strong blizzards showed no significant trend.

5.2 Updated CRHM-AHM validation

The 1995 to 2015 Nash–Sutcliffe Efficiency (NSE) and mean bias were found to be 0.40 and 6 %, respectively, suggesting that the model's streamflow performance is consistent with that showed by Krogh et al. (2017), and changing vegetation dynamic parameterization has a small impact on the short-term model's streamflow performance.

5.3 Trends comparison between modelling scenarios

5.3.1 Sub-basin scale

Figure 6 presents trends in annual (water year) evapotranspiration and sublimation for various HRUs. Evapotranspiration (ET) refers to the actual wetted surface and canopy intercepted rain evaporation and plant transpiration as calculated by Penman–Monteith (P–M) and Priestley–Taylor (P–T; wetlands and lakes) methods (Krogh et al., 2017), but restricted by not only stomatal conductance in P–M but also by available storage of intercepted rainfall, ponded surface water and soil moisture content and the soil moisture withdrawal curve in CRHM. ET in ΔC and ΔCV has been significantly decreasing between 2 and $5\,\mathrm{mm\,decade^{-1}}$ for some HRUs, whereas in ΔV it has been increasing from HRU no. 3 (upper gully/drift). Evaporation from canopy rainfall inter-

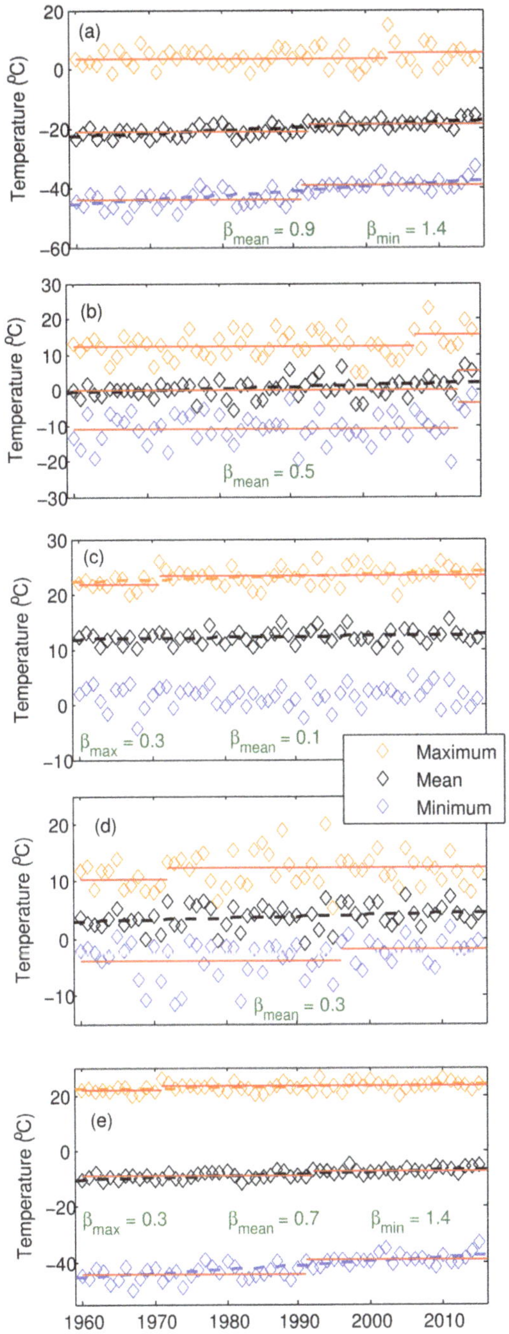

Figure 5. Observed seasonal and annual maximum, mean and minimum temperature for each water year (October–September) calculated from mean daily temperature, between 1960 and 2015. **(a)** Winter (October–April), **(b)** spring (May), **(c)** summer (June–August), **(d)** fall (September) and **(e)** annual. The dashed line is the linear regression using Sen (1968). The slope (β) in $^{\circ}\mathrm{C\,decade^{-1}}$ for statistically significant trends at the $p \leq 0.05$ significance threshold is shown. The solid red line shows the annual change point.

Table 2. Slope for statistically significant weather trends at the $p \leq 0.05$ significance threshold using the Mann–Kendall test.

Period	Short-wave irradiance ($W m^{-2} decade^{-1}$)	Long-wave irradiance ($W m^{-2} decade^{-1}$)	All-wave irradiance ($W m^{-2} decade^{-1}$)	Wind speed ($m s^{-1} decade^{-1}$)	Relative humidity (% $decade^{-1}$)
Winter	0.8	2.8	3.9	–	0.6
Spring	−4.3	2.4	–	–	–
Summer	−6	3.3	−2.9	−0.1	1.4
Fall	–	3.7	2.4	–	1.2
Annual	−1.4	2.9	1.5	–	0.8

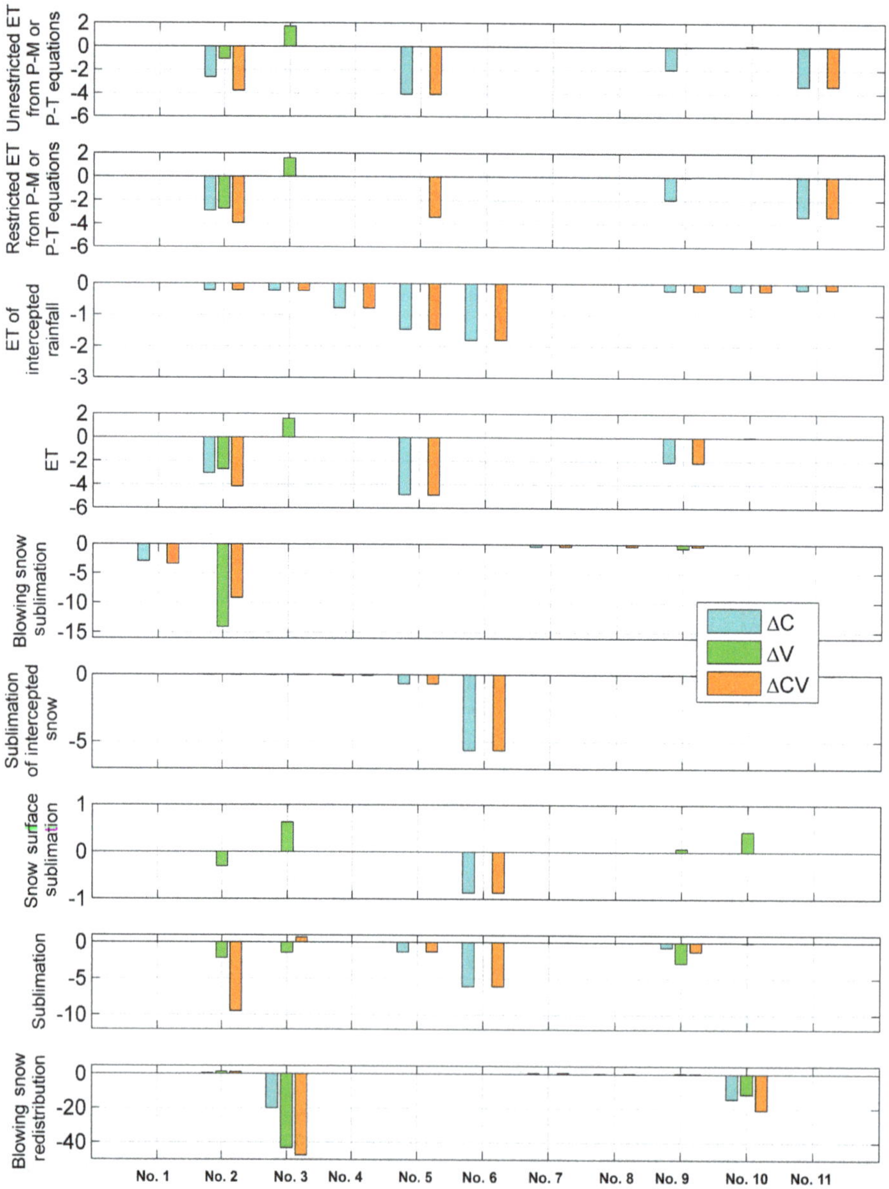

Figure 6. Units in mm $decade^{-1}$. Scenario comparison of significant trends ($p \leq 0.05$) for selected mass fluxes at an HRU-scale. The x axis is as follows: upper tundra (no. 1), upper sparse shrubs (no. 2), upper gully/drift (no. 3), close shrubs (no. 4), taiga forest (no. 5), forest (no. 6), lower tundra (no. 7), open water (no. 8), lower sparse shrubs (no. 9), lower gully/drift (no. 10) and wetland (no. 11).

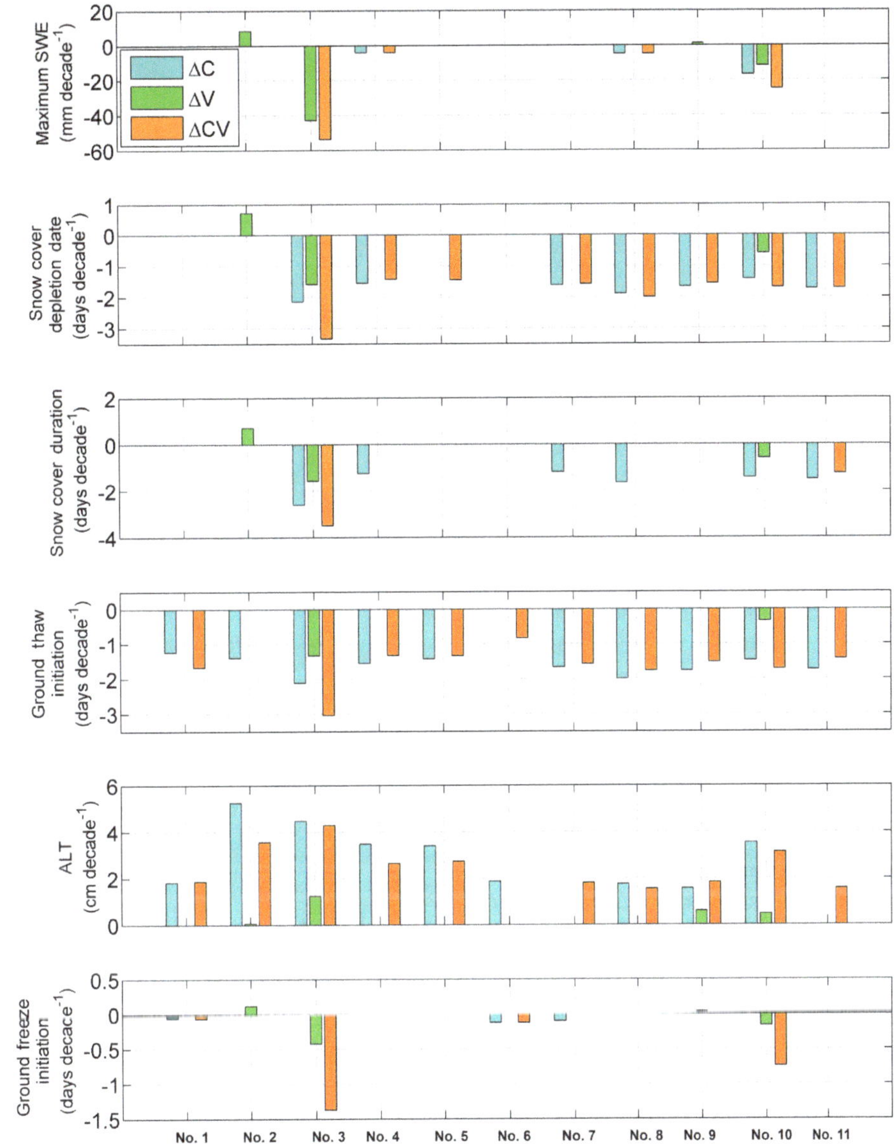

Figure 7. Comparison of significant trends ($p \leq 0.05$) for snow and ground freeze/thaw-related variables at HRU scale for the three scenarios. Note that trends for snow cover depletion date, snow cover duration and ground thaw initiation are in dates, and for maximum SWE, ALT and snow ablation are shown in rates. The x axis is as follows: upper tundra (no. 1), upper sparse shrubs (no. 2), upper gully/drift (no. 3), close shrubs (no. 4), taiga forest (no. 5), forest (no. 6), lower tundra (no. 7), open water (no. 8), lower sparse shrubs (no. 9), lower gully/drift (no. 10) and wetland (no. 11). ALT: active layer thickness. SWE: snow water equivalent.

ception has been decreasing by up to 2 mm decade^{-1} in ΔC and ΔCV for most HRUs, but has no trend in ΔV where only vegetation increases. Soil-moisture-restricted and soil-moisture-unrestricted ET from P–M and P–T equations has virtually the same trends, except from taiga forest (HRU no. 5), suggesting that soil moisture content has had little effect on ET. Blowing snow sublimation has a decreasing trend in the upper and lower shrub HRUs for ΔV and ΔCV where vegetation increases, with the largest trend in the upper basin (~ -14 mm decade^{-1}). Decreasing blowing snow sublimation by 3 mm decade^{-1} was found in the upper tundra HRU for ΔC and ΔCV. Sublimation from canopy-intercepted snow has a decreasing trend for all of the vegetated HRUs in ΔC and ΔCV, with the largest trend in the forest HRU (roughly 6 mm decade^{-1}). Sublimation at the snow surface has a decreasing trend in the forest (HRU no. 6) in ΔC and ΔCV (~ -1 mm decade^{-1}), whereas in ΔV it has an increasing trend in the upper and lower gully/drift (HRU no. 3 and no. 10) and a decreasing trend in the upper shrub HRU. Annual sublimation, defined as the sum of the previous three sublimation terms, has a decreasing trend in ΔV for the upper and lower shrubs and upper Gully/drift (about

Table 3. Mean change point analysis of the atmospheric forcing variables.

Atmospheric variable	Mean annual change	Year
Precipitation (mm)	369 to 321	1972
Air temperature (°C)	−9.1 to −7.1	1992
Short-wave irradiance ($W\,m^{-2}$)	112 to 104	1969
Long-wave irradiance ($W\,m^{-2}$)	230 to 242	1969
All-wave irradiance ($W\,m^{-2}$)	344 to 348	1997
Wind speed ($m\,s^{-1}$)	n/a	n/a
Relative humidity (%)	69 to 75	2013

n/a means not applicable.

2 to 3 mm $decade^{-1}$, respectively). In ΔC and ΔCV it has a decreasing trend in the forested HRUs and lower shrub HRU, driven by the decreasing sublimation from canopy interception, which is the dominant sublimation term over the basin (Krogh et al., 2017). Blowing snow redistribution, defined as the divergence between incoming and outgoing blowing snow transport, decreased in the upper and lower gully/drift HRU for all scenarios, between −20 and −45 mm $decade^{-1}$ in the upper basin and −10 and −20 mm $decade^{-1}$ in the lower basin.

Figure 7 present a series of trends related to snow cover and ground freeze/thaw. Maximum SWE for ΔC decreased in some HRUs, with the largest trend in the lower sparse shrub HRU (−17 mm $decade^{-1}$), whereas for ΔCV the largest decreasing trend was found in the upper gully/drift HRU (−54 mm $decade^{-1}$). Maximum SWE for ΔV showed increasing and decreasing trends in the sparse shrub and gully/drift HRUs, respectively, with the largest changes found in the upper basin. Note that increasing vegetation cover and density hampered blowing snow transport from sparse shrub towards gully/drift HRUs (Fig. 6). The snow cover depletion date for ΔC and ΔCV advanced in almost all HRUs, around −1 and −3 days $decade^{-1}$, whereas for ΔV, both advance and retreat were found in the upper sparse shrub and gully/drift HRUs, respectively. Snow cover duration for ΔC and ΔCV declined for some HRUs (around −1 and −3 days $decade^{-1}$), whereas for ΔV, both extension and decline was found in the upper sparse shrub and gully/drift HRUs (roughly 1 and −1 days $decade^{-1}$). Ground thaw initiation had similar changes as the snow cover depletion timing, which is expected as ground thaw typically starts once the ground is snow-free and temperatures are above 0 °C. Active layer thickness (ALT) for ΔC and ΔCV deepened throughout the basin at a rate of between 2 and 5 cm $decade^{-1}$, whereas for ΔV it deepened in the sparse shrub and gully/drift HRUs (< 2 cm $decade^{-1}$). Snow ablation rate, here defined as the ratio between the maximum SWE and the number of days between maximum SWE and

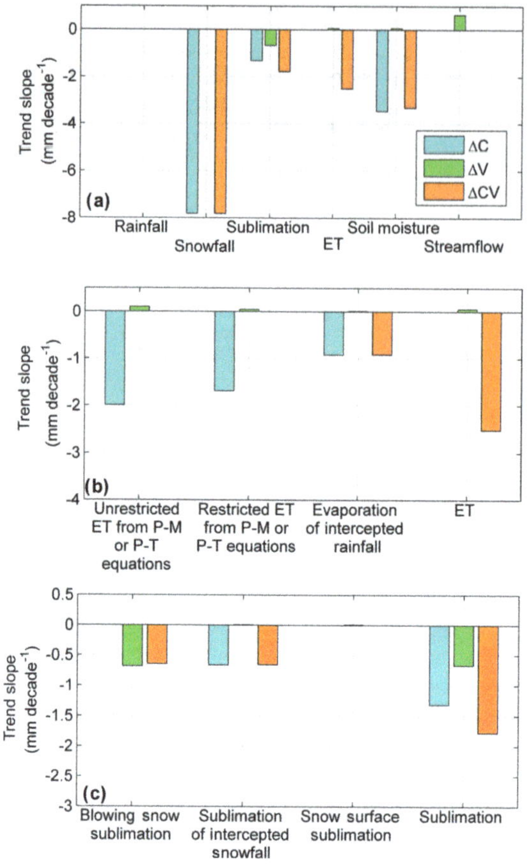

Figure 8. Comparison of basin-scale annual mass fluxes trends ($p \leq 0.05$) over the water years from 1960 to 2015, for the three scenarios. **(a)** Main mass fluxes. **(b)** Evapotranspiration components. **(c)** Sublimation components.

the depletion of snow cover, decreased for ΔC and ΔCV in some HRUs by between −0.1 and −1.5 mm day^{-1} $decade^{-1}$, whereas for ΔV it increased in the sparse shrub and decreased in the gully/drift HRUs.

5.3.2 Basin scale

The primary annual mass flux trends from the three modelling scenarios are presented in Fig. 8 at the basin scale. No trend was found for annual rainfall depths; however, a decreasing trend of −7.8 mm $decade^{-1}$ was found for snowfall depths in ΔC and ΔCV. The rainfall ratio (rainfall divided by total precipitation) exhibited no trend. Annual sublimation losses decreased by −1.3, −0.7 and −1.8 mm $decade^{-1}$ in scenarios ΔC, ΔV and ΔCV, respectively. The sublimation trend in ΔC was driven by decreasing sublimation from canopy interception, likely due to decreasing snowfall. Decreasing sublimation in the ΔV scenario was driven by decreasing blowing snow sublimation caused by expanding and densifying tundra shrubs, whereas for the ΔCV scenario, both drove sublimation trends. Annual ET losses decreased by −2.5 mm $decade^{-1}$ in ΔCV, in contrast to the trend to increase by 0.06 mm decade^{-1} for ΔV, driven by positive trends in all ET components. ET in ΔC showed no trend. Decreasing ET in ΔCV was driven by a decreasing trend in evaporation of rain intercepted in the canopy. To investigate the potential impact of changes in stomata resistance on evapotranspiration, trends in mean annual stomata resistance were also calculated. For both scenarios with changing climate (ΔC and ΔCV), no trend was found; however, for the changing vegetation-only scenario (ΔV) a positive trend of 1.6 s m^{-1} decade^{-1} was found, which agrees with the small increase in ET found for this scenario. Annual streamflow shows an increasing trend of 0.6 mm decade^{-1} only for ΔV, likely due to the increasing snow accumulation at some HRUs (Fig. 6) as a result of reduced blowing snow transport.

Table 4 presents the change point analysis for selected annual mass fluxes at the basin scale for the three scenarios. Rainfall shows an increase from 131 to 196 mm in 2013, whereas snowfall decreases from 211 to 169 mm in 1997. Similarly to the trend analysis, sublimation shows a decrease in all the modelling scenarios, from 39 to 28 mm, 37 to 35 mm and 42 to 36 mm, for ΔC, ΔV and ΔCV, respectively. ET, which showed no significant trend for ΔC, presents a decreasing change point from 160 to 144 mm for ΔC, driven by the drier conditions. ET for ΔV shows no change point, despite the small significant trend in ET (0.06 mm decade^{-1}). For the combined scenario (ΔCV) ET shows a decrease change point in 1977 from 160 to 144 mm, driven by drier conditions and the decrease in radiative energy for ET (all-wave irradiance). Streamflow for ΔC has a decreasing change point from 180 to 140 mm in 1973, despite the lack of significant trend (Fig. 8). For ΔV, streamflow has a small increase from 133 to 135 mm in 1992, which somewhat counteracts the effect of changing climate (180 to 140 mm), resulting in a smaller change from 178 to 140 mm in 1973 for ΔCV.

5.4 Streamflow regime change

The ΔCV scenario most comprehensively represents historical change in climate and vegetation in the Havikpak Creek basin; therefore, it was used to estimate and diagnose changes in streamflow. Figure 9 presents annual time series of variables associated with annual streamflow and peak streamflow for the water years between 1960 and 2015. These time series are annual streamflow volume (Fig. 9a), annual peak daily streamflow discharge (Fig. 9b), date (day of the year, DOY) of peak discharge (Fig. 9c), the DOY of the centre of mass (50 % of volume passed) of streamflow discharge (Fig. 9d) and daily streamflow discharge associated with different exceedance probabilities: 5, 25, 50, 75 and 95 %, using a Weibull distribution function (Fig. 9e). The Weibull distribution was used as it successfully represents daily streamflow probability distribution at HPC (not shown). The DOY of peak daily annual streamflow and the DOY

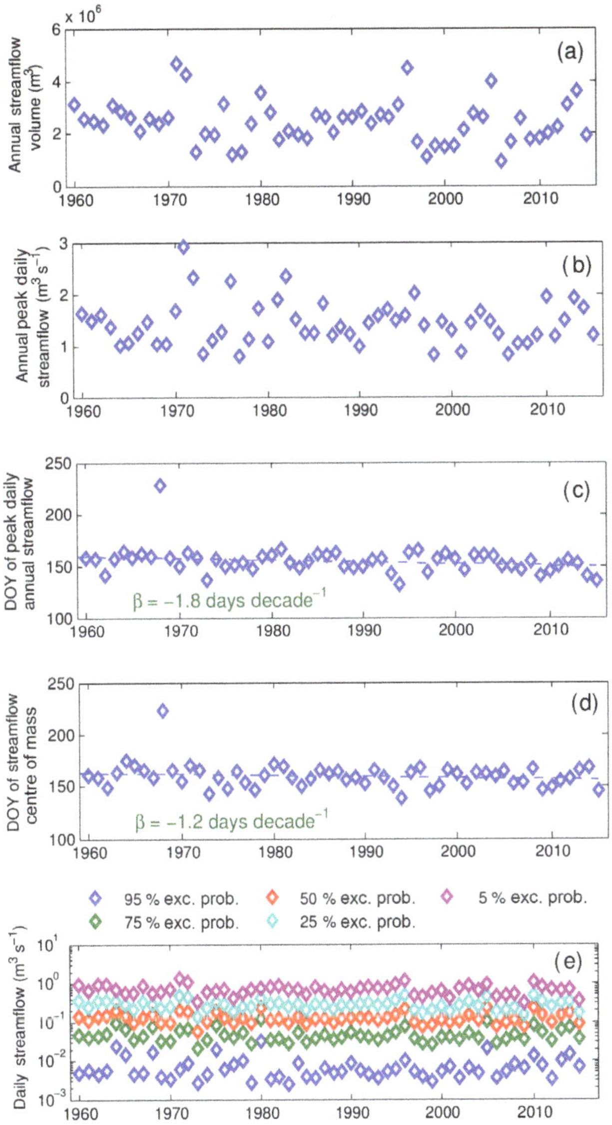

Figure 9. (a) Annual streamflow volume. **(b)** Annual peak daily streamflow. **(c)** Day of the year (DOY) of peak daily streamflow. **(d)** DOY of streamflow volume discharge centre of mass. **(e)** Streamflow discharge associated with various exceedance probabilities. The *x* axis of all subplots is the water year starting in October.

of the streamflow's centre of mass decreased by −1.8 and 1.2 days decade^{-1}, respectively. This finding is consistent with the earlier snow depletion date shown in Fig. 7. The abnormally high value for the DOY peak daily annual streamflow and of the streamflow's centre of mass (Fig. 9c and d; DOY = 226, mid-August) for the water year 1968 is associated with a water year with abnormally high rainfall-runoff compared to snowmelt runoff. No trends were found in monthly streamflow volumes for each month between May and October (not shown), except for September, which decreased at about −47.1 m^3 decade^{-1}.

Table 4. Change point analysis for selected annual basin-scale mass fluxes for the three modelling scenarios.

Mass fluxes	ΔC: Δ climate-only		ΔV: Δ vegetation-only		ΔCV: Δ climate and vegetation	
	Mean change (mm)	Year	Mean change (mm)	Year	Mean change (mm)	Year
Rainfall	131 to 196	2013	n/a	n/a	131 to 196	2013
Snowfall	211 to 169	1997	n/a	n/a	211 to 169	1997
Sublimation	39 to 28	2013	37 to 35	1988	42 to 36	1980
ET	160 to 144	1977	n/a	n/a	160 to 144	1977
Soil moisture	80 to 48	1968	n/a	n/a	82 to 49	1968
Streamflow	180 to 140	1973	133 to 135	1992	178 to 140	1973

n/a means not applicable.

Figure 10 presents the mean daily streamflow discharge for observed streamflow (1995–2015) and the three modelling scenarios over the period 1960–2016. The ΔC and ΔCV scenarios show very similar mean hydrographs, with streamflow discharge starting in mid-April reaching the peak discharge at $0.7\,m^3\,s^{-1}$ on 8 June and ending by mid-November. The ΔV scenario presents a very different mean discharge response, which is not surprising as meteorological drivers largely control the mean conditions and these were kept constant in this scenario. Under this scenario (ΔV), streamflow starts in mid-May reaching the peak discharge at $1.7\,m^3\,s^{-1}$ on 22 May, and it ends in mid-August, having a much shorter discharge season. The current mean hydrological regime, discussed in detail by Krogh et al. (2017), shows an earlier peak flow compared with the long-term ΔC or ΔCV scenarios, which is consistent with the reduction in the date of peakflow presented in Fig. 9c. Also, larger late-fall streamflow discharge is present under current conditions.

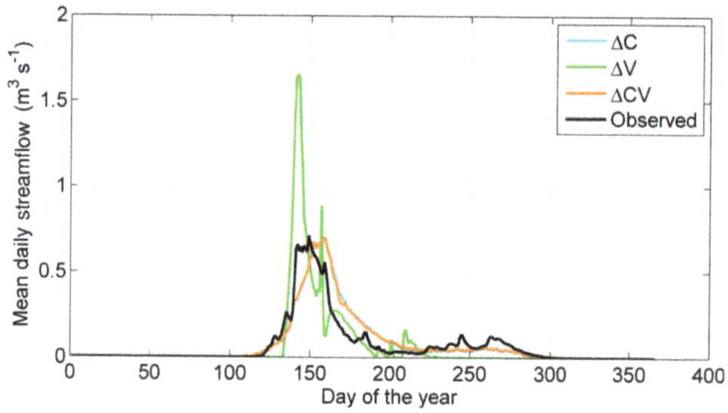

Figure 10. Mean annual hydrograph for the observed streamflow (1995–2015), and the three modelling scenarios (1960–2016): changing climate-only (ΔC), changing vegetation-only (ΔV) and changing climate and vegetation (ΔCV). Note the overlap between the ΔC and ΔCV scenarios.

5.5 Teleconnections

Table 5 lists Pearson correlations coefficients between annual basin-scale mass fluxes and five climatic indices. Statistically significant correlation coefficients with p values ≤ 0.05 are in bold. Significant correlations were found between some mass fluxes and the North Pacific Index (NPI), Southern Oscillation Index (SOI) and Pacific Decadal Oscillation (PDO); however, even significant Pearson coefficients were relatively low (≤ 0.4), suggesting that large-scale climatic oscillations do not have an important effect on Havikpak Creek basin hydrology. The same analysis on a seasonal scale provided similarly low correlation coefficients (not shown).

6 Discussion

6.1 Changing climate

The increasing air temperature trends at Inuvik found in this study (Fig. 5) qualitatively agree with those trends found by other studies using gridded data products (DeBeer et al., 2016; Vincent et al., 2015). Inuvik winters have warmed to the greatest degree; minimum and mean air temperature have increased by 8.0 and 5.2 °C, respectively, over 1960–2016. No temporal trend in precipitation was found at Inuvik (Fig. 4), except for a decrease in the spring ($-2.7\,mm\,decade^{-1}$; Fig. 4); however, the change point analysis showed an important decrease in the year 1972 from 369 to $321\,mm\,yr^{-1}$ for the mean annual precipitation (Table 3). Vincent et al. (2015) investigated long-term trends in precipitation records over Canada for the period between 1948 and 2012 using the gridded and spatially interpolated CANGRD dataset (Rapaic et al., 2015). For the region around Havikpak Creek Vincent et al. (2015) showed significant spatial variability with a small increase of less than 10 % in annual precipitation. The CANGRD dataset is a spatially interpolated 50 km product that is based on the AHCCD dataset and has shown problems when compared against weather station data, particularly north of 60° N (Milewska and Hogg, 2001). Different trends found in this study and Vincent et al. (2015) can be explained by interpolation errors

Table 5. Pearson correlation coefficient between basin-scale mass fluxes and climatic indexes, using water year values (October–September). Correlation coefficients with p values ≤ 0.05 are in bold.

Climatic index	AO	NAO	NPI	SOI	PDO
Rainfall	0.134	0.002	0.207	−0.007	−0.110
Snowfall	−0.013	0.151	0.116	−0.168	0.022
Precipitation	0.075	0.114	0.219	−0.130	−0.054
Sublimation	−0.021	0.125	0.256	0.185	**−0.340**
Blowing snow sublimation	0.044	0.020	**0.336**	0.200	**−0.397**
Snowpack sublimation	−0.049	0.077	0.108	0.203	−0.249
Sublimation of intercepted snowfall	−0.014	0.172	0.143	−0.071	−0.058
Restricted ET from P–M or P–T equations	0.161	−0.200	0.253	**0.268**	**−0.332**
Evaporation from canopy interception	−0.034	0.125	−0.017	−0.078	0.061
ET	0.156	−0.183	0.250	0.256	**−0.323**
Soil moisture	−0.010	−0.158	0.191	−0.065	−0.073
Streamflow	−0.002	0.062	0.083	−0.262	0.141

in the CANGRD dataset and the different period of analysis. This suggests that careful assessment of regional climate products needs to be performed when looking at individual sites, particularly in the Arctic where there are few stations.

As presented in Sect. 3.2, the precipitation time series was produced using mostly the AHCCD dataset and corrected records from automated weather stations (AWSs) for wind undercatch, producing a discontinuity in the time series in the mid-1990s. Although uncertainty exists in the precipitation records, there is a relatively high confidence in the accuracy of precipitation, supported by the typically low wind speed limiting wind undercatch losses, the meticulous quality control and corrections used in the AHCCD dataset (Mekis and Vincent, 2011), the well-established wind undercatch correction used for the AWS snow gauge and the snow surveys from small clearing with minimal snow distribution and sublimation that allows a good estimation of seasonal snowfall. Comparing this precipitation dataset with another nearby station is challenging, as there is no station with similar long-term records close to Inuvik. Nevertheless, the impacts of such uncertainty on the presented results are expected to be small and should not change the core discussions and conclusions of this study.

Mean annual short-wave irradiance from combined ERA-40 and ERA-I decreased by $-1.4\,\mathrm{W\,m^{-2}\,decade^{-1}}$ (Table 2) or −7.4 % over 1960–2016 with respect to 1960. Other studies have also found that measured solar irradiance in the Arctic has decreased. For example, Weston et al. (2007) found a decreasing trend in solar irradiance at two Canadian Arctic sites: Alert and Resolute Bay, Nunavut Territory, for the period 1964–2002 and 1957–2003, respectively. They argued that decreases in short-wave irradiance are driven by changes in atmospheric composition, such as aerosols and greenhouse gases, producing a decreasing in the calculated daily Clearness Index. However, the ERA-I irradiance model calculation (Saunders et al., 1999) does not include the effect of aerosol scattering, but it does include the effect of greenhouse gasses, such as water vapour and carbon monoxide. Mean annual long-wave irradiance shows an increasing trend of $2.9\,\mathrm{W\,m^{-2}\,decade^{-1}}$ (Table 2) or 7.3 % over 1960–2016 with respect to 1960. This result agrees with global observations, showing an increase in long-wave radiation (Ohmura, 2009), particularly over the Canadian Arctic, for which observed net long-wave is also increasing (Weston et al., 2007), and is consistent with an increase in cloud cover and/or water vapour in the atmosphere with resulting increasing atmospheric emissivity and/or increasing air temperatures. The annual modelled all-wave irradiance is increasing by 2.6 %, but with seasonal variations. Winter all-wave irradiance has been increasing by 10 %, providing more energy to snowmelt and sublimation, whereas summer all-wave irradiance has been decreasing by 3 %, which decreases the energy for ET and ground thaw.

6.2 Changes to the hydrological cycle

The precipitation phase shifted from snowfall to rainfall in the scenarios including climate change (ΔC and ΔCV; Fig. 8) by 22.7 % from 1960 to 2016; this was driven by the increase in mean annual air temperature of 3.7 °C (Table 1). Snow cover duration decreased for ΔC (some HRUs; Fig. 7), whereas for ΔV both decreased and increased over the HRUs; however, the ΔCV resulted in a shortened snow season (most HRUs). This shortening was mostly driven by changing climate with reduced snowfall and snow redistribution to drifts by wind. Similarly to the snow season duration, the snow cover depletion date for ΔCV decreased between 8 and 17 days over 1960–2016 (Fig. 7), with the greatest decrease in the upper gully/drift HRU, due to decreasing blowing snow redistribution and hence peak SWE. As peak streamflow in HPC is dominated by snowmelt events, these changes are consistent with the 10-day advance in peak streamflow date. Peak SWE decreased between 12 and 33 % in the ΔC, whereas for ΔV it increased in the sparse shrub

HRUs by 3 to 30 % and decreased in the gully/drift HRUs by 22 to 40 %. The ΔCV scenario resulted in diminishing peak SWE by 12 to 50 %, due to the combination of decreasing snowfall and blowing snow redistribution from sparse shrubs to gully/drift HRUs. Snow ablation rates for ΔC decreased by 0.3 to 1.1 % over 1960–2016, whereas for ΔV they decreased in the sparse shrub HRUs by 0.3 to 0.4 % and increased by 0.31 to 0.4 % in the gully/drift HRUs. Changes in snow ablation rates due to a warmer climate have been investigated in other cold regions. Rasouli et al. (2014) and Pomeroy et al. (2015) modelled snow hydrology in mountain basins in Yukon and Alberta, Canada, respectively, and attributed the lower snow ablation rates under climate change to an earlier snowmelt season, occurring when lower solar radiation inputs are available. Using snow accumulation records in the western USA, Musselman et al. (2017) reached a similar conclusion. López-Moreno et al. (2012) also found a reduction in ablation rates in the Spanish Pyrenees under a scenario of warmer temperatures. However, here some snow ablation rates increased for ΔV, suggesting climatic factors are not the only control in ablation rates, but that vegetation dynamics can compensate or even reverse trends in ablation rates due to changing climate.

Sublimation decreased in ΔC by 23 %, due to a decrease in sublimation of intercepted snow by 19 %. Factors decreasing sublimation of intercepted snow are warmer temperatures, causing accelerated snow unloading from the canopy, and decreasing snowfall. Sublimation for ΔV decreased by 10 %, due to blowing snow sublimation dropping by 44 %, mostly in the upper basin. Decreasing blowing snow in this scenario is driven by shrub densification, increasing the aerodynamic roughness height and wind speeds required to initiate blowing snow transport. Shrub densification intensified the effect of changing climate on sublimation, decreasing sublimation by 29 % over the study period. ET did not show a significant trend in the changing climate-only scenario; however, the ET of intercepted rainfall and soil moisture-restricted ET from the P–M or P–T equations decreased by 51 and 6 %, respectively. This is explained by the different inter-annual variability of the two ET terms hampering the individual trends; nevertheless, the change point analysis of mean annual ET shows a decrease of 16 mm yr^{-1} in 1977 (9.5 % with respect to 1960 for ΔC and ΔCV), consistently with the simulated trends of each ET component. ET for ΔV increased by a marginal 0.2 %, due to the 1.5 % increase in soil moisture content and 0.8 % increase in ET of intercepted rainfall due to shrub expansion. The combined effect of changing climate and vegetation decreased ET by 8.5 %, driven largely by changing climate. Decreasing summer all-wave irradiance (3 %) and soil moisture content (19 %) were driving decreasing ET in the scenario with combined climate and vegetation changes.

Active layer thickness (ALT) for ΔC increased by 11 to 28 cm over 1960–2016 for most HRUs, caused by the earlier snow depletion date (8 to 11 days) and ground thaw initiation (6 to 11 days), and warmer ground-surface temperatures due to warmer air temperatures. ALT increased up to 7 cm for ΔV in some HRUs, driven by the earlier snow cover depletion date (3 to 8 days) and ground thaw initiation (2 to 6 days). The effect of changing vegetation dampened the deepening in ALT found in the ΔC scenario for most HRUs; nevertheless, in the ΔCV scenario, ALT increased by 11 to 22 cm.

Annual streamflow volume from HPC has dropped by 38 mm (21 %) whilst annual precipitation has dropped by 48 mm (13 %) since 1960. We argue that the 10 mm (21 %) difference between the decrease in precipitation and streamflow discharge from HPC suggests a small degree of hydrological resiliency, here defined as the capacity of a basin to actively counteract the impact of changes in weather on streamflow discharge, which is explained by the declining ET and sublimation. This result emphasizes the need for a full physically based representation of the hydrological cycle in models so that the processes driving this resiliency can be used to diagnose its function.

6.3 Havikpak Creek basin changes versus other Arctic studies

The ΔCV scenario best represents historical change in Havikpak Creek basin; therefore, it is used to compare with other Arctic studies. Snow cover depletion dates in HPC accelerated between 1.5 and 3.2 days decade^{-1} (Fig. 7), which are higher than the average trend presented by Liston and Hiemstra (2011) for the entire Arctic (−1.28 days decade^{-1}), but smaller than their largest trend in the Arctic (−9.89 days decade^{-1}). The maximum ALT depth increased by 1.8 to 4.2 cm decade^{-1} (Fig. 7), which is smaller than the average trend of 4.7 cm decade^{-1} modelled by Oelke et al. (2004) over the Mackenzie River basin. Differences in ALT change simulations can be due to (1) differences in the model's spatial representation (Oelke et al., 2004 used grids of 25 km, with which small-scale features are not well represented); (2) differences in the ground freeze/thaw method algorithm (Oelke et al. used a one-dimensional heat conduction, i.e. lateral flow is neglected), and (3) the driving meteorology (Oelke et al. used the NCEP/NCAR reanalysis, which has shown some problems in representing Arctic climate) (Serreze et al., 1998; Serreze and Hurst, 2000). However, the average permafrost conditions of the Mackenzie River basin are thinner and warmer compared with those in HPC, and so average changes in ALT are expected to be larger for the Mackenzie River basin than for HPC.

Annual streamflow volume at HPC has dropped (Table 4); unfortunately there are no long-term studies of small streams that originate in the Arctic to compare this result with. There are studies showing increasing large river basin streamflow into the Arctic (McClelland et al., 2006; Overeem and Syvitski, 2010; Peterson et al., 2002; Rood et al., 2017; Yang et al., 2002). However, a significant portion of the runoff in these basins originates south of the Arctic Circle (e.g. the Mackenzie and the Lena River basins in Canada and Russia, respectively), and therefore these trends are not representative of changes in Arctic hydrology. Previous studies have argued that the increase in the streamflow of large rivers flowing into the Arctic is driven by increasing baseflow due to permafrost thaw and increasing precipitation. However, HPC annual precipitation and streamflow have both dropped and the earlier shifts in the hydrograph are inconsistent with such mechanism. Instead, baseflow during the end of the summer is minimal, streamflow has been decreasing during September and no winter flow has been observed. Only a few similarities can be found between results of studies of large river basins flowing to the Arctic and HPC, such as increasing ALT and accelerating snow-free dates; however, most processes, such as evapotranspiration and streamflow, depend on the local-scale interaction between several physical processes, which are undergoing distinct changes that are not evident in rivers flowing into the Arctic. Therefore, the results of studies of these large river basins should not be confused with the results for an Arctic hydrology study.

This study considered changing climate and transient vegetation change separately to identify their individual effects; nevertheless, they are strongly coupled in the historical record. Warming temperatures are well correlated with shrub growth (Myers-Smith et al., 2011), which has a positive feedback to atmospheric heating by decreasing surface albedo, generating greater sensible heat flux to the atmosphere (Pomeroy et al., 2006), and a negative feedback by consuming more atmospheric CO_2 (Myers-Smith et al., 2011). The modelling scenario experiments here revealed that most simulated trends in the water balance are attributable to changes in climate; however, the effect of transient vegetation, as expressed in shrub expansion and densification, was shown to further reduce blowing snow redistribution and sublimation, which intensified climate-change-driven trends produced by the reduced snow accumulation. This emphasizes the need to include transient vegetation changes in hydrological simulations, which are typically neglected in hydrological models. Reliable rates of change in vegetation species, height and density need to be available for this purpose; therefore, comprehensive studies investigating these changes in other transitioning environments are needed.

7 Conclusion

This study diagnosed changes in the hydrology of a small Arctic basin in the tundra–taiga transition using a spatially distributed and physically based hydrological model. It considered both transient climate and vegetation changes for the first time. There was no evidence for intensification of the hydrological cycle as instead, most processes slowed. In the changing climate-only scenario, statistically significant changes were found for diminishing snow accumulation, sublimation, blowing snow redistribution, snow cover duration, snow ablation rate and evapotranspiration, deepening active layer thickness and earlier snow cover depletion and ground thaw initiation. These, along with warming temperatures, declining summer net radiation and declining precipitation, resulted in diminished annual streamflow volume of 38 mm over the 56 years. However the decline in streamflow did not match the larger decline in precipitation (48 mm), providing some evidence of resilience to climate change, as despite rising temperatures, both evapotranspiration and sublimation dropped with declining precipitation and this attenuated the streamflow volume decline. Transient vegetation change further decreased blowing snow sublimation by reducing blowing snow transport. The combination of changing climate and transient vegetation change resulted in annual streamflow volume dropping by 38 mm over 56 years – a change that is not substantially different from that due to climate change alone. These results suggest that historical changes in vegetation and a degree of hydrological resiliency have not compensated for the effects of climate change on the hydrological regime of Havikpak Creek. They provide the first estimates of long-term change for a drainage basin located completely within the Arctic Circle, and demonstrate the large, complex and recent hydrological changes that have occurred, which can be used as a reference to inform other studies of Arctic climate change impacts.

Author contributions. SAK and JWP designed the study. JWP developed the CRHM modelling platform. SAK performed the simulations and the statistical analyses and prepared the manuscript with contributions from JWP to the manuscript structure, readability and analysis and discussion of the results.

Competing interests. The authors declare that they have no conflict of interest.

Special issue statement. This article is part of the special issue "Understanding and predicting Earth system and hydrological change in cold regions". It is not associated with a conference.

Acknowledgements. Financial support for this study was provided by CONYCIT under the BECAS CHILE scholarship program, Global Water Futures, Changing Cold Regions Network, NSERC Discovery Grants, Canada Research Chairs and Yukon Environment. The authors thank Tom Brown of the Centre for Hydrology for providing technical support with the CRHM platform. Comments from Joseph Shea, Paul Whitfield, Lucia Scaff and two anonymous reviewers are greatly appreciated.

Edited by: Bettina Schaefli

References

Boé, J., Terray, L., Habets, F., and Martin, E.: Statistical and dynamical downscaling of the Seine basin climate for hydro-meteorological studies, Int. J. Climatol., 27, 1643–1655, https://doi.org/10.1002/joc.1602, 2007.

Bond, W. J. and Keeley, J. E.: Fire as a global 'herbivore': the ecology and evolution of flammable ecosystems, Trends Ecol. Evol., 20, 387–394, https://doi.org/10.1016/j.tree.2005.04.025, 2005.

Bonsal, B. R., Prowse, T. D., Duguay, C. R., and Lacroix, M. P.: Impacts of large-scale teleconnections on freshwater-ice break/freeze-up dates over Canada, J. Hydrol., 330, 340–353, https://doi.org/10.1016/j.jhydrol.2006.03.022, 2006.

Brown, R., Derksen, C., and Wang, L.: A multi-data set analysis of variability and change in Arctic spring snow cover extent, 1967–2008, J. Geophys. Res., 115, D16111, https://doi.org/10.1029/2010JD013975, 2010.

Burn, D. H. and Hag Elnur, M. A.: Detection of hydrologic trends and variability, J. Hydrol., 255, 107–122, https://doi.org/10.1016/S0022-1694(01)00514-5, 2002.

Cederstrom, D. J., Johnston, P. M., and Subitzky, S.: Occurrence and development of ground water in permafrost regions, Washington, D.C., available at: https://pubs.usgs.gov/circ/1953/0275/report.pdf (last access: 12 July 2018), 1953.

Chow, D. and Levermore, G. J.: New algorithm for generating hourly temperature values using daily maximum, minimum and average values from climate models, Build. Serv. Eng. Res. Technol., 28, 237–248, https://doi.org/10.1177/0143624407078642, 2007.

Colbeck, S. C.: A theory of water percolation in snow, J. Glaciol., 11, 369–385, 1972.

Connon, R. F., Quinton, W. L., Craig, J. R., and Hayashi, M.: Changing hydrologic connectivity due to permafrost thaw in the lower Liard River valley, NWT, Canada, Hydrol. Process., 28, 4163–4178, https://doi.org/10.1002/hyp.10206, 2014.

DeBeer, C. M., Wheater, H. S., Carey, S. K., and Chun, K. P.: Recent climatic, cryospheric, and hydrological changes over the interior of western Canada: A review and synthesis, Hydrol. Earth Syst. Sci., 20, 1573–1598, https://doi.org/10.5194/hess-20-1573-2016, 2016.

Dee, D. P., Uppala, S. M., Simmons, A. J., Berrisford, P., Poli, P., Kobayashi, S., Andrae, U., Balmaseda, M. A., Balsamo, G., Bauer, P., Bechtold, P., Beljaars, A. C. M., van de Berg, L., Bidlot, J., Bormann, N., Delsol, C., Dragani, R., Fuentes, M., Geer, A. J., Haimberger, L., Healy, S. B., Hersbach, H., Hólm, E. V., Isaksen, L., Kållberg, P., Köhler, M., Matricardi, M., McNally, A. P., Monge-Sanz, B. M., Morcrette, J.-J., Park, B.-K., Peubey, C., de Rosnay, P., Tavolato, C., Thépaut, J.-N., and Vitart, F.: The ERA-Interim reanalysis: configuration and performance of the data assimilation system, Q. J. Roy. Meteorol. Soc., 137, 553–597, https://doi.org/10.1002/qj.828, 2011.

Déry, S. J. and Wood, E. F.: Teleconnection between the Arctic Oscillation and Hudson Bay river discharge, Geophys. Res. Lett., 31, 2–5, https://doi.org/10.1029/2004GL020729, 2004.

Ellis, C. R., Pomeroy, J. W., and Link, T. E.: Modeling increases in snowmelt yield and desynchronization resulting from forest gap-thinning treatments in a northern mountain headwater basin, Water Resour. Res., 49, 936–949, https://doi.org/10.1002/wrcr.20089, 2013.

Essery, R. L. H. and Pomeroy, J. W.: Vegetation and Topographic Control of Wind-Blown Snow Distributions in Distributed and Aggregated Simulations for an Arctic Tundra Basin, J. Hydrometeorol., 5, 735–744, https://doi.org/10.1175/1525-7541(2004)005<0735:VATCOW>2.0.CO;2, 2004.

Flügel, W.-A.: Delineating hydrological response units by geographycal information system analyses for regional hydrological modelling using PRMS/MMS in the drainage basin of the River Bröl, Germany, Hydrol. Process., 9, 423–436, 1995.

Goodison, B. E., Louie, P. Y., and Yang, D.: WMO Solid Precipitation Measurement Intercomparison, Report No. 67, World Meteorological Organization, available at: http://www.wmo.int/pages/prog/www/reports/WMOtd872.pdf (last access: 12 July 2018), 1998.

Gray, D. M., Toth, B., Zhao, L., Pomeroy, J. W., and Granger, R. J.: Estimating areal snowmelt infiltration into frozen soils, Hydrol. Process., 15, 3095–3111, https://doi.org/10.1002/hyp.320, 2001.

Güntner, A., Olsson, J., Calver, A., and Gannon, B.: Cascade-based disaggregation of continuous rainfall time series: the influence of climate, Hydrol. Earth Syst. Sci., 5, 145–164, https://doi.org/10.5194/hess-5-145-2001, 2001.

Hamed, K. H.: Trend detection in hydrologic data: The Mann-Kendall trend test under the scaling hypothesis, J. Hydrol., 349, 350–363, https://doi.org/10.1016/j.jhydrol.2007.11.009, 2008.

Hamed, K. H. and Rao, A. R.: A modified Mann-Kendall trend test for autocorrelated data, J. Hydrol., 204, 182–196, https://doi.org/10.1016/S0022-1694(97)00125-X, 1998.

Hansen, J., Ruedy, R., Sato, M., and Lo, K.: Global surface temperature change, Rev. Geophys., 48, RG4004, https://doi.org/10.1029/2010RG000345, 2010.

Hedstrom, N. R. and Pomeroy, J. W.: Measurements and modelling of snow interception in the boreal forest, Hydrol. Process., 12, 1611–1625, https://doi.org/10.1002/(SICI)1099-1085(199808/09)12:10/11<1611::AID-HYP684>3.0.CO;2-4, 1998.

Hess, A., Iyer, H., and Malm, W.: Linear trend analysis: a comparison of methods, Atmos. Environ., 35, 5211–5222, https://doi.org/10.1016/S1352-2310(01)00342-9, 2001.

Hinkley, D. V: Inference About the Change-Point in a Sequence of Random Variables, Biometrika, 57, 1–17, https://doi.org/10.2307/2334932, 1970.

Hinzman, L. D., Bettez, N. D., Bolton, W. R., Chapin, F. S., Dyurgerov, M. B., Fastie, C. L., Griffith, B., Hollister, R. D., Hope, A., Huntington, H. P., Jensen, A. M., Jia, G. J., Jorgenson, T., Kane, D. L., Klein, D. R., Kofinas, G., Lynch, A. H., Lloyd, A. H., McGuire, A. D., Nelson, F. E., Oechel, W. C., Osterkamp, T. E., Racine, C. H., Romanovsky, V. E., Stone, R. S., Stow, D. A., Sturm, M., Tweedie, C. E., Vourlitis, G. L., Walker, M. D., Walker, D. A., Webber, P. J., Welker, J. M., Winker, K. S., and Yoshikawa, K.: Evidence and Implications of Recent Climate Change in Northern Alaska and Other Arctic Regions, Climatic Change, 72, 251–298, https://doi.org/10.1007/s10584-005-5352-2, 2005.

Hurrell, J. W., Kushnir, Y., and Visbeck, M. H.: The North Atlantic Oscillation, Science, 291, 603–605, https://doi.org/10.1126/science.1058761, 2001.

Janowicz, J. R.: Observed trends in the river ice regimes of northwest Canada, Hydrol. Res., 41, 462–470, https://doi.org/10.2166/nh.2010.145, 2010.

Ju, J. and Masek, J. G.: The vegetation greenness trend in Canada and US Alaska from 1984–2012 Landsat data, Remote Sens. Environ., 176, 1–16, https://doi.org/10.1016/j.rse.2016.01.001, 2016.

Juminikis, A. R.: Thermal Geotechnics, Rutgers University Press, New Brunswick, New Jersey, 1977.

Kane, D. L.: Snowmelt infiltration into seasonally frozen soils, Cold Reg. Sci. Technol., 3, 153–161, https://doi.org/10.1016/0165-232X(80)90020-8, 1980.

Kane, D. L. and Stein, J.: Water movement into seasonally frozen soils, Water Resour. Res., 19, 1547–1557, https://doi.org/10.1029/WR019i006p01547, 1983.

Kane, D. L., Hinzman, L. D., Benson, C. S., and Liston, G. E.: Snow hydrology of a headwater arctic basin, 1. Physical measurements and process studies, Water Resour. Res., 27, 1099–1109, https://doi.org/10.1029/91WR00262, 1991.

Kendall, M. G.: Rank Correlation Methods, Griffin, London, 1975.

Killick, R., Haynes, K., Eckley, I., Fearnhead, P., and Lee, J.: Methods for Changepoint Detection, available at: https://cran.r-project.org/web/packages/changepoint/changepoint.pdf (last access: 12 July 2018), 2016.

Krogh, S. A., Pomeroy, J. W., and McPhee, J.: Physically Based Mountain Hydrological Modeling Using Reanalysis Data in Patagonia, J. Hydrometeorol., 16, 172–193, https://doi.org/10.1175/JHM-D-13-0178.1, 2015.

Krogh, S. A., Pomeroy, J. W., and Marsh, P.: Diagnosis of the hydrology of a small Arctic basin at the tundra-taiga transition using a physically based hydrological model, J. Hydrol., 550, 685–703, https://doi.org/10.1016/j.jhydrol.2017.05.042, 2017.

Lantz, T. C., Marsh, P., and Kokelj, S. V.: Recent Shrub Proliferation in the Mackenzie Delta Uplands and Microclimatic Implications, Ecosystems, 16, 47–59, https://doi.org/10.1007/s10021-012-9595-2, 2013.

Laudon, H., Spence, C., Buttle, J., Carey, S. K., McDonnell, J. J., McNamara, J. P., Soulsby, C., and Tetzlaff, D.: Save northern high-latitude catchments, Nat. Geosci., 10, 324–325, https://doi.org/10.1038/ngeo2947, 2017.

Lawrence, M. G.: The relationship between relative humidity and the dewpoint temperature in moist air: A simple conversion and applications, B. Am. Meteorol. Soc., 86, 225–233, https://doi.org/10.1175/BAMS-86-2-225, 2005.

Liljedahl, A. K., Boike, J., Daanen, R. P., Fedorov, A. N., Frost, G. V., Grosse, G., Hinzman, L. D., Iijma, Y., Jorgenson, J. C., Matveyeva, N., Necsoiu, M., Raynolds, M. K., Romanovsky, V., Schulla, J., Tape, K. D., Walker, D. A., Wilson, C., Yabuki, H., and Zona, D.: Pan-Arctic ice-wedge degradation in warming permafrost and influence on tundra hydrology, Nat. Geosci., 9, 1–8, https://doi.org/10.1038/ngeo2674, 2016.

Lindsay, R., Wensnahan, M., Schweiger, A., and Zhang, J.: Evaluation of seven different atmospheric reanalysis products in the arctic, J. Climate, 27, 2588–2606, https://doi.org/10.1175/JCLI-D-13-00014.1, 2014.

Liston, G. E. and Hiemstra, C. A.: The changing cryosphere: Pan-Arctic snow trends (1979–2009), J. Climate, 24, 5691–5712, https://doi.org/10.1175/JCLI-D-11-00081.1, 2011.

López-Moreno, J. I., Pomeroy, J. W., Revuelto, J., and Vicente-Serrano, S. M.: Response of snow processes to climate change: spatial variability in a small basin in the Spanish Pyrenees, Hydrol. Process., 27, 2637–2650, https://doi.org/10.1002/hyp.9408, 2012.

López-Moreno, J. I., Boike, J., Sanchez-Lorenzo, A., and Pomeroy, J. W.: Impact of climate warming on snow processes in Ny-Ålesund, a polar maritime site at Svalbard, Global Planet. Change, 146, 10–21, https://doi.org/10.1016/j.gloplacha.2016.09.006, 2016.

Mann, H. B.: Nonparametric Test against Trend, Econometrica, 13, 245–259, https://doi.org/10.2307/1907187, 1945.

Mantua, N. J. and Hare, S. R.: The Pacific Decadal Oscillation, J. Oceanogr., 58, 35–44, https://doi.org/10.1023/A:1015820616384, 2002.

Marsh, P. and Woo, M.-K.: Wetting front advance and freezing of meltwater within a snow cover: 1. Observations in the Canadian Arctic, Water Resour. Res., 20, 1853–1864, https://doi.org/10.1029/WR020i012p01865, 1984a.

Marsh, P. and Woo, M.-K.: Wetting front advance and freezing of meltwater within a snow cover: 2. A simulation model, Water Resour. Res., 20, 1865–1874, https://doi.org/10.1029/WR020i012p01865, 1984b.

Marsh, P., Bartlett, P., MacKay, M., Pohl, S., and Lantz, T.: Snowmelt energetics at a shrub tundra site in the western Canadian Arctic, Hydrol. Process., 24, 3603–3620, https://doi.org/10.1002/hyp.7786, 2010.

Martin, A. C., Jeffers, E. S., Petrokofsky, G., Myers-Smith, I. H., and Macias-Fauria, M.: Shrub growth and expansion in the Arctic tundra: an assessment of controlling factors using an evidence-based approach, Environ. Res. Lett., 12, 085007, https://doi.org/10.1088/1748-9326/aa7989, 2017.

McClelland, J. W., Déry, S. J., Peterson, B. J., Holmes, R. M., and Wood, E. F.: A pan-arctic evaluation of changes in river discharge during the latter half of the 20th century, Geophys. Res. Lett., 33, 2–5, https://doi.org/10.1029/2006GL025753, 2006.

Mekis, É. and Vincent, L. A.: An Overview of the Second Generation Adjusted Daily Precipitation Dataset for Trend Analysis in Canada, Atmosphere-Ocean, 49, 163–177, https://doi.org/10.1080/07055900.2011.583910, 2011.

Ménard, C. B., Essery, R., Pomeroy, J., Marsh, P. and Clark, D. B.: A shrub bending model to calculate the albedo of shrub-tundra, Hydrol. Process., 28, 341–351, https://doi.org/10.1002/hyp.9582, 2012.

Ménard, C. B., Essery, R., and Pomeroy, J. W.: Modelled sensitivity of the snow regime to topography, shrub fraction and shrub height, Hydrol. Earth Syst. Sci., 18, 2375–2392, https://doi.org/10.5194/hess-18-2375-2014, 2014.

Milewska, E. and Hogg, W. D.: Spatial representativeness of a long-term climate network in Canada, Atmosphere-Ocean, 39, 145–161, https://doi.org/10.1080/07055900.2001.9649671, 2001.

Musselman, K. N., Clark, M. P., Liu, C., Ikeda, K., and Rasmussen, R.: Slower snowmelt in a warmer world, Nat. Clim. Change, 7, 214–219, https://doi.org/10.1038/nclimate3225, 2017.

Myers-Smith, I. H. and Hik, D. S.: Climate warming as a driver of tundra shrubline advance, J. Ecol., 106, 547–560, https://doi.org/10.1111/1365-2745.12817, 2018.

Myers-Smith, I. H., Forbes, B. C., Wilmking, M., Hallinger, M., Lantz, T., Blok, D., Tape, K. D., Macias-Fauria, M., Sass-Klaassen, U., Lévesque, E., Boudreau, S., Ropars, P., Hermanutz, L., Trant, A., Collier, L. S., Weijers, S., Rozema, J., Rayback, S. A., Schmidt, N. M., Schaepman-Strub, G., Wipf, S., Rixen, C., Ménard, C. B., Venn, S., Goetz, S., Andreu-Hayles, L., Elmendorf, S., Ravolainen, V., Welker, J., Grogan, P., Epstein, H. E., and Hik, D. S.: Shrub expansion in tundra ecosystems: dynamics, impacts and research priorities, Environ. Res. Lett., 6, 045509, https://doi.org/10.1088/1748-9326/6/4/045509, 2011.

Oelke, C., Zhang, T., and Serreze, M. C.: Modeling evidence for recent warming of the Arctic soil thermal regime, Geophys. Res.

Lett., 31, 4–7, https://doi.org/10.1029/2003GL019300, 2004.

Ohmura, A.: Observed decadal variations in surface solar radiation and their causes, J. Geophys. Res., 114, D00D05, https://doi.org/10.1029/2008JD011290, 2009.

Overeem, I. and Syvitski, J. P. M.: Shifting Discharge Peaks in Arctic Rivers, 1977–2007, Geogr. Ann., 92, 285–296, https://doi.org/10.1111/j.1468-0459.2010.00395.x, 2010.

Pan, X., Yang, D., Li, Y., Barr, A., Helgason, W., Hayashi, M., Marsh, P., Pomeroy, J., and Janowicz, R. J.: Bias corrections of precipitation measurements across experimental sites in different ecoclimatic regions of western Canada, The Cryosphere, 10, 2347–2360, https://doi.org/10.5194/tc-10-2347-2016, 2016.

Payette, S. and Filion, L.: White spruce expansion at the tree line and recent climatic change, Can. J. Forest. Res., 15, 241–251, https://doi.org/10.1139/x85-042, 1985.

Peterson, B. J., Holmes, R. M., McClelland, J. W., Vorosmarty, C. J., Lammers, R., Shiklomanov, A. I., Shiklomanov, I. A., and Rahmstorf, S.: Increasing River Discharge to the Arctic Ocean, Science, 298, 2171–2173, https://doi.org/10.1126/science.1077445, 2002.

Pomeroy, J. W. and Brun, E.: Physical Properties of Snow, in: Snow Ecol. an Interdiscip. Exam. Snow-covered Ecosyst., chap. 2, edited by: Jones, H. G., Pomeroy, J. W., Walker, D. A., and Hoham, R. W., Cambridge University Press, New York, USA, 45–118, http://www.usask.ca/hydrology/papers/Pomeroy_et_al_2001.pdf (last access: 12 July 2018), 2001.

Pomeroy, J. W. and Goodison, B. E.: Winter and Snow, in: The Surface Climates of Canada, chap. 4, edited by: Bailey, W. G., Oke, T. R., and Rouse, W. R., McGill-Queen's University Press, Montreal, Kingston, Canada, 68–100, 1997.

Pomeroy, J. W. and Marsh, P.: The application of remote sensing and a blowing snow model to determine snow water equivalent over northern basins, in: Applications of remote sensing in hydrology: NHRI Symposium No. 17, edited by: Kite, G. W., Pietroniro, A., and Pultz, T. J., National Hydrology Research Institute, Environment Canada, Saskatoon, SK, 253–270, 1996.

Pomeroy, J. W., Marsh, P., and Gray, D. M.: Application of a distributed blowing snow model to the Arctic, Hydrol. Process., 11, 1451–1464, https://doi.org/10.1002/(SICI)1099-1085(199709)11:11<1451::AID-HYP449>3.0.CO;2-Q, 1997.

Pomeroy, J. W., Parviainen, J., Hedstrom, N., and Gray, D. M.: Coupled modelling of forest snow interception and sublimation, Hydrol. Process., 12, 2317–2337, https://doi.org/10.1002/(SICI)1099-1085(199812)12:15<2317::AID-HYP799>3.0.CO;2-X, 1998.

Pomeroy, J. W., Bewley, D. S., Essery, R. L. H., Hedstrom, N. R., Link, T., Granger, R. J., Sicart, J. E., Ellis, C. R., and Janowicz, J. R.: Shrub tundra snowmelt, Hydrol. Process., 20, 923–941, https://doi.org/10.1002/hyp.6124, 2006.

Pomeroy, J. W., Fang, X., Shook, K., and Whitfield, P. H.: Predicting in Ungauged Basins using Physical Principles Obtained using The Deductive, Inductive, and Abductive Reasoning Approach, in: Putting Prediction in Ungauged Basins into Practice, Canadian Water Resources Association (CWRA) and the International Association of Hydrological Sciences (IAHS), Ottawa, 41–62, 2013a.

Pomeroy, J. W., Spence, C., Whitfield, P. H., and Spence, C.: Putting Prediction in Ungauged Basins into Practice, Canadian Water Resources Association, International Association of Hydrological Sciences, available at: https://cwra.org/en/resource-center/publications/bookstore/20-publications/245-putting-prediction-in-ungauged-basins-into-practice (last access: 12 July 2018), 2013b.

Pomeroy, J. W., Fang, X., and Rasouli, K.: Sensitivity of snow processes to warming in the Canadian Rockies, 72nd East. Snow Conf. Sherbrooke, Quebec, Canada, 22–33, 2015.

Prowse, T., Alfredsen, K., Beltaos, S., Bonsal, B., Duguay, C., Korhola, A., McNamara, J., Pienitz, R., Vincent, W. F., Vuglinsky, V., and Weyhenmeyer, G. A.: Past and future changes in Arctic lake and river ice, Ambio, 40, 53–62, https://doi.org/10.1007/s13280-011-0216-7, 2011.

Przybylak, R., Vízi, Z., and Wyszyński, P.: Air temperature changes in the Arctic from 1801 to 1920, Int. J. Climatol., 30, 791–812, https://doi.org/10.1002/joc.1918, 2010.

Quinton, W. L. and Carey, S. K.: Towards an energy-based runoff generation theory for tundra landscapes, Hydrol. Process., 22, 4649–4653, https://doi.org/10.1002/hyp.7164, 2008.

Quinton, W. L. and Gray, D. M.: Estimating subsurface drainage from organic-covered hillslopes underlain by permafrost?: toward a combined heat and mass flux model, in: Sixth IAHS Scientific Assmble, Maastricht, the Netherlands, 2001.

Quinton, W. L. and Marsh, P.: A conceptual framework for runoff generation in a permafrost environment, Hydrol. Process., 13, 2563–2581, https://doi.org/10.1002/(SICI)1099-1085(199911)13:16<2563::AID-HYP942>3.0.CO;2-D, 1999.

Rapaic, M., Brown, R., Markovic, M., and Chaumont, D.: An Evaluation of Temperature and Precipitation Surface-Based and Reanalysis Datasets for the Canadian Arctic, 1950–2010, Atmosphere-Ocean, 42, 283–303, https://doi.org/10.1080/07055900.2015.1045825, 2015.

Rasmusson, E. M. and Carpenter, T. H.: Variations in Tropical Sea Surface Temperature and Surface Wind Fields Associated with the Southern Oscillation/El Niño, Mon. Weather Rev., 110, 354–384, https://doi.org/10.1175/1520-0493(1982)110<0354:VITSST>2.0.CO;2, 1982.

Rasouli, K., Pomeroy, J. W., Janowicz, J. R., Carey, S. K., and Williams, T. J.: Hydrological sensitivity of a northern mountain basin to climate change, Hydrol. Process., 28, 4191–5208, https://doi.org/10.1002/hyp.10244, 2014.

Rasouli, K., Pomeroy, J. W., and Marks, D. G.: Snowpack sensitivity to perturbed climate in a cool mid-latitude mountain catchment, Hydrol. Process., 29, 3925–3940, https://doi.org/10.1002/hyp.10587, 2015.

Romero-Lankao, P., Smith, J. B., Davidson, D. J., Diffenbaugh, N. S., Kinney, P. L., Kirshen, P., Kovacs, P., and Villers-Ruiz, L.: North America, in: Climate Change 2014: Impacts, Adaptation, and Vulnerability, Part B: Regional Aspects, Contribution of Working Group II to the Fifth Assessment Report of the Intergovernmental Panel of Climate Change, edited by: Barros, V. R., Field, C. B., Dokken, D. J., Mastrandrea, M. D., Mach, K. J., Bilir, T. E., Chatterjee, M., Ebi, K. L., Estrada, Y. O., Genova, R. C., Girma, B., Kissel, E. S., Levy, A. N., MacCracken, S., Mastrandrea, P. R., and White, L. L., Cambridge University Press, Cambridge, UK and New York, NY, USA, 1439–1498, 2014.

Rood, S. B., Kaluthota, S., Philipsen, L. J., Rood, N. J., and Zanewich, K. P.: Increasing discharge from the Mackenzie River system to the Arctic Ocean, Hydrol. Process., 31, 150–160, https://doi.org/10.1002/hyp.10986, 2017.

Saunders, R., Matricardi, M., and Brunel, P.: An improved fast radiative transfer model for assimilation of satellite radiance observations, Q. J. Roy. Meteorol. Soc., 125, 1407–1425, https://doi.org/10.1002/qj.1999.49712555615, 1999.

Schmidt, R. A.: Properties of blowing snow, Rev. Geophys., 20, 39–44, https://doi.org/10.1029/RG020i001p00039, 1982.

Schmidt, R. A. and Gluns, D. R.: Snowfall interception on branches of three conifer species, Can. J. Forest. Res., 21, 1262–1269, https://doi.org/10.1139/x91-176, 1991.

Sen, P. K.: Estimates of the regression coefficient based on Kendall's Tau, J. Am. Stat. Assoc., 63, 1379–1389, https://doi.org/10.2307/2285891, 1968.

Serreze, M. C. and Hurst, C. M.: Representation of Mean Arctic Precipitation from NCEP – NCAR and ERA Reanalyses, J. Climate, 13, 182–201, 2000.

Serreze, M. C., Key, J. R., Box, J. E., Maslanik, J. A., and Steffen, K.: A new monthly climatology of global radiation for the arctic and comparisons with NCEP-NCAR reanalysis and ISCCP-C2 fields, J. Climate, 11, 121–136, https://doi.org/10.1175/1520-0442(1998)011<0121:ANMCOG>2.0.CO;2, 1998.

Serreze, M. C., Bromwich, D. H., Clark, M. P., Etringer, A. J., Zhang, T., and Lammers, R.: Large-scale hydro-climatology of the terrestrial Arctic drainage system, J. Geophys. Res., 108, 8160, https://doi.org/10.1029/2001JD000919, 2002.

Serreze, M. C., Clark, M. P., and Bromwich, D. H.: Monitoring Precipitation over the Arctic Terrestrial Drainage System?: Data Requirements, Shortcomings, and Applications of Atmospheric Reanalysis, J. Hydrometeorol., 4, 387–407, 2003.

Smith, C.: Correcting the wind bias in snowfall measurements made with a Geonor T-200B precipitation gauge and alter wind shield, in: CMOS Bulletin SCMO, vol. 36, Canadian Meteorological and Oceanographic Society (CMOS), Ottawa, Canada, 162–167, 2008.

Sturm, M., Racine, C., and Tape, K.: Increasing shrub abundance in the Arctic, Nature, 411, 456–457, 2001.

Sturm, M. E. A., Mcfadden, J. P., Liston, G. E., Chapin, F. S., Racine, C. H., and Holmgre, J.: Snow–Shrub Interactions in Arctic Tundra: A Hypothesis with Climatic Implications, J. Climate, 14, 336–344, 2000.

Suarez, F., Binkley, D., Kaye, M. W., and Stottlemyer, R.: Expansion of forest stands into tundra in the Noatak National Preserve, northwest Alaska, Ecoscience, 6, 465–470, https://doi.org/10.1080/11956860.1999.11682538, 1999.

Tape, K., Sturm, M., and Racine, C.: The evidence for shrub expansion in Northern Alaska and the Pan-Arctic, Global Chang. Biol., 12, 686–702, https://doi.org/10.1111/j.1365-2486.2006.01128.x, 2006.

Thompson, D. W. J. and Wallace, J. M.: The Arctic Oscillation signature in the wintertime geopotential height and temperature fields, Geophys. Res. Lett., 25, 1297, https://doi.org/10.1029/98GL00950, 1998.

Trenberth, K. E. and Hurrell, J. W.: Decadal atmosphere-ocean variations in the Pacific, Clim. Dynam., 9, 303–319, https://doi.org/10.1007/BF00204745, 1994.

Uppala, S. M., Kallberg, P. W., Simmons, A. J., Andrae, U., Bechtold, V. D. C., Fiorino, M., Gibson, J. K., Haseler, J., Hernandez, A., Kelly, G. A., Li, X., Onogi, K., Saarinen, S., Sokka, N., Allan, R. P., Andersson, E., Arpe, K., Balmaseda, M. A., Beljaars, A. C. M., Van De Berg, L., Bidlot, J., Bormann, N., Caires, S., Chevallier, F., Dethof, A., Dragosavac, M., Fisher, M., Fuentes, M., Hagemann, S., Hólm, E., Hoskins, B. J., Isaksen, L., Janssen, P. A. E. M., Jenne, R., Mcnally, a. P., Mahfouf, J.-F., Morcrette, J.-J., Rayner, N. A., Saunders, R. W., Simon, P., Sterl, A., Trenberth, K. E., Untch, A., Vasiljevic, D., Viterbo, P., and Woollen, J.: The ERA-40 re-analysis, Q. J. Roy. Meteorol. Soc., 131, 2961–3012, https://doi.org/10.1256/qj.04.176, 2005.

Vincent, L. A., Zhang, X., Brown, R. D., Feng, Y., Mekis, E., Milewska, E. J., Wan, H. and Wang, X. L.: Observed Trends in Canada's Climate and Influence of Low-Frequency Variability Modes, J. Climate, 28, 4545–4560, https://doi.org/10.1175/JCLI-D-14-00697.1, 2015.

Walsh, J. E.: Cryosphere and Hydrology, in: Arctic Climate Impact Assessment – Scientific Report, edited by: Symon, C., Arris, L., and Heal, B., Cambridge University Press, Cambridge, 183–242, 2005.

Walvoord, M. A. and Kurylyk, B. L.: Hydrologic impacts of thawing permafrost – a review, Vadose Zone J., 15, 1–20, https://doi.org/10.2136/vzj2016.01.0010, 2016.

Wanishsakpong, W., McNeil, N., and Notodiputro, K. A.: Trend and pattern classification of surface air temperature change in the Arctic region, Atmos. Sci. Lett., 17, 378–383, https://doi.org/10.1002/asl.668, 2016.

Wessel, D. A. and Rouse, W. R.: Modelling Evaporation From Wetland Tundra, Bound.-Lay. Meteorol., 41, 109–130, https://doi.org/10.1007/BF00712666, 1994.

Westerling, A. L., Hidalgo, H. G., Cayan, D. R., and Swetnam, T. W.: Warming and Earlier Spring Increase Western U.S. Forest Wildfire Activity, Science, 313, 940–943, https://doi.org/10.1126/science.1128834, 2006.

Weston, S. T., Bailey, W. G., McArthur, L. J. B., and Hertzman, O.: Interannual solar and net radiation trends in the Canadian Arctic, J. Geophys. Res.-Atmos., 112, 1–8, https://doi.org/10.1029/2006JD008000, 2007.

Whitfield, P. H., Hall, A. W., and Cannon, A. J.: Changes in the seasonal cycle in the circumpolar Arctic, 1976–95: temperature and precipitation, Arctic, 57, 80–93, 2004.

Williamson, T., Colombo, S., Duinker, P., Gray, P., Hennessey, R., Houle, D., Johnston, M., Ogden, A., and Spittlehouse, D.: Climate change and Canada's forests: from impacts to adaptation, Sustainable Forest Management Network and Natural Resources Canada, Canadian Forest Service, Northern Forestry Centre, Edmonton, AB, 2009.

Woo, M.-K. and Sauriol, J.: Channel Development in Snow-Filled Valleys, Resolute, N.W.T., Canada, Geogr. Ann., 62, 37–56, 1980.

Woo, M.-K., Kane, D. L., Carey, S. K., and Yang, D.: Progress in permafrost hydrology in the new millennium, Permafrost Periglac. Process., 19, 237–254, https://doi.org/10.1002/ppp.613, 2008.

Xu, L., Myneni, R. B., Chapin III, F. S., Callaghan, T. V., Pinzon, J. E., Tucker, C. J., Zhu, Z., Bi, J., Ciais, P., Tømmervik, H., Euskirchen, E. S., Forbes, B. C., Piao, S. L., Anderson, B. T., Ganguly, S., Nemani, R. R., Goetz, S. J., Beck, P. S. A., Bunn, A. G., Cao, C., and Stroeve, J. C.: Temperature and vegetation seasonality diminishment over northern lands, Nat. Clim. Change, 6, 581–586, https://doi.org/10.1038/nclimate1836, 2013.

Yang, D., Kane, D. L., Hinzman, L. D., Zhang, X., Zhang, T., and Ye, H.: Siberian Lena River hydrologic regime and recent change, J. Geophys. Res.-Atmos., 107, 1–10, https://doi.org/10.1029/2002JD002542, 2002.

Yip, Q. K. Y., Burn, D. H., Seglenieks, F., Pietroniro, A., and Soulis, E. D.: Climate Impacts on Hydrological Variables in the Mackenzie River Basin, Can. Water Resour. J., 37, 209–230, https://doi.org/10.4296/cwrj2011-899, 2012.

Zhang, Z., Kane, D. L., and Hinzman, L. D.: Development and application of a spatially-distributed Arctic hydrological and thermal process model (ARHYTHM), Hydrol. Process., 14, 1017–1044, https://doi.org/10.1002/(SICI)1099-1085(20000430)14:6<1017::AID-HYP982>3.0.CO;2-G, 2000.

6

Using distributed temperature sensing for Bowen ratio evaporation measurements

Bart Schilperoort[1], **Miriam Coenders-Gerrits**[1], **Willem Luxemburg**[1], **César Jiménez Rodríguez**[1,3], **César Cisneros Vaca**[2], **and Hubert Savenije**[1]

[1]Delft University of Technology, Water Resources Section, Stevinweg 1, 2628 CN Delft, the Netherlands
[2]University of Twente, Faculty of Geo-Information Science and Earth Observation (ITC), Hengelosestraat 99, 7514 AE, Enschede, the Netherlands
[3]Tecnológico de Costa Rica, Escuela de Ingeniería Forestal. 159-7050, Cartago, Costa Rica

Correspondence: Bart Schilperoort (b.schilperoort@tudelft.nl)

Abstract. Rapid improvements in the precision and spatial resolution of distributed temperature sensing (DTS) technology now allow its use in hydrological and atmospheric sciences. Introduced by Euser et al. (2014) is the use of DTS for measuring the Bowen ratio (BR-DTS), to estimate the sensible and latent heat flux. The Bowen ratio is derived from DTS-measured vertical profiles of the air temperature and wet-bulb temperature. However, in previous research the measured temperatures were not validated, and the cables were not shielded from solar radiation. Additionally, the BR-DTS method has not been tested above a forest before, where temperature gradients are small and energy storage in the air column becomes important.

In this paper the accuracy of the wet-bulb and air temperature measurements of the DTS are verified, and the resulting Bowen ratio and heat fluxes are compared to eddy covariance data. The performance of BR-DTS was tested on a 46 m high tower in a mixed forest in the centre of the Netherlands in August 2016. The average tree height is 26 to 30 m, and the temperatures are measured below, in, and above the canopy. Using the vertical temperature profiles the storage of latent and sensible heat in the air column was calculated.

We found a significant effect of solar radiation on the temperature measurements, leading to a deviation of up to 3 K. By installing screens, the error caused by sunlight is reduced to under 1 K. Wind speed seems to have a minimal effect on the measured wet-bulb temperature, both below and above the canopy. After a simple quality control, the Bowen ratio measured by DTS correlates well with eddy covariance (EC) estimates ($r^2 = 0.59$). The average energy balance closure between BR-DTS and EC is good, with a mean underestimation of 3.4 W m^{-2} by the BR-DTS method. However, during daytime the BR-DTS method overestimates the available energy, and during night-time the BR-DTS method estimates the available energy to be more negative. This difference could be related to the biomass heat storage, which is neglected in this study.

The BR-DTS method overestimates the latent heat flux on average by 18.7 W m^{-2}, with RMSE = 90 W m^{-2}. The sensible heat flux is underestimated on average by 10.6 W m^{-2}, with RMSE = 76 W m^{-2}. Estimates of the BR-DTS can be improved once the uncertainties in the energy balance are reduced. However, applying, for example, Monin–Obukhov similarity theory could provide independent estimates for the sensible heat flux. This would make the determination of the highly uncertain and difficult to determine net available energy redundant.

1 Introduction

In recent years distributed temperature sensing (DTS) technology has quickly improved (Bao and Chen, 2012). The precision and spatial resolution now allow its widespread use in hydrological and atmospheric sciences (Selker et al., 2006; Thomas et al., 2012), from measuring groundwater flow (Blume et al., 2013) and seepage into streams (Westhoff et al., 2007) to soil moisture (Steele-Dunne et al., 2010), soil heat flux (Bense et al., 2016), and wind speed (Sayde et al., 2015). First introduced by Euser et al. (2014), DTS can also be used for measuring the Bowen ratio, to estimate the evaporation flux. A dry and wet stretch of the same fibre optic cable are installed vertically to obtain the so-called dry- and wet-bulb temperature gradient, respectively. This method mitigates some problems of the conventional Bowen ratio, since usually at least two different sensors are used to measure the temperature and vapour pressure gradients, of which each has its own independent error (Angus and Watts, 1984; Fuchs and Tanner, 1970). The DTS-based Bowen ratio does not suffer from this drawback, by having a large amount of data points over the height (up to 8 per metre) with only a single sensor. It also has a resolution of 0.06 K for 1 min averages (Silixa machine calibration), and will be more accurate when measuring over a longer time period, allowing for very small temperature gradients to be measured.

In addition to estimating the latent and sensible heat flux, the measurements can also be used to get a better understanding of the processes taking place in complex ecosystems, such as forests. A vertical temperature and humidity profile is available in high resolution and precision, both above, inside, and under the canopy. DTS can also estimate different components of the energy balance, such as the heat storage in the air column, and the soil heat flux (Jansen et al., 2011). Finally, it can be used to increase our understanding of the energy exchange between the canopy and undergrowth layers by looking at the air temperature gradient under the canopy.

This paper elaborates on the method of Euser et al. (2014), by considering more energy balance components like the latent and sensible heat storage in the air column, including a data-quality system, and using the potential air temperature. The performance of the method is tested in a mixed forest in the Netherlands by looking at the accuracy of the DTS-measured air temperature and wet-bulb temperature, compared to reference temperature and humidity sensors. It appears that solar radiation can have a significant influence on the cable temperature, which can be mitigated by providing artificial shadow. Lastly, the fluxes resulting from the method are compared to an eddy covariance (EC) system, and the sources of differences between the methods are shown.

2 Materials and methods

2.1 Theory

The Bowen ratio energy balance method (BREB) combines the energy balance with the Bowen ratio (Oliphant et al., 2004). The energy balance can be described by

$$R_\mathrm{N} + A = \rho\lambda E + H + G_\mathrm{S} + \frac{\mathrm{d}Q}{\mathrm{d}t}, \tag{1}$$

where R_N is the net radiation ($\mathrm{W\,m^{-2}}$), $\rho\lambda E$ the latent heat flux ($\mathrm{W\,m^{-2}}$), H the sensible heat flux ($\mathrm{W\,m^{-2}}$), G_S the soil heat flux ($\mathrm{W\,m^{-2}}$), and $\frac{\mathrm{d}Q}{\mathrm{d}t}$ is the change of energy storage in the system ($\mathrm{W\,m^{-2}}$). A represents a net advection of energy into the system ($\mathrm{W\,m^{-2}}$), but is assumed to be 0. The energy flux associated with photosynthesis (G_P) was not measured, and is therefore not included in the equation. The Bowen ratio (β) is the ratio of the sensible heat flux to the latent heat flux and can be approximated using the air temperature gradient and the vapour pressure difference over the height (Bowen, 1926)

$$\beta = \frac{H}{\rho\lambda E} \approx \gamma \frac{\Delta T_\mathrm{a}}{\Delta e_\mathrm{a}}, \tag{2}$$

where γ is the psychrometric constant ($\mathrm{kPa\,K^{-1}}$) (see Eq. 10), ΔT_a the difference in air temperature between two heights (K) and Δe_a the difference in actual vapour pressure between the two heights (kPa). However, when gradients are very small, the adiabatic lapse rate cannot be neglected (Barr et al., 1994). Therefore the potential temperature should be used instead:

$$\beta = \frac{H}{\rho\lambda E} = \frac{c_p}{\lambda}\frac{\partial\Theta/\partial z}{\partial q/\partial z} = \gamma \frac{\partial\Theta/\partial z}{\partial e_\mathrm{a}/\partial z}, \tag{3}$$

where c_p is the specific heat of air ($\mathrm{MJ\,kg^{-1}}$) (see Eq. 6), λ the latent heat of vaporization ($2.45\,\mathrm{MJ\,kg^{-1}\,K^{-1}}$), Θ the potential temperature (K), q the specific humidity ($\mathrm{kg\,kg^{-1}}$) (see Eq. 7) and z the height above the ground (m). The potential temperature gradient can be approximated by the right-hand side of Eq. (4), as the ratio $\frac{\Theta}{T_\mathrm{a}}$ is nearly 1 (Pal Arya, 1988):

$$\frac{\partial\Theta}{\partial z} = \frac{\Theta}{T_\mathrm{a}}\left(\frac{\partial T_\mathrm{a}}{\partial z} + \Gamma\right) \approx \frac{\partial T_\mathrm{a}}{\partial z} + \Gamma, \tag{4}$$

where T_a is the air temperature (K), and Γ is the adiabatic lapse rate (typically around $0.01\,\mathrm{K\,m^{-1}}$). The numerical implementations of Eqs. (3) and (4) are explained in Sect. 3.2. Under dry and unsaturated conditions the lapse rate is equal to (Pal Arya, 1988)

$$\Gamma = \frac{g}{c_p}, \tag{5}$$

where g is the gravitational acceleration ($9.81\,\mathrm{m\,s^{-2}}$). The specific heat capacity of air is determined by (Stull, 2015)

$$c_p = 1.004 + 1.84q \tag{6}$$

and the specific humidity by (Pal Arya, 1988)

$$q = \varepsilon\frac{e_\mathrm{a}}{P}, \tag{7}$$

where ε is the ratio of molecular mass of water vapour to dry air (0.622), and P the atmospheric pressure (kPa). The actual vapour pressure is determined by (Allen et al., 1998)

$$e_\mathrm{a}(T_\mathrm{a}) = e_\mathrm{s}(T_\mathrm{w}) - \gamma(T_\mathrm{a} - T_\mathrm{w}), \tag{8}$$

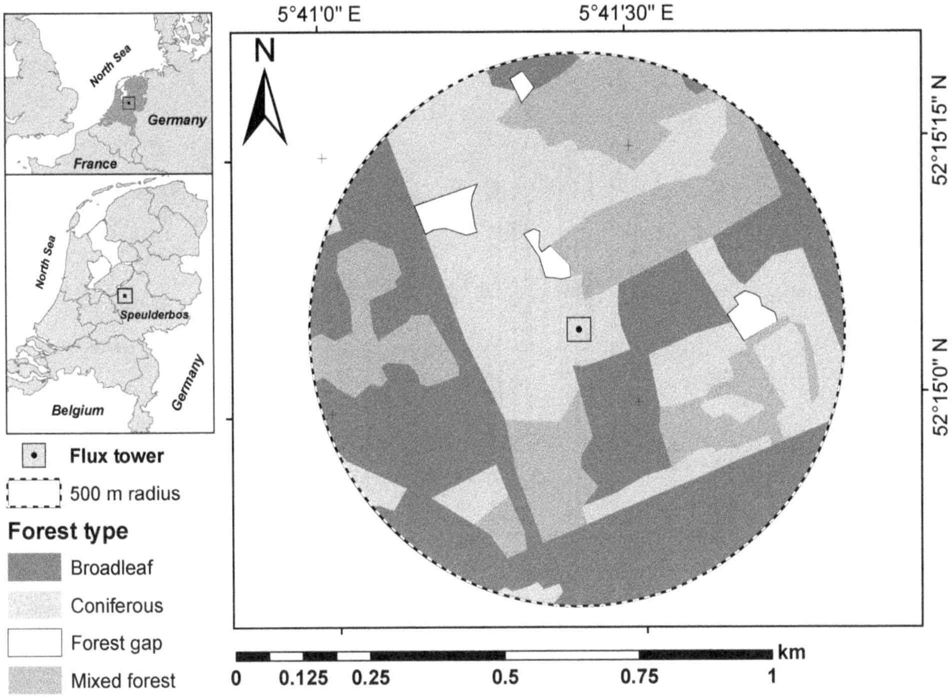

Figure 1. Forest type distribution within 500 m of the tower site at Speulderbos Forest, the Netherlands.

where T_w is the wet-bulb temperature (K), and e_s the saturation vapour pressure (kPa) given by (Koutsoyiannis, 2012)

$$e_\mathrm{s}(T_\mathrm{w}) = 0.61 \cdot \exp\left(\frac{19.9 \cdot T_\mathrm{w}}{273 + T_\mathrm{w}}\right). \tag{9}$$

The psychrometer constant is related to the air pressure and ventilation of the psychrometer (Harrison and Wood, 2012; Allen et al., 1998). If sufficiently ventilated, the psychrometric constant is defined by (Allen et al., 1998)

$$\gamma = \frac{c_p P}{\varepsilon \lambda} = 0.665 \times 10^{-3} \cdot P. \tag{10}$$

As the air pressure also varies over height, the measurements have to be corrected for elevation using the following approximation (Stull, 2015, p. 8):

$$P(z) = P_0 \cdot \exp(-z/7290) \tag{11}$$

with P_0 being the pressure at sea level (kPa). By combining the Bowen ratio (Eq. 3) with the energy balance (Eq. 1), the latent heat flux and sensible heat flux can be determined:

$$H = \frac{R_\mathrm{N} - G_\mathrm{S} - \frac{\mathrm{d}Q}{\mathrm{d}t}}{1 + \frac{1}{\beta}}, \tag{12}$$

$$\rho \lambda E = \frac{R_\mathrm{N} - G_\mathrm{S} - \frac{\mathrm{d}Q}{\mathrm{d}t}}{1 + \beta}. \tag{13}$$

The storage component in the energy balance has multiple parts, ranging from the storage of heat in the soil, to the storage of heat in the form of water vapour in the air column:

$$\frac{\mathrm{d}Q}{\mathrm{d}t} = \frac{\mathrm{d}Q_H}{\mathrm{d}t} + \frac{\mathrm{d}Q_E}{\mathrm{d}t}. \tag{14}$$

The changes in storage of heat and water vapour in the air column below the height at which the energy fluxes (R_N, H and $\rho\lambda E$) are measured are represented by $\frac{\mathrm{d}Q_H}{\mathrm{d}t}$ and $\frac{\mathrm{d}Q_E}{\mathrm{d}t}$ respectively ($\mathrm{W\,m^{-2}}$). The change in biomass heat storage ($\frac{\mathrm{d}Q_B}{\mathrm{d}t}$) was not measured, and is therefore not included in this equation. $\frac{\mathrm{d}Q_H}{\mathrm{d}t}$ and $\frac{\mathrm{d}Q_E}{\mathrm{d}t}$ are defined as (Barr et al., 1994)

$$\frac{\mathrm{d}Q_H}{\mathrm{d}t} = \int_0^z \rho_\mathrm{a} c_p \frac{\mathrm{d}T_\mathrm{a}}{\mathrm{d}t} \mathrm{d}z, \tag{15}$$

$$\frac{\mathrm{d}Q_E}{\mathrm{d}t} = \int_0^z \rho_\mathrm{a} \lambda \frac{\mathrm{d}q}{\mathrm{d}t} \mathrm{d}z. \tag{16}$$

3 Study site

The measurements were carried out at the Speulderbos mixed forest (52°15′4″N, 5°41′24″E), on a tower located within a patch of Douglas fir trees (*Pseudotsuga menziesii* (Mirb.) Franco) of 2.5 ha in Garderen, the Netherlands (Fig. 1). The surrounding area is characterized by the presence of broadleaved and coniferous tree species, distributed in blocks around the tower site (Bosveld and Bouten, 2001). Within a

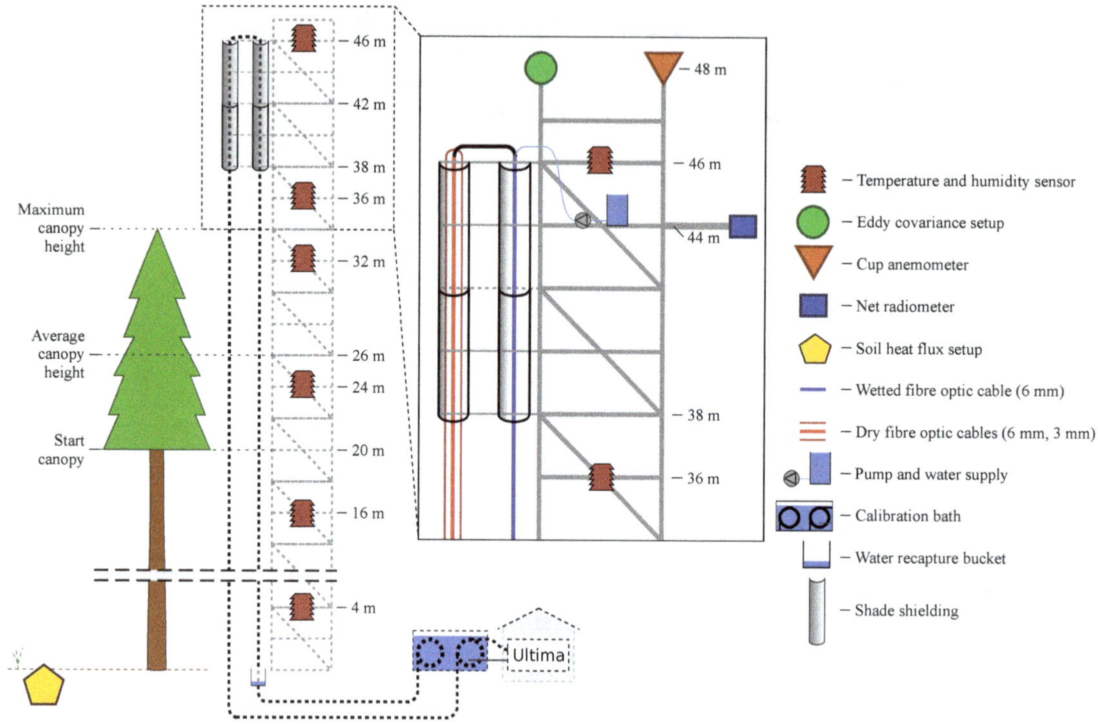

Figure 2. Schematic overview of the measurement setup at the tower.

500 m radius it is possible to find native tree species such as beech (*Fagus sylvatica* L.), pedunculate oak (*Quercus robur* L.) and Scots pine (*Pinus sylvestris* L.), as well as the introduced species hemlock (*Tsuga heterophylla* (Rafinesque) Sargent) and Japanese larch (*Larix kaempferi* (Lambert) Carriére) (Erisman et al., 1998; Raj et al., 2014; Su et al., 2009; Bosveld and Bouten, 2001; Tietema et al., 2002; Van Wijk et al., 2000; Weligepolage et al., 2013). Canopy heights differ between cover types depending on species and growing stage. Some coniferous canopies like the Douglas fir have a canopy height between 26 and 30 m, while the broadleaved stands can reach up to 30 m height for old-growth beech trees (Weligepolage et al., 2012; Wilkes et al., 2017), or heights under 10 m for smaller pedunculate oak trees.

The study site has an oceanic climate (Cfb) under the Köpen classification system, with a yearly average temperature of 9.8 °C and an average precipitation of 910 mm yr^{-1} (Sluijter, 2011). The topography is slightly undulating with smooth height differences (Raj et al., 2014), a well-drained soil, and a groundwater table below 40 m depth (Tiktak and Bouten, 1994). The soil texture ranges from fine sand to sandy loam (Weligepolage et al., 2012; Tietema et al., 2002; Van Wijk et al., 2000).

3.1 Setup

The temperature of fibre optic cables is measured using the DTS technique (Selker et al., 2006). In the setup, two cables with different diameters were used. The first cable has a diameter of 6 mm and has both a dry and a wetted stretch. To wet the cable it was wrapped in cloth, and water was supplied to it continuously. A second cable with a diameter of 3 mm was used to study the effects of solar radiation, as a thinner cable will warm up less (De Jong et al., 2015). However, this method added additional uncertainties due to the required extrapolation and the 3 mm cable was not used in this study. (While correlation with reference sensors improved, the uncertainty of extrapolation caused extra noise in the Bowen ratio calculations). Both cables were connected to the same DTS machine (in single-ended mode) and calibrated in a calibration bath (see Fig. 2)

The DTS machine used was the Silixa Ultima (Silixa Ltd, 2017), which has a sampling resolution of 12.5 cm, measurement resolution of 35 cm, and a measurement standard deviation of 0.06 K at a 1 min time resolution.

The fibre optic cable with a diameter of 6 mm was secured at the top of the tower, with the dry stretch hanging 1.2 m away from the tower, and the wet stretch 0.25 m away. The cable with a diameter of 3 mm was secured next to the dry 6 mm cable. The response times of the cables are of the order of 2–3 min for the 6 mm cable, and 20 to 40 s for the 3 mm cable. The cables were secured at multiple locations distributed over the height (in and above the canopy; see Fig. 2), using loops (with a diameter of 5 cm) to prevent direct contact with the support structure. For both cables a stretch of 10 m at both the start and end was placed in a calibration bath, an enclosed styrofoam box filled with water, along with two

Pt100 temperature probes that were connected to the DTS machine. An air bubbler was installed in the styrofoam box to ensure a homogeneous temperature distribution. The cables were shielded from direct solar radiation using screen gauze secured onto PVC rings; see Fig. 3. Only the southern 180° of the cables was shielded, to allow for sufficient ventilation. The screen gauze had holes 1.5 mm wide, and the mesh material had a diameter of 0.3 mm. Two layers of the gauze were used. Each segment of shield was 2 m long, and was secured to the tower with a horizontal beam. Due to the angle of the incident sunlight the gauze was able to block most direct sunlight, except during the early morning. To supply the wet cable with water, a reservoir was installed near the top of the tower, along with a pump. The pump speed was set to 1500 mL h^{-1} during sunny days without rainfall, and to 800 mL h^{-1} on other days, which was enough to keep the cable wet over the entire height, while keeping the influence of relatively warm water at the top of the cable at a minimum. As water supplied at the top has a higher temperature than the wet-bulb temperature, the top 2 m of wet cable data was excluded from the data analysis to allow the slowly flowing water to reach the wet-bulb temperature.

A net radiometer (Kipp & Zonen CNR4) was located on the top of the tower (48 m), measuring both incoming and outgoing short- and longwave radiation. One-minute averages were logged. On the tower six humidity and temperature sensors were located over the height, at 4, 16, 24, 32, 36 and 46 m a.g.l. (above ground level). The lower four were Rotronic HC2-S3C03 sensors (with active ventilation), and the top two were Campbell CS216 sensors with passive ventilation. The sensors were inter-calibrated to the sensor at 24 m. The temperature and humidity was logged at 1 min averages.

At the top of the tower an EC system was installed to measure the sensible and latent heat fluxes. It consisted of a Campbell CSAT3 sonic anemometer and a LI-COR Biosciences LI7500 gas analyser connected to a CR5000 Campbell data logger, to which the data were logged at 20 Hz.

Two cup anemometers (Onset S-WSB-M003) were used to measure the wind speed, one at the top of the tower (48 m), and one below the canopy (4 m). The data from the lower anemometer lacks the resolution to properly measure the low wind speeds below the canopy, which are at times too low to be registered. One-minute average wind speeds, along with the maximum gust speeds, were logged.

The biomass heat storage change and the photosynthesis energy flux were not measured. The biomass heat storage change is estimated to have a maximum of 45 W m^{-2}, and the photosynthesis energy flux is estimated to be of the order of 5 W m^{-2} (Barr et al., 1994; Michiles and Gielow, 2008). For the soil heat flux, the soil temperature was measured at different depths (1, 3, 4, 8, 20, 50 cm). Soil moisture was measured using Campbell Sci. Inc. CS616 water content reflectometers. Thermal conductivity was fitted to soil heat flux

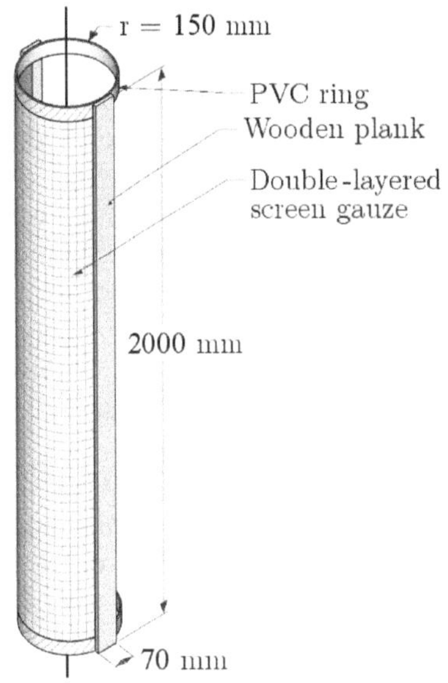

Figure 3. Schematic of one 2 m segment of the solar screen construction.

measurements done at 8 cm. The soil heat flux was then determined using the harmonics method (van der Tol, 2012).

3.2 Data processing

The DTS machine was set to measure the cable temperature at 1 min averaging intervals. For the comparison with reference temperature sensors, these 1 min resolution data are used. To compare the wet-bulb temperature measured by the fibre optic cable to the reference sensors, the reference wet-bulb temperature is iteratively derived from the reference air temperature and relative humidity. For the purpose of calculating the Bowen ratio, the temperature and actual vapour pressure are averaged over time for 15 min time periods. For DTS Bowen ratio calculations, the temperatures between 38.5 and 44 m are used. This area is shaded from the sun by the screen gauze, and at the top of the stretch the new water on the wet cable has reached the wet-bulb temperature.

When calculating the gradients for the Bowen ratio, the 15 min average temperature and vapour pressure are fit to the natural logarithm of the height, in the following form:

$$T_{\mathrm{a,\,fit}} = a \cdot \ln(z) + b. \tag{17}$$

A logarithmic shape of the profiles was assumed based on Monin–Obukhov similarity theory. A linear fit was also looked at, but it resulted in a minimal difference in the resulting fit. From the fits the temperature difference over height is then calculated:

$$\frac{\partial \Theta}{\partial z} \approx \frac{\partial T_{\mathrm{a}}}{\partial z} + \Gamma(z) \approx \frac{\Delta T_{a,\mathrm{fit}}}{\Delta z} + \Gamma(z) \approx \frac{T_{\mathrm{a,\,fit}}(z=44) - T_{\mathrm{a,\,fit}}(z=38.5)}{44-38.5} + \Gamma(\bar{z}=41.25), \tag{18}$$

$$\frac{\partial e_\mathrm{a}}{\partial z} \approx \frac{\Delta e_{a,\mathrm{fit}}}{\Delta z} = \frac{e_{\mathrm{a,\ fit}}(z=44) - e_{\mathrm{a,\ fit}}(z=38.5)}{44-38.5}, \tag{19}$$

where $\Delta T_{\mathrm{a,\ fit}}$ is the difference in air temperature (K) of the fitted temperature curve, between the top and bottom of the height range used for the Bowen ratio. $\Delta e_{\mathrm{a,\ fit}}$ is the difference in vapour pressure (kPa) of the fitted vapour pressure curve between those heights. Δz is the difference in height (m). The coefficients of determination of the regressions of the temperature and vapour pressure, $r_{T_\mathrm{a},z}$ and $r_{e_\mathrm{a},z}$, can be used for determining the goodness of fit. A high (positive or negative) regression means that the logarithmic slope (of the 15 min average) is very well defined.

To calculate the air column storage terms $\frac{\mathrm{d}Q_H}{\mathrm{d}t}$ and $\frac{\mathrm{d}Q_E}{\mathrm{d}t}$ (Eqs. 15 and 16), the DTS-measured temperature and vapour pressure are used, except for the centre of the canopy where DTS data are not accurate due to the sunlight and lack of screens in the canopy. The temperature and specific humidity are integrated over the height from 0 to 41 m, up to the height of the Bowen ratio measurements.

As quality control scheme for the DTS-measured Bowen ratio, two flags are used. The first flag tests the correlation coefficient of the actual vapour pressure over height, for which we chose a lower limit of 0.20 (Eq. 20). We do not consider $r_{T_\mathrm{a},z}$ of the air temperature gradient as it is always higher than $r_{e_\mathrm{a},z}$ (as the uncertainty in e_a is higher due to the propagation of errors in T_a and T_w). The second flag is for the case where the Bowen ratio approaches −1, which causes the uncertainty in the BREB fluxes to be very high, as the denominator of Eqs. (12) and (13) approaches 0 (Payero et al., 2003):

$$\text{Flag 1}: r^2_{e_\mathrm{a},z} > 0.20, \tag{20}$$

$$\text{Flag 2}: \beta < -1.1 \ \text{ or } \ \beta > -0.9. \tag{21}$$

If flag 1 is true, the outcome of the Bowen ratio calculation is considered reliable. The other data points are removed from further analysis. If flag 2 is also true, then the Bowen ratio can be used for calculating the atmospheric heat fluxes.

After processing the EC data using LI-COR's EddyPro® software (LI-COR Inc., 2016), several quality flags are available. The quality flag system used is from Mauder and Foken (2006), ranging from 0 (best) to 2 (worst). The EC fluxes with a quality flag of 0 or 1 are used in this research.

To summarize, the method of this paper differs in a few points from Euser et al. (2014). The fit of the Bowen ratio temperature and vapour pressure profiles is done separately, to get the correct ratio, as $\frac{\partial T}{\partial z} / \frac{\partial e_\mathrm{a}}{\partial z} \neq \frac{\partial T}{\partial e_\mathrm{a}}$. More energy balance storage terms are taken into account, namely the latent and specific heat storage in the air column. The potential temperature is used instead of the air temperature, to correct for the lapse rate. The local air pressure is taken into account in the calculations, as it has an influence on the psychrometric constant, specific heat capacity and specific humidity. Lastly, a system for simple quality flags is introduced to allow for simple objective quality control.

4 Results and discussion

4.1 Meteorological conditions

For the comparison of the DTS temperature with the reference temperature data (Sect. 4.2), the days 10–22 August 2016 are used.

For a good comparison between DTS and EC, both devices should work properly. Due to several technical problems with data collection, only 11 days within the measurement campaign have both EC and DTS data available, namely 10, 12–14, 19–22, and 28–30 August 2016. On the other days data are missing in either the EC or the DTS. The meteorological conditions of these days are shown in Fig. 4. All days were partially clouded, or completely clouded. The wind direction was mainly west and northeast. Above the canopy the wind speed varied between 2 and 6 m s^{-1}, while under the canopy the wind speed was often too low to be measured with the cup anemometer (under 0.4 m s^{-1}).

4.2 Temperature validation

In Fig. 5 the comparison between the 6 mm DTS cable and the reference sensor is shown. For the above-canopy comparison, the 46 m reference sensor is compared to the cable temperatures at 44 m height, as the temperatures at the top are unreliable due to influence from the sun and the warm water from the reservoir. Below the canopy the dry cable temperature correlates perfectly with the reference sensor temperature (Fig. 5e). In and above the canopy incoming solar radiation warms up the fibre optic cable (Fig. 5a, c), which causes an error at 34 m, where no screen was installed. This error is a deviation of up to 3 K from the reference sensor temperature (for 1 min temperature averages). The comparison at 34 m also has an offset – this is a constant error of about 1 K, due to the reference temperature sensor drift and inter-calibration problems. The addition of screens above the canopy largely reduces the error from solar radiation to under 1 K, leading to a very good agreement between the two sensor types (Fig. 5a).

Below the canopy the wet cable temperature is in good agreement with the reference wet-bulb temperature (Fig. 5f), even though wind speeds were often low. This shows that the wet cable gives a good estimate of the wet-bulb temperature. At 34 m, where no screens were placed, the error in the wet-bulb temperature is larger than the error in the air temperature. Deviations of up to 4 K occur in the measurement period. The shielded top part of the wet cable performs much better (Fig. 5b), and errors are small (under 1 K).

4.3 Bowen ratio verification

The Bowen ratio resulting from the BR-DTS method (β_DTS) is compared to the EC Bowen ratio (β_EC), at a 15 min averaging interval. In Fig. 6 the correlation between the EC Bowen ratio estimate and the BR-DTS is shown. It shows a grouping

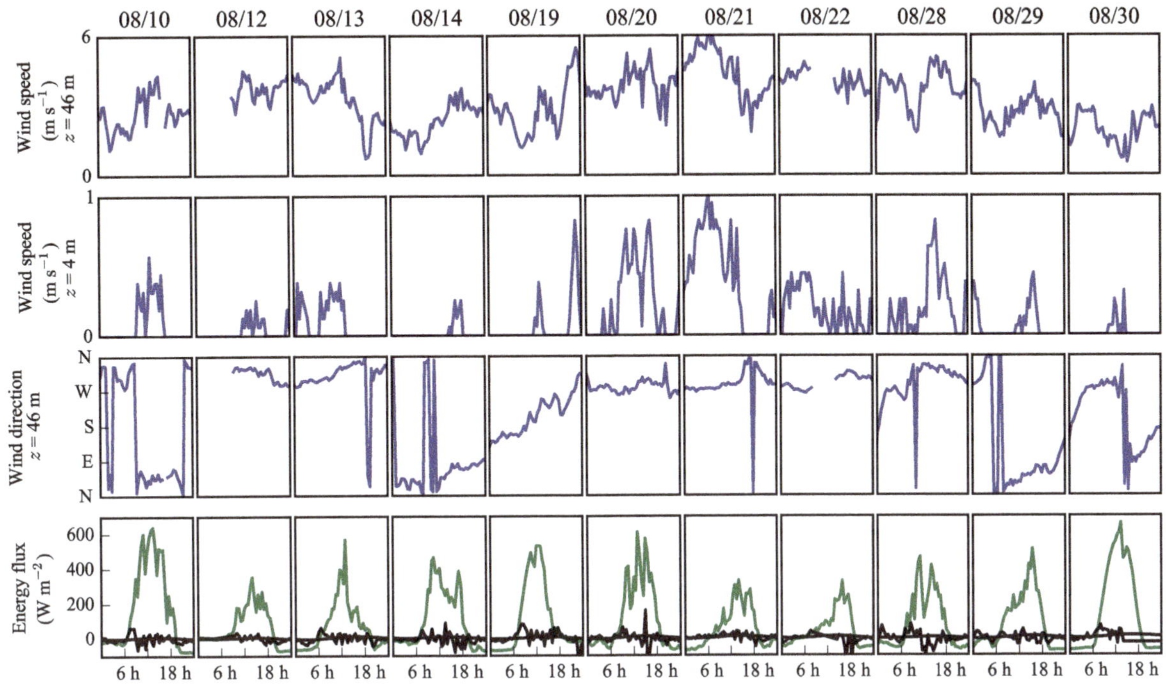

Figure 4. Meteorological conditions during the days that both DTS and EC data were available. From top to bottom: wind speed at the top of the tower, wind speed at the bottom of the tower, wind direction at the top of the tower, and the measured energy fluxes (green: net radiation; red: soil heat flux; black: energy storage change $\frac{dQ}{dt}$).

around the 1 : 1 line, and a good correlation ($r^2 = 0.59$). The EC Bowen ratio was only calculated for fluxes with an absolute value larger than $10\,W\,m^{-2}$, as the uncertainty of the EC Bowen ratio is very high when the fluxes are small. Even the negative (night-time) values seemed to be accurate, since they passed the quality control flags. However, both EC and BR-DTS have problems measuring the night-time Bowen ratio. For EC this is due to the lower friction velocity at night (Wilson, 2002), while for the BR-DTS method the gradients are very small due to the small fluxes.

One drawback of the DTS-based Bowen ratio is the assumption that the eddy diffusivity of heat and water vapour are the same. In reality these eddy diffusivities can be dissimilar (Irmak et al., 2014). This can cause an error (both a bias and extra noise) in the Bowen ratio as measured by the temperature and vapour pressure gradients compared to the EC Bowen ratio. Another source of differences between β_{DTS} and β_{EC} is that the two are measured at different heights.

During the measurement period the 80 % fetch of the EC system was between 200 and 300 m. By applying the findings of Stannard (1997), the Bowen ratio 80% equilibrium ratio would be reached at a fetch-to-height ratio of 20 to 40. This corresponds to a distance of 350 to 700 m. The fetch of the Bowen ratio will therefore not be equal to the EC fetch, which could cause some differences in measured fluxes.

4.4 Energy balance closure

A known problem in measuring fluxes is that the energy balance often does not close well. This is caused by differences in fetch between the used devices, device inaccuracies, and possibly problems with the EC method (Wilson, 2002). Part of the difference between the BR-DTS method and the EC method may be explained by this energy balance closure problem. EC measurements have a fetch which does not include the area close to the flux tower. The available energy in the BR-DTS method depends on measurements of net radiation, ground heat flux and heat storage change ($\frac{dQ}{dt}$) close to the tower. Heterogeneity in the fetch may cause differences between the two methods. In addition, the biomass heat storage change ($\frac{dQ_B}{dt}$) was not measured for the BR-DTS method, and was assumed to be $0\,W\,m^{-2}$. The photosynthesis energy flux (G_P) was also assumed to be $0\,W\,m^{-2}$.

To investigate the energy balance closure for the two methods, we summed up the available fluxes in the following equations, where $\frac{dQ}{dt}$ is the storage term from Eq. (14):

$$B_{DTS} = R_N - G_S - \frac{dQ}{dt}, \tag{22}$$

$$B_{EC} = H_{EC} + \rho\lambda E_{EC}, \tag{23}$$

where B_{DTS} is the energy available for heat fluxes in the BR-DTS method ($W\,m^{-2}$) and B_{EC} is the sum of the EC-measured heat fluxes ($W\,m^{-2}$).

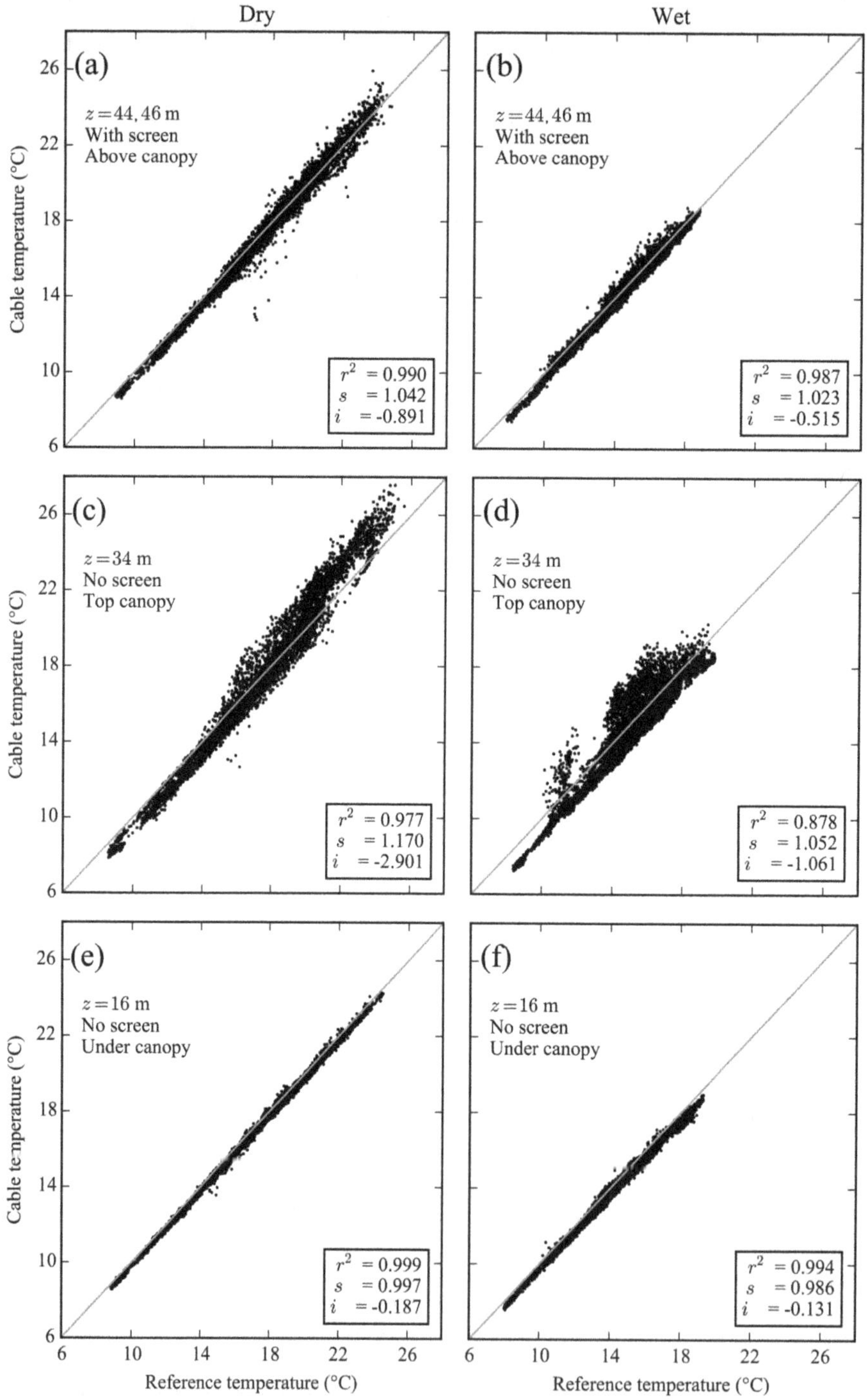

Figure 5. Comparison between the 6 mm DTS cable and reference temperatures. Grey line shows 1:1 correlation. Data from 10–23 August 2016. **(a)** Dry cable at 44 m and reference air temperature at 46 m; the cable is shielded by the screen. **(b)** Wet cable at 44 m and reference wet-bulb temperature at 46 m; the cable is shielded by the screen. **(c)** Dry cable and reference air temperature at 34 m; the cable is exposed to direct sunlight. **(d)** Wet cable and reference wet-bulb temperature at 34 m; the fibre optic cable is exposed to direct sunlight. **(e)** Dry cable and reference air temperature at 16 m, under the canopy so less direct sunlight hits the fibre optic cable. **(f)** Wet cable and reference wet-bulb temperature at 16 m, under the canopy so less direct sunlight hits the fibre optic cable. Shown are the linear correlation coefficients: the coefficient of determination (r^2), the slope (s) and the intercept (i).

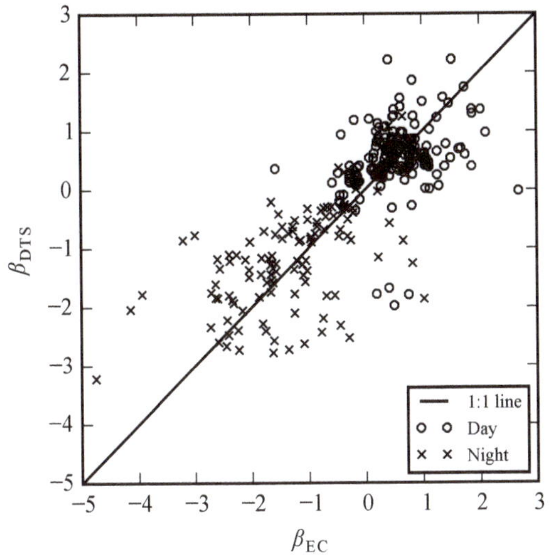

Figure 6. Correlation between the DTS-measured (β_{DTS}) and EC-measured (β_{EC}) Bowen ratios. Daytime data are between 07:00 and 18:00. Data from 10, 12–14, 19–22, and 28–30 August. $R^2 = 0.59$. RMSE = 0.81. $n = 319$ data points.

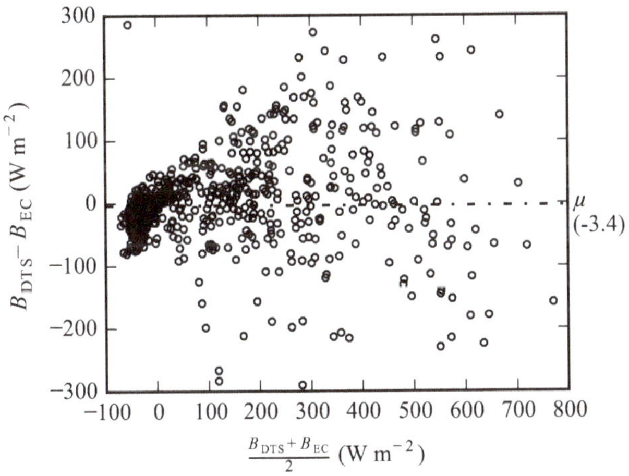

Figure 7. Tukey mean-difference plot comparing B_{DTS} and B_{EC}. With $\mu = -3.4$ W m^{-2}, RMSE = 76 W m^{-2}, $n = 741$ data points (15 min averages). Data from 10, 12–14, 19–22, and 28–30 August 2016.

To compare the two measurement methods, a Tukey mean-difference (or Bland–Altman) plot was made (Fig. 7) (Altman and Bland, 1983). The mean of the two measurement methods is plotted against the difference between them. The mean difference (μ) between B_{DTS} and B_{EC} is a 3.4 W m^{-2} underestimation by the BR-DTS method. At low fluxes (below 100 W m^{-2}), the BR-DTS method measures less energy available for fluxes compared to EC. At high fluxes (over 400 W m^{-2}) the opposite is visible. One possible reason for this is that the biomass heat flux ($\frac{dQ_B}{dt}$) was not measured,

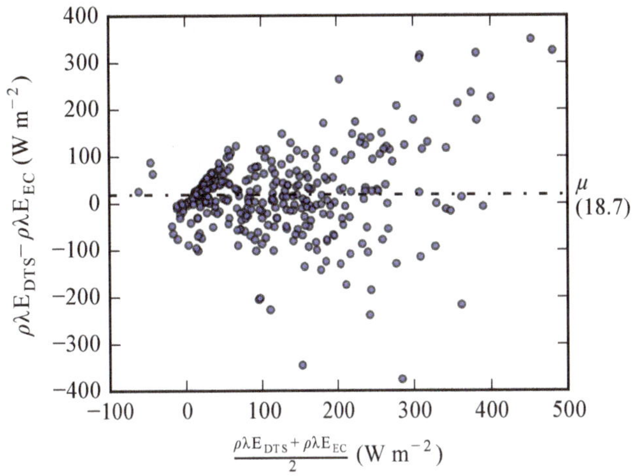

Figure 8. Tukey mean-difference plot comparing $\rho\lambda E_{EC}$ and $\rho\lambda E_{DTS}$. With $\mu = 18.7$ W m^{-2}, RMSE = 90 W m^{-2} (15 min averages). Data from 10, 12–14, 19–22, and 28–30 August 2016.

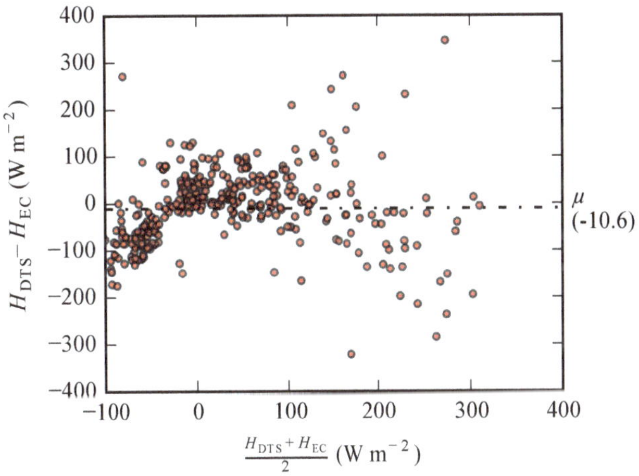

Figure 9. Tukey mean-difference plot comparing H_{EC} and H_{DTS}. With $\mu = -10.6$ W m^{-2}, RMSE = 82 W m^{-2}. (15 min averages). Data from 10, 12–14, 19–22, and 28–30 August 2016.

which causes an underestimation of the available energy in B_{DTS} during the night, and an overestimation during the day.

4.5 Energy fluxes

Figures 8 and 9 show the mean difference plots comparing the latent and sensible heat fluxes of the EC method to the BR-DTS method. The BR-DTS fluxes are calculated above the canopy, using only temperature data from the shielded cables. The Tukey mean-difference plot for the latent heat flux shows no large bias when comparing the BR-DTS method to EC, with the mean difference being a 18.7 W m^{-2} overestimation by the BR-DTS method (Fig. 8).

The Tukey mean-difference plot comparing the sensible heat flux (Fig. 9) shows a strong negative bias for negative

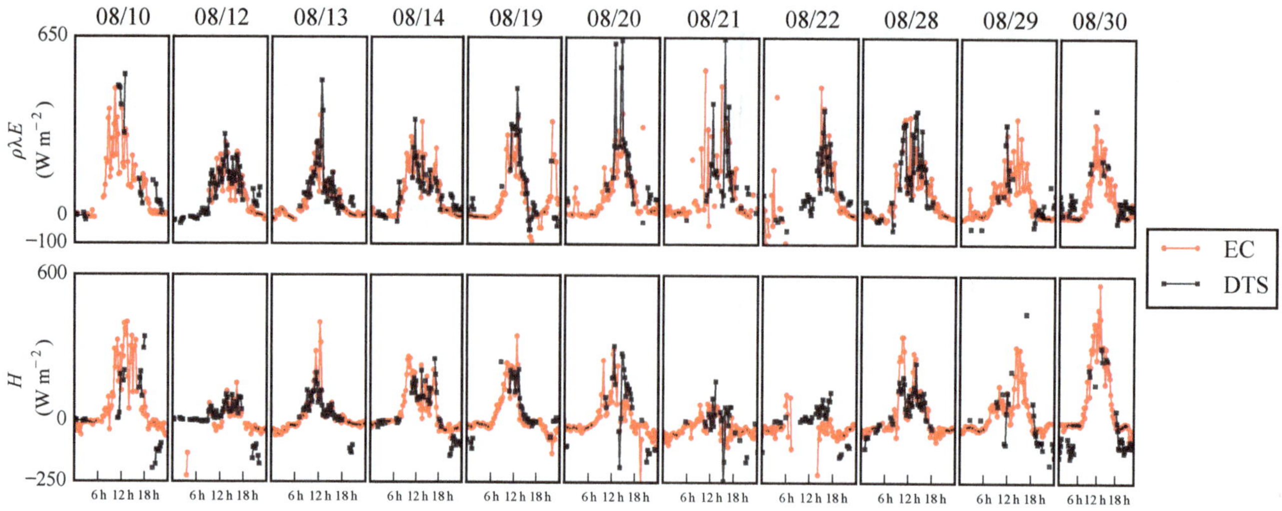

Figure 10. Plot comparing the BR-DTS and EC measured sensible (H) and latent ($\rho\lambda E$) heat fluxes over time (15 min averages). Data from 10, 12–14, 19–22, and 28–30 August 2016.

fluxes, resulting from the negative bias in the energy balance comparison (Fig. 7). At positive fluxes there seems to be a positive bias ($H_{DTS} > H_{EC}$). The mean difference is small, being a $10.6\,W\,m^{-2}$ underestimation by the BR-DTS method.

Figure 10 shows the time series of the BR-DTS and EC measured heat fluxes. The daytime flux estimates correspond well, and follow the same trends. The night-time BR-DTS estimates of the sensible heat flux are more negative than the EC estimates, one possible reason being the energy balance differences discussed before. On many days, during the early morning and start of the evening, the BR-DTS has missing values, which is mainly due to the inversion of the gradient, as the temperature gradients changes from negative (stable conditions) to positive (unstable conditions) and vice versa. This inversion causes uncertainty, which is filtered out by the quality control flags.

5 Conclusions and recommendations

This technical note investigates the use of the BR-DTS method above a forest canopy, and introduces a number of improvements on the method as presented by Euser et al. (2014). The performance is investigated by comparing the measured DTS cable temperatures to reference sensors, looking at energy balance closure, and comparing the measured Bowen ratio, sensible heat flux and latent heat flux to EC measurements.

When comparing the fibre optic cable temperature to reference sensors, it can be seen that the wet-bulb and air temperatures can be well represented. Under the canopy, where the cables are shaded from direct sunlight, the DTS cable and reference sensors are in near-perfect agreement. However, above the canopy direct sunlight may cause a large error, up to 3 K. This error can be largely mitigated by placing screens to block the sunlight, reducing the error to less than 1 K. Hence screens are effective and should also be placed in the canopy.

The Bowen ratio measured by DTS correlates well with EC estimates ($r^2 = 0.59$). A simple quality control method, using the goodness of fit of the vapour pressure gradient, also works well, and filters out most outliers and errors. The small gradients above the forest canopy are hard to measure accurately, which increases the uncertainty during days where fluxes (and thus gradients) are small. The Bowen ratio assumption that the eddy diffusivities of heat and vapour are equal was not studied, but can be a source of differences between the BR-DTS and EC methods. The difference in fetch for the two methods can also be a cause for differences.

The energy balance closure between the BR-DTS method and EC is in good agreement, with the mean difference being a $3.4\,W\,m^{-2}$ underestimation by the BR-DTS method, and an uncertainty of RMSE $= 76\,W\,m^{-2}$. However, the BR-DTS method estimates a more negative amount of available energy during night-time, and a more positive amount during daytime compared to EC. One cause could be the lack of biomass heat storage change measurements, which is of the order of $45\,W\,m^{-2}$. Another source for the difference is that the energy balance components of the BR-DTS method are generally point measurements, while EC and the Bowen ratio both have a large fetch. As a result, heterogeneity can cause large differences in the available energy for latent and sensible heat fluxes.

When comparing the latent heat flux of the two methods, they are in agreement, although the uncertainty is high (RMSE $= 90\,W\,m^{-2}$). The BR-DTS method slightly overestimates the latent heat flux, with a mean difference of $18.7\,W\,m^{-2}$. The results for the sensible heat flux are similar, with an uncertainty of RMSE = $82\,\mathrm{W\,m^{-2}}$, and the BR-DTS method underestimating the sensible heat flux by $10.6\,\mathrm{W\,m^{-2}}$. However, the underestimation mainly takes place during night-time, which can be caused by differences in available energy.

While the average profiles can be useful and valuable, extra information could be gained by opting for a smaller diameter fibre optic cable, and measuring at a high frequency (1 Hz). This could give new insights into surface interactions and could show convective cells transporting heat upwards.

A way to improve the performance of the BR-DTS method is to find an independent estimate for the sensible heat flux (H), to avoid the uncertainties in the energy balance components (R_N, $\frac{\mathrm{d}Q}{\mathrm{d}t}$). Through the universal functions of the Monin–Obukhov similarity theory estimates of the sensible heat flux can be made. This could be done either by measuring the wind speed over height (Stricker and Brutsaert, 1978) using DTS (Sayde et al., 2015) or by applying the flux-variance method (Katul et al., 1995). The Bowen ratio can then be used to calculate the latent heat flux.

Competing interests. The authors declare that they have no conflict of interest.

Acknowledgements. This research was funded by NWO Earth and Life Sciences (ALW), veni-project 863.15.022, the Netherlands. We would like to thank Murat Ucer and Christiaan van der Tol (University of Twente) for providing us with access to their measurement location and reference data, and the students Tara van Iersel and Tom Oostdijk for assisting with setting up the DTS measurements.

Edited by: Pierre Gentine

References

Allen, R. G., Pereira, L. S., Raes, D., and Smith, M.: FAO Irrig. Drain. Paper No. 56, Irrig. Drain., 300, 300 pp., http://www.kimberly.uidaho.edu/water/fao56/fao56.pdf, 1998.

Altman, D. G. and Bland, J. M.: Measurement in Medicine: The Analysis of Method Comparison Studies, J. R. Stat. Soc. D-Sta., 32, 307–317, https://doi.org/10.2307/2987937, 1983.

Angus, D. E. and Watts, P. J.: Evapotranspiration – How good is the Bowen ratio method?, Agr. Water Manage., 8, 133–150, https://doi.org/10.1016/0378-3774(84)90050-7, 1984.

Bao, X. and Chen, L.: Recent progress in distributed fiber optic sensors, Sensors (Basel), 12, 8601–8639, https://doi.org/10.3390/s120708601, 2012.

Barr, A. G., King, K. M., Gillespie, T. J., Den Hartog, G., and Neumann, H. H.: A comparison of bowen ratio and eddy correlation sensible and latent heat flux measurements above deciduous forest, Bound.-Lay. Meteorol., 71, 21–41, https://doi.org/10.1007/BF00709218, 1994.

Bense, V., Read, T., and Verhoef, A.: Using distributed temperature sensing to monitor field scale dynamics of ground surface temperature and related substrate heat flux, Agr. Forest Meteorol., 220, 207–215, https://doi.org/10.1016/j.agrformet.2016.01.138, 2016.

Blume, T., Krause, S., Meinikmann, K., and Lewandowski, J.: Upscaling lacustrine groundwater discharge rates by fiber-optic distributed temperature sensing, Water Resour. Res., 49, 7929–7944, https://doi.org/10.1002/2012WR013215, 2013.

Bosveld, F. and Bouten, W.: Evaluation of transpiration models with observations over a Douglas-fir forest, Agr. Forest Meteorol., 108, 247–264, https://doi.org/10.1016/S0168-1923(01)00251-9, 2001.

Bowen, I. S.: The ratio of heat losses by conduction and by evaporation from any water surface, Phys. Rev., 27, 779–787, https://doi.org/10.1103/PhysRev.27.779, 1926.

de Jong, S. A. P., Slingerland, J. D., and van de Giesen, N. C.: Fiber optic distributed temperature sensing for the determination of air temperature, Atmos. Meas. Tech., 8, 335–339, https://doi.org/10.5194/amt-8-335-2015, 2015.

Erisman, J., Draaijers, G., Steingröver, E., Van Dijk, H., Boxman, A., and De Vries, W.: Assessment of the exposure and loads of acidifying and eutrophying pollutants and ozone, as well as their harmful influence on the vitality of the trees and the Speulder forest ecosystem as a whole, Water Air Soil Poll., 105, 539–571, 1998.

Euser, T., Luxemburg, W. M. J., Everson, C. S., Mengistu, M. G., Clulow, A. D., and Bastiaanssen, W. G. M.: A new method to measure Bowen ratios using high-resolution vertical dry and wet bulb temperature profiles, Hydrol. Earth Syst. Sci., 18, 2021–2032, https://doi.org/10.5194/hess-18-2021-2014, 2014.

Fuchs, M. and Tanner, C.: Error analysis of bowen ratios measured by differential psychrometry, Agr. Meteorol., 7, 329–334, https://doi.org/10.1016/0002-1571(70)90027-0, 1970.

Harrison, R. G. and Wood, C. R.: Ventilation effects on humidity measurements in thermometer screens, Q. J. Roy. Meteor. Soc., 138, 1114–1120, https://doi.org/10.1002/qj.985, 2012.

Irmak, S., Kilic, A., and Chatterjee, S.: On the Equality Assumption of Latent and Sensible Heat Energy Transfer Coefficients of the Bowen Ratio Theory for Evapotranspiration Estimations: Another Look at the Potential Causes of Inequalities, Climate, 2, 181–205, https://doi.org/10.3390/cli2030181, 2014.

Jansen, J., Stive, P. M., van de Giesen, N., Tyler, S., Steele-Dunne, S. C., and Williamson, L.: Estimating soil heat flux using Distributed Temperature Sensing, GRACE, Remote Sensing and Ground-based Methods in Multi-Scale Hydrology, 140–144, 2011.

Katul, G., Goltz, S. M., Hsieh, C. I., Cheng, Y., Mowry, F., and Sigmon, J.: Estimation of surface heat and momentum fluxes using the flux-variance method above uniform and non-uniform terrain, Bound.-Lay. Meteorol., 74, 237–260, https://doi.org/10.1007/BF00712120, 1995.

Koutsoyiannis, D.: Clausius-Clapeyron equation and saturation vapour pressure: simple theory reconciled with practice, Eur. J. Phys., 33, 295–305, https://doi.org/10.1088/0143-0807/33/2/295, 2012.

LI-COR Inc.: EddyPro® Version 6.2 [Computer Software], 2016.

Mauder, M. and Foken, T.: Impact of post-field data processing on eddy covariance flux estimates and energy balance closure, Meteorol. Z., 15, 597–609, https://doi.org/10.1127/0941-2948/2006/0167, 2006.

Michiles, A. A. and Gielow, R.: Above-ground thermal energy storage rates, trunk heat fluxes and surface energy balance in a central Amazonian rainforest, Agr. Forest Meteorol., 148, 917–930, https://doi.org/10.1016/j.agrformet.2008.01.001, 2008.

Oliphant, A. J., Grimmond, C. S. B., Zutter, H. N., Schmid, H. P., Su, H. B., Scott, S. L., Offerle, B., Randolph, J. C., and

Ehman, J.: Heat storage and energy balance fluxes for a temperate deciduous forest, Agr. Forest Meteorol., 126, 185–201, https://doi.org/10.1016/j.agrformet.2004.07.003, 2004.

Pal Arya, S.: Introduction to Micrometeorology, Academic Press, San Diego, California, 1988.

Payero, J. O., Neale, C. M. U., Wright, J. L., and Allen, R. G.: Guidelines for validating Bowen ratio data, T. ASAE, 46, 1051, https://doi.org/10.13031/2013.13967, 2003.

Raj, R., Hamm, N. A., van der Tol, C., and Stein, A.: Variance-based sensitivity analysis of BIOME-BGC for gross and net primary production, Ecol. Model., 292, 26–36, 2014.

Sayde, C., Thomas, C. K., Wagner, J., and Selker, J.: High-resolution wind speed measurements using actively heated fiber optics, Geophys. Res. Lett., 42, 10064–10073, https://doi.org/10.1002/2015GL066729, 2015.

Schilperoort, B., Cisneros Vaca, C. R., and Jiménez Rodríguez, C. D.: Speulderbos flux, storage, and temperature profile measurements, https://doi.org/10.4121/uuid:5c81f10a-1249-4b85-8dec-2b029dd88b99, 2017.

Selker, J. S., Thévenaz, L., Huwald, H., Mallet, A., Luxemburg, W., Van De Giesen, N., Stejskal, M., Zeman, J., Westhoff, M., and Parlange, M. B.: Distributed fiber-optic temperature sensing for hydrologic systems, Water Resour. Res., 42, 1–8, https://doi.org/10.1029/2006WR005326, 2006.

Silixa Ltd: ULTIMA® DTS, available at: http://silixa.com/technology/ultima-dts/, last access: 26 June 2017.

Sluijter, R.: De Bosatlas van het klimaat, Noordhoff Uitgevers, Groningen, 2011.

Stannard, D. I.: A Theoretically Based Determination of Bowen-ratio Fetch Requirements, Bound.-Lay. Meteorol., 83, 375–406, https://doi.org/10.1023/A:1000286829849, 1997.

Steele-Dunne, S. C., Rutten, M. M., Krzeminska, D. M., Hausner, M., Tyler, S. W., Selker, J., Bogaard, T. A., and Van De Giesen, N. C.: Feasibility of soil moisture estimation using passive distributed temperature sensing, Water Resour. Res., 46, 1–12, https://doi.org/10.1029/2009WR008272, 2010.

Stricker, H. and Brutsaert, W.: Actual evapotranspiration over a summer period in the "Hupsel catchment", J. Hydrol., 39, 139–157, https://doi.org/10.1016/0022-1694(78)90119-1, 1978.

Stull, R.: Practical Meteorology – An Algebra-based Survey of Atmospheric Science, Department of Earth, Ocean & Atmospheric Sciences, University of British Columbia, Vancouver, BC, https://doi.org/10.14288/1.0300441, 2015.

Su, Z., Timmermans, W. J., van der Tol, C., Dost, R., Bianchi, R., Gómez, J. A., House, A., Hajnsek, I., Menenti, M., Magliulo, V., Esposito, M., Haarbrink, R., Bosveld, F., Rothe, R., Baltink, H. K., Vekerdy, Z., Sobrino, J. A., Timmermans, J., van Laake, P., Salama, S., van der Kwast, H., Claassen, E., Stolk, A., Jia, L., Moors, E., Hartogensis, O., and Gillespie, A.: EAGLE 2006 – Multi-purpose, multi-angle and multi-sensor in-situ and airborne campaigns over grassland and forest, Hydrol. Earth Syst. Sci., 13, 833–845, https://doi.org/10.5194/hess-13-833-2009, 2009.

Thomas, C. K., Kennedy, A. M., Selker, J. S., Moretti, A., Schroth, M. H., Smoot, A. R., Tufillaro, N. B., and Zeeman, M. J.: High-Resolution Fibre-Optic Temperature Sensing: A New Tool to Study the Two-Dimensional Structure of Atmospheric Surface-Layer Flow, Bound.-Lay. Meteorol., 142, 177–192, https://doi.org/10.1007/s10546-011-9672-7, 2012.

Tietema, A., Mol-Dijkstra, J. P., Kros, J., and De Vries, W.: Dynamic nitrogen deposition thresholds during forest stand development in a Douglas fir forest analysed with two nitrogen models SMART2 and MERLIN, Hydrol. Earth Syst. Sci., 6, 375–382, https://doi.org/10.5194/hess-6-375-2002, 2002.

Tiktak, A. and Bouten, W.: Soil water dynamics and long-term water balances of a Douglas fir stand in the Netherlands, J. Hydrol., 156, 265–283, 1994.

van der Tol, C.: Validation of remote sensing of bare soil ground heat flux, Remote Sens. Environ., 121, 275–286, https://doi.org/10.1016/j.rse.2012.02.009, 2012.

Van Wijk, M., Dekker, S., Bouten, W., Bosveld, F., Kohsiek, W., Kramer, K., and Mohren, G.: Modeling daily gas exchange of a Douglas-fir forest: comparison of three stomatal conductance models with and without a soil water stress function, Tree Physiol., 20, 115–122, 2000.

Weligepolage, K., Gieske, A., and Su, Z.: Surface roughness analysis of a conifer forest canopy with airborne and terrestrial laser scanning techniques, Int. J. Appl. Earth Obs., 14, 192–203, https://doi.org/10.1016/j.jag.2011.08.014, 2012.

Weligepolage, K., Gieske, A., and Su, Z.: Effect of spatial resolution on estimating surface albedo: A case study in Speulderbos forest in The Netherlands, Int. J. Appl. Earth Obs., 23, 18–28, 2013.

Westhoff, M. C., Savenije, H. H. G., Luxemburg, W. M. J., Stelling, G. S., van de Giesen, N. C., Selker, J. S., Pfister, L., and Uhlenbrook, S.: A distributed stream temperature model using high resolution temperature observations, Hydrol. Earth Syst. Sci., 11, 1469–1480, https://doi.org/10.5194/hess-11-1469-2007, 2007.

Wilkes, P., Lau, A., Disney, M., Calders, K., Burt, A., de Tanago, J. G., Bartholomeus, H., Brede, B., and Herold, M.: Data acquisition considerations for Terrestrial Laser Scanning of forest plots, Remote Sens. Environ., 196, 140–153, https://doi.org/10.1016/j.rse.2017.04.030, 2017.

Wilson, K.: Energy balance closure at FLUXNET sites, Agr. Forest Meteorol., 113, 223–243, https://doi.org/10.1016/S0168-1923(02)00109-0, 2002.

7

An improved Grassberger–Procaccia algorithm for analysis of climate system complexity

Chongli Di[1], **Tiejun Wang**[1], **Xiaohua Yang**[2], **and Siliang Li**[1]

[1]Institute of Surface-Earth System Science, Tianjin University, Tianjin, 300072, P. R. China
[2]State Key Laboratory of Water Environment Simulation, School of Environment, Beijing Normal University, Beijing, 100875, P. R. China

Correspondence: Tiejun Wang (tiejun.wang@tju.edu.cn)

Abstract. Understanding the complexity of natural systems, such as climate systems, is critical for various research and application purposes. A range of techniques have been developed to quantify system complexity, among which the Grassberger–Procaccia (G-P) algorithm has been used the most. However, the use of this method is still not adaptive and the choice of scaling regions relies heavily on subjective criteria. To this end, an improved G-P algorithm was proposed, which integrated the normal-based K-means clustering technique and random sample consensus (RANSAC) algorithm for computing correlation dimensions. To test its effectiveness for computing correlation dimensions, the proposed algorithm was compared with traditional methods using the classical Lorenz and Henon chaotic systems. The results revealed that the new method outperformed traditional algorithms in computing correlation dimensions for both chaotic systems, demonstrating the improvement made by the new method. Based on the new algorithm, the complexity of precipitation, and air temperature in the Hai River basin (HRB) in northeastern China was further evaluated. The results showed that there existed considerable regional differences in the complexity of both climatic variables across the HRB. Specifically, precipitation was shown to become progressively more complex from the mountainous area in the northwest to the plain area in the southeast, whereas the complexity of air temperature exhibited an opposite trend, with less complexity in the plain area. Overall, the spatial patterns of the complexity of precipitation and air temperature reflected the influence of the dominant climate system in the region.

1 Introduction

There are increasing interests in understanding system complexity, ranging from natural phenomena to social behaviors (Bras, 2015; Lin et al., 2015; Wang et al., 2016). As an open system with random external forcings and nonlinear dissipation, climate systems are highly complex (Nicolis and Nicolis, 1984; Jayawardena and Lai, 1994; Rind, 1999; Wang et al., 2015). Owing to nonlinear interactions among atmosphere, hydrosphere, and biosphere, climatic variables exhibit highly nonlinear and dynamic characteristics, which reflect the complexity of climate systems (Palmer, 1999; Rial et al., 2004; Sivakumar, 2005; Wu et al., 2010). It is thus imperative to quantitatively measure the complexity of climatic variables for understanding underlying processes. However, no common definition of system complexity exists in scientific communities, particularly from a mathematical perspective (Carbone et al., 2016). To resolve this issue, numerous concepts and methods, including chaos theory, wavelet analysis, and dynamical analysis, have been proposed to describe the complexity of climate systems (Lorenz, 1963; Di et al., 2014; Feldhoff et al., 2015; Sivakumar, 2017; Meseguer-Ruiz et al., 2017). For instance, the chaos theory has been extensively used to characterize the chaotic and nonlinear features of climate systems (Sivakumar, 2001). Overall, previous studies based on the chaos theory revealed that the time series of air temperature and precipitation is nonstationary with abundant information. The complexity of rainfall and temperature dynamics has been widely used to indicate the extent of the complexity of climate systems (Dhanya and Kumar, 2010; Gan et al., 2002).

One of the important parameters in the chaos theory is correlation dimension, which can be used to measure the complexity and chaotic properties of variables, including precipitation and streamflow (Sivakumar et al., 2002; Dhanya and Kumar, 2011; Kyoung et al., 2011; Lana et al., 2016). Conceptually, the correlation dimension of a variable indicates the number of primary controls of the variable and thus determines the degree of freedom of the underlying process (Sivakumar and Singh, 2012). Despite the wide applications in various scientific fields, the use of the correlation dimension method is still hindered by certain limitations. For instance, the dimension method proposed by Grassberger and Procaccia (1983b) (denoted as the G-P method hereafter) is commonly used in the fields of hydrology and atmospheric science; however, its calculation procedures are still problematic (Ji et al., 2011). Specifically, the G-P method utilizes phase space reconstruction (Packard et al., 1980) and the embedding theorem (Takens, 1981) to compute correlation dimensions, which requires selection of an appropriate scaling region. The scaling region is a domain, over which an object exhibits self-similarity across a range of scales. However, the G-P method relies on visual inspections for choosing scaling regions, which is subject to human errors (Sprott and Rowlands, 2001). To tackle this problem, alternative methods have been developed to improve the original G-P method (Maragos and Sun, 1993). For example, Jothiprakash and Fathima (2013) utilized empirical equations to calculate the upper limit of scaling regions. Ji et al. (2011) applied the clustering analysis technique to determine scaling regions. However, these existing methods for identifying scaling regions are still not adaptive and the choice of scaling regions relies heavily on subjective criteria, and the use of the least squares method for fitting straight lines to determine correlation exponents can include outliers (Cantrell, 2008) and thus is not optimal. Therefore, studies are still warranted to seek more objective and adaptive algorithms for identifying scaling regions to obtain more accurate estimates of correlation dimensions.

The primary aims of this study were twofold. First, a new algorithm was proposed to improve the original G-P method, which integrated the methods of normal estimation, K-means clustering (Lloyd, 1982) and random sample consensus (RANSAC; Fischler and Bolles, 1981). The classical Lorenz and Henon chaotic systems were chosen to test the effectiveness of the proposed algorithm for estimating correlation dimensions. Afterwards, the newly developed algorithm was utilized to investigate the nonlinear characteristics of precipitation and air temperature across the Hai River basin (HRB) in northeastern China. The HRB has been facing serious water shortage issues due to climate change and increasing water demand. Although previous studies have investigated climate variability (e.g., precipitation, air temperature, and evaporation) in the HRB from different perspectives (Bao et al., 2012; Sang et al., 2012; Chu et al., 2010a), to our best knowledge, there have been no attempts made so far to quantify the nonlinear characteristics of climatic variables, especially regarding their chaotic behaviors in the HRB, which is essential for understanding the nonlinearity of the climate system in the region. Furthermore, the HRB is a diverse hydroclimatic region with many subwatersheds of varying geographical and hydroclimatic conditions, which makes the region ideal for understanding the climate system complexity. The rest of this paper is organized as follows: Sect. 2 describes the calculation procedures of the proposed algorithm, which is then tested using classical mathematical models in Sect. 3. Section 4 describes the data obtained from the HRB and presents the results and analysis. Conclusions are made in the last part of this paper.

2 Methodology

2.1 Algorithm for computing correlation dimension

Correlation dimensions can be used to identify the complexity of dynamical systems with varying complexity degrees (e.g., low-dimensional vs. high-dimensional systems). A wealth of algorithms have been developed for computing correlation dimensions, among which the G-P algorithm has been used most and is also adopted in this study. The G-P algorithm uses the concept of phase space reconstruction (Packard et al., 1980) from a single-variable time series. Here, the method of delays (Takens, 1981) was employed for reconstructing phase space. Given a time series $\boldsymbol{X}_i$ ($i = 1, 2, \ldots, N$), a multi-dimensional phase space can be reconstructed as follows:

$$\boldsymbol{Y}_j = (\boldsymbol{X}_j, \boldsymbol{X}_{j+\tau}, \boldsymbol{X}_{j+2\tau}, \cdots, \boldsymbol{X}_{j+(m-1)\tau}), \tag{1}$$

where $j = 1, 2, \cdots, N-(m-1)\tau$, m is the dimension of $\boldsymbol{Y}_j$ called embedding dimension, τ is delay time, and $\boldsymbol{X}_j$ is the reconstructed phase space vector.

For the m-dimensional reconstructed phase space, the correlation function $C(r, m)$ is defined as follows:

$$C(r, m) = \lim_{N\to\infty} \frac{2}{N(N-1)} \sum_{i,j=1}^{N} H(r - \|\boldsymbol{Y}_i - \boldsymbol{Y}_j\|),$$
$$1 \le i \le j \le N, \tag{2}$$

where $\|\boldsymbol{Y}_i - \boldsymbol{Y}_j\|$ is the Euclidean distance between the vectors $\boldsymbol{Y}_i$ and $\boldsymbol{Y}_j$. $H(x)$ is the Heaviside function with $H(x) = 1$ for $x>0$ and $H(x) = 0$ for $x \le 0$, where $x = r - \|\boldsymbol{Y}_i - \boldsymbol{Y}_j\|$ and r is the vector norm (i.e., radius of a sphere) centered on $\boldsymbol{Y}_i$ or $\boldsymbol{Y}_j$; $r_{\min}$ and $r_{\max}$ were set as the minimum and maximum distances between points, respectively (Ji et al., 2011; Lai and Lerner, 1998). If $r \le r_{\min}$, none of the vector points falls within the volume element and $C(r, m) = 0$. Otherwise, if $r \ge r_{\max}$, all vector points fall within the volume element and $C(r, m) = 1$. If there exists an attractor in the reconstructed system, $C(r, m)$ and r are related through the following relationship:

$$C(r, m) \underset{\substack{r\to 0\\ N\to\infty}}{\approx} \alpha r^{D_2(m)}, \tag{3}$$

where α is a constant and $D_2(m)$ is the correlation exponent.

$D_2(m)$ is usually estimated using the least squares method by fitting a straight line through $\ln r$ vs. $\ln C(r, m)$. According to the relationship between $D_2(m)$ and m, the saturation value of $D_2(m)$ is defined as the correlation dimension. If the saturation value is low (e.g., a low correlation dimension), the system is considered to exhibit low-dimensional deterministic dynamics (i.e., a chaotic system); otherwise, the system is a stochastic one. The range over which the straight line is fitted through $\ln r$ vs. $\ln C(r, m)$ is called the scaling region, where the slope is defined. Clearly, choosing an appropriate scaling region is critical for computing correlation dimensions. In previous studies, scaling regions are usually determined by visual inspections, and this will be prone to individual preferences and thus not objective. Therefore, an objective method with adaptive procedures for computing correlation dimensions is still desired.

2.2 Scaling region identification

To overcome the limitation of the original G-P algorithm for selecting scaling regions, we propose an adaptive identification algorithm of scaling regions, which utilizes the normal-based K-means clustering technique and the RANSAC algorithm. The use of the normal-based K-means clustering technique is to partition all normals of the scatter points into K clusters with high similarity and to remove the points that are outside of the range of the scaling region. The RANSAC algorithm was introduced to fit a straight line through the log-transformed points obtained by the normal-based K-means clustering technique, which had been shown to outperform the traditional least squares method for fitting straight lines (Kyoung, 2011; Ji et al., 2011). To illustrate the advantages of using the RANSAC algorithm for linear fitting, a hypothetical example is shown in Fig. 1, which compares the fitting results obtained from the RANSAC algorithm and the traditional least squares method. The input data are sampled from a line $y = 0.5x$, with added noises and outliers. Here, for the RANSAC algorithm, the inliers are the points used to fit the line, whereas the outliers are removed from the line fitting. It can be seen from Fig. 1 that the fitting line ($y = 0.60x - 0.068$; $R^2 = 0.854$) obtained from the least squares method is seriously affected by outliers and deviated from the original line $y = 0.5x$. By contrast, the RANSAC method is able to distinguish the inliers from outliers effectively and results in a satisfactory fitting line ($y = 0.49x + 0.007$; R^2 =0.990), demonstrating the advantage of using the RANSAC algorithm for linear fitting.

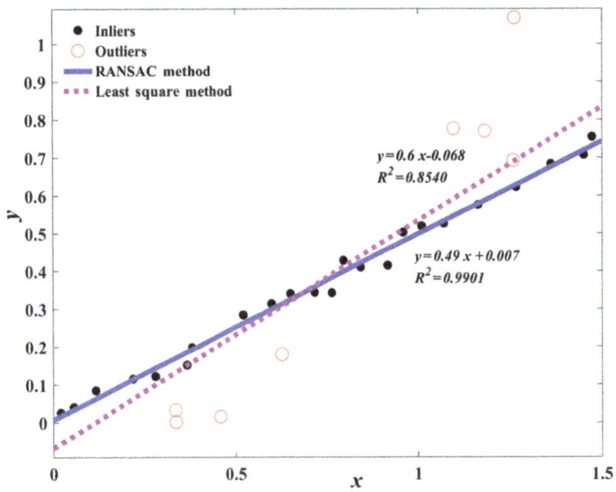

Figure 1. Comparison of the fitted lines obtained from the RANSAC algorithm and the least squares method.

The flow chart of the proposed procedures for calculating correlation dimensions is given in Fig. 2, which consists of five major steps:

1. For the time series $x(t)$, the time delay τ is computed by an autocorrelation function (Liebert and Schuster, 1989). Then the minimum embedding dimension $m_{\min} = 2$ was set and the phase space was reconstructed by increasing m to obtain the correlation exponent function $C(r, m)$.

2. The normals of the scatter points on the $\ln r \sim \ln C(r, m)$ line are estimated via principal component analysis (Mitra et al., 2004).

3. The K-means clustering technique is performed on the normal set N with $K = 2$ to obtain two different clusters. Set a threshold value T to determine the angle α between the two clusters. If $\alpha > T$, the data set with larger differences in normals is discarded. Then, the K-means clustering technique is repeated on the remaining data set until $\alpha \leq T$.

4. The RANSAC algorithm is used to fit a straight line through the set of remaining scatter points.

5. The slope of the line obtained from the RANSAC method is computed to acquire the correlation dimension $D_2(m)$ for each m. Finally, the saturation correlation dimension is determined using the plot $D_2(m)$ vs. m.

3 Verification of the proposed algorithm

To test the effectiveness of the proposed algorithm, the classical chaotic models of Lorenz (1963) in Eq. (4) and

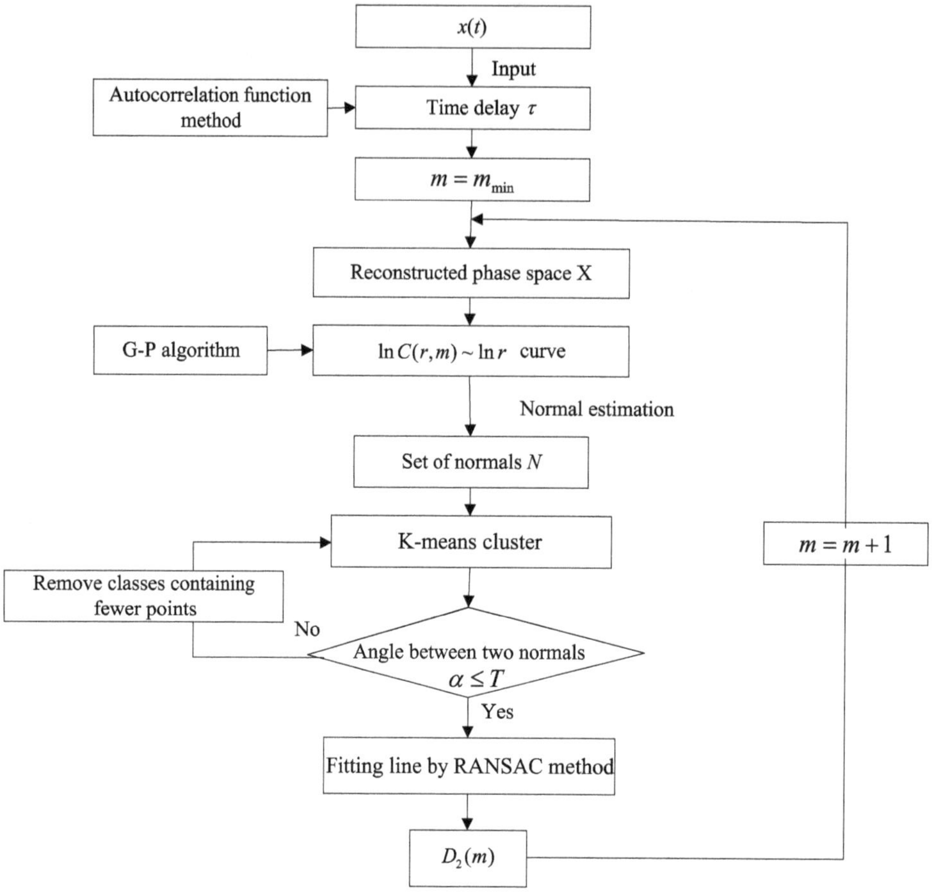

Figure 2. Flow chart of the proposed algorithm for computing correlation dimensions (the details are listed in the text).

Henon (1976) in Eq. (5) were used. The Lorenz and Henon systems with existing theoretical correlation dimensions have been studied the most in the past and thus widely used to analyze the chaotic behavior in climate systems and to test the effectiveness of algorithms for computing climate system complexity (e.g., Grassberger and Procaccia, 1983a; Lai and Lerner, 1998; Ji et al., 2011).

$$\dot{x} = \sigma(-x + y),\ \dot{y} = -xz + rx - y,\ \dot{z} = xy - bz, \tag{4}$$

$$x_{n+1} = y_n + 1 - ax_n^2,\ y_{n+1} = bx_n\ , \tag{5}$$

where $\sigma = 10$, $b = 28$, and $r = 8/3$ in Eq. (4), and $a = 1.4$ and $b = 0.3$ in Eq. (5). The theoretical dimensions of the Lorenz and the Henon systems are 2.05 ± 0.01 and 1.25 ± 0.02, respectively (Grassberger and Procaccia, 1983a). As a comparison, the results obtained by our proposed method were compared with the theoretical dimensions and the values obtained by another two commonly used algorithms, including the intuitive judgment method (IJM) and the point-based K-means clustering method (PKC).

According to the autocorrelation function, the time delay τ was determined to be 10 for the Lorenz system, with m varying from 2 to 20. Figure 3a shows the relationship between $\ln C(r, m)$ and $\ln r$, with m ranging from 2 to 20. Figure 3b shows the slopes of $\ln C(r, m)$ against $\ln r$ by increasing the embedding dimension m (i.e., the bottom curves are associated with smaller m values in Fig. 3b). The threshold value T was set as 5° and K was set as 2. The scaling regions of the curves in Fig. 3a were determined using the normal-based K-means clustering technique. As an example, an arbitrary curve was first selected from Fig. 3a, and the results are presented in Fig. 4. It can be seen from Fig. 4 that the process of the proposed method for determining the scaling region is adaptive. Specifically, for the selected curve shown in Fig. 4a, the normals of the curve were first computed based on Step 2 and the results are plotted in Fig. 4b. Different from previous K-means methods (e.g., the point-based K-means clustering method), we measured the similarity of points using the diversity between normals of different points. The reason for using the normal-based method is that the directions of normals for different points may vary considerably (see Fig. 4b), whereas for the point-based K-means method, the distance between different points might be small, making it difficult to separate the points into different clusters (Fig. 3a). The obtained two separate clusters of the normals (in red and blue)

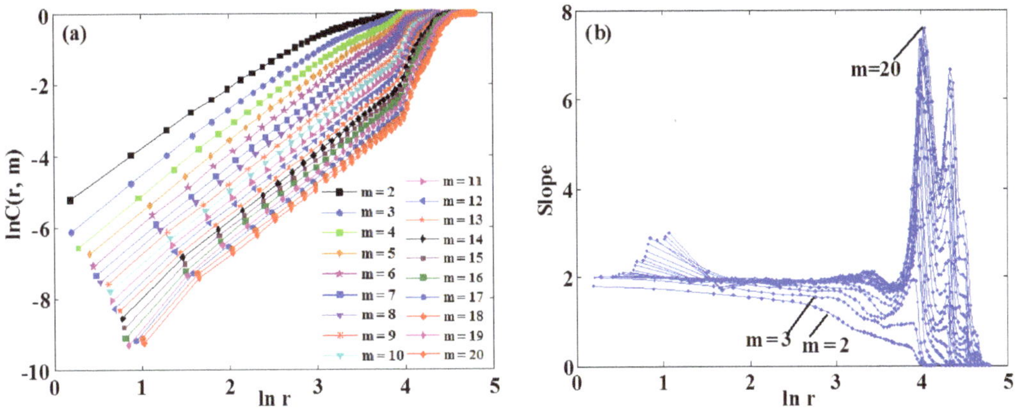

Figure 3. Correlation integral as a function of r, with embedding dimension m ranging from 2 to 20 for the Lorenz attractor: **(a)** $\ln C(r, m)$ vs. $\ln r$, **(b)** the slopes of $\ln C(r, m)$ vs. $\ln r$.

are shown in Fig. 4c. If the angle α between the two clusters was larger than T, the one with larger differences in normals was discarded. Then, the K-means clustering technique was performed again on the remaining data set. This process was usually repeated for 2–3 times until $\alpha \leq T$ (e.g., Fig. 4c to e). The final scaling region was determined as shown in Fig. 4f.

Figure 5a shows the final fitted lines through the scaling regions using the RANSAC method. The slope of the fitted line is the correlation dimension for each corresponding m. Figure 5b presents the graph of $D_2(m)$ against m with the value of m varying from 2 to 20. From Fig. 5b, we can see that $D_2(m)$ was saturated when $m>5$ with the saturation value approximately equal to 2.054, which was comparable to the theoretical value of the correlation dimension for the Lorenz attractor (i.e., 2.05 ± 0.01). Following the same procedures, the obtained correlation dimension for the Henon attractor was 1.243, which was also close to its theoretical value (i.e., 1.25 ± 0.02).

To verify the accuracy of our algorithm for computing correlation dimensions, the results derived from the proposed algorithm were compared with the ones obtained from the IJM and PKC methods. The IJM method was based on visual inspections to determine scaling regions (Jothiprakash and Fathima, 2013), while the PKC method integrated the K-means algorithm and the point-slope-error technique to determine scaling regions (Ji et al., 2011). The obtained correlation dimensions are reported in Table 1. For the Lorenz system, the differences in the correlation dimensions between the theoretical value and the ones obtained from IJM and PKC were 0.18 ± 0.01 and 0.014 ± 0.01, respectively, whereas the difference was much smaller for the newly proposed algorithm (i.e., 0.004 ± 0.01). Similar conclusions can be also made for the Henon system, demonstrating the improved performance of the proposed algorithm for determining correlation dimensions. It should be stressed that despite the improvement made by our proposed algorithm, further studies are still needed to address the issues in the computation of correlation dimensions. For example, estimation of correlation dimensions is partly dependent on the proper selection of time delay and embedding dimension; therefore, the impacts of their uncertainties should be further assessed.

Table 1. Comparison of the correlation dimensions derived from different methods. TCD: theoretical correlation dimension; IJM: intuitive judgment method; PKC: point-based K-means clustering; NPA: newly proposed algorithm.

Attractor	TCD	IJM	PKC	NPA
Lorenz	2.05 ± 0.01	2.23 ± 0.02	2.064	2.054
Henon	1.25 ± 0.02	1.354 ± 0.02	1.240	1.243

4 Application, results, and analysis

The correlation dimension method is an important diagnostic tool for understanding the complexity of natural systems with chaotic characteristics. In this section, a case study is presented to illustrate the use of the newly developed algorithm for studying the complexity of climate systems. Specifically, the algorithm was first utilized to compute the correlation dimensions of precipitation and air temperature using time series obtained from the HRB. Afterwards, the regional patterns of correlation dimensions for precipitation and air temperature in the HRB were analyzed.

4.1 Study area and data

The HRB is located in northeastern China (112–120° E, 35–43° N; Fig. 6), which hosts one of the most important economic zones in China (White et al., 2015). Under the influences of climate change and human activities, complex water issues have become increasingly prominent in the HRB (Liu and Xia, 2004). Topography varies considerably across the

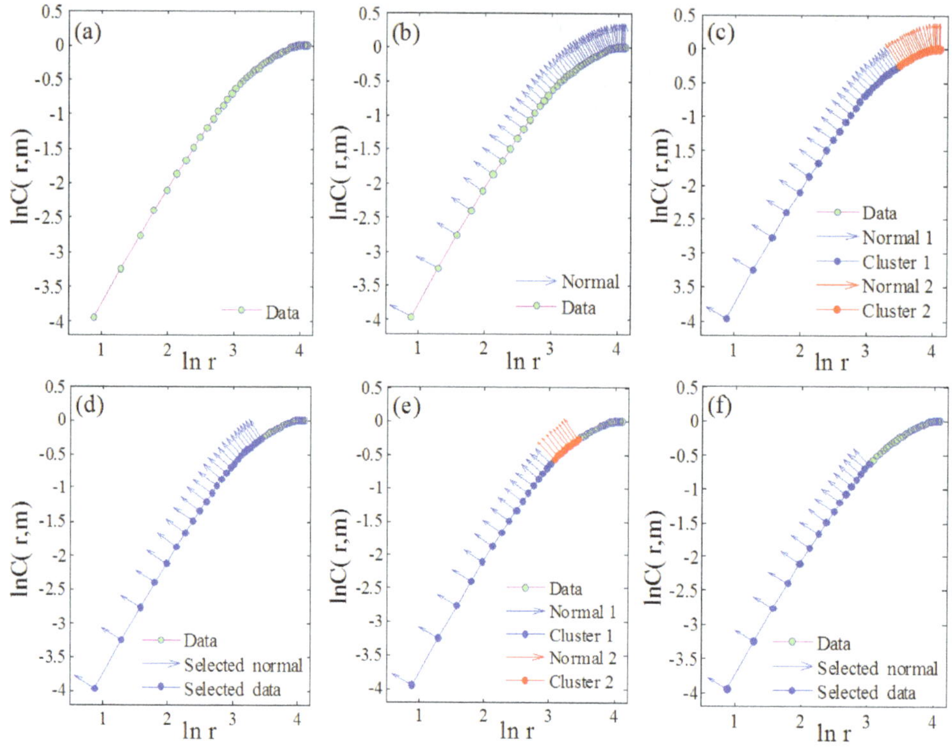

Figure 4. Illustration of using the normal-based K-means clustering technique for determining the scaling region. The curve shown here was randomly selected from Fig. 3.

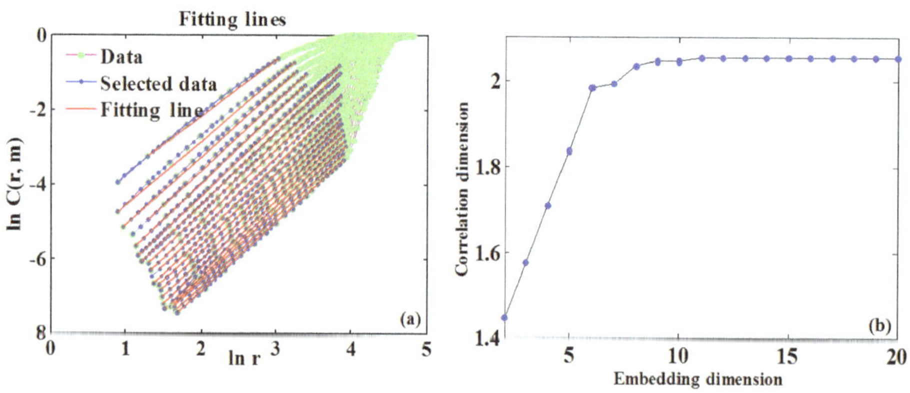

Figure 5. The final fitted lines and the correlation dimension of the Lorenz system: **(a)** the final fitted lines through the scaling regions and **(b)** correlation dimensions as a function of embedding dimension.

area, with 22 % of the total area for mountains in the western and northern parts, 40 % for plains in the eastern and southern parts, and 38 % for hilly areas in the central part. The regional climate in the HRB is of a semiarid or subhumid type, with mean annual precipitation of 539.0 mm year^{-1} and mean annual temperature of 10.2 °C. Mean annual precipitation increases from the mountainous areas in the west to the plains in the east, while mean annual temperature decreases from south to north. In addition, precipitation in the HRB exhibits significant interdecadal and interannual variations. To apply the proposed algorithm for computing correlation dimensions, monthly precipitation and air temperature data spanning from 1951 to 2016 were retrieved from 40 meteorological stations in the HRB and nearby areas (Fig. 6), which were operated by the China Meteorological Administration http://data.cma.cn/site/index.html, last access: 28 September 2018).

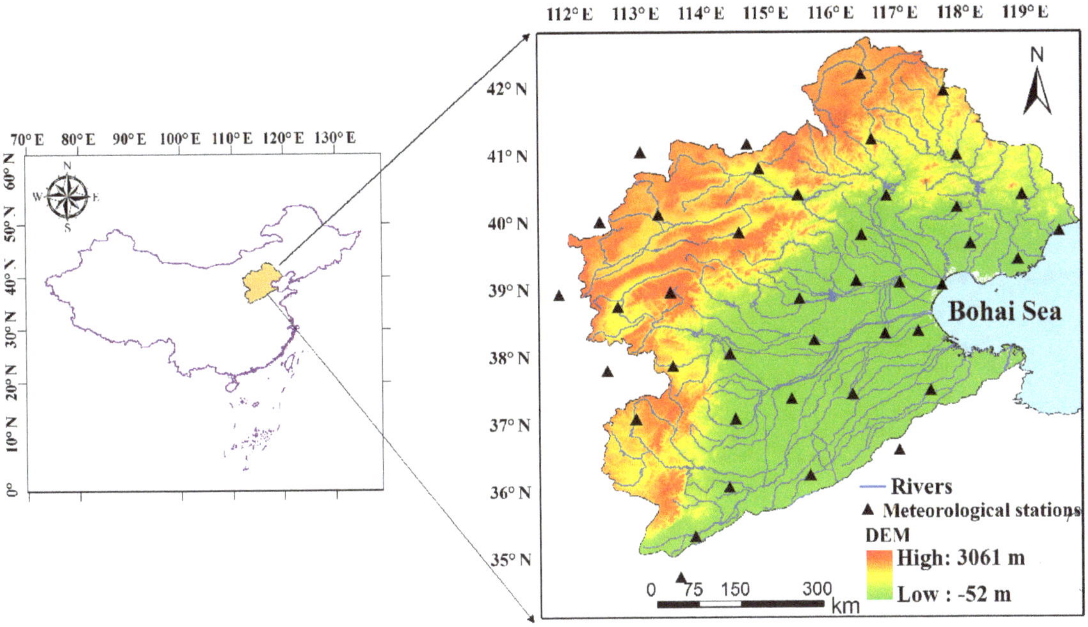

Figure 6. Locations of meteorological stations in the Hai River basin.

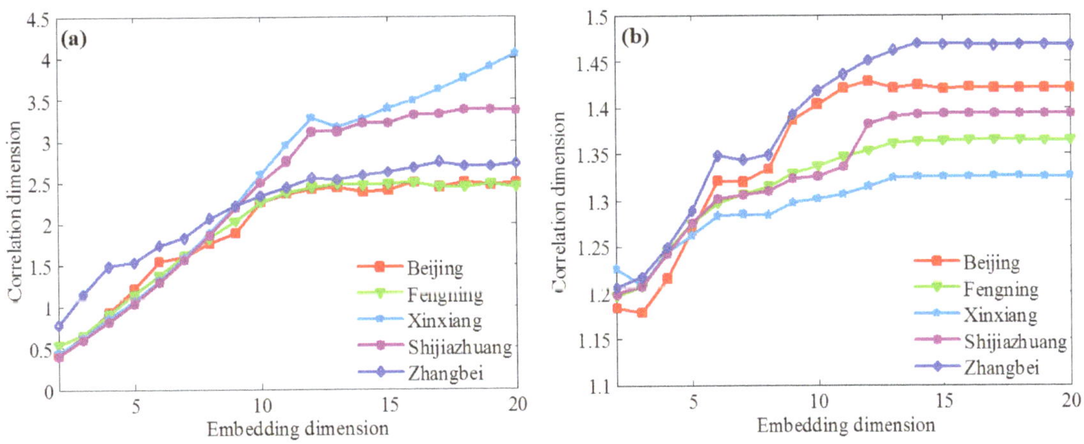

Figure 7. Variation of correlation dimension vs. embedding dimension of climate variables: **(a)** precipitation and **(b)** air temperature.

4.2 Results and analysis

The correlation dimensions of precipitation and air temperature at all 40 meteorological stations were computed using the algorithm proposed in this study. Figure 7 shows the relationships between correlation dimension and embedding dimension for precipitation and air temperature at five representative stations across the HRB (i.e., Beijing, Fengning, Shijiazhuang, Xinxiang, and Zhangbei). The embedding dimensions of precipitation and air temperature for the five stations varied between 10 and 12. It is evident that the relationship between correlation dimension and embedding dimension for precipitation and air temperature differed among the selected stations. In general, correlation dimensions for precipitation showed gradual saturation processes with respective saturation values of 2.378, 2.407, 3.055, and 2.550 for Beijing, Fengning, Shijiazhuang, and Zhangbei stations, respectively (Fig. 7a), indicating chaotic dynamical characteristics of precipitation. By comparison, the correlation dimension for precipitation at the Xinxiang station increased with increasing embedding dimensions, suggesting random characteristics of precipitation. For air temperature, the correlation dimensions at the five stations also showed gradual saturation processes (Fig. 7b), suggesting low dimensional chaotic characteristics for air temperature.

Figure 8 presents the spatial distributions of the saturated correlation dimensions at the 40 meteorological stations for precipitation and air temperature in the HRB. For both pre-

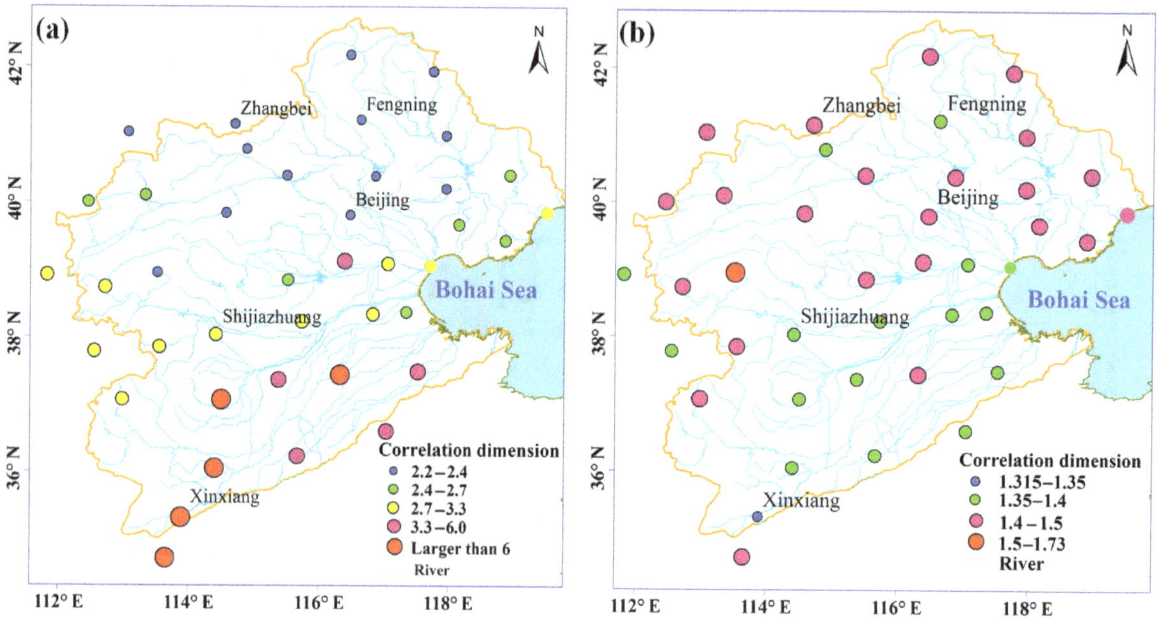

Figure 8. The spatial distribution of the correlation dimension values for all the 40 stations: **(a)** precipitation and **(b)** temperature.

cipitation and air temperature, the correlation dimensions varied markedly across the area. The correlation dimension for precipitation ranged from less than 3 to more than 6, while the correlation dimension was much lower for temperature (i.e., less than 2). Overall, the ranges of the correlation dimensions for precipitation and air temperature were comparable to previously reported values in other regions with similar climatic conditions (Kyoung et al., 2011; Sivakumar and Singh, 2012; Sivakumar et al., 2014). More importantly, the considerable spatial variations in the dimensionality for both climatic variables suggest the regional differences in the complexity of the climate system in the HRB. Specifically, the correlation dimension for precipitation tended to be smaller in the northwestern mountainous area, with values of less than 2.5. In the central area, the correlation dimension for precipitation became larger, with values of greater than 3, while precipitation in the southeastern plain area showed very high correlation dimensions, with values of larger than 6. Given that correlation dimensions indicate the number of controls on the underlying process (Sivakumar and Singh, 2012), Fig. 8a suggests that precipitation processes become progressively more complex from the mountainous area to the plain area in the HRB. Interestingly, the regional pattern of the correlation dimension for air temperature showed an opposite trend with smaller values mainly located in the northern HRB, indicating more complex temporal dynamics of air temperature in the area.

The spatial pattern of the correlation dimension for precipitation in the HRB may be largely attributed to the regional flow pathway of moisture flux, which is mainly controlled by the East Asian Summer Monsoon (EASM). The HRB is located in a monsoon-dominated region, where the EASM plays a leading role in the regional meteorological system. Chen et al. (2013) showed that the EASM had significant impacts on the spatiotemporal distribution of precipitation in eastern China. Li et al. (2017) further suggested that there was a significant correlation between precipitation and the EASM index in the HRB. Wang et al. (2011) revealed that large-scale atmospheric circulations had close relationships with precipitation patterns in the HRB by analyzing the moisture flux derived from NCAR/NCEP reanalysis data. Influenced by the large-scale atmospheric circulation, precipitation in the middle and southeast parts of the HRB is more sensitive to climate variability due to their locations closer to the ocean. This leads to the decreasing trend of precipitation from the southeast to the northwest in the HRB, suggesting that the supply of moisture for precipitation in the region mainly comes from the ocean.

Partly owing to the closer geographical proximity to the ocean (Fig. 8), the EASM has a stronger impact on precipitation in the southern and central areas than in the northern part of the HRB. Furthermore, at the north corner of the HRB, the westerlies primarily affect the hydrometeorological system and thus weaken the impact of the EASM on precipitation (Li et al., 2017). In addition, other factors (e.g., topography, vegetation distribution, and human activity) may also have impacts on regional patterns of climate variables. In particular, the Yan and Taihang mountain range located in the northwestern HRB obstructs the vapor transport driven by the EASM, resulting in lower spatiotemporal variability in precipitation in the northern part of the HRB. As a result, precipitation had higher degrees of complexity in the southern HRB, while its complexity was lower in the mountainous area in the northwestern HRB. As for air temperature, the orographic effect on air temperature might be stronger in the mountainous area (Chu et al., 2010b), resulting in the higher complexity of temperature in this area. However, it should be noted that the range of the correlation dimension for air temperature from 1.0 to 2.0 suggests that two primary controls on temperature exist at all stations across the region.

5 Conclusions

In this study, the original G-P algorithm for calculating correlation dimensions was modified by incorporating the normal-based K-means clustering technique and the RANSAC algorithm. Using the proposed method, the spatial patterns of the complexity of precipitation and air temperature in the HRB were analyzed. The following conclusions were reached:

1. The effectiveness of the proposed method for calculating correlation dimensions was illustrated using the classical Lorenz and Henon chaotic systems. The results showed that the new method outperformed the traditional intuitive judgment and point-based K-means clustering method for computing correlation dimensions.

2. Except for few stations in the northern region, precipitation at most of the meteorological stations in the HRB showed chaotic behaviors. Specifically, the correlation dimension for precipitation showed an increasing trend from the mountainous region in the northwest to the plain area in the southeast, indicating that precipitation processes became progressively more complex from the mountainous area to the plain area. The spatial pattern of the complexity of precipitation reflected the influence of the dominant climate system in the region. Meanwhile, air temperature at all meteorological stations showed chaotic characteristics. In contrast to precipitation, the complexity of air temperature exhibited an opposite trend, with less complexity in the plain area.

The modified G-P algorithm proposed in this study can be used more objectively to characterize the complexity of climate systems (and other hydrological systems, such as streamflow, soil moisture, and groundwater) and thus provide a more reliable estimate of the number of dominant factors governing climate systems. Theoretically, it can provide valuable information for optimizing the number of parameters in climate models to reduce computational demands and model parameter uncertainties. Furthermore, the findings of this study can be used for the regionalization of hydrometeorological systems in the HRB, which has important significance in prediction in ungaged areas (Lebecherel et al., 2016). It should be noted that more studies are still required to verify the present results using other nonlinear techniques, such as the Lyapunov exponent (Wolf et al., 1985) and approximate entropy (Pincus, 1995), which might provide additional insights into climate complexity analysis.

Author contributions. CD performed the improved method, experiments and algorithm coding. CD and TW conceived the framework and analysis of the paper. XY contributed to data collection. SL took part in the data analysis. All the authors contributed to writing and revising the paper.

Competing interests. The authors declare that they have no conflict of interest.

Acknowledgements. The work was supported by the National Key R&D Program of China (no. 2016YFA0601002, no. 2016YFC0401305, and no. 2017YFC0506603), and by the National Natural Scientific Foundation of China (no. 51679007 and no. U1612441). The authors would also like to acknowledge the financial support from the Tianjin University, and Tiejun Wang also acknowledges the financial support from the Thousand Talent Program for Young Outstanding Scientists for this study.

Edited by: Dimitri Solomatine

References

Bao, Z. X., Zhang, J. Y., Wang, G. Q., Fu, G. B, He, R. M., Yan, X. L., Jin, J. L., Liu, Y. L., and Zhang, A. J.: Attribution for decreasing streamflow of the Haihe River basin, northern China: Climate variability or human activities?, J. Hydrol., 460, 117–129, https://doi.org/10.1016/j.jhydrol.2012.06.054, 2012.

Bras, R. L.: Complexity and organization in hydrology: A personal view, Water Resour. Res., 51, 6532–6548, https://doi.org/10.1002/2015WR016958, 2015.

Cantrell, C. A.: Technical Note: Review of methods for linear least-squares fitting of data and application to atmospheric chemistry problems, Atmos. Chem. Phys., 8, 5477–5487, https://doi.org/10.5194/acp-8-5477-2008, 2008.

Carbone, A., Jensen, M., and Sato, A. H.: Challenges in data science: a complex systems perspective, Chaos Soliton. Fract., 90, 1–7, https://doi.org/10.1016/j.chaos.2016.04.020, 2016.

Chen, F., Yuan, Y. J., Wei, W. S., Fan, Z. A., Yu, S. L., Zhang, T. W., Zhang, R. B., Shang, H. M., and Qin, L.: Reconstructed precipitation for the north-central China over the past 380 years and its linkages to East Asian summer monsoon variability, Quat. Int., 283, 36–45, https://doi.org/10.1016/j.quaint.2012.05.047, 2013.

Chu, J. T., Xia, J., Xu, C. Y., Li, L., and Wang, Z. G.: Spatial and temporal variability of daily precipitation in Haihe River basin, 1958–2007, J. Geogr. Sci., 20, 248–260, https://doi.org/10.1007/s11442-010-0248-0, 2010a.

Chu, J. T., Xia, J., Xu, C. Y., and Singh, V. P.: Statistical downscaling of daily mean temperature, pan evaporation and precipitation for climate change scenarios in Haihe River, China, Theor. Appl. Climatol., 99, 149–161, https://doi.org/10.1007/s00704-009-0129-6, 2010b.

Dhanya, C. T. and Kumar, D. N.: Nonlinear ensemble prediction of chaotic daily rainfall, Adv. Water Resour., 33, 327–347, https://doi.org/10.1016/j.advwatres.2010.01.001, 2010.

Dhanya, C. T. and Kumar, D. N.: Multivariate nonlinear ensemble prediction of daily chaotic rainfall with climate inputs, J. Hydrol., 403, 292–306, https://doi.org/10.1016/j.jhydrol.2011.04.009, 2011.

Di, C. L., Yang, X. H., and Wang, X. C.: A four-stage hybrid model for hydrological time series forecasting, Plos One, 9, e104663, https://doi.org/10.1371/journal.pone.0104663, 2014.

Feldhoff, J. H., Lange, S., Volkholz, J., Donges, J. F., Kurths, J., and Gerstengarbe, F.: Complex networks for climate model evaluation with application to statistical versus dynamical modeling of South American climate, Clim. Dynam., 44, 1567–1581, https://doi.org/10.1007/s00382-014-2182-9, 2015.

Fischler, M. A. and Bolles, R. C.: Random sample consensus: a paradigm for model fitting with applications to image analysis and automated cartography, Commun. ACM, 24, 381–395, 1981.

Gan, T. Y., Wang, Q., and Seneka, M.: Correlation dimensions of climate sub-systems and their geographic variability, J. Geophys. Res.-Atmos., 107, 4728, https://doi.org/10.1029/2001JD001268, 2002.

Grassberger, P. and Procaccia, I.: Characterization of strange attractors, Phys. Rev. Lett., 50, 346–349, https://doi.org/10.1103/PhysRevLett.50.346, 1983a.

Grassberger, P. and Procaccia, I.: Measuring the strangeness of strange attractors, Physica D, 9, 189–208, https://doi.org/10.1016/0167-2789(83)90298-1, 1983b.

Henon, M.: A two-dimensional mapping with a strange attractor, Commun. Math. Phys., 50, 69–77, https://doi.org/10.1007/BF01608556, 1976.

Jayawardena, A. W. and Lai, F.: Analysis and prediction of chaos in rainfall and stream flow time series, J. Hydrol., 153, 23–52, https://doi.org/10.1016/0022-1694(94)90185-6, 1994.

Ji, C. C., Zhu, H., and Jiang, W.: A novel method to identify the scaling region for chaotic time series correlation dimension calculation, Chinese Sci. Bull., 56, 925–932, https://doi.org/10.1007/s11434-010-4180-6, 2011.

Jothiprakash, V. and Fathima, T. A.: Chaotic analysis of daily rainfall series in Koyna reservoir catchment area, India, Stoch. Env. Res. Risk A., 27, 1371–1381, https://doi.org/10.1007/s00477-012-0673-y, 2013.

Kyoung, M. S., Kim, H. S., Sivakumar, B., Singh, V. P., and Ahn, K. S.: Dynamic characteristics of monthly rainfall in the Korean Peninsula under climate change, Stoch. Env. Res. Risk A., 25, 613–625, https://doi.org/10.1007/s00477-010-0425-9, 2011.

Lai, Y. C. and Lerner, D.: Effective scaling regime for computing the correlation dimension from chaotic time series, Physica D, 115, 1–18, https://doi.org/10.1016/S0167-2789(97)00230-3, 1998.

Lana, X., Burgueno, A., Martinez, M. D., and Serra, C.: Complexity and predictability of the monthly Western Mediterranean Oscillation index, Int. J. Climatol., 36, 2435–2450, https://doi.org/10.1002/joc.4503, 2016.

Lebecherel, L., Andreassian, V., and Charles, P.: On evaluating the robustness of spatialproximity-based regionalization methods, J. Hydrol., 539, 196–203, https://doi.org/10.1016/j.jhydrol.2016.05.031, 2016.

Li, F. X., Zhang, S. Y., Chen, D., He, L., and Gu, L. L.: Interdecadal variability of the east Asian summer monsoon and its impact on hydrologic variables in the Haihe River Basin, China, J. Resour. Ecol., 8, 174–184, https://doi.org/10.5814/j.issn.1674-764X.2017.02.008, 2017.

Liebert, W. and Schuster, H. G.: Proper choice of the time delay for the analysis of chaotic time series, Phys. Lett. A, 142, 107–111, https://doi.org/10.1016/0375-9601(89)90169-2, 1989.

Lin, H., Vogel, H., Phillips, J., and Fath, B. D.: Complexity of soils and hydrology in ecosystems, Ecol. Model., 298, 1–3, https://doi.org/10.1016/j.ecolmodel.2014.11.016, 2015.

Liu, C. and Xia, J.: Water problems and hydrological research in the Yellow River and the Huai and Hai River basins of China, Hydrol. Process., 18, 2197–2210, https://doi.org/10.1002/hyp.5524, 2004.

Lloyd, S. P.: Least squares quantization in PCM, IEEE Trans. Inf. Theory, 28, 129–137, https://doi.org/10.1109/TIT.1982.1056489, 1982.

Lorenz, E. N.: Deterministic nonperiodic flow, J. Atmos. Sci., 20, 130–141, https://doi.org/10.1175/1520-0469(1963)020<0448:TMOV<2.0.CO;2, 1963.

Maragos, P. and Sun, F.: Measuring the fractal dimension of signals: Morphological covers and iterative optimization, IEEE Trans. Signal Process., 41, 108–121, https://doi.org/10.1109/TSP.1993.193131, 1993.

Meseguer-Ruiz, O., Olcina Cantos, J., Sarricolea, P., and Martin-Vide, J.: The temporal fractality of precipitation in mainland Spain and the Balearic Islands and its relation to other precipitation variability indices, Int. J. Climatol., 37, 849–860, https://doi.org/10.1002/joc.4744, 2017.

Mitra, N. J., Nguyen, A. N., and Guibas, L.: Estimating surface normals in noisy point cloud data, J. Comput. Geom. Appl., 14, 261–276, https://doi.org/10.1145/777792.777840, 2004.

Nicolis, C. and Nicolis, G.: Is there a climatic attractor?, Nature, 311, 529–532, https://doi.org/10.1038/311529a0, 1984.

Packard, N. H., Crutchfield, J. P., Farmer, J. D., and Shaw, R. S.: Geometry from a time series, Phys. Rev. Lett., 45, 712–716, https://doi.org/10.1103/PhysRevLett.45.712, 1980.

Palmer, T. N.: A Nonlinear Dynamical Perspective on Climate Prediction, J. Climate, 12, 575–591, https://doi.org/10.1175/1520-0442(1999)012<0575:ANDPOC>2.0.CO;2, 1999.

Pincus, S.: Approximate entropy (ApEn) as a complexity measure, Chaos, 5, 110–117, https://doi.org/10.1063/1.166092, 1995.

Rial, J. A., Pielke, R. A., Beniston, M., Claussen, M., Canadell, J., Cox, P., Held, H., De Noblet-Ducoudre, N., Prinn, R., Reynolds, J. F., and Salas, J. D.: Nonlinearities, feedbacks and critical thresholds within the Earth's climate system, Climate Change, 65, 11–38, https://doi.org/10.1023/B:CLIM.0000037493.89489.3f, 2004.

Rind, D.: Complexity and climate, Science, 284, 105–107, https://doi.org/10.1126/science.284.5411.105, 1999.

Sang, Y., Wang, Z., and Li, Z.: Discrete wavelet entropy aided detection of abrupt change: A case study in the haihe river basin, china, Entropy-Switz., 14, 1274–1284, https://doi.org/10.3390/e14071274, 2012.

Sivakumar, B.: Rainfall dynamics at different temporal scales: A chaotic perspective, Hydrol. Earth Syst. Sci., 5, 645–652, https://doi.org/10.5194/hess-5-645-2001, 2001.

Sivakumar, B.: Chaos in rainfall: variability, temporal scale and zeros, J. Hydroinform., 7, 175–184, https://doi.org/10.2166/hydro.2005.0015, 2005.

Sivakumar, B.: Chaos in Hydrology: Bridging determinism and stochasticity, Springer, the Netherlands, https://doi.org/10.1007/978-90-481-2552-4, 2017.

Sivakumar, B. and Singh, V. P.: Hydrologic system complexity and nonlinear dynamic concepts for a catchment classification framework, Hydrol. Earth Syst. Sci., 16, 4119–4131, https://doi.org/10.5194/hess-16-4119-2012, 2012.

Sivakumar, B., Persson, M., Berndtsson, R., and Uvo, C. B.: Is correlation dimension a reliable indicator of low-dimensional chaos in short hydrological time series?, Water Resour. Res., 38, 3–1, https://doi.org/10.1029/2001WR000333, 2002.

Sivakumar, B., Woldemeskel, F. M., and Puente, C. E.: Nonlinear analysis of rainfall variability in Australia, Stoch. Env. Res. Risk A., 28, 17–27, https://doi.org/10.1007/s00477-013-0689-y, 2014.

Sprott, J. C. and Rowlands, G.: Improved correlation dimension calculation, Int. J. Bifurcat. Chaos, 11, 1865–1880, https://doi.org/10.1142/S021812740100305X, 2001.

Takens, F.: Detecting strange attractors in turbulence. Dynamical Systems and Turbulence, Warwick 1980, Springer, Berlin, Heidelberg, 366–381, 1981.

Wang, W. G., Shao, Q. X., Peng, S. Z., Zhang, Z. X., Xing, W. Q., An, G. Y., and Yong, B.: Spatial and temporal characteristics of changes in precipitation during 1957–2007 in the Haihe River basin, China, Stoch. Env. Res. Risk A., 25, 881–895, https://doi.org/10.1007/s00477-011-0469-5, 2011.

Wang, W., Wei, J., Shao, Q., Xing, W., Yong, B., Yu, Z., and Jiao, X.: Spatial and temporal variations in hydro-climatic variables and runoff in response to climate change in the Luanhe River basin, China, Stoch. Env. Res. Risk A., 29, 1117–1133, https://doi.org/10.1007/s00477-014-1003-3, 2015.

Wang, W., Lai, Y., and Grebogi, C.: Data based identification and prediction of nonlinear and complex dynamical systems, Phys. Rep., 644, 1–76, https://doi.org/10.1016/j.physrep.2016.06.004, 2016.

White, D. J., Feng, K. S., Sun, L. X., and Hubacek, K.: A hydro-economic MRIO analysis of the Haihe River Basin's water footprint and water stress, Ecol. Model., 318, 157–167, https://doi.org/10.1016/j.ecolmodel.2015.01.017, 2015.

Wolf, A., Swift, J. B., and Swinney, H. L.: Determining Lyapunov exponents from a time series, Physica D, 16, 285–317, https://doi.org/10.1016/0167-2789(85)90011-9, 1985.

Wu, C. L., Chau, K. W., and Fan, C.: Prediction of rainfall time series using modular artificial neural networks coupled with data-preprocessing techniques, J. Hydrol., 389, 146–167, https://doi.org/10.1016/j.jhydrol.2010.05.040, 2010.

8

Sensitivity of young water fractions to hydro-climatic forcing and landscape properties across 22 Swiss catchments

Jana von Freyberg[1,2], **Scott T. Allen**[1], **Stefan Seeger**[3], **Markus Weiler**[3], **and James W. Kirchner**[1,2]

[1]Department of Environmental Systems Science, ETH Zurich, Zurich, Switzerland
[2]Swiss Federal Institute for Forest, Snow and Landscape Research (WSL), Birmensdorf, Switzerland
[3]Faculty of Environment and Natural Resources, University of Freiburg, Freiburg im Breisgau, Germany

Correspondence: Jana von Freyberg (jana.vonfreyberg@usys.ethz.ch)

Abstract. The young water fraction F_{yw}, defined as the proportion of catchment outflow younger than approximately 2–3 months, can be estimated directly from the amplitudes of seasonal cycles of stable water isotopes in precipitation and streamflow. Thus, F_{yw} may be a useful metric in catchment inter-comparison studies that investigate landscape and hydro-climatic controls on streamflow generation. Here, we explore how F_{yw} varies with catchment characteristics and climatic forcing, using an extensive isotope data set from 22 small- to medium-sized (0.7–351 km^2) Swiss catchments. We find that flow-weighting the tracer concentrations in streamwater resulted in roughly 26 % larger young water fractions compared to the corresponding unweighted values, reflecting the fact that young water fractions tend to be larger when catchments are wet and discharge is correspondingly higher. However, flow-weighted and unweighted young water fractions are strongly correlated with each other among the catchments. They also correlate with terrain, soil, and land-use indices, as well as with mean precipitation and measures of hydrologic response. Within individual catchments, young water fractions increase with discharge, indicating an increase in the proportional contribution of faster flow paths at higher flows. We present a new method to quantify the discharge sensitivity of F_{yw}, which we estimate as the linear slope of the relationship between the young water fraction and flow. Among the 22 catchments, discharge sensitivities of F_{yw} are highly variable and only weakly correlated with F_{yw} itself, implying that these two measures reflect catchment behaviour differently. Based on strong correlations between the discharge sensitivity of F_{yw} and several catchment characteristics, we suggest that low discharge sensitivities imply greater persistence in the proportions of fast and slow runoff flow paths as catchment wetness changes. In contrast, high discharge sensitivities imply the activation of different dominant flow paths during precipitation events, such as when subsurface water tables rise into more permeable layers and/or the river network expands further into the landscape.

1 Introduction

Naturally occurring variations in stable water isotopes ($\delta^{18}O$, δ^2H) or chemically passive solutes (e.g. chloride) are commonly used in catchment studies to track the flow of water and to gain insight into catchment storage and mixing behaviour (Buttle, 1994; Kendall and McDonnell, 1998; Klaus and McDonnell, 2013). Many catchment studies use these tracers to estimate time-averaged transit time distributions, to characterize the heterogeneity of flow pathways, and to estimate mobile catchment storage (e.g. Benettin et al., 2015; Birkel et al., 2011; Hrachowitz et al., 2009; Staudinger et al., 2017). Transit time distributions are often inferred from conservative tracers using lumped-parameter models (McGuire and McDonnell, 2006). Because the mean transit time expresses the ratio between mobile catchment storage and the average flow rate, it is widely used in catchment inter-comparison studies (e.g. Hrachowitz et al., 2009; McGuire et al., 2005; Staudinger et al., 2017). However, estimates of mean transit time can be biased and unreliable, especially for spatially heterogeneous catchments (Kirchner, 2016b; Seeger and Weiler, 2014). Instead, the young water fraction F_{yw} – i.e. the average fraction of streamflow that is younger than a specified threshold age – has recently been proposed as a more reliable measure of water age in heterogeneous catchments (Kirchner, 2016a, b). Young water fractions with a threshold age of roughly 2–3 months can be estimated directly from the amplitude ratio of the seasonal cycles in stable water isotopes in precipitation and streamwater.

The amplitudes of the seasonal isotopic cycles in precipitation and streamwater can be estimated directly from the isotope measurements themselves, or by volume-weighting these measurements by the corresponding precipitation or discharge rates. Precipitation isotopes should generally be volume-weighted to prevent small precipitation events, potentially with anomalous isotope values, from substantially influencing the calculated seasonal precipitation isotope cycle. Streamwater isotope values can also be flow-weighted, using stream discharges as weights. Higher streamflows should typically correspond to larger young water fractions, for the simple reason that flow peaks typically follow intense rainfall and contain more recent precipitation than base flows (e.g. Kirchner, 2016b; von Freyberg et al., 2017). Hence, the flow-weighted average young water fraction (here denoted F^*_{yw}) is expected to be higher than the unweighted average young water fraction (F_{yw}). Both F_{yw} and F^*_{yw} are calculated over periods of a year or longer, and represent the average catchment behaviour over that time. In calculating the unweighted F_{yw}, each unit of time counts equally, and benchmark tests using a nonstationary lumped catchment model confirm that the calculated F_{yw} should accurately reflect the time-averaged fraction of young water in discharge (Kirchner, 2016b). By contrast, in calculating the flow-weighted F^*_{yw}, each unit of flow counts equally, and benchmark tests confirm that the calculated F^*_{yw} reflects the cumulative volume of young water, as a fraction of the cumulative volume of discharge, over the corresponding period (Kirchner, 2016b). Although F^*_{yw} and F_{yw} have previously been compared in benchmark tests, a systematic evaluation based on tracer data from natural catchments has not yet been carried out.

At sites where precipitation isotopes are not measured directly, catchment isotopic inputs can be estimated from nearby long-term monitoring stations using various spatial interpolation methods. These interpolation methods differ in their assumptions about temperature- and elevation-dependent isotope fractionation effects, and their treatment of seasonal snowpack storage. Based on a global database of $\delta^{18}O$ in precipitation, Jasechko et al. (2016) calculated the seasonal cycle amplitudes and their standard errors for each station and interpolated them to generate a global grid of the seasonal cycle amplitudes. These interpolated coefficients were volume-weighted by the spatial pattern of precipitation over each catchment. An alternative approach (Allen et al., 2018; see Supplement) builds on the Jasechko et al. (2016) method with an additional step that accounts for the residuals of the observations from the fitted seasonal cycles. To generate a high-resolution precipitation isotope map for Switzerland, Seeger and Weiler (2014) interpolated $\delta^{18}O$ in monthly precipitation from long-term monitoring stations in central Europe, using an elevation-gradient approach. They combined their interpolation method with an energy-balance-based snow model to estimate the liquid input to the soil surface at monthly temporal resolution. The latter two interpolation methods have been rigorously tested with real-world isotope measurements, and thus may be particularly useful for estimating young water fractions in catchments where no long-term precipitation isotope measurements exist.

Another analytical decision that affects the interpretation of F^*_{yw} and F_{yw} relates to whether snowpack storage is considered to be part of catchment storage, or not. If one measures precipitation to the snow surface as the catchment input, then snowpack accumulation and melt are implicitly included in catchment storage (e.g. Staudinger et al., 2017). In this case, comparisons of seasonal cycles in precipitation and streamflow should reflect the young water fraction resulting from the combination of snowpack and subsurface storage. Alternatively, if one uses precipitation and snowmelt arriving at the soil surface as the catchment input (for example, with melt pan lysimeters, or modelled snowpack outflows), then snowpack accumulation and melt are implicitly excluded from catchment storage. In this case, comparisons of seasonal cycles in streamflow and sub-snowpack catchment input should reflect the young water fraction resulting from subsurface storage alone. Because the total catchment storage in the first case (including snowpack storage) is larger than the subsurface storage alone, the resulting young water fractions are expected to be smaller. Previous studies that estimated young water fractions in snow-dominated watersheds (Jasechko et al., 2016; Song et al., 2017) did not differentiate between these two concepts of catchment storage and simply used incoming precipitation in the young water fraction calculations, thus implicitly considering snowpack storage as part of catchment storage (as in the first case outlined above). This approach is practical in view of the challenges of measuring or modelling snowmelt and its isotopic composition. However, it is still unclear whether, in cases where snowmelt can be modelled or measured, explicitly considering snowmelt as a catchment input would significantly alter young water fraction estimates.

Because the young water fraction can be estimated from sparse and irregular tracer data, it has been suggested as a useful metric for catchment inter-comparison studies (Kirchner, 2016a). To date, however, most catchment inter-comparison studies have investigated controls on mean transit times instead. Mean transit times have been variably found to be correlated with (for example) flow path lengths and gradients (McGuire et al., 2005), drainage density (Soulsby et al., 2010), the areal fraction of hydrologically responsive soils (Tetzlaff et al., 2009), bedrock permeability (Hale and McDonnell, 2016), or combinations of multiple factors (Hrachowitz et al., 2009; Seeger and Weiler, 2014). So far, only a few catchment inter-comparison studies have used young water fractions (Song et al., 2017; Stockinger et al., 2016; Jasechko et al., 2016). A global analysis of 254 watersheds, revealed large spatial variability in young streamflow, which correlated inversely with average topographic gradients and water table depths (Jasechko et al., 2016). Jasechko et al. (2016) hypothesized that steeper landscapes are associated with more pervasive rock fracturing, deeper infiltration, and reduced shallow lateral flow, all of which would reduce the young water fraction in steep terrain. However, the correlation between topographic steepness and young water fractions was highly scattered, indicating that other factors are also involved. Jasechko et al.'s (2016) study sites were mostly larger than 1000 km^2 (25th percentile 1753 km^2, median 10 800 km^2) and thus were probably affected by a complex interplay of landscape characteristics, climatic variability, and human impacts. Identifying landscape and climatic drivers that potentially control catchment storage behaviour may be easier in small- to medium-sized catchments with near-natural streamflow regimes (Holko et al., 2015).

In the present study, we use seasonal cycles in $\delta^{18}O$ to estimate young water fractions for 22 sites in Switzerland with catchment areas between 0.7 and 351 km^2. In a first step, we evaluate how choices of methodology affect the young water fraction estimates, with emphasis on (i) the spatial interpolation method for precipitation isotopes, (ii) the conceptual representation of snow storage, and (iii) flow-weighting the streamwater isotope data. Because the 22 study catchments cover a wide range of landscape and hydro-climatic characteristics, in the second part of this study, we test for correlations between the young water fraction and a wide range of landscape and hydro-climatic indices. Finally, we present a method for estimating the linear dependence of the young water fraction on the streamflow regime and propose that the slope of this relationship may be a diagnostic indicator of streamflow generation processes.

2 Theoretical background: young water fractions from seasonal cycles of stable water isotopes in precipitation and streamwater

The isotopic composition of precipitation follows a seasonal cycle (Feng et al., 2009). The damping and phase shift of this seasonal cycle as it is transmitted through catchments (Fig. 1) can be used to infer timescales of catchment storage and transport (e.g. DeWalle et al., 1997; Soulsby et al., 2006). Sine-wave fitting can quantify the amplitude ratio A_S/A_P and phase shift $\varphi_S - \varphi_P$ between the seasonal isotope cycles in precipitation and streamflow (the indices P and S refer to precipitation and streamwater, respectively). The seasonal isotope cycles in precipitation and streamwater can be described by:

$$c_P(t) = A_P \sin(2\pi f t - \varphi_P) + k_P \quad (1)$$

and

$$c_S(t) = A_S \sin(2\pi f t - \varphi_S) + k_S. \quad (2)$$

In Eqs. (1) and (2), A is the amplitude (‰), φ is the phase of the seasonal cycle (in radians, with 2π rad equalling 1 year), t is the time (decimal years), f is the frequency (yr^{-1}), and k (‰) is a constant describing the vertical offset of the isotope signal.

If one assumes that the transit times of water through the catchment follow a particular transit time distribution, the mean transit time can be calculated as a function of the amplitude ratio A_S/A_P. However, mean transit times inferred from seasonal tracer cycles in runoff from heterogeneous catchments are potentially subject to severe aggregation bias (Kirchner, 2016a). Alternatively, the amplitude ratio A_S/A_P can be used to estimate the fraction of water younger than a specified threshold age. Compared to the mean transit time, this "young water fraction" (F_{yw}) is markedly less vulnerable to aggregation bias, and less sensitive to the assumed shape of the catchment transit time distribution (Kirchner, 2016a, b). For a wide range of transit time distributions, the young water threshold age is approximately 2.3 ± 0.8 months (Kirchner, 2016a).

We can estimate the amplitudes A_S and A_P of the seasonal isotope cycles in Eqs. (1) and (2) by using multiple linear regression to obtain the coefficients a and b in

$$c_P(t) = a_P \cos(2\pi f t) + b_P \sin(2\pi f t) + k_P \quad (3)$$

and

$$c_S(t) = a_S \cos(2\pi f t) + b_S \sin(2\pi f t) + k_S. \quad (4)$$

The amplitudes A_S and A_P are then determined by

$$A_P = \sqrt{a_P^2 + b_P^2}$$

and

$$A_S = \sqrt{a_S^2 + b_S^2}. \quad (5)$$

Following Kirchner (2016a), we calculate young water fractions as the amplitude ratio A_S/A_P. We estimate the coefficients a_S, b_S, a_P, and b_P by fitting Eqs. (3) and (4) using iteratively re-weighted least squares (IRLS), a robust estimation method that minimizes the influence of any potential outliers (an *R* script with our IRLS code is provided in the Supplement). In estimating a_P and b_P, we volume-weight Eq. (3) to avoid giving undue leverage to low-precipitation periods. To calculate the unweighted young water fraction F_{yw}, we estimate a_S and b_S from Eq. (4) using unweighted IRLS. For the flow-weighted young water fraction (F^*_{yw}), we estimate a_S and b_S from Eq. (4) using discharge-weighted IRLS (see the *R* script provided in the Supplement). Uncertainties in the calculated unweighted and flow-weighted young water fractions are expressed as standard errors (SEs) and are estimated using Gaussian error propagation.

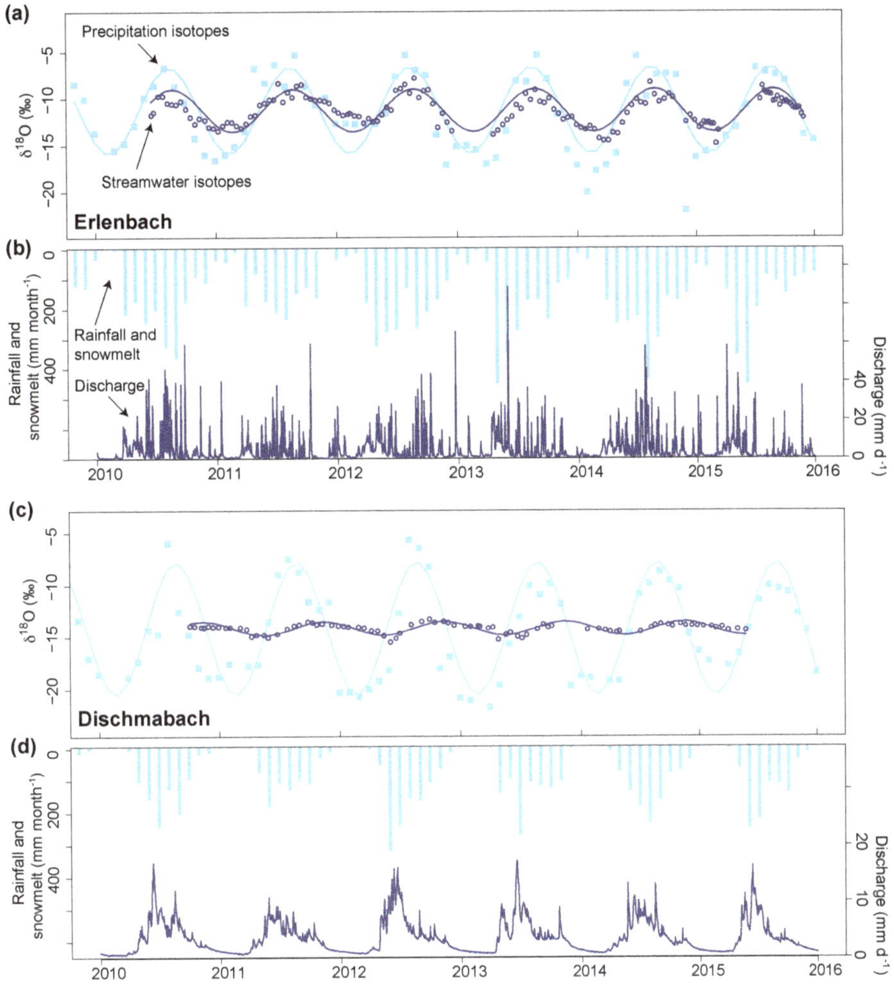

Figure 1. Hydrologic and isotopic seasonality of precipitation and streamflow for the Erlenbach and Dischmabach catchments. Precipitation isotopes were interpolated with the method of Seeger and Weiler (2014). Sinusoidal cycles were fitted to the isotope data using iteratively re-weighted least squares regression. The seasonal cycles of the streamwater isotopes exhibit damping and phase shifts relative to the precipitation isotopic cycles. Stronger damping of the seasonal isotope cycle, implying a smaller fraction of young water in streamflow, can be observed in the Dischmabach catchment.

3 Data set

The 22 study catchments cover areas between 0.7 and 351 km^2 and have mean elevations between 472 and 2369 m a.s.l. (Table 1). Most of the sites are located in the Swiss Plateau and in the northern Alps, where the geology is characterized by sedimentary rocks (limestones, sandstones, marls, marly shales, conglomerates, breccias) and unconsolidated sediments (clay, silts, sands). In the southern Alps, two high-elevation catchments (Dischmabach and Riale di Calneggia) are predominantly underlain by metamorphic rock (mica schist, gneiss), and Ova da Cluozza is the only study catchment underlain by dolomite (Fig. 2a and b).

Land use at lower elevations (400–800 m) is predominantly agriculture, while grassland and forests can be found at elevations up to around 1400 m. Much of the area above 1700 m is characterized by grasses, shrubs, and sparse vegetation. At two high-elevation sites, Dischmabach and Ova da Cluozza, up to $\sim 2\,\%$ of the drainage area is covered by glaciers. At all sites, the human influence on river discharge is small, resulting in near-natural streamflow regimes.

Switzerland is characterized by a humid to temperate continental climate, with the Alps creating climatically distinct subregions. The wettest regions can be found in the northern pre-Alps and Alps, as well as in the canton of Ticino south of the Alps. The driest regions are located in inner Alpine valleys in the cantons of Valais and Grisons (Fig. 2c). Average annual precipitation rates for the 22 catchments range from 887 to 1853 mm based on observations from 2000 to 2015 (Table 1). To differentiate between the hydro-climatic regimes of the catchments, we grouped them into three classes (snow-dominated, rainfall-dominated, and hybrid) proposed by Staudinger et al. (2017). Precipitation is

Table 1. General properties of the 22 study catchments and streamwater isotope time series.

Catchment name	Gauging station	Longitude (WGS84)	Latitude (WGS84)	Area (km^2)	Mean elevation (range) (m)	Average annual precipitation* (mm)	Hydro-climatic regime	Sampling period for $\delta^{18}O$ in streamwater (number of samples)
Aabach	Mönchaltdorf	8.7206	47.3110	49.0	635 (519–1092)	1358	Rainfall-dominated	Sep 2010–Feb 2013 (62)
Aach	Salmsach, Hungerbühl	9.3572	47.5505	50.0	472 (408–560)	1141	Rainfall-dominated	Jul 2010–Dec 2011 (26)
Allenbach	Adelboden	7.5521	46.4860	28.8	1852 (1293–2742)	1338	Snow-dominated	Sep 2010–May 2015 (87)
Alp	Einsiedeln	8.7393	47.1508	46.5	1154 (845–1894)	1776	Hybrid	May 2010–2011/2015 (131)
Biber	Biberbrugg	8.7209	47.1534	31.6	999 (827–1495)	1658	Rainfall-dominated	May 2010–Nov 2015 (140)
Dischmabach	Davos, Kriegsmatte	9.8772	46.7754	43.2	2369 (1663–3139)	1072	Snow-dominated	Nov 2010–May 2015 (81)
Emme	Eggiwil, Heidbüel	7.8047	46.8711	127.0	1285 (743–2216)	1372	Hybrid	Jun 2010–Nov 2013 (71)
Ergolz	Liestal	7.7342	47.4882	261.2	584 (305–1165)	1081	Rainfall-dominated	Jun 2010–Nov 2015 (140)
Erlenbach	Erlenbach	8.7089	47.0452	0.7	1359 (1117–1650)	1853	Hybrid	Jul 2010–May 2015 (140)
Guerbe	Belp, Mülimatt	7.5155	46.7888	55.4	1037 (556–2152)	1236	Hybrid	Jul 2010–Dec 2012 (64)
Ilfis	Langnau	7.7975	46.9379	187.9	1037 (681–2087)	1443	Hybrid	Jul 2010–May 2015 (128)
Langeten	Huttwil, Häberenbad	7.8282	47.1225	60.3	760 (598–1100)	1297	Rainfall-dominated	Jul 2010–May 2015 (103)
Lümpenenbach	Lümpenenbach	8.7052	47.0467	0.9	1336 (1092–1508)	1803	Hybrid	Oct 2010–Nov 2015 (132)
Mentue	Yvonand, La Mauguettaz	6.7242	46.7768	105.0	679 (447–926)	1097	Rainfall-dominated	Jul 2010–Feb 2013 (63)
Murg	Waengi	8.9529	47.4963	76.8	648 (467–1036)	1314	Rainfall-dominated	Jul 2010–Nov 2014 (95)
Ova da Cluozza	Zernez	10.1183	46.6932	26.9	2364 (1519–3160)	887	Snow-dominated	Aug 2010–Sep 2013 (65)
Riale di Calneggia	Cavergno, Pontit	8.5429	46.3696	23.9	1986 (881–2908)	1686	Snow-dominated	Jul 2010–Dec 2012 (55)
Rietholzbach	Mosnang, Rietholz	9.0123	47.3761	3.2	794 (671–938)	1415	Rainfall-dominated	Jul 2010–Feb 2013 (68)
Schaechen	Bürglen, Galgenwäldli	8.6517	46.8710	107.9	1719 (487–3260)	1565	Snow-dominated	Apr 2011–May 2015 (66)
Sense	Thoerishaus, Sensematt	7.3514	46.8883	351.2	1068 (554–2184)	1258	Hybrid	Oct 2010–Mar 2013 (47)
Sitter	Appenzell	9.4104	47.3319	88.2	1301 (768–2500)	1771	Hybrid	Nov 2010–May 2015 (97)
Vogelbach	Vogelbach	8.7161	47.0761	1.6	1335 (1038–1540)	1800	Hybrid	Jun 2010–Nov 2015 (139)

* Based on interpolated data from PREVAH and the time period January 2000–December 2015.

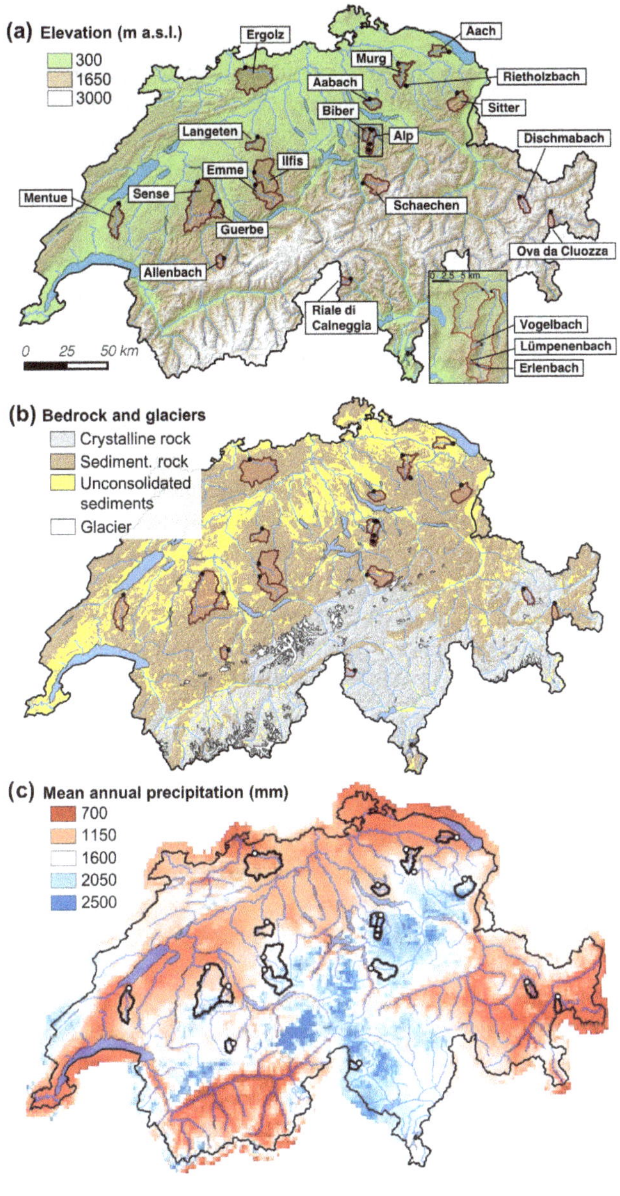

Figure 2. Locations of the 22 study catchments in Switzerland **(a)**, bedrock geology **(b)**, and mean annual precipitation based on the observation period 1991–2010 **(c)**. The expanded panel in **(a)** shows the sub-catchments of the Alp basin.

distributed more-or-less evenly throughout the year, although peak inputs to the soil surface (melt and precipitation) are shifted towards spring and summer in all snow-dominated sites and some hybrid sites.

3.1 Hydro-climatic data

Daily discharge data for 18 of the 22 sites were provided by the Swiss Federal Office for the Environment. Discharge measurements for the Aabach catchment were made available by the Office for Waste, Water, Energy and Air (WWEA) of the canton of Zurich. Discharge data for the Erlenbach, Vogelbach, and Lümpenenbach catchments were provided by the Swiss Federal Institute for Forest, Snow and Landscape Research (WSL), Birmensdorf, Switzerland.

Meteorological data for each site and each 100 m elevation band were interpolated from measurements taken by the national meteorological service of Switzerland (MeteoSwiss), using the PREVAH (Precipitation-Runoff-EVApotranspiration HRU) model (Viviroli et al., 2009). Mean precipitation for each 100 m elevation band was aggregated to obtain area-weighted catchment average values.

3.2 Catchment properties

The average hydro-climatic properties at the sites were described by various indices, such as mean monthly values of discharge $\overline{Q}$ and precipitation Q, as well as mean daily precipitation intensity $\overline{P}_{\text{intensity}}$. To quantify the variability of the flow regimes, we determined the average coefficient of variation of daily discharge (CV_Q) and the quick-flow index (QFI). The QFI is the average ratio between $(Q - Q_{\text{bf}})$ and Q, where Q is daily discharge and Q_{bf} is daily baseflow; Q_{bf} was calculated with the "BaseflowSeparation" function in the *EcoHydRology* package (version 0.4.12) in *R* using a recursive digital filter parameter of 0.925, as recommended by Nathan and McMahon (1990). All of these hydro-climatic indices were calculated for each site and for the duration of the site-specific streamwater isotope sampling campaigns, which varied between approximately 1 and 5 years (Table 1).

The seasonal variability of monthly precipitation for the years 2000–2015 was expressed through the amplitude and the phase shift of a fitted sinusoidal function (Berghuijs et al., 2014):

$$P(t) = \overline{P}\left[1 + A_{\text{precip}} \sin\left(2\pi\left(t - \varphi_{\text{precip}}\right)/\tau\right)\right], \tag{6}$$

where P is the precipitation volume (mm month^{-1}), $\overline{P}$ is the average of P (mm month^{-1}), A_{precip} is the seasonal amplitude of precipitation (–), t is the time (months), τ is the duration of a full seasonal cycle (12 months), and φ_{precip} is the phase (months). The phase describes the offset from the beginning of the seasonal cycle, which is defined here as 1 January. The parameters A_{precip} and φ_{precip} were obtained by non-linear fitting to the monthly precipitation data using Newton's method. Strong precipitation seasonality would be expressed in a high A_{precip} value.

The hydro-climatic indices are to some extent redundant with one another. Unsurprisingly, mean monthly discharge ($\overline{Q}$) and mean monthly precipitation ($\overline{P}$) were significantly correlated with each other across the 22 sites. Furthermore, $\overline{Q}$ was significantly correlated with the seasonality of precipitation (A_{precip}), and the QFI was significantly correlated with the coefficient of variation of daily discharge (CV_Q) (Table 4).

To quantify the geomorphological characteristics of the study catchments, we used terrain indices (median flow path length L, median flow gradient G, the ratio L/G, and median topographic wetness index – TWI) which were calculated previously by Seeger and Weiler (2014) for all 22 study sites using a digital elevation model with 25 m spatial resolution. The indices L/G and TWI were previously applied in numerous catchment inter-comparison studies, e.g. Hrachowitz et al. (2009), McGuire et al. (2005), and Tetzlaff et al. (2009). In addition, we calculated the drainage density DD (the total channel length divided by the catchment area) based on the official river network from the topographical landscape model of Switzerland (swissTLM3D, ©2017 swisstopo; resolution 8 m or better).

Hydrologic soil properties and vegetation cover information were extracted from geospatial data provided by the Swiss Federal Office for Agriculture (BLW, 2012) and the Swiss Federal Statistical Office (BFS, 2004), respectively. The data product "land suitability" uses six soil properties – soil depth, large particle fraction, water storage capacity, nutrient storage capacity, permeability, and soil wetness index – to generate a map of 144 different soil classes. Each soil property is ranked from 0 (very low) to 5 or 6 (very high). For our analysis, we calculated the areal fractions of aggregated soil properties that are usually associated with fast runoff processes, i.e. low water storage capacity (ranks 1–3), low permeability (rank 1–3), and high soil wetness index (i.e. saturated soils, ranks 4–5). From the data product "forest diversity" we extracted the fraction of forested area for each catchment.

The hydrogeological properties of the study sites were obtained from the official geotechnical map of Switzerland (1 : 200 000, ©2017 swisstopo). We extracted the areal fractions of low, intermediate, and high groundwater productivity for each catchment. Representative groundwater table depths could not be determined for all sites due to their complex small-scale topographic and geologic heterogeneity. The hydrologic soil properties, as well as the hydrogeological properties of the individual sites, are provided in the Supplement (Table S1).

Correlations between the catchments' young water fractions, hydro-climatic conditions, and landscape properties were assessed with the Spearman rank correlation coefficient ρ (Spearman, 1987). Following conventional practice, we consider correlations with $p < 0.05$ to be statistically significant.

3.3 Streamwater isotope data

Streamwater grab samples were collected approximately fortnightly at 21 sites between mid-2010 and mid-2011 or later (see Table 1 for exact dates). Oxygen isotope ratios ($\delta^{18}O$) were measured with a Picarro isotope analyser (Picarro Inc., Santa Clara, CA, USA) at the University of Freiburg, Germany, and are reported here as δ values relative to the VSMOW standard. For the Rietholzbach catchment, fortnightly streamwater $\delta^{18}O$ data were provided by the Institute for Atmospheric and Climate Science at ETH Zurich.

3.4 Precipitation isotope data

Values of $\delta^{18}O$ in precipitation were not measured directly at the 22 study catchments. Instead, $\delta^{18}O$ values from monthly cumulative precipitation samples were interpolated from long-term observations at nearby monitoring stations (the Swiss network for Observations of Isotopes in the Water Cycle – NAQUA-ISOT, the Global Network of Isotopes in Precipitation – GNIP, and the Austrian Network of Isotopes in Precipitation – ANIP). We used two different interpolation approaches that we summarize below: method 1 after Seeger and Weiler (2014) and method 2 similar to that of Allen et al. (2018). More detailed descriptions of both interpolation methods 1 and 2 can be found in Seeger and Weiler (2014) and in the Supplement, respectively.

In method 1, we adjusted a kriging interpolation of the available precipitation isotope values from 26 long-term monitoring stations for local differences in elevation. For this, we used the monthly average elevation gradient of $\delta^{18}O$ in precipitation, estimated from three isotope monitoring stations in central Switzerland (Meiringen, Guttannen, and Grimsel, Fig. S1 in the Supplement) that cover a similar elevation range as the 22 study catchments. Method 1 can be extended using an energy-balance-based model to explicitly simulate the storage of winter precipitation in the snowpack. The energy-balance-based model uses PREVAH simulations of air temperature, wind speed, incoming shortwave radiation, and precipitation amount to predict the meltwater amounts and their average isotopic compositions for each 100 m elevation band (without considering isotopic fractionation of the snowpack and snowmelt).

In method 2, we fitted isotope data from 19 long-term monitoring stations to sine curves using least squares. We then constructed a multiple linear regression model to explain the best-fit sine parameters as functions of latitude, longitude, and elevation. These spatially varying sine parameters were used to construct interpolated seasonal cycle maps for all of Switzerland. These seasonal cycles were then adjusted using kriged interpolations of the monthly residuals of station measurements from their fitted seasonal patterns, to account for non-sinusoidal isotope dynamics. For both interpolation methods 1 and 2, monthly isotope values were mass-weighted based on the monthly elevation-dependent precipitation volumes obtained from the PREVAH model (Viviroli et al., 2009).

4 Methodological evaluation of the young water fraction framework

4.1 Comparing two methods for spatial interpolation of $\delta^{18}O$ in precipitation

Here, we apply two different methods 1 and 2 (Sect. 3.4) for interpolating monthly precipitation isotopes from nearby

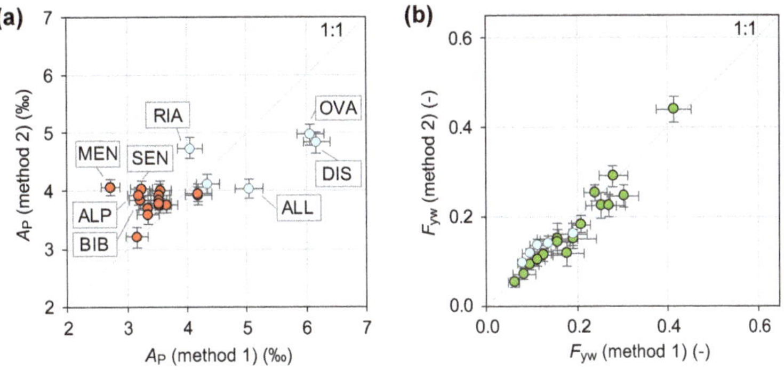

Figure 3. (a) Comparison of flow-weighted seasonal amplitudes of precipitation δ^{18}O cycles (A_P) obtained with two different interpolation methods, method 1 (Seeger and Weiler, 2014) and method 2 (based on Allen et al., 2018; Supplement). Differences in A_P between the two interpolation methods were significant for the catchments highlighted with their abbreviated names. The abbreviations for the study sites stand for Allenbach (ALL), Alp (ALP), Biber (BIB), Dischmabach (DIS), Mentue (MEN), Ova da Clouzza (OVA), Riale di Calneggia (RIA), and Sense (SEN). **(b)** Comparison of young water fractions derived from the two interpolation methods. High-elevation, snow-dominated catchments are marked in light blue colour. Error bars show ±1 SE.

long-term monitoring stations and compare the resulting seasonal cycles of precipitation isotopes and their effects on the calculated young water fractions. In this comparison, method 1 is used without the snow module because method 2 does not allow for explicit simulation of snow accumulation and melt.

Figure 3a shows that the seasonal precipitation isotope cycle amplitudes (A_P) obtained with both methods are similar for most catchments; the differences range from -1.34 ± 0.21 ‰ (±SE, Mentue) to 1.35 ± 0.29 ‰ (Dischmabach). Method 2 results in larger A_P values for five sites (Alp, Biber, Mentue, Sense, and Riale di Caneggia), compared to the results of method 1 (Fig. 3a). However, smaller A_P values are obtained with method 2 for three high-elevation sites: Allenbach, Dischmabach, and Ova de Cluozza. Overall, A_P spanned a range of 1.77 ‰ with method 2, compared to a larger range of 3.47 ‰ with method 1. Nevertheless, for most sites the differences in A_P between the two methods are small compared to the absolute values of A_P, and thus the choice of the interpolation method only marginally affects the estimated young water fractions F_{yw}. For all sites, the absolute differences between the values of F_{yw} calculated with the two interpolation methods are below 0.06 and statistically insignificant (i.e. smaller than twice their pooled uncertainties, Fig. 3b).

A systematic test of both interpolation methods using on-site, long-term precipitation isotope measurements would go beyond the scope of this study. However, method 1 was tested with isotope measurements from six stations (Seeger and Weiler, 2014), and we evaluated the performance of method 2 as described in the Supplement. Results from the two methods are likely to differ because they make different assumptions about the changes in precipitation isotopic composition with elevation. For our objectives, however, it is helpful that these two different approaches yield different A_P in several cases, because it allows us to show that this level of variability in A_P has only minor effects on the calculated young water fractions. Our comparison thus demonstrates that both approaches for spatially interpolating δ^{18}O in precipitation yield consistent young water fraction estimates for the 22 study catchments.

4.2 The effect of snow storage on the seasonal cycle amplitudes and phases of precipitation isotopes

At high-elevation sites with seasonally cold climates, precipitation (and its isotopic signature) will be stored temporarily in the snowpack in winter, and will be released during the melt season. Thus, significant volumes of isotopically depleted snow meltwater may reach the river system during spring and early summer, when the isotope signal of incoming precipitation is more enriched. As a result, the seasonal isotopic variation in water reaching the soil surface (rainwater and snowmelt) is likely to be smaller than the seasonal variation in precipitation alone.

In order to investigate the effect of snow storage on the amplitudes and phases of precipitation isotope cycles, we applied method 1 with the snow module, so that the input to the soil surface, and its isotope signal, can be described by a mixture of rainwater and snowmelt (the "delayed input" scenario in Fig. 4). Alternatively, method 1 can also be applied without the snow module, i.e. by ignoring snowpack as a separate storage, such that the catchment input is taken directly from the incoming precipitation and its isotopic composition (the "direct input" scenario in Fig. 4). Figure 4a shows, as an example, the time series of input water flux and δ^{18}O (not volume-weighted) at the Dischmabach catchment for both scenarios. The delayed release of depleted winter precipitation from the snowpack ("delayed input" scenario) results in a smaller seasonal amplitude of the input tracer signal. However, when this input tracer signal is volume-weighted,

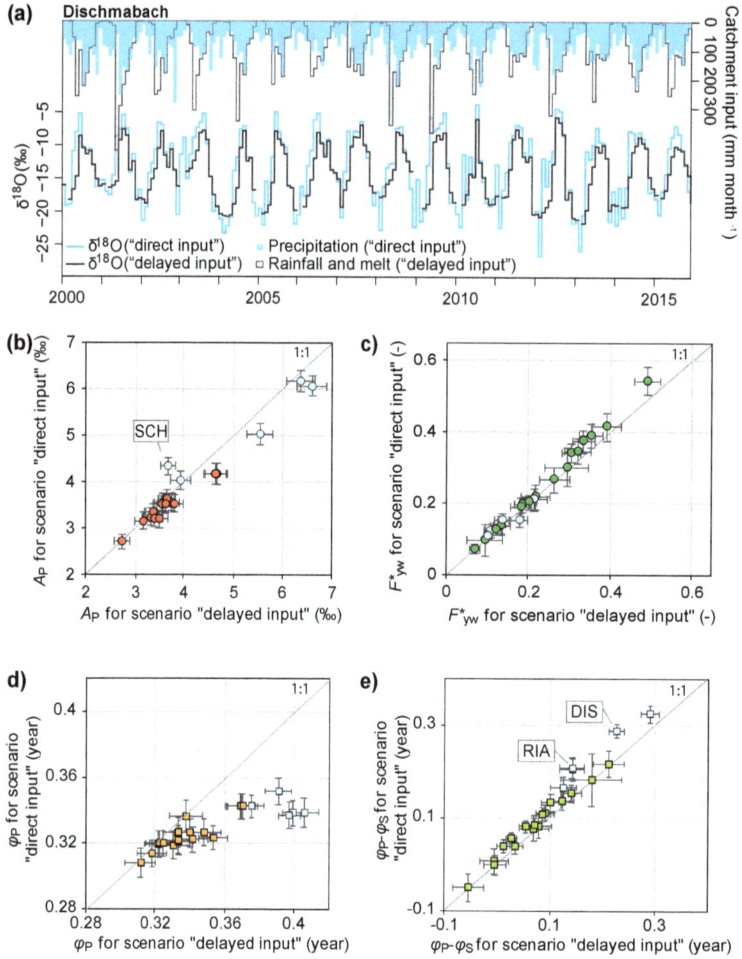

Figure 4. **(a)** Time series of catchment input volumes and $\delta^{18}O$ values (not volume-weighted) for the Dischmabach catchment, calculated using the interpolation method of Seeger and Weiler (2014), with and without modelling of snow accumulation and melt ("delayed input" and "direct input", respectively). Panels **(b)** and **(c)** compare the seasonal amplitudes of the precipitation isotope cycles (volume-weighted), and the resulting flow-weighted young water fractions, with and without modelling of snow accumulation and melt. Panels **(d)** and **(e)** compare the phases of the seasonal precipitation isotope cycles, and the resulting phase shifts, with and without modelling of snow accumulation and melt. High-elevation, snowmelt-dominated sites are marked in light blue. The abbreviations for the study sites stand for Dischmabach (DIS), Riale di Calneggia (RIA), and Schaechen (SCH). Error bars show ±1 SE.

the fitted seasonal amplitudes (A_P) are statistically indistinguishable between the "direct" and "delayed" input scenarios for 21 of the 22 sites (Fig. 4b). This result arises because the "delayed" input scenario gives very little weight to winter inputs in snow-dominated catchments (because snowmelt volumes during winter conditions are small), allowing the fitted cycles to deviate from the winter isotope values. The difference in A_P for both scenarios is statistically significant only at the Schaechen catchment, which contains the highest-elevation snowpacks in our data set (elevation up to 3260 m a.s.l., Table S1). As a consequence, snowmelt at the Schaechen site is isotopically more depleted compared to the other, lower-elevation sites. For the hybrid and rain-dominated sites, the A_P values are almost indistinguishable between the two scenarios, either because snowmelt occurs early in the season when rainwater and snowmelt have similar isotopic signatures (i.e. hybrid catchments), or because the contribution of snowmelt is small compared to that of rainfall (rain-dominated catchments). As a consequence, the young water fractions F_{yw} are virtually identical between the "direct input" and "delayed input" scenarios (Fig. 4c).

As can be seen in Fig. 4a, the delayed meltwater input shifts the seasonal isotope pattern toward later in the season. Thus the "delayed input" scenario results in later cycle phases (φ_P) compared to the "direct input" scenario (Fig. 4d), with statistically significant differences for the five high-elevation, snow-dominated sites and for four hybrid catchments (Erlenbach, Lümpenbach, Vogelbach, and Sitter). However, the "delayed input" scenario had a statistically significant effect on the phase shift between input and output ($\varphi_S - \varphi_P$) only at Dischmabach (where it altered the phase shift by 0.06 years) and Riale di Calneggia (where it altered the phase shift by 0.07 years; Fig. 4e). In the analysis presented below, we use interpolated precipitation isotope values obtained with method 1 that explicitly account for snowpack accumulation and melt (i.e. the "delayed input" scenario) in order to be consistent with previous studies where this data set has been used (Seeger and Weiler, 2014; Staudinger et al., 2017).

4.3 Comparing unweighted and flow-weighted young water fractions

We use the isotope and discharge data sets of the 22 catchments to estimate young water fractions from the ratios of the seasonal cycle amplitudes A_S and A_P, with and without discharge-weighting (F^*_{yw} and F_{yw}, respectively). Figure 5a shows that flow-weighting the streamwater isotope values results in a roughly 25 % increase in the fitted seasonal streamwater isotope cycle amplitudes A_S, relative to the unweighted A_S values for the same sites. Statistically significant differences between unweighted and flow-weighted values of A_S were found for Dischmabach, Emme, Mentue, Rietholzbach, and Sense, as well as Alp, Erlenbach, Lümpenenbach, Vogelbach, and Biber (which are all located nearby one another and share similar catchment characteristics). Perhaps unsurprisingly, the effect of flow-weighting on A_S is largest in catchments with highly variable flow regimes, i.e. at sites with relatively large coefficients of variation of daily discharge (CV_Q) and QFIs (Table 2). In such catchments, robust estimation of the flow-weighted F^*_{yw} may require a smart sampling strategy that captures a representative range of hydrologic conditions.

The flow-weighted F^*_{yw} values range from 0.07 ± 0.01 to 0.49 ± 0.03 (±SE), whereas the unweighted F_{yw} values range from 0.06 ± 0.01 to 0.37 ± 0.03. Thus, flow-weighting the streamwater isotope values yields young water fractions (F^*_{yw}) that are around 26 % larger than those calculated from unweighted streamwater isotope values (F_{yw}; Fig. 5b, Table 3), because high flows generally contain more young water than base flows. The average values of F^*_{yw} and F_{yw} are 0.22 ± 0.02 and 0.17 ± 0.02, respectively, meaning that approximately one-fifth of total discharge was younger than roughly 2.3 ± 0.8 months (assuming that the catchment transit times can be described by gamma distributions with shape factors α ranging from 0.3 to 2; Kirchner, 2016a). Our F_{yw} results are within the range of young water fractions reported for rivers in mountainous regions in North America and central Europe by Jasechko et al. (2016).

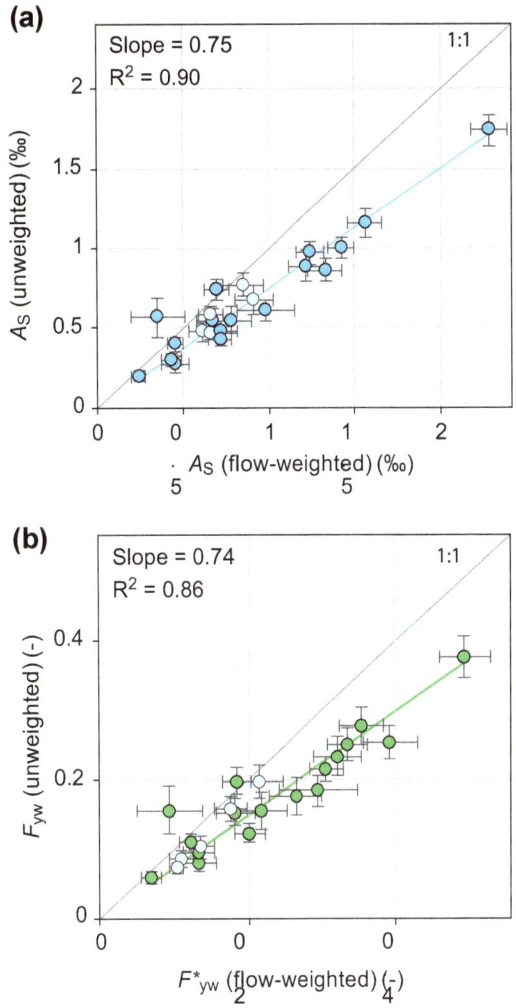

Figure 5. Panel **(a)** compares the seasonal amplitudes of streamwater isotope cycles (A_S) with and without flow weighting. High-elevation, snowmelt-dominated sites are marked in light blue. Panel **(b)** compares flow-weighted young water fractions F^*_{yw} with unweighted young water fractions (F_{yw}). Error bars show ±1 SE. Unweighted young water fractions are roughly 26 % smaller than flow-weighted young water fractions across these catchments.

5 Relationships between young water fractions, hydro-climatic conditions, and landscape characteristics

By examining how the catchments' young water fractions correlate with their landscape and hydro-climatic characteristics, we aim to identify dominant controls on their hydrological behaviour. Below, we present our results for flow-weighted young water fractions (F^*_{yw}); however, the unweighted young water fractions (F_{yw}) yield very similar results, as both values are significantly correlated with each other ($\rho = 0.9$, $p < 0.001$; Table 4).

Table 4 and Fig. 6 show that young water fractions exhibit statistically significant positive correlations with five hydro-climatic indices: mean monthly discharge ($\overline{Q}$), mean monthly precipitation ($\overline{P}$), mean daily precipitation intensity ($\overline{P}_{intensity}$), coefficient of variation of daily discharge (CV_Q), and QFI. These correlations suggest that

Table 2. Hydro-climatic and topographic indices, as well as soil properties of the 22 study catchments.

Catchment name	Quick-flow index QFI (–)	Coefficient of variation of Q (%)	Average discharge $\overline{Q}$ (mm month^{-1})	Average precipitation $\overline{P}$ (mm month^{-1})	Average precip. intensity (mm day^{-1})	Precipitation amplitude A_{precip} (mm month^{-1})	Median flow path length L (m)	Median flow path gradient G (m m^{-1})	L/G (m)	Drainage density, DD (km km^{-2})	Mean topo-graphic gradient (%)	Median topographic wetness index TWI (–)	Fraction forested area (%)	Fraction low-permeab. soils (%)	Fraction saturated soils (%)
Aabach	0.62	126.39	54.9	106.1	5.4	0.4	407	0.04	10 175	1.97	5.5	9.22	14.8	0	17
Aach	0.63	137.20	33.3	85.1	4.8	0.4	481	0.02	24 050	1.46	2.1	9.61	14.3	0	13
Allenbach	0.40	100.77	109.3	99.4	4.7	1.0	423	0.31	1365	2.45	45.1	9.67	15.9	57	20
Alp	0.66	138.90	126.3	158.2	6.3	0.6	196	0.21	933	3.88	25.7	10.92	50.6	48	23
Biber	0.72	149.36	96.2	150.2	5.8	0.5	207	0.16	1294	3.77	18.2	9.48	43.2	30	15
Dischmabach	0.31	85.61	99.5	76.4	3.8	1.3	647	0.33	1961	1.74	46.1	10.07	2.8	59	0
Emme	0.72	142.06	88.7	116.6	4.8	0.6	286	0.27	1059	3.38	33.1	9.76	38.8	49	27
Ergolz	0.59	118.42	39.5	87.7	4.1	0.2	421	0.15	2807	1.34	19.5	9.99	39.9	41	1
Erlenbach	0.81	169.73	138.9	162.4	6.6	0.8	169	0.20	845	6.61	23.9	9.27	62.3	4	95
Guerbe	0.49	90.33	100.3	94.9	4.3	0.5	258	0.19	1358	2.59	27.6	11.67	30.6	35	13
Ilfis	0.53	113.61	79.2	127.5	5.2	0.5	157	0.30	523	3.31	28.6	8.87	50.0	28	6
Langeten	0.30	61.72	53.9	118.2	4.7	0.3	308	0.11	2800	1.70	12.6	9.52	18.4	0	3
Lümpenenbach	0.68	141.41	152.0	157.1	6.0	0.7	155	0.17	912	6.57	19.6	9.85	29.3	4	96
Mentue	0.52	154.74	29.7	89.3	3.9	0.2	364	0.08	4550	1.47	8.9	9.10	23.0	0	0
Murg	0.52	97.33	62.8	116.6	5.1	0.3	219	0.10	2190	2.16	12.0	10.83	32.0	0	8
Ova da Cluozza	0.39	88.68	72.8	61.3	3.5	1.2	616	0.46	1339	0.72	59.3	9.43	13.8	34	0
Riale di Calneggia	0.52	154.05	143.7	129.3	6.4	1.1	647	0.46	1407	1.03	64.4	9.51	15.3	44	0
Rietholzbach	0.69	140.73	87.2	121.1	5.6	0.4	194	0.18	1078	2.09	14.6	9.88	22.3	0	0
Schaechen	0.40	86.35	120.3	140.0	5.9	0.9	646	0.38	1700	1.36	54.3	9.38	19.2	67	5
Sense	0.53	101.76	54.2	95.2	4.3	0.5	227	0.20	1135	2.84	24.1	9.00	32.9	24	19
Sitter	0.62	120.79	94.2	168.7	6.4	0.6	329	0.27	1219	2.65	35.2	9.96	30.8	61	14
Vogelbach	0.70	162.67	117.3	162.2	6.3	0.8	193	0.28	689	6.55	28.9	9.69	82.1	51	49

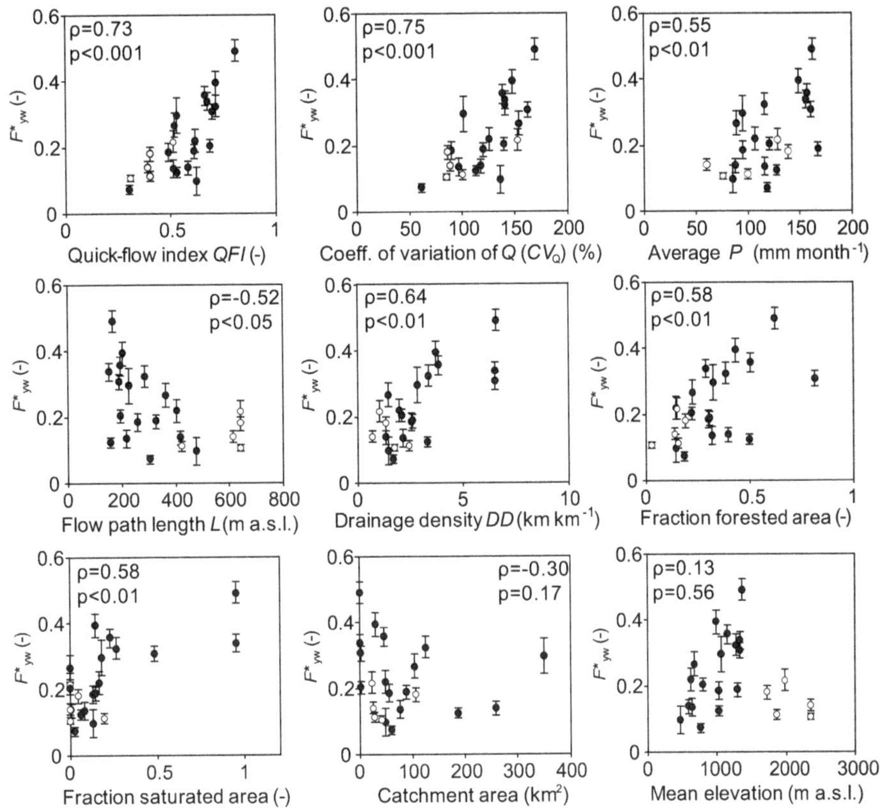

Figure 6. Scatter plots showing how young water fractions correlate with climatic and landscape indices. High-elevation, snowmelt-dominated sites are marked in light blue. Error bars show ±1 SE. Spearman rank correlation coefficients (ρ) and corresponding p values are provided in the individual figures.

young water fractions tend to be highest in humid catchments where prompt runoff response is facilitated by fast flow paths and/or high-intensity precipitation events. F^*_{yw} was also significantly correlated with high values of drainage density (DD) and low values of flow path length (L) (Table 4). There was also a significant negative correlation with the ratio of the flow path length to gradient (L/G), but as there is nearly zero correlation with G itself, the correlation with L/G apparently arises through L alone. Drainage density is inversely proportional to median flow path length, so the strong positive correlation of F^*_{yw} with DD and negative correlation with L can be viewed as two sides of the same coin. All else equal, high values of DD, and thus small values of L, facilitate faster runoff, which is directly linked to higher values of CV_Q and QFI.

A statistically significant inverse correlation ($\rho = -0.36$, $p < 0.0001$) between F_{yw} and the logarithm of the topographic gradient was found by Jasechko et al. (2016) for 254 sites across Europe and North America, with the surprising implication that steeper catchments have less (not more) young streamflow. Among our individual catchments, however, we find no correlation between F_{yw} (or F^*_{yw}) and topographic gradient. This may be partly explained by the lack of low gradient catchments among our study sites; our gradients span a range of 0.02–0.64 compared to ~ 0.0007–0.11 in Jasechko et al. (2016), and the correlation that they observe appears to be largely driven by sites with gradients less than roughly 0.01. Nevertheless, our data set fits within the global pattern found by Jasechko et al. (2016), and the median F_{yw} of our 22 mostly high-gradient study catchments (0.16, 95 % confidence interval 0.10–0.21) is smaller than the global median (0.21, 95 % confidence interval 0.19–0.24) consistent with the gradient-dependence hypothesized by Jasechko et al. (2016).

Some studies have identified catchment area as a major control on mean transit times (e.g. DeWalle et al., 1997; Soulsby et al., 2000), however, the inverse correlation of F^*_{yw} and F_{yw} with catchment area only becomes significant ($\rho = -0.49$, $p < 0.05$) when the five high-elevation, snow-dominated sites are omitted from the analysis (Fig. 6). The young water fractions of the remaining 17 sites were also strongly correlated with mean catchment elevation ($\rho = 0.65$, $p < 0.005$, Fig. 6), which in turn is a major control on other hydro-climatic indices ($\overline{Q}$, $\overline{P}$) and topographic indices (DD, G, L, L/G, and TWI).

Table 3. Values ± SE (standard errors) of flow-weighted seasonal amplitude coefficients of precipitation isotopes (A_P), unweighted and flow-weighted seasonal amplitude coefficients of streamwater isotopes (A_S), unweighted and flow-weighted young water fractions, and the discharge sensitivity of the young water fraction (estimated as the linear slope of the Q–F_{yw} relationship; see Sect. 6).

Catchment name	$A_P \pm$ SE (‰)	$A_S \pm$ SE (‰)		$F_{yw} \pm$ SE (–)		Discharge
	Volume-weighted	Unweighted	Flow-weighted	Unweighted	Flow-weighted	sensitivity of $F_{yw} \pm$ SE (day mm^{-1})
Aabach	3.57 ± 0.18	0.55 ± 0.09	0.77 ± 0.12	0.15 ± 0.03	0.22 ± 0.04	0.0530 ± 0.0247
Aach	3.65 ± 0.18	0.57 ± 0.12	0.35 ± 0.16	0.16 ± 0.03	0.10 ± 0.04	–*
Allenbach	5.54 ± 0.27	0.48 ± 0.06	0.61 ± 0.08	0.09 ± 0.01	0.11 ± 0.02	0.0185 ± 0.0065
Alp	3.50 ± 0.19	0.97 ± 0.07	1.24 ± 0.08	0.28 ± 0.03	0.35 ± 0.03	0.0119 ± 0.0063
Biber	3.39 ± 0.19	0.86 ± 0.07	1.33 ± 0.10	0.25 ± 0.03	0.39 ± 0.04	0.0299 ± 0.0074
Dischmabach	6.36 ± 0.29	0.46 ± 0.04	0.66 ± 0.05	0.07 ± 0.01	0.10 ± 0.01	0.0169 ± 0.0021
Emme	3.80 ± 0.20	0.88 ± 0.10	1.22 ± 0.11	0.23 ± 0.03	0.32 ± 0.03	0.0237 ± 0.0107
Ergolz	3.15 ± 0.18	0.30 ± 0.04	0.43 ± 0.06	0.09 ± 0.01	0.14 ± 0.02	0.0651 ± 0.0186
Erlenbach	4.63 ± 0.22	1.74 ± 0.09	2.27 ± 0.10	0.37 ± 0.03	0.49 ± 0.03	0.0066 ± 0.0029
Guerbe	3.63 ± 0.19	0.55 ± 0.07	0.67 ± 0.09	0.15 ± 0.02	0.18 ± 0.03	0.0214 ± 0.0113
Ilfis	3.63 ± 0.19	0.40 ± 0.05	0.45 ± 0.05	0.11 ± 0.01	0.12 ± 0.02	0.0061 ± 0.0067
Langeten	3.36 ± 0.18	0.20 ± 0.03	0.24 ± 0.04	0.06 ± 0.01	0.07 ± 0.01	0.0503 ± 0.0166
Lümpenenbach	4.66 ± 0.22	1.16 ± 0.09	1.56 ± 0.10	0.25 ± 0.02	0.33 ± 0.03	0.0174 ± 0.0049
Mentue	2.72 ± 0.16	0.48 ± 0.07	0.72 ± 0.09	0.18 ± 0.03	0.26 ± 0.04	0.0732 ± 0.0360
Murg	3.39 ± 0.18	0.27 ± 0.05	0.45 ± 0.08	0.08 ± 0.01	0.13 ± 0.03	0.0304 ± 0.0114
Ova da Cluozza	6.60 ± 0.30	0.68 ± 0.09	0.91 ± 0.11	0.10 ± 0.01	0.14 ± 0.02	0.0328 ± 0.0077
Riale di Calneggia	3.94 ± 0.20	0.77 ± 0.07	0.85 ± 0.12	0.20 ± 0.02	0.22 ± 0.03	0.0451 ± 0.0145
Rietholzbach	3.54 ± 0.18	0.43 ± 0.04	0.71 ± 0.06	0.12 ± 0.01	0.20 ± 0.02	0.0132 ± 0.0045
Schaechen	3.67 ± 0.16	0.58 ± 0.05	0.66 ± 0.07	0.16 ± 0.02	0.18 ± 0.02	0.0174 ± 0.0050
Sense	3.32 ± 0.19	0.61 ± 0.07	0.98 ± 0.17	0.18 ± 0.02	0.29 ± 0.05	0.0463 ± 0.0105
Sitter	3.75 ± 0.18	0.74 ± 0.06	0.69 ± 0.07	0.20 ± 0.02	0.19 ± 0.02	−0.0085 ± 0.0090
Vogelbach	4.64 ± 0.22	1.00 ± 0.06	1.42 ± 0.08	0.21 ± 0.02	0.31 ± 0.02	0.0107 ± 0.0034

* The catchment Aach was omitted from the analysis because its isotope data set contained only two data points during high-flow conditions.

Across the 22 catchments, F^*_{yw} is positively correlated with the areal fraction of saturated soils ($\rho = 0.58$, $p < 0.01$) and low-permeability soils ($\rho = 0.52$, $p < 0.05$). These relationships remain significant when the snow-dominated sites are omitted from the analysis. A strong positive relationship with F^*_{yw} can be expected because saturated soils and low-permeability soils are often associated with overland flow and/or fast subsurface flow mechanisms triggered by exceedance of soil water storage thresholds (saturation excess; Dunne and Black, 1970) or precipitation intensity (infiltration excess; Horton, 1933). Particularly high fractions of saturated soils occur at three neighbouring catchments (Erlenbach, Lümpenenbach and Vogelbach) that are characterized by shallow gleysols (Feyen et al., 1996; Fischer et al., 2015). Together with the nearby Biber catchment, these four sites exhibit the largest young water fractions in our data set. No correlation was evident between F^*_{yw} and the fraction of soils with low water storage capacity, likely due to the strong influence of six sites where this fraction was zero.

F^*_{yw} is not significantly correlated with the areal fractions mapped as having high, intermediate, or low groundwater productivity, here used as a proxy for the catchments' hydrogeologic properties. This result is perhaps unsurprising; most groundwater is probably older than the threshold age that defines young water, so the young water fraction will not be sensitive to how much older the groundwater is. Instead, the fraction of young water should primarily reflect mechanisms that control flow processes and routing near the land surface (shallow groundwater, soil water, overland flow) rather than groundwater flow in deep aquifers where flow velocities can be several orders of magnitude slower.

Across our study catchments, the young water fraction is strongly correlated with the areal fraction of forest ($\rho = 0.58$, $p < 0.01$; Tables 1 and 4). Excluding the snow-dominated sites from the analysis slightly weakens this relationship although it remains statistically significant ($\rho = 0.51$, $p < 0.05$). One would normally expect tree roots to increase soil permeability, resulting in greater infiltration and groundwater recharge (Brantley et al., 2017). However, on steep forested slopes, abundant lateral preferential flow pathways (e.g. macropores) may facilitate rapid transport of water (Whipkey, 1965). Thus, the correlation we observe may be artefactual, since across our sites, forest cover is also correlated with higher drainage densities and shorter mean

Table 4. Spearman rank correlation coefficients relating the flow-weighted (F^*_{yw}) and unweighted (F_{yw}) young water fractions, and the discharge sensitivity of F_{yw}, to selected hydro-climatic indices and landscape properties of the 22 Swiss catchments. The corresponding p values are indicated by regular font in grey fields ($p < 0.05$), bold font in grey fields ($p < 0.01$), as well as italic and underlined font in grey fields ($p < 0.001$); fields without grey shading indicate $p > 0.05$.

					Hydro-climatic indices						Topographic indices									Soils	
		F^*_{yw}	F_{yw}	Discharge sensitivity [a]	$\overline{Q}$	$\overline{P}$	A_{precip}	Mean $P_{intensity}$	*QFI*	CV_Q	Catchm. area	Mean Elevation	Topogr. gradient	*DD*	*TWI*	*L*	*G*	*L/G*	Fraction forested area	Fraction low-permeab. soils	Fraction saturated soils
	Flow-weighted young water fraction F^*_{yw}		*0.90*	-0.14	0.44	**0.55**	0.14	0.52	*0.73*	*0.75*	-0.30	0.13	-0.03	**0.64**	0.01	-0.52	-0.02	**-0.55**	**0.58**	0.52	**0.58**
	Unweighted young water fraction F_{yw}	*0.90*		-0.32	0.50	**0.65**	0.23	*0.67*	*0.76*	*0.77*	-0.26	0.17	0.04	**0.62**	-0.04	-0.43	0.04	**-0.54**	**0.54**	0.43	**0.63**
	Discharge sensitivity of F_{yw} [a]	-0.14	-0.32		**-0.61**	**-0.64**	-0.44	**-0.56**	-0.33	-0.13	0.32	-0.44	-0.38	**-0.59**	0.52	0.46	-0.46	*0.75*	-0.38	-0.38	-0.33
Hydro-climatic indices	Average discharge $\overline{Q}$	0.44	0.50	**-0.61**		**0.62**	*0.75*	*0.65*	0.20	0.25	**-0.60**	*0.72*	**0.58**	0.46	-0.52	-0.21	**0.58**	**-0.57**	0.20	0.32	0.38
	Average precipitation $\overline{P}$	**0.55**	**0.65**	**-0.64**	**0.62**		0.20	*0.91*	0.52	0.47	-0.32	0.20	0.11	**0.64**	-0.27	**-0.57**	0.15	**-0.62**	**0.58**	0.49	0.50
	Precipitation amplitude A_{precip}	0.14	0.23	-0.44	*0.75*	0.20		0.26	-0.11	-0.03	-0.50	*0.95*	*0.84*	0.11	*-0.70*	0.25	*0.83*	-0.36	-0.18	-0.08	0.13
	Mean $P_{intensity}$	0.52	*0.67*	**-0.56**	*0.65*	*0.91*	0.26		**0.56**	**0.56**	-0.43	0.18	0.09	0.48	-0.21	-0.40	0.12	-0.47	0.40	0.44	0.45
	Quick-flow index *QFI*	*0.73*	*0.76*	-0.33	0.20	0.52	-0.11	**0.56**		*0.82*	-0.27	-0.20	-0.30	*0.66*	0.11	**-0.62**	-0.26	-0.51	**0.62**	*0.70*	**0.62**
	Coefficient of variation CV_Q	*0.75*	*0.77*	-0.13	0.25	0.47	-0.03	**0.56**	*0.82*		-0.39	-0.06	-0.20	0.44	0.04	-0.42	-0.16	-0.36	0.47	0.46	0.41
Topographic indices	Catchment area	-0.30	-0.26	0.32	**-0.60**	-0.32	-0.50	-0.43	-0.27	-0.39		-0.43	-0.11	-0.24	0.25	0.19	-0.17	0.29	0.12	-0.26	-0.18
	Mean elevation	0.13	0.17	-0.44	*0.72*	0.20	*0.95*	0.18	-0.20	-0.06	-0.43		*0.88*	0.10	*-0.79*	0.19	*0.89*	-0.43	-0.15	-0.10	0.05
	Topographic gradient	-0.03	0.04	-0.38	**0.58**	0.11	*0.84*	0.09	-0.30	-0.20	-0.11	*0.88*		-0.07	*-0.86*	0.33	*0.97*	-0.37	-0.09	-0.19	-0.11
	Drainage density *DD*	**0.64**	**0.62**	**-0.59**	0.46	**0.64**	0.11	0.48	*0.66*	0.44	-0.24	0.10	-0.07		-0.06	*-0.84*	-0.04	*-0.77*	*0.73*	**0.63**	*0.83*
	Topogr. wetness index *TWI*	0.01	-0.04	0.52	-0.52	-0.27	*-0.70*	-0.21	0.11	0.04	0.25	*-0.79*	*-0.86*	-0.06		-0.02	*-0.91*	**0.59**	-0.02	0.03	0.07
	Flow path length *L*	-0.52	-0.43	0.46	-0.21	**-0.57**	0.25	-0.40	**-0.62**	-0.42	0.19	0.19	0.33	*-0.84*	-0.02		0.25	*0.73*	*-0.75*	**-0.65**	**-0.57**
	Flow gradient *G*	-0.02	0.04	-0.46	**0.58**	0.15	*0.83*	0.12	-0.26	-0.16	-0.17	*0.89*	*0.97*	-0.04	*-0.91*	0.25		-0.45	-0.06	-0.16	-0.13
	L/G	**-0.55**	**-0.54**	*0.75*	**-0.57**	**-0.62**	-0.36	-0.47	-0.51	-0.36	0.29	-0.43	-0.37	*-0.77*	**0.59**	*0.73*	-0.45		**-0.65**	-0.51	-0.50
	Fraction forested area	**0.58**	**0.54**	-0.38	0.20	**0.58**	-0.18	0.40	**0.62**	0.47	0.12	-0.15	-0.09	*0.73*	-0.02	*-0.75*	-0.06	**-0.65**		*0.66*	**0.55**
Soils	Fraction low-permeab. soils	0.52	0.43	-0.38	0.32	0.49	-0.08	0.44	*0.70*	0.46	-0.26	-0.10	-0.19	**0.63**	0.03	**-0.65**	-0.16	-0.51	*0.66*		**0.60**
	Fraction saturated soils	**0.58**	**0.63**	-0.33	0.38	0.50	0.13	0.45	**0.62**	0.41	-0.18	0.05	-0.11	*0.83*	0.07	**-0.57**	-0.13	-0.50	0.55	**0.60**	

[a] The catchment Aach was omitted from the analysis because its isotope data set contained only two data points during high-flow conditions.

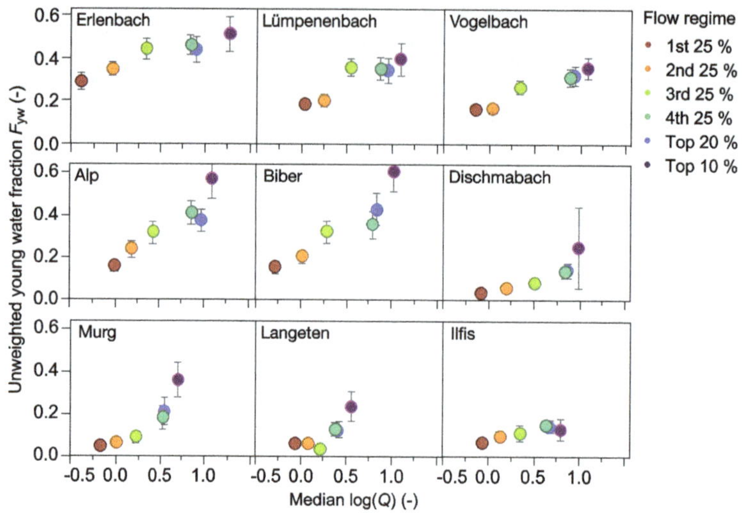

Figure 7. Variation in unweighted young water fractions with flow regime (log-transformed) for the nine Swiss catchments that have sufficiently long time series of streamwater isotope measurements. Error bars show ±1 SE. The young water fraction increases with discharge differently at different sites, suggesting different degrees of activation of fast flow paths at high flows.

flow paths, as well as higher fractions of saturated and low-permeability soils, all of which can plausibly increase the young water fraction. More generally, among our 22 study sites, hydro-climatic characteristics are correlated with landscape properties, making it challenging to clearly identify individual controls on the young water fraction. Broadly, however, we can conclude that high young water fractions are generally associated with hydro-climatic factors (e.g. humid climate and high precipitation intensity) and landscape characteristics (e.g. low soil permeability and high drainage density) that facilitate fast streamflow responses.

5.1 Discharge sensitivity of the young water fraction as a diagnostic indicator of runoff generation processes

The catchment inter-comparison analysis presented in Sect. 5 suggests that wetter catchments, and those with shorter and faster flow paths, have larger young water fractions. In individual catchments, one would also expect young water fractions (and thus seasonal isotope cycles) to be variable in time, i.e. to be larger during periods of stronger precipitation forcing and wetter antecedent conditions, as shallower, faster flow paths become more dominant, and as the stream network extends farther into the landscape, shortening the average path length of subsurface flow (Godsey and Kirchner, 2014). In this section, we examine how young water fractions respond to changes in catchment wetness, as reflected in stream discharge.

5.2 Young water fractions of distinct flow regimes

Our expectation that the young water fraction should be higher under wetter conditions (and thus during higher stream discharges) is borne out by the observation that flow-weighted young water fractions are systematically higher than unweighted young water fractions (Sect. 4.3). We can visualize the relationship between F_{yw} and stream discharge (as a proxy for catchment wetness) by separating the streamwater isotope time series into different discharge ranges and calculating the seasonal isotope cycles and F_{yw} values individually for each of these flow regimes. These flow regimes comprise the 1st to 4th quartiles, as well as the upper 20 and 10 % of daily discharges at the day of sampling. For instance, from the 140 streamwater isotope samples at the Erlenbach site, each quartile of Q comprised 35 samples, while the upper 20 and 10 %, of daily discharges comprised 28 and 14 samples, respectively. In Fig. 7, we plot F_{yw} in relation to the median discharge values of the six flow regimes at nine of our study sites. These sites have the longest isotope time series in our data set, allowing us to estimate robust seasonal cycle coefficients A_S for each individual flow regime. At our sites with shorter time series, sub-sampling individual flow regimes would result in highly uncertain A_S estimates.

The visual patterns shown in Fig. 7 are similar for catchments located close to each other, such as for Alp and Biber, or for Lümpenenbach, Vogelbach, and Erlenbach. However, young water fractions vary substantially among the sites in Fig. 7, with F_{yw} in the lowest flow regime ranging from 0.03 at Dischmabach to 0.29 at Erlenbach and F_{yw} in the highest flow regime ranging from 0.13 at Ilfis to 0.60 at Biber. Figure 7 suggests that the relationship between discharge and F_{yw} may be a diagnostic fingerprint linked to hydrological properties that control the storage and release of young water. However, the nine catchments shown in Fig. 7 are too small of a sample to draw any robust conclusions concerning

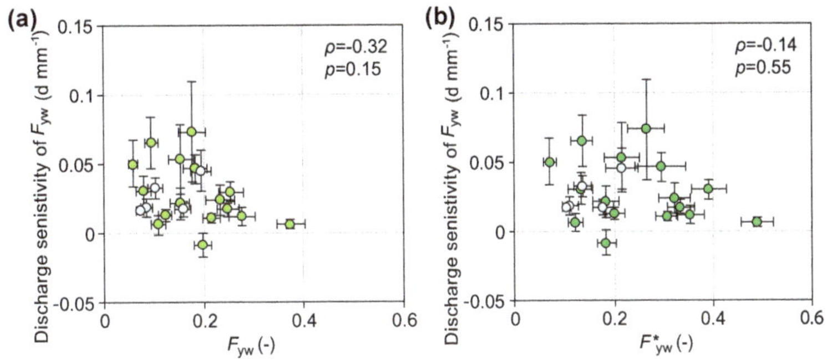

Figure 8. Scatter plots of the unweighted and flow-weighted young water fractions versus the discharge sensitivity of F_{yw} calculated for 21 of the 22 Swiss catchments (no discharge sensitivity was calculated for the Aach catchment because only two isotope values existed for high-flow conditions). High-elevation, snowmelt-dominated sites are marked in light blue. Error bars show ±1 SE. There is no systematic relationship between the young water fractions and their discharge sensitivities.

how this fingerprint may vary with catchments' landscape characteristics and hydro-climatic conditions.

5.3 Estimating the discharge sensitivity of F_{yw} and linking it to catchments' landscape and hydro-climatic characteristics

As a first-order estimate of the sensitivity of F_{yw} to discharge across all 22 study catchments, we calculated the linear slope of the relationship between Q and F_{yw}, using a method that does not require breaking the streamwater isotope time series into separate flow regimes (and thus has more modest data requirements than plots like Fig. 7). Instead of fitting a linear slope to the few data points shown in Fig. 7, we estimated the linear slope of the Q–F_{yw} relationship directly from the tracer time series $c_S(t)$ and $c_P(t)$. For each site, we assume that the seasonal amplitude of precipitation isotopes (A_P) is independent of Q, leaving the seasonal amplitude of streamwater isotopes A_S as the only flow-rate-dependent variable. If A_S varies with discharge but A_P does not, then the young water fraction F_{yw} varies with Q as follows:

$$F_{yw}(Q) = A_S(Q)/A_P. \tag{7}$$

If we approximate A_S as a linear function of Q,

$$A_S(Q) = n_S + m_S Q, \tag{8}$$

we can estimate the linear slope (m_S) and the intercept (n_S) through nonlinear fitting (analytic Gauss–Newton algorithm) by replacing A_S in Eq. (2) with $A_S(Q)$ from Eq. (6), yielding the following:

$$c_S(t) = (n_S + m_S Q) \cdot \sin(2\pi f t - \varphi_S) + k_S. \tag{9}$$

In Eq. (8), φ_S is the phase of the seasonal streamwater isotope cycle (rad), t is the time (decimal year), f is the frequency (yr^{-1}), and k_S (‰) is a constant describing the vertical offset of the streamwater isotope signal. For the sake of simplicity, Eq. (8) assumes that the amplitude of the seasonal cycle varies with Q but the phase φ_S does not. Numerical experiments (e.g. Fig. 8 in Kirchner, 2016b) suggest that the change in streamwater isotope cycle phase φ_S between high and low flows should have only a minor influence on the estimate of the parameters in Eq. (8), because the change in φ_S can only be large when the cycle is strongly damped (i.e. during low-flow conditions), and the phase of such a strongly damped cycle will have little effect on the fit to the data.

Combining Eqs. (6) and (7) yields

$$F_{yw}(Q) = \frac{n_S + m_S Q}{A_P} = \frac{n_S}{A_P} + \frac{m_S}{A_P} Q \tag{10}$$

and thus, the linear slope of the dependence of F_{yw} on Q can be approximated as m_S/A_P, which has units of Q^{-1}. The uncertainty in this slope was estimated through Gaussian error propagation. Please note that Eq. (9) quantifies discharge sensitivity based on the linear slope of the relationship between F_{yw} and Q, whereas Fig. 7 shows how F_{yw} varies with $\log(Q)$ for different fractions of the discharge distribution. By replacing Q with $\log(Q)$ in Eqs. (6)–(9), one could easily determine the linear slope of the relationship between F_{yw} and $\log(Q)$ instead.

For convenience, we term this linear slope of the Q–F_{yw} relationship the "discharge sensitivity" of F_{yw}. Our use of this term should not be interpreted to mean that F_{yw} depends, in a mechanistic sense, on discharge per se. Instead, we use the term to indicate the statistical sensitivity of F_{yw} to discharge, where discharge is a proxy indicator of catchment wetness conditions and hydro-climatic forcing. Catchments with high discharge sensitivity of F_{yw} (steep linear slope in Eq. 9) are ones in which the young water fraction varies greatly between low and high flows, suggesting that faster flow paths are more predominant in larger events. Conversely, catchments with low discharge sensitivity (shallower linear slopes in Eq. 9) are ones in which young water fractions are broadly similar between low and high flows, suggesting that the same predominant flow paths are activated in similar proportions in both large and small runoff events.

On average, we find that every 1 mm day^{-1} increase in discharge is associated with an increase of 0.0202 ± 0.0046 in F_{yw}. From this analysis, we excluded the Aach catchment because only two streamwater samples were collected during high-flow conditions, resulting in an unrealistic and highly uncertain value for m_S. At the remaining 21 sites, the discharge sensitivities of F_{yw} range between zero (within error) at Ilfis and Sitter, and 0.0732 ± 0.0360 day mm^{-1} at Mentue. A similar analysis was carried out by Wilusz et al. (2017) for two neighbouring catchments in Plynlimon, Wales. For those two sites, Wilusz et al. (2017) combined a rainfall–runoff model with a rank storage selection (rSAS) transit time model and estimated an increase in F_{yw} of 0.031 to 0.040, respectively, with every 1 mm day^{-1} increase in average annual precipitation. Multiplying their "precipitation sensitivities of F_{yw}" by the site-specific runoff ratios (0.78 and 0.90) yields average discharge sensitivities of F_{yw} of 0.0242 and 0.0360 day mm^{-1}, respectively, which are within the range of values we obtained for our 22 Swiss study sites. Even though the methods, tracers, and timescales Wilusz et al. used to estimate F_{yw} differed from ours, the similarity in the discharge sensitivities between their sites and ours suggests that this may be a robust and reproducible metric that could be useful in future catchment studies.

For our study catchments, there was no systematic relationship between the young water fraction (either F_{yw} or F^*_{yw}) and the discharge sensitivity, indicating that they are different and largely independent measures of catchment behaviour (Figs. 8 and 9). The discharge sensitivity of F_{yw} is, however, strongly correlated to a range of landscape and hydro-climatic conditions, including $\overline{P}$ ($\rho = -0.64$, see also Fig. 9b), $\overline{P}_{\text{intensity}}$ ($\rho = -0.56$), $\overline{Q}$ ($\rho = -0.61$), DD ($\rho = -0.59$), L/G ($\rho = 0.75$), L ($\rho = 0.46$), G ($\rho = -0.46$), TWI ($\rho = 0.52$), A_{precip} ($\rho = -0.44$), and mean catchment elevation ($\rho = 0.44$). All of these correlations remain statistically significant (and many become stronger) when the snow-dominated sites are excluded from the analysis.

In contrast, calculating linear slopes between F_{yw} and $\log(Q)$, instead of Q, yields no significant correlations with any of the variables in Tables 2 or S1. It should be noted that calculations based on $\log(Q)$ will be more strongly influenced by small discharges, whereas calculations based on Q will be more strongly influenced by the upper tail of the Q distribution. Thus, since our primary focus is storm runoff generation, we interpret the discharge sensitivities of F_{yw} based on Q instead of $\log(Q)$.

Our results suggest that catchments with low discharge sensitivity of F_{yw} are characterized by high elevations, dense river networks (high DD, low L/G) and/or generally humid conditions (high $\overline{P}$). These catchment properties are generally associated with predominantly shallow runoff flow paths during both large and small precipitation events, such that the fraction of young water remains relatively high under widely varying flow regimes. In contrast, in catchments characterized by lower drainage density and less humid conditions, larger or higher-intensity storms are likely to strongly alter the proportions of different dominant flow paths, leading to bigger variations in F_{yw} (i.e. higher discharge sensitivity). For example, the dynamic extension of the stream network (e.g. Godsey and Kirchner, 2014; Jensen et al., 2017) and/or the increase in hydrologic connectivity between the stream network and the surrounding landscape (e.g. Detty and McGuire, 2010; Phillips et al., 2011; von Freyberg et al., 2015) should more strongly influence the relative proportion of young streamflow in catchments where drainage density is not already high. Likewise, the activation of shallow flow paths during larger storm events will have a bigger influence on F_{yw} in drier catchments than in wetter ones, where shallow flow paths are likely to be activated during both large and small events.

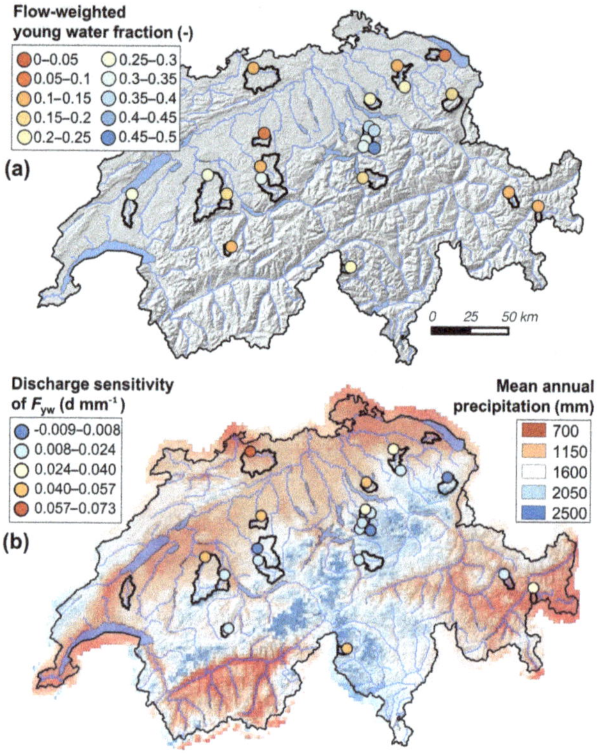

Figure 9. **(a)** Flow-weighted young water fractions at the 22 Swiss study catchments; **(b)** discharge sensitivity of F_{yw} at the same sites (mean annual precipitation for the period 1991–2010 is shown for comparison).

Interestingly, although F_{yw} and its discharge sensitivity are not significantly correlated with each other, they are often correlated with catchment characteristics in opposite ways (Table 4). For example, DD, $\overline{Q}$, $\overline{P}$, $\overline{P}_{\text{intensity}}$, QFI, and CV_Q exhibit positive correlations with F_{yw} but also exhibit negative correlations with the discharge sensitivity of F_{yw}. In catchments with dense river networks and/or generally hu-

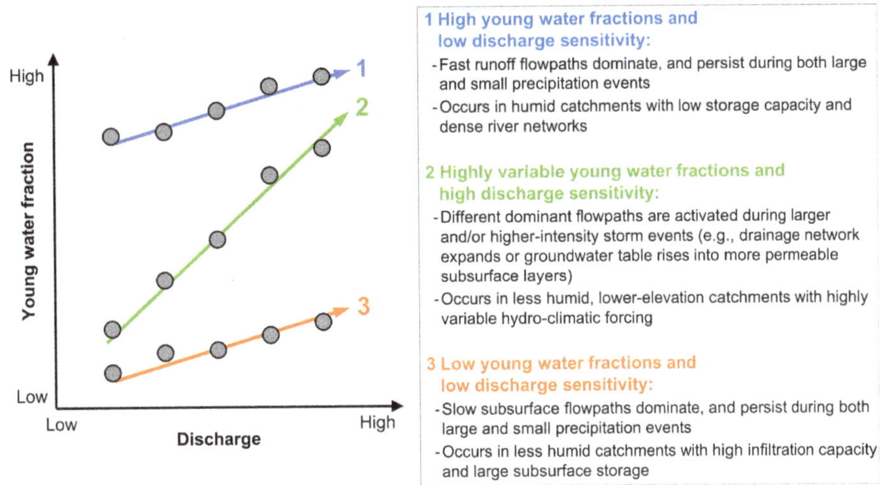

Figure 10. Conceptual description of the mechanistic relationship between young water fractions and discharge, which is used here as a proxy for catchment wetness and hydro-climatic forcing. The three colours of the arrows represent three individual hypothetical catchments.

mid climates, fast runoff flow paths will dominate (and thus F_{yw} and F^{*}_{yw} will be high). These same conditions should also make fast runoff flow paths more persistent, with the result that the young water fraction will not be strongly dependent on catchment wetness conditions or hydro-climatic forcing (and thus discharge sensitivity will be low).

5.4 A conceptual model of the mechanistic relationship between young water fractions and discharge

Figure 10 presents a conceptual summary of the relationships between the young water fraction, its discharge sensitivity, and landscape and hydro-climatic characteristics that control streamflow generation. We suggest that the general trend of the Q–F_{yw} relationship is positive because high-flow periods during precipitation events are likely to contain larger fractions of young water travelling by quick flow paths, while low-flow conditions are primarily sustained by older groundwater. In Fig. 10, the steepness of the linear slope expresses how extensively fast flow paths are activated during high flows. In theory, a linear slope of zero (i.e. F_{yw} insensitive to discharge) would represent strictly linear rainfall–runoff behaviour with a constant mixing fraction of young and old water. In natural systems, however, the relative proportions of streamflow generation mechanisms are likely to vary between high and low flows, making F_{yw} sensitive to discharge. From our analyses in Sect. 6.1. and 6.2, we find that low discharge sensitivities of F_{yw} can occur at sites with either high or low young water fractions (cases 1 and 3, respectively, in Fig. 10; e.g. Erlenbach and Ilfis, respectively, in Fig. 7). Case 1 might be found in humid catchments with frequent precipitation, low storage capacity, and dense river networks, where shallow runoff flow paths dominate both during and between events (e.g. triggered by saturation excess). Case 3 is more likely to occur in catchments with high infiltration capacity and large subsurface storage, where slow subsurface flow paths dominate both during events and between them, leading to consistently low young water fractions. A steep linear slope (case 2 in Fig. 10; e.g. Alp, Biber or Murg in Fig. 7) is likely to occur in catchments where the relative contributions of fast and slow flow paths vary dramatically in response to hydro-climatic forcing or antecedent wetness conditions, for example through drainage network expansion, or shifts in hydrological connectivity due to groundwater tables rising into more permeable layers.

The hydrological concepts presented in Fig. 10 are based on the young water fraction analysis for 21 Swiss catchments that share several landscape and hydro-climatic characteristics, such as similar vegetation cover, relatively humid climate, and (partly) mountainous terrain. Hence, we must be cautious about extending this conceptual model to regions characterized by (semi-)arid or arctic climates, very different vegetation cover or predominantly flat terrain. In addition, linking young water fractions to catchment wetness conditions and hydro-climatic forcing may be difficult in catchments with streamflow regimes that are discontinuous or strongly affected by lakes, water management (e.g. groundwater pumping, artificial groundwater recharge, irrigation, or water diversion) or land-use change (e.g. urban development, soil degradation, or forest clear cutting). Nevertheless, long-term tracer data sets from other catchments could be used to expand our analysis beyond the Swiss study sites and to test the transferability of the conceptual model presented in Fig. 10.

6 Summary and conclusions

The fraction of streamflow younger than roughly 2–3 months has recently been proposed as a robust measure of water age which can be estimated directly from the seasonal cycles of stable water isotopes in precipitation and streamflow (Kirchner, 2016a, b). Here, we have leveraged an extensive isotope data set from 22 small- to medium-sized Swiss catchments to explore how the young water fraction (F_{yw}) varies with catchment characteristics and climatic forcing.

Catchment inter-comparison studies require application of consistent procedures across sites, so we quantified how choices of methodology may affect estimates of F_{yw}. Across the 22 sites, F_{yw} values were not particularly sensitive to the spatial interpolation methods used to estimate precipitation isotope signatures (Sect. 4.1), or sensitive to whether one accounts for snow accumulation and melt in estimating isotopic inputs to the catchment (Sect. 4.2). Flow-weighting the streamwater isotope measurements, however, yielded flow-weighted young water fractions (F^*_{yw}) that were roughly 26 % larger than their unweighted counterparts (F_{yw}; Sect. 4.3, Fig. 5). This result is not surprising, because flow peaks typically follow intense rainfall and thus should contain more recent precipitation than base flows. Here we quantify, for the first time, how flow-weighting affects young water fractions, using real-world data.

The flow-weighted young water fractions of the 22 Swiss catchments ranged from 0.07 ± 0.01 to 0.49 ± 0.03 ($\pm$SE), whereas the unweighted F_{yw} were slightly smaller, ranging from 0.06 ± 0.01 to 0.37 ± 0.03. The F_{yw} values from our study sites span roughly the 10th to 80th percentiles of the F_{yw} values estimated by Jasechko et al. (2016) for 254 rivers around the world. The median F_{yw} among the 22 Swiss catchments was 0.16 (95 % confidence interval 0.10–0.21), somewhat less than the global median of 0.21 (95 % confidence interval 0.19–0.24; Jasechko et al., 2016), consistent with Jasechko et al.'s (2016) observation that young water fractions tend to be smaller in steeper landscapes. Among the 22 Swiss catchments, F_{yw} and F^*_{yw} were positively correlated with catchment characteristics that control wetness conditions (e.g. mean monthly precipitation and mean precipitation intensity) and near-surface flow routing (e.g. drainage density and areal fractions of saturated soils; Sect. 5).

By calculating young water fractions for individual ranges of streamflow, we demonstrated that young water fractions generally increase with discharge (Q), and that this sensitivity of F_{yw} to Q varies from site to site (Sect. 6.1, Fig. 8). We developed a method to quantify the discharge sensitivity of F_{yw} through calculating the linear slope of the Q–F_{yw} relationship (Eqs. 6 to 9). The discharge sensitivity expresses how F_{yw} responds to changes in river discharge, which is used here as a proxy for catchment wetness and hydroclimatic forcing. Across our study catchments, the young water fraction and its discharge sensitivity were not correlated with each other, suggesting that these metrics represent different diagnostic indicators of catchment hydrologic behaviour (Sect. 6.2, Fig. 8). We hypothesize that low discharge sensitivities imply greater persistence in the relative contributions of fast and slow flow paths to streamflow during both high and low flows. High discharge sensitivities, however, imply shifts in flow path dominance during higher flows, such as when subsurface water tables rise into more permeable layers or the river network expands further into the landscape.

Based on our analysis, we developed a generalized conceptual description that relates F_{yw} and its discharge sensitivity to dominant streamflow generation mechanisms (Sect. 6.3, Fig. 10), which could be useful for analysing the effects of future climate change on catchment hydrological behaviour. It remains to be tested whether this conceptual description is transferable to other sites with landscape features and hydro-climatic forcing that are substantially different from our 22 Swiss study catchments.

Author contributions. MW provided stream water isotope measurements. SS and SA modeled precipitation isotopes. JF and JK analyzed the data set. JF prepared the paper with contributions from JK, MW, SS and SA.

Competing interests. The authors declare that they have no conflict of interest.

Acknowledgements. The collection and analysis of the streamwater isotope data were mainly funded as part of the National Research Programme NRP 61 by the Swiss National Science Foundation within the project Drought-CH. We thank Massimiliano Zappa from the Swiss Federal Research Institute WSL for providing interpolated meteorological data for the 22 study catchments, and Wouter Berghuijs for helpful discussions. Comments by Markus Hrachowitz and Stefanie Lutz helped to improve the paper.

Edited by: Ralf Merz

References

Allen, S. T., Kirchner, J. W., and Goldsmith, G. R.: Predicting Spatial Patterns in Precipitation Isotope (δ^2H and δ^{18}O) Seasonality Using Sinusoidal Isoscapes, Geophys. Res. Lett., 45, 4859–4868, https://doi.org/10.1029/2018GL077458, 2018.

Benettin, P., Kirchner, J. W., Rinaldo, A., and Botter, G.: Modeling chloride transport using travel time distributions at Plynlimon, Wales, Water Resour. Res., 51, 3259–3276, https://doi.org/10.1002/2014WR016600, 2015.

Berghuijs, W. R., Sivapalan, M., Woods, R. A., and Savenije, H. H. G.: Patterns of similarity of seasonal water balances: A window into streamflow variability over a range of time scales, Water Resour. Res., 50, 5638–5661, https://doi.org/10.1002/2014WR015692, 2014.

BFS – Bundesamt für Statistik: GEOSTAT: Waldmischungsgrad der Schweiz, Neuchâtel, 1–12, 2004.

Birkel, C., Soulsby, C., and Tetzlaff, D.: Modelling catchment-scale water storage dynamics: reconciling dynamic storage with tracer-inferred passive storage, Hydrol. Process., 25, 3924–3936, https://doi.org/10.1002/Hyp.8201, 2011.

BLW – Bundesamt für: Minimales Geodatenmodell, 77.2 Digitale Bodeneignungskarte der Schweiz, Berne, 1–23, 2012.

Brantley, S. L., Eissenstat, D. M., Marshall, J. A., Godsey, S. E., Balogh-Brunstad, Z., Karwan, D. L., Papuga, S. A., Roering, J., Dawson, T. E., Evaristo, J., Chadwick, O., McDonnell, J. J., and Weathers, K. C.: Reviews and syntheses: on the roles trees play in building and plumbing the critical zone, Biogeosciences, 14, 5115–5142, https://doi.org/10.5194/bg-14-5115-2017, 2017.

Buttle, J. M.: Isotope hydrograph separations and rapid delivery of pre-event water from drainage basins, Prog. Phys. Geogr., 18, 16–41, https://doi.org/10.1177/030913339401800102, 1994.

Detty, J. M. and McGuire, K. J.: Topographic controls on shallow groundwater dynamics: implications of hydrologic connectivity between hillslopes and riparian zones in a till mantled catchment, Hydrol. Process., 24, 2222–2236, https://doi.org/10.1002/hyp.7656, 2010.

DeWalle, D. R., Edwards, P. J., Swistock, B. R., Aravena, R., and Drimmie, R. J.: Seasonal isotope hydrology of three Appalachian forest catchments, Hydrol. Process., 11, 1895–1906, https://doi.org/10.1002/(SICI)1099-1085(199712)11:15<1895::AID-HYP538>3.0.CO;2-#, 1997.

Dunne, T.,and Black, R. D.: Partial Area Contributions to Storm Runoff in a Small New-England Watershed, Water Resour. Res., 6, 1296–1311, https://doi.org/10.1029/Wr006i005p01296, 1970.

Feng, X. H., Faiia, A. M., and Posmentier, E. S.: Seasonality of isotopes in precipitation: A global perspective, J. Geophys. Res.-Atmos., 114, D08116, https://doi.org/10.1029/2008jd011279, 2009.

Feyen, H., Leuenberger, J., Papritz, A., Gysi, M., Fluehler, H., and Schleppi, P.: Runoff processes in catchments with a small scale topography, Phys. Chem. Earth, 21, 177–181, https://doi.org/10.1016/S0079-1946(97)85581-4, 1996.

Fischer, B. M. C., Rinderer, M., Schneider, P., Ewen, T., and Seibert, J.: Contributing sources to baseflow in pre-alpine headwaters using spatial snapshot sampling, Hydrol. Process., 29, 5321–5336, https://doi.org/10.1002/hyp.10529, 2015.

Godsey, S. E. and Kirchner, J. W.: Dynamic, discontinuous stream networks: Hydrologically driven variations in active drainage density, flowing channels and stream order, Hydrol. Process., 28, 5791–5803, https://doi.org/10.1002/hyp.10310, 2014.

Hale, V. C. and McDonnell, J. J.: Effect of bedrock permeability on stream base flow mean transit time scaling relations: 1. A multi-scale catchment intercomparison, Water Resour. Res., 52, 1358–1374, https://doi.org/10.1002/2014WR016124, 2016.

Holko, L., Holzmann, H., de Lima, M. I. P., and de Lima, J. L. M. P.: Hydrological research in small catchments – an approach to improve knowledge on hydrological processes and global change impacts, J. Hydrol. Hydromech., 63, 181–182, 2015.

Horton, R. E.: The role of infiltration in the hydrologic cycle, Eos T. Am. Geophys. Un., 14, 446–460, 1933.

Hrachowitz, M., Soulsby, C., Tetzlaff, D., Dawson, J. J. C., and Malcolm, I. A.: Regionalization of transit time estimates in montane catchments by integrating landscape controls, Water Resour. Res., 45, W05421, https://doi.org/10.1029/2008WR007496, 2009.

Jasechko, S., Kirchner, J. W., Welker, J. M., and McDonnell, J. J.: Substantial proportion of global streamflow less than three months old, Nat. Geosci., 9, 126–129, https://doi.org/10.1038/Ngeo2636, 2016.

Jensen, C. K., McGuire, K. J., and Prince, P. S.: Headwater stream length dynamics across four physiographic provinces of the Appalachian Highlands, Hydrol. Process., 31, 3350–3363, 2017.

Kendall, C. and McDonnell, J. J.: Isotope tracers in catchment hydrology, Elsevier, Amsterdam, New York, 1–839, 1998.

Kirchner, J. W.: Aggregation in environmental systems – Part 1: Seasonal tracer cycles quantify young water fractions, but not mean transit times, in spatially heterogeneous catchments, Hydrol. Earth Syst. Sci., 20, 279–297, https://doi.org/10.5194/hess-20-279-2016, 2016a.

Kirchner, J. W.: Aggregation in environmental systems – Part 2: Catchment mean transit times and young water fractions under hydrologic nonstationarity, Hydrol. Earth Syst. Sci., 20, 299–328, https://doi.org/10.5194/hess-20-299-2016, 2016b.

Klaus, J. and McDonnell, J. J.: Hydrograph separation using stable isotopes: Review and evaluation, J. Hydrol., 505, 47–64, https://doi.org/10.1016/j.jhydrol.2013.09.006, 2013.

McGuire, K. J. and McDonnell, J. J.: A review and evaluation of catchment transit time modeling, J. Hydrol., 330, 543–563, https://doi.org/10.1016/j.jhydrol.2006.04.020, 2006.

McGuire, K. J., McDonnell, J. J., Weiler, M., Kendall, C., McGlynn, B. L., Welker, J. M., and Seibert, J.: The role of topography on catchment-scale water residence time, Water Resour. Res., 41, W05002, https://doi.org/10.1029/2004WR003657, 2005.

Nathan, R. J. and McMahon, T. A.: Evaluation of Automated Techniques for Base-Flow and Recession Analyses, Water Resour. Res., 26, 1465–1473, https://doi.org/10.1029/Wr026i007p01465, 1990.

Phillips, R. W., Spence, C., and Pomeroy, J. W.: Connectivity and runoff dynamics in heterogeneous basins, Hydrol. Process., 25, 3061–3075, https://doi.org/10.1002/hyp.8123, 2011.

Seeger, S. and Weiler, M.: Reevaluation of transit time distributions, mean transit times and their relation to catchment topography, Hydrol. Earth Syst. Sci., 18, 4751–4771, https://doi.org/10.5194/hess-18-4751-2014, 2014.

Song, C., Wang, G., Liu, G., Mao, T., Sun, X., and Chen, X.: Stable isotope variations of precipitation and streamflow reveal the young water fraction of a permafrost watershed, Hydrol. Process., 31, 935–947, https://doi.org/10.1002/hyp.11077, 2017.

Soulsby, C., Malcolm, R., Helliwell, R., Ferrier, R. C., and Jenkins, A.: Isotope hydrology of the Allt a' Mharcaidh catchment, Cairngorms, Scotland: implications for hydrological pathways and residence times, Hydrol. Process., 14, 747–762, 2000.

Soulsby, C., Tetzlaff, D., Rodgers, P., Dunn, S., and Waldron, S.: Runoff processes, stream water residence times and controlling landscape characteristics in a mesoscale catchment: An initial evaluation, J. Hydrol., 325, 197–221, https://doi.org/10.1016/j.jhydrol.2005.10.024, 2006

Soulsby, C., Tetzlaff, D., and Hrachowitz, M.: Are transit times useful process-based tools for flow prediction and classification in ungauged basins in montane regions?, Hydrol. Process., 24, 1685–1696, https://doi.org/10.1002/hyp.7578, 2010.

Spearman, C.: The proof and measurement of association between two things. By C. Spearman, 1904, Am. J. Psychol., 100, 441–471, 1987.

Staudinger, M., Stoelzle, M., Seeger, S., Seibert, J., Weiler, M., and Stahl, K.: Catchment water storage variation with elevation, Hydrol. Process., 31, 2000–2015, https://doi.org/10.1002/hyp.11158, 2017.

Stockinger, M. P., Bogena, H. R., Lücke, A., Diekkrüger, B., Cornelissen, T., and Vereecken, H.: Tracer sampling frequency influences estimates of young water fraction and streamwater transit time distribution, J. Hydrol., 541, 952–964, https://doi.org/10.1016/j.jhydrol.2016.08.007, 2016.

Tetzlaff, D., Seibert, J., and Soulsby, C.: Inter-catchment comparison to assess the influence of topography and soils on catchment transit times in a geomorphic province; the Cairngorm mountains, Scotland, Hydrol. Process., 23, 1874–1886, 2009.

Viviroli, D., Zappa, M., Gurtz, J., and Weingartner, R.: An introduction to the hydrological modelling system PREVAH and its pre- and post-processing-tools, Einviron. Model. Softw., 24, 1209–1222, https://doi.org/10.1016/j.envsoft.2009.04.001, 2009.

von Freyberg, J., Rao, P. S. C., Radny, D., and Schirmer, M.: The impact of hillslope groundwater dynamics and landscape functioning in event-flow generation: a field study in the Rietholzbach catchment, Switzerland, Hydrogeol. J., 23, 935–948, https://doi.org/10.1007/s10040-015-1238-1, 2015.

von Freyberg, J., Studer, B., and Kirchner, J. W.: A lab in the field: high-frequency analysis of water quality and stable isotopes in stream water and precipitation, Hydrol. Earth Syst. Sci., 21, 1721–1739, https://doi.org/10.5194/hess-21-1721-2017, 2017.

Whipkey, R. Z.: Subsurface stormflow from forested slopes, Int. Assoc. Scient. Hydrol., 10, 74–85, 1965.

Wilusz, D. C., Harman, C. J., and Ball, W. P.: Sensitivity of Catchment Transit Times to Rainfall Variability Under Present and Future Climates, Water Resour. Res., 53, 10231–10256, https://doi.org/10.1002/2017WR020894, 2017.

9

Seasonal drought predictability and forecast skill in the semi-arid endorheic Heihe River basin in northwestern China

Feng Ma[1,2], **Lifeng Luo**[2], **Aizhong Ye**[1], **and Qingyun Duan**[1]

[1]State Key Laboratory of Earth Surface and Ecological Resources, Faculty of Geographical Science, Beijing Normal University, Beijing 100875, China
[2]Department of Geography, Environment, and Spatial Sciences, Michigan State University, East Lansing, Michigan, USA

Correspondence: Aizhong Ye (azye@bnu.edu.cn)

Abstract. Endorheic and arid regions around the world are suffering from serious drought problems. In this study, a drought forecasting system based on eight state-of-the-art climate models from the North American Multi-Model Ensemble (NMME) and a Distributed Time-Variant Gain Hydrological Model (DTVGM) was established and assessed over the upstream and midstream of Heihe River basin (UHRB and MHRB), a typical arid endorheic basin. The 3-month Standardized Precipitation Index (SPI3) and 1-month Standardized Streamflow Index (SSI1) were used to capture meteorological and hydrological drought, and values below − 1 indicate drought events. The skill of the forecasting systems was evaluated in terms of anomaly correlation (AC) and Brier score (BS) or Brier skill score (BSS). The predictability for meteorological drought was quantified using AC and BS with a "perfect model" assumption, referring to the upper limit of forecast skill. The hydrological predictability was to distinguish the role of initial hydrological conditions (ICs) and meteorological forcings, which was quantified by root-mean-square error (RMSE) within the ESP (Ensemble Streamflow Prediction) and reverse ESP framework. The UHRB and MHRB showed season-dependent meteorological drought predictability and forecast skill, with higher values during winter and autumn than that during spring. For hydrological forecasts, the forecast skill in the UHRB was higher than that in MHRB. Predicting meteorological droughts more than 2 months in advance became difficult because of complex climate mechanisms. However, the hydrological drought forecasts could show some skills up to 3–6 lead months due to memory of ICs during cold and dry seasons. During wet seasons, there are no skillful hydrological predictions from lead month 2 onwards because of the dominant role of meteorological forcings. During spring, the improvement of hydrological drought predictions was the most significant as more streamflow was generated by seasonal snowmelt. Besides meteorological forcings and ICs, human activities have reduced the hydrological variability and increased hydrological drought predictability during the wet seasons in the MHRB.

1 Introduction

Drought is among the most costly natural hazards in many parts of the world. It is defined as a prolonged period of below-average rainfall, leading to water shortages in soil and the hydrologic system. Drought can have a substantial effect on many sectors, such as agriculture, ecosystem and economy, and its impacts can vary from region to region. Mitigation of drought impact requires improved understanding of the predictability of drought and the capability to predict drought at sufficient lead times (Below et al., 2007; Sheffield and Wood, 2012; Smith and Katz, 2013). Statistical, dynamic and hybrid (statistical–dynamic) methods have been used for drought predictions (Mariotti et al., 2013; Hao et al., 2018; Mishra and Singh, 2011; Pozzi et al., 2013; Luo and Wood, 2007; Luo et al., 2008). The statistical method is based on the historical relationship between some aspects of drought and a number of predictors (e.g., large-scale climate signals). The dynamic method relies on the skill of state-of-the-art general circulation models (GCMs) or hydrologic models that represent physical processes linked with drought development. A hybrid method combines the statistical and dynamic methods, which has been shown to improve drought prediction in certain case studies (Pan et al., 2013; Schepen et al., 2016). However, drought remains one of the least understood natural hazards that are affected by many contributing factors, including meteorological anomalies, land–atmosphere interaction and human activities (Van Loon et al., 2016a, b), which makes accurate drought prediction a challenge (Hao et al., 2018).

Recently, climate forecasts from the North American Multi-Model Ensemble (NMME; Kirtman et al., 2014) have been widely applied to drought predictions globally and regionally (Ma et al., 2015, 2017; Mo and Lyon, 2015; Thober and Kumar, 2015; Yuan and Wood, 2013; Yuan, 2016). The NMME-based climate forecasts (e.g., precipitation and temperature) for hydrometeorological forecasts exhibit some improvement in skills over the reference forecasts such as climatology forecasts, persistence forecasts or ensemble streamflow prediction (ESP) (e.g., Ma et al., 2015, 2017; Mo and Lettenmaier, 2014; Shukla et al., 2016; Yuan, 2016). However, the improvement varies with different regions and seasons, and the understanding of its application in endorheic and semi-arid and arid basins remains poor. Endorheic regions cover $\sim 11.4\,\%$ of global land, which are mostly located in semi-arid and arid regions (Li et al., 2018). The semi-arid and arid regions occupy approximately 40 % of earth's land surface, and show an accelerated expansion trend (Huang et al., 2016b). The fragile ecosystems in such regions are sensitive to climate change and human activities (Huang et al., 2016a). Therefore, the aim of this study is to fill the gaps in understanding drought predictions and predictability in endorheic and arid basins by addressing the following questions:

How do climate forecasts perform for meteorological and hydrological drought forecasting in an endorheic river basin?

How do meteorological forcings, initial hydrologic conditions and human activities influence hydrological predictability?

Here, predictability is considered as the possible maximum forecast skill that a forecast system can achieve (Luo and Wood, 2006). The Heihe River basin, which is a typical endorheic river basin in the semi-arid and arid region of northwestern China, is selected in this study to address the above questions. The Heihe River basin is an important part of the historic Silk Road and an important breadbasket in northwestern China (Zhang et al., 2015). However, the basin is subject to serious drought problems historically and in recent decades related to climate change and intensified human activities (Zhang et al., 2016). Therefore, it is crucial to develop a drought forecast system to promote the development of adapting strategies for sustainable water resource and ecological management in the Heihe River basin.

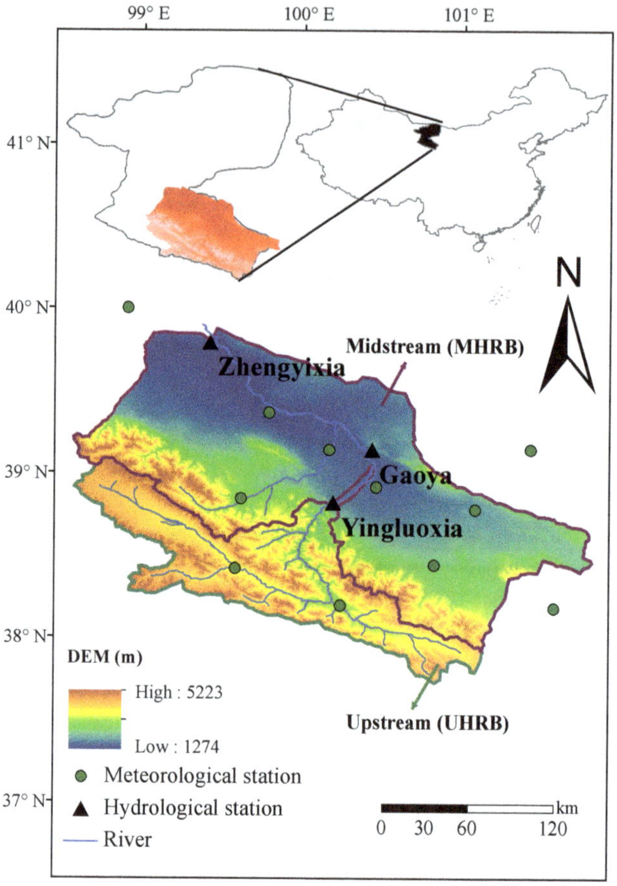

Figure 1. Geographical location of the study area (upstream and midstream of Heihe River basin) used in this study. The top diagram shows the location of the entire Heihe River basin. The bottom diagram shows the geographic distribution of hydrometeorological stations in the UHRB and MHRB.

The study is organized as follows: in Sect. 2, we provide a brief description of the study region and data used in this study; in Sect. 3, we introduce the framework of the study and methods used; in Sect. 4, we present the analysis results and discussion, followed by conclusions in Sect. 5. Because few studies have focused on dynamic drought predictions based on GCMs and/or hydrological models in the region, this study will offer new clues for the development of more accurate drought monitoring and forecasting systems.

2 Study area and data

2.1 The Heihe River basin

The Heihe River basin (HRB; Fig. 1) is the second largest inland river basin of China, located deep in the hinterland of Eurasia. The river originates from the north side of Qilian Mountain, with a drainage area of $128\,900\,\mathrm{km}^2$ (37°41′–42°42′ N, 96°42′–02°00′ E; Ma and Frank, 2006). The HRB has an apparent landscape, ecological and climate gradient from upstream to downstream. The landscape varies from glaciers and alpine biomes in the upstream to oases with irrigated agriculture in the midstream, to riparian ecosystems and the vast Gobi desert in the downstream. Most precipitation is concentrated in the upstream during wet season (June–September). During spring and summer, as temperature rises, snowmelt and glacier melt and permafrost thaw generally occur. Most of the streamflow in the HRB originates from precipitation, snowmelt, glacier melt and permafrost thaw in the upstream mountains (Wang et al., 2008), which contribute approximately 71 %, 25 % and 4 % of the total runoff (Li et al., 2018). Most water consumption happens in the midstream for human activities (e.g., agricultural irrigation). In this study, we focus on the upstream (UHRB) and midstream (MHRB) of HRB, as these two subbasins show drastic differences in terms of impacts of human activities on hydrological processes.

2.2 Data

Daily temperature and precipitation data at 0.5° spatial resolution (Zhao and Zhu, 2015) are obtained for 1961–2016, which were interpolated using thin plate spline (TPS) and 3-D geospatial information from 2472 meteorological stations by the National Meteorological Information Center, China Meteorological Administration (CMA) (Hutchinson, 1998a, b). Hydrological data (1982–2011) used in this study were monthly streamflow datasets from Yingluoxia (YLX) and Zhengyixia (ZYX) hydrologic stations that are located at the outlet of UHRB and MHRB. The data for the hydrological model (the Distributed Time-Variant Gain Hydrological Model, DTVGM, in this study) setup and calibration were mainly obtained from the Chinese Academy of Sciences, Gansu Water Resources Bulletin, and Statistical Yearbooks, which are presented in Ma et al. (2018) in detail.

Climate hindcast data with 1° × 1°grids (Table 1) were obtained from the North American Multi-Model Ensemble (NMME; Kirtman et al., 2014) archive (http://iridl.ldeo.columbia.edu/SOURCES/.Models/.NMME/; last access: 13 February 2018). Monthly precipitation, maximum, mean and minimum temperature data covering 1982–2010 were used for this study. The climate models with real-time forecast were selected for drought forecasting. In this study, "lead month 1" refers to forecasts initialized at the beginning of one month for itself, and "lead month 2" is that for the next month.

3 Methodology

3.1 Meteorological and hydrological drought index

To analyze meteorological and hydrological drought, the Standardized Precipitation Index (SPI, McKee et al., 1993) and Standardized Streamflow Index (SSI; Vicente-Serrano et al., 2012) were used in this study. In the calculation, a probability distribution of monthly precipitation or streamflow for each month was generated and standardized using the empirical Gringorten plotting position (Farahmand and AghaKouchak, 2015; Gringorten, 1963):

$$p(x_i) = \frac{i-0.44}{n+0.12}, \tag{1}$$

where n is the time span, i is the position of precipitation or streamflow time series sorted from smallest to largest and $p(x_i)$ is the corresponding empirical probability. Finally, normalization was calculated to make the index comparable over time and space. For SPI in the upstream and midstream, catchment average precipitation in the upstream and midstream were used respectively. SSI in the upstream and midstream was calculated using streamflow at YLX (the outlet of the upstream) and the streamflow difference between ZYX (the outlet of the midstream) and YLX. SPI-3 and SSI-1 were selected to characterize the seasonal meteorological and hydrological droughts, respectively. We select SSI-1 for its good description of hydrological drought (e.g., Barker et al., 2016; Gustard et al., 1992; Huang et al., 2017; Ma et al., 2018). A total of 9 different lead months' forecasts are combined with observation to construct 3 months' accumulated precipitation for computing SPI3. For example, for SPI3 in August, lead month 1 uses the forecast at lead month 1 (August) combined with 2 months' observation (June to July). Lead month 2 means the sum of forecasts at lead month 1–2 (July and August) and 1 month of observation in June. Lead month 3 refers to forecasts at lead month 1–3 (June to August). In this study, drought is defined when the drought index value is below −1.

3.2 Seasonal drought forecasting system

In this study, meteorological drought forecasting was produced using NMME climate forecasts, and hydrological drought forecasting makes use of a hydrological model forced by NMME climate forecasts (Fig. 2). To improve the forecast skill and drive the hydrological model, the NMME hindcasts were bias-corrected and downscaled using the quantile mapping method (Wood et al., 2002). The 1 ° monthly NMME precipitation and temperature hindcasts were interpolated into 0.5 ° with bilinear interpolation over the Heihe River basin. Then the cumulative distribution functions (CDFs) of NMME hindcasts and observations were constructed with all years except the target year (leave-one-out), and matched to remove model forecast bias. The daily hindcasts were then generated using a temporally downscaled technique by matching the monthly hindcasts with the daily samples from observations. In this approach (Luo and Wood, 2008), a randomly selected daily observation time series from the entire historical period (1961–2016) is used as a candidate for each member, and they are adjusted to match the predicted monthly values from the distributions obtained

Table 1. Information of NMME models.

NMME models	Spatial resolution	Hindcast	Member	Max lead months
CMC1-CanCM3	Global, 1 × 1	1981–2010	10	12
CMC2-CanCM4	Global, 1 × 1	1981–2010	10	12
COLA-RSMAS-CCSM3	Global, 1 × 1	1982–2010	6	12
COLA-RSMAS-CCSM4	Global, 1 × 1	1982–2010	9	12
GFDL-CM2p1-aer04	Global, 1 × 1	1982–2010	10	12
GFDL-CM2p5-FLOR-A06	Global, 1 × 1	1980–2010	12	12
NASA-GMAO-062012	Global, 1 × 1	1981–2010	7	9
NCEP-CFSv2	Global, 1 × 1	1982–2010	24	10

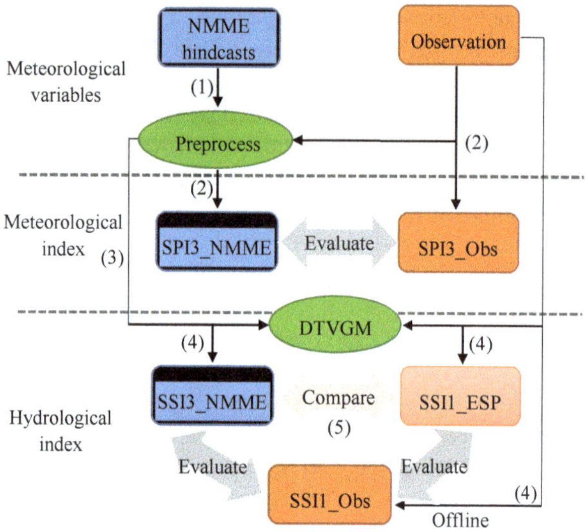

Figure 2. Flow chart explaining the seasonal meteorological and hydrological drought forecasting system. The numbers (1)–(5) refer to the steps for the development and assessment of a seasonal drought prediction system.

in the previous step. Finally, the 0.5 ° bias-corrected daily hindcasts were bilinearly interpolated into 140 subbasins to drive the hydrological model over the HRB. The last step is only necessary in this study as our hydrological model as described below runs on subbasin scales instead of regular latitude–longitude grids.

In this study, the Distributed Time-Variant Gain Hydrological Model (DTVGM; Mao et al., 2016) was used to simulate and predict streamflow. It is a distributed, catchment-based hydrological model with modules to simulate snow, runoff, streamflow routing, water use and reservoir operation. The degree-day factor (DDF) method is used to compute snowmelt. The runoff module is based on the water balance equation, and the routing is based on a kinematic scheme (Ye et al., 2010, 2013). Three runoff components for each subbasin, i.e., surface runoff, subsurface runoff and base flow, are computed based on precipitation, soil hydraulic parameters, and land cover parameters and sets of calibrated parameters (Ma et al., 2018; Ye et al., 2010). The human activities module, i.e., water use and reservoir models, can be switched on or off to simulate real streamflow and naturalized streamflow. Here, water use includes irrigation water, industrial water and domestic water, which are derived from irrigation areas and irrigation quota, industrial GDP and population distribution, respectively. The reservoir regulation rules are defined according to reservoir storage capacity and ecological flow during wet (June–September) and dry seasons. The DTVGM model has been calibrated with the human activities module turned on using observed streamflow at YLX, Gaoya and ZYX stations. The DTVGM could capture the variations of streamflow well, with Nash–Sutcliffe efficiency coefficient (NSE) values greater than 0.86 and 0.52 for the UHRB and MHRB during both calibration and validation periods, respectively. In the MHRB, except for input and structural errors, the unrefined human activities module also increases the uncertainties of the hydrology model, leading to a NSE value lower than that in the UHRB. Detailed description of the DTVGM model and its calibration process can be found in Mao et al. (2016) and Ma et al. (2018). Before making forecasts, the DTVGM was continuously run for the period of 1961–1981 to spin up the hydrological model, and continuously driven by observed meteorological forcings from 1982 to 2010 to generate the offline initial hydrological conditions (ICs) for NMME-based and ESP-type forecast experiments.

In this study, ESP (Ensemble Streamflow Prediction) forecasts (see Sect. 3.4), which were based on ensemble of historical meteorological forcings, were used as a reference hydrological forecast.

3.3 Forecast verification

The meteorological and hydrological drought forecast skills for 9 different lead months were evaluated using both deterministic and probabilistic metrics. The deterministic metric that we use is the anomaly correlation (AC; Wilks, 2011), which is calculated as

$$AC = \frac{\sum_{i}^{n} F'_i O'_i}{\left\langle \sum_{i}^{n} F'^2_i \times \sum_{i}^{n} O'^2_i \right\rangle^{1/2}}, \quad (2)$$

where F'_i is the anomaly for NMME hindcasts and O'_i is the anomaly for observations; for a given lead and target month/season and member, i is the target month/season and n is the time span (29 years in this study).

The probabilistic metric that we use is the Brier score (BS; Wilks, 2011), which is the mean squared error of probability forecasts, considering that the observation is $o_k = 1$ if drought occurs, otherwise $o_k = 0$. The BS can be defined as

$$BS = \frac{1}{n}\sum_{k=1}^{n}(y_k - o_k)^2, \quad (3)$$

where k means a numbering of the n forecast–event pairs and y_k is the forecast probability. The BS is negatively oriented $(0 \leq BS \leq 1)$, with perfect forecast exhibiting $BS = 0$. The Brier skill score (BSS; Wilks, 2011) is then calculated as

$$BSS = 1 - \frac{BS}{BS^*}, \quad (4)$$

where BS* is the BS for the reference system. A positive BSS value indicates a better forecast, while a negative value indicates a worse forecast than the reference system.

3.4 Experiment design for understanding the hydrological predictability

With the calibrated DTVGM hydrological model, two experiments were carried out to distinguish the importance of initial hydrological conditions (ICs) and meteorological forcings in the hydrological forecasting. The first experiment is the Ensemble Streamflow Prediction (ESP), which was initialized at the beginning of each month during 1982–2010, with 28 ensemble members of 9-month meteorological forcings taken from the same period without the target year. For example, the ESP simulation starting in January 1982 was initialized on the first day in January, and driven by 28 9-month meteorological forcings during January–September of 1983, 1984, ..., 2010. The second experiment is the so-called reverse-ESP (revESP; Shukla and Lettenmaier, 2011; Shukla et al., 2013; Wood and Lettenmaier, 2008), which was driven by the observed meteorological forcings of the target year, but with 28 different ICs except the target year. For example, the revESP simulation starting in January 1982 was driven by the meteorological forcings during January–September of 1982, but initialized with hydrological conditions in January of 1982, 1984, ..., 2010.

To determine whether the meteorological forcings or ICs are more important in the hydrologic cycle, the RMSE ratio was calculated, which is defined as

$$\text{RMSE ratio} = \frac{\text{RMSE}_{\text{ESP}}}{\text{RMSE}_{\text{revESP}}}, \quad (5)$$

where RMSE_{ESP} and $\text{RMSE}_{\text{revESP}}$ are the root-mean-square error (RMSE) for ESP and revESP. Here, the RMSE was calculated using the ensemble means of ESP and revESP. The RMSE ratio is lower than 1 when ICs play a more important role than the meteorological forcings in the hydrologic predictability, and the ratio is larger than 1 when the meteorological forcings dominate.

4 Results and discussion

4.1 The predictability and forecast skill for meteorological droughts

To evaluate the performance of seasonal drought prediction system, we first examined the predictability and forecast skill of NMME meteorological predictions based on SPI3 series in terms of AC metric (Fig. 3). The red box refers to the AC for predictability, and the blue box is the AC for forecast skill. Here, predictability is defined by using a "perfect model" assumption (Luo and Wood, 2006; Wang and Yuan, 2018), which considers one member of the ensemble as the "observation", and the average of remaining members as prediction. The analysis is rotated through using all 88 ensemble members as the observations, and 88 values of AC are then calculated as the ensemble of predictability. For each season (3 months), there are 29 year hindcasts leading to a sample size of 87, so a correlation of 0.21 is statistically significant using a 95 % significance interval.

Meteorological predictability is higher than the forecast skill in terms of AC in most cases (Fig. 3), indicating some room for improving the meteorological predictions. The meteorological predictability and forecast skill depend on the target season because of a strong seasonality for climate in the HRB. The predictability is higher in autumn and winter than that in summer and spring, which corresponds to a higher forecast skill in autumn and winter. Most climate anomalies (i.e., sea surface temperature (SST) anomaly) occur in winter and autumn, and SST is also a potentially important predictor (Becker et al., 2014). In addition, the climatic noise of monthly precipitation over China has obvious seasonal variation and it is greater in summer than in winter (Liu et al., 2000). The multi-model ensemble mean skill, shown by the blue diamonds, is generally located at the upper part of the distribution, indicating that the forecast skill of the grand ensemble mean is higher than that of most members. It is not surprising that the SPI3 predictions perform well in the first 2 lead months, where 1 or 2 of the three months come from observations. However, as the lead month increases beyond 2 months, both the predictability and forecast skill decrease significantly, for which correlations of most of members are lower than 0.21.

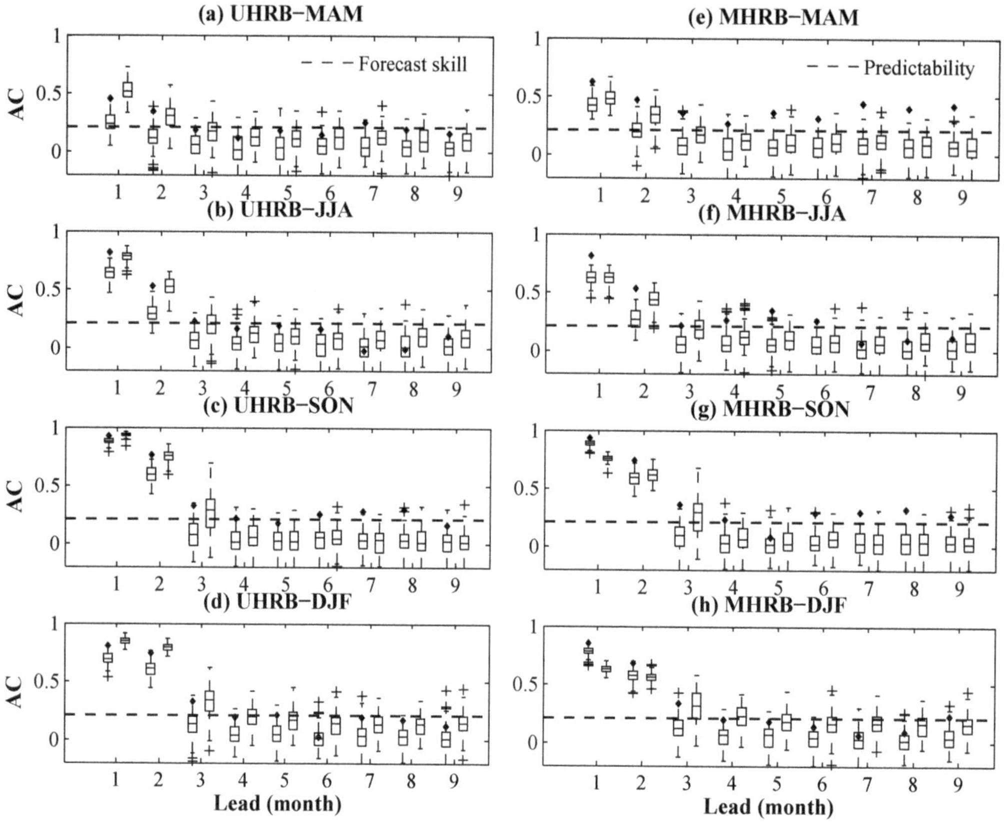

Figure 3. Anomaly correlation (AC) of the forecast of seasonal SPI3. The red box plots show the spread of predictability, and the blue box plots show the spread of forecast skill for each ensemble member. The blue diamonds show the AC of the grand ensemble mean. The blue (red) crosses show the outliers for forecast skill (predictability). The dashed black line indicates the threshold (AC = 0.21) of 95 % confidence intervals calculated from the *t* test.

In fact, NMME climate predictions have lower predictability and forecast skill in the northwest inland areas of China, in comparison with southeast monsoon regions (Ma et al., 2016). The HRB is located far from oceans and in mid-latitudes, and is little affected by SST from oceans, especially equatorial oceans, which are the major source of predictability at a seasonal timescale. Topographic influence on regional and local weather and climate cannot be resolved by GCMs, for example, local ascending motion affected by the Qinghai–Tibet Plateau exists and has a considerable impact on precipitation over the HRB (Sun et al., 2011). In addition, the joint extreme phases of climate oscillations instead of a single one could trigger extremes (e.g., drought) over the arid endorheic basin, and almost no climate models can capture the complex and multiple teleconnections (Shi et al., 2016).

When defining a meteorological drought as SPI3 below −1, the results of predictability and forecast skill for meteorological drought predictions are similar to that for SPI3 predictions. Figure 4 shows the BS for meteorological drought events in the upstream and midstream of Heihe River basin. AC of SPI3 reflects the forecast skill for both the dry spells and wet spells. The BS for the different months shows the forecast skill, which primarily aims at predicting drought events. It can be seen that the NMME-based meteorological droughts show higher predictability and forecast skill in the first 2 lead months, especially during October–January. The results indicate that the NMME-based SPI3 predictions could reasonably capture meteorological drought conditions in the first 2 lead months.

4.2 The predictability and forecast skill for hydrological droughts

Hydrological processes and predictions are more complex, especially in the MHRB. Figure 5 shows the performance of NMME for SSI1 predictions compared to ESP, which is assessed using the AC metric. The predictions from NMME and ESP experiments are verified against DTVGM offline simulations, driven by observed meteorological forcings and calibrated against observed streamflow at Yingluoxia, Gaoya and Zhengyixia stations. Ensemble hydrological predictions from NMME show less spread than ensemble meteorological predictions, especially in the cold seasons. Due to the memory of initial hydrologic conditions, hydrological predictions show less uncertainty than the corresponding meteorological forcings. There are notable differences in hy-

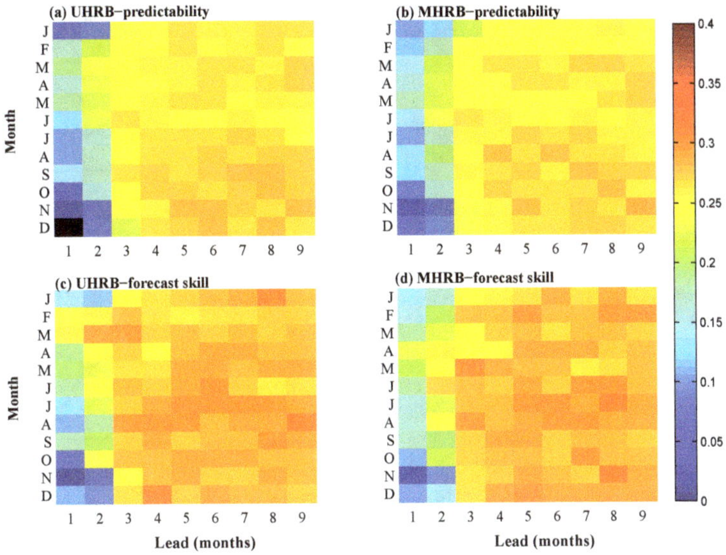

Figure 4. Brier score (BS) of NMME forecast for meteorological drought events. **(a–b)** Meteorological drought predictability in the upstream **(a)** and upstream **(b)**. **(c–d)** Meteorological drought forecast skill in the upstream **(c)** and midstream **(d)**. Here, a meteorological drought event happens when the SPI3 value is below −1. The BS is negatively oriented ($0 \leq BS \leq 1$), with perfect forecast exhibiting $BS = 0$. The color from deep blue to deep red (0–0.4) means increasing BS values, i.e., decreasing predictability or forecast skill.

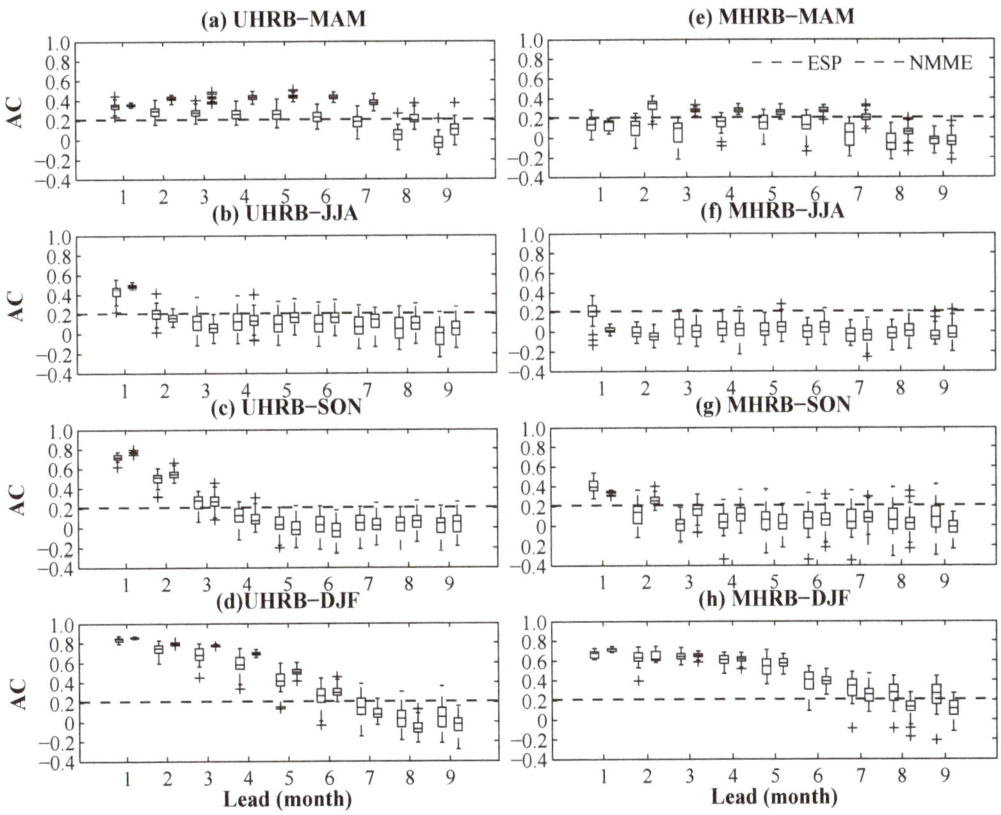

Figure 5. Anomaly correlation (AC) of the forecast of seasonal SSI1. The red box plots show the spread of AC of each member from NMME, and the blue box plots show that from ESP. The blue (red) crosses show the outliers for NMME (ESP) forecast skill. The dashed black line indicates the threshold ($AC = 0.21$) of 95 % confidence intervals calculated from the t test. The predictions and simulations are carried out with the human activities module switched on.

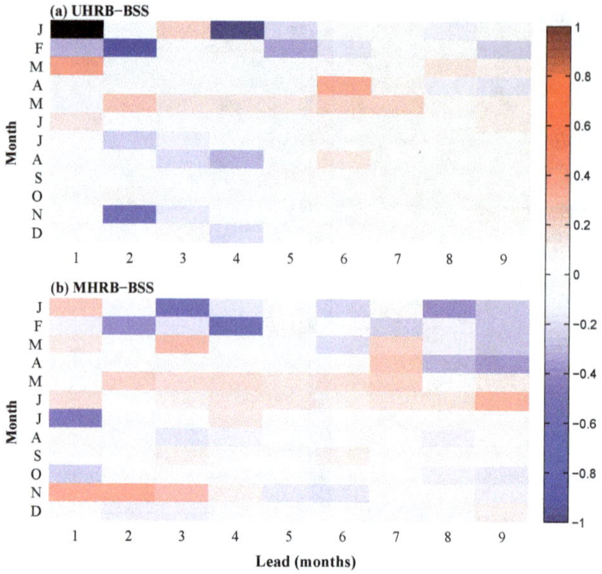

Figure 6. Brier skill score (BSS) of NMME forecast for hydrological drought events. Here, a hydrological drought event happens when the SSI1 value is below −1. The reference forecasts are simulations from the ESP experiment. The predictions and simulations are carried out with the human activities module switched on.

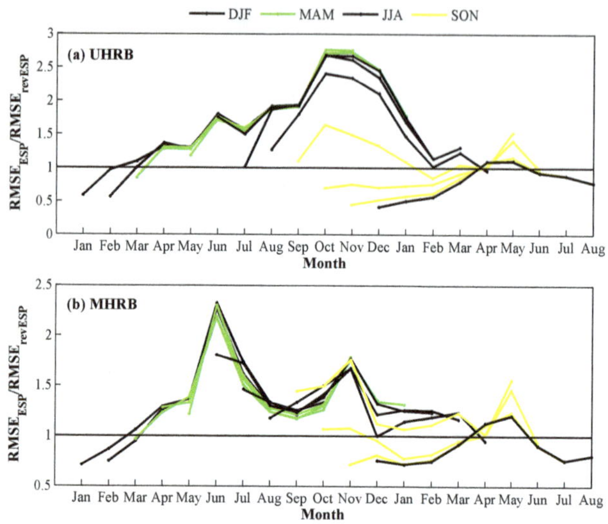

Figure 7. The RMSE ratio ($RMSE_{ESP}/RMSE_{revESP}$) as a function of start month and lead time over the upstream and midstream. The $RMSE_{ESP}$ ($RMSE_{revESP}$) is calculated between the SSI1 series from the offline simulation and that from the ESP (revESP) experiments with the human activities module switched on.

drological predictions between the upstream and midstream, with higher overall forecast skill in the upstream. The winter season shows the highest forecast skill, followed by autumn, spring and summer. During spring, NMME hydrological predictions show the most significant improvement over ESP, in spite of low forecast skill for precipitation predictions. During March–June, approximately 70 % of streamflow is generated by seasonal snowmelt (Wang and Li, 1999). Therefore, hydrological predictions skill may also rely more on temperature predictions, which are generally more skillful than precipitation predictions (Becker et al., 2014; Shukla et al., 2016), and the accuracy of the initial hydrological condition in terms of snow amount. In summer, NMME hydrological predictions show some improvement compared to ESP in the upstream, especially in the first lead month. However, in the midstream, low forecast skills were detected in all lead months, which are likely due to poor precipitation predictions and effects from human activities. Predicting seasonal streamflow during summer in advance is difficult and both NMME and ESP exhibit weak skills. In autumn, lead time with good forecast skill could be up to 3 months in the upstream and 2 months in the midstream, which are similar to meteorological predictions. In addition, NMME hydrological predictions also show improvement over ESP in the first 3–4 lead months.

In winter, until lead month 6, both ESP and NMME show skillful hydrology predictions due to the more important role of initial conditions, which will be discussed in the later part of this section.

To further evaluate the performance of NMME-based hydrological predictions compared with ESP for droughts (i.e., SSI1 $<$ 1), the BSS metric is used and shown in Fig. 6. The skill for hydrological drought predictions from NMME is higher than that from ESP during late spring. The improvement is even more clear for longer lead times (6–7 lead months), which may be linked with the higher skill in temperature forecasting. In general, NMME outperforms ESP for the 1–4 lead months (with some exceptions), but the improvement is not obvious due to a long memory of initial conditions during the cold season and poor meteorological predictions during summer.

As mentioned above, the forecast skill for NMME shows a notable difference between meteorological and hydrological droughts during different seasons, that is because besides meteorological forcings, initial conditions also play an important role in hydrological predictions. Figure 7 shows the relative role of initial hydrological conditions (ICs) in hydrological predictability for different months and lead times over the upstream and midstream of Heihe River basin. During cold and dry seasons (October–March), the RMSE ratio is lower than 1 in the first 2–7 months, indicating that the hydrological initial conditions play a more important role in hydrological predictability up to 2–7 lead months. The maximum lead times in which the ICs prevail over the meteorological forcings in the hydrological predictability could even be up to 5–7 months during October–December. In general, as the lead time increases, the contribution of initial conditions decreases, while that of the meteorological forcings gradually increases over the ICs. For the forecasts in April–August, the influence of ICs could not persist for 1 month, and the meteo-

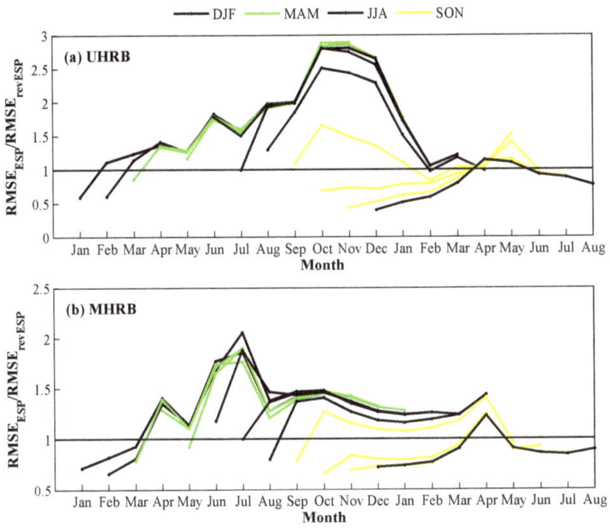

Figure 8. The RMSE ratio ($RMSE_{ESP}/RMSE_{revESP}$) as a function of start month and lead time over the upstream and midstream. The $RMSE_{ESP}$ ($RMSE_{revESP}$) is calculated between the SSI1 series from the offline simulation and that from the ESP (revESP) experiments with the human activities module switched off.

rological forcings significantly contribute to the hydrological predictability. This means that the ICs make more contribution to hydrological predictability during the cold and dry season than that during the warm and wet season. This helps to explain why hydrological predictions are more skillful in the dry season than that in the wet season.

From a hydrological perspective, the MHRB is a human-dominant basin (Ma et al., 2018). To explore the influence of human activities on hydrological predictability, an additional experiment is conducted by turning off the human activities module in the hydrological model. The RMSE ratios of ESP and revESP without human activities are then calculated, and the results are shown in Fig. 8. The impact of human activities is less noticeable in the upstream because there is less human activity there. As for the midstream, without the impact of human activities, the RMSE ratio is less than 1 in the first lead month initialized in May and July–September (Fig. 8b). This indicates that the initial hydrological conditions have less variability in the wet seasons due to more human activities (e.g., irrigation and reservoir regulation), and human activities reduce the effect of ICs on hydrological predictability for 1 month. Therefore, the memory from ICs could only last for less than 1 month and would not contribute much to the hydrological prediction when human activities are the interference, which makes hydrological predictability rely more on meteorological prediction. In addition, the RMSE ratios in the midstream have a smaller spread for all forecast lead times when human activities are considered (Figs. 7b, 8b). This indicates that human activities reduce hydrologic variability between years, and could potentially increase the hydrological predictability. Considering droughts (i.e., dry conditions), human activities could also increase hydrological drought predictability, mainly by reasonable reservoir regulation. When droughts happen, discharge from reservoir will be increased, aiming to maintain ecological flow and human (domestic, agricultural and industrial) water demand, which will decrease the hydrological variability during dry periods. Therefore, human activities can outweigh meteorological variability and play a more important role in hydrological predictability. The results are similar to Yuan et al. (2017), which found human interventions can outweigh the climate variability for the hydrological drought forecasting over the Yellow River basin.

How do human activities influence hydrological drought forecast skill? Figures 9–10 show NMME-based hydrological drought forecast skill against ESP in terms of AC and BSS, when the human activities module is switched off. The forecast skill for NMME-based and ESP-based hydrological forecasts without influence of human activities (Fig. 9) is higher than that with human intervention (Fig. 5), especially in the midstream. The influence of human activities mainly occurs in the spring and early summer. Comparing Figs. 6 and 10 shows that NMME-based drought predictions have more skill improvement over the ESP-based predictions when human activities are involved. The improvement can be attained at lead times of 1–4 months in the winter, and longer lead times during April–September in the midstream. That means human activities have reduced the influence of ICs on hydrological drought predictions.

5 Conclusions

Understanding the performance of climate predictions at regional or global scales provides an important basis for the utility and improvement of these products. In recent decades, drought prediction based on climate predictions at seasonal scales has improved significantly due to a range of global climate models (Hao et al., 2018). However, hydroclimatic drought prediction and predictability over an endorheic and arid basin that is affected by complex climate mechanisms remains a great challenge. A breakthrough in predicting hydroclimatic drought over endorheic basins can bring major improvements in the development of reliable drought monitoring and prediction systems at regional and global scales. In this study, the seasonal meteorological and hydrological drought predictability and forecast skill in the Heihe River basin (HRB), a typical endorheic and arid basin with distinctive characteristics from upstream to midstream, were presented in detail. Here, meteorological predictability refers to the upper limit of forecast skill using a perfect model assumption, while hydrological predictability is to quantify the role of initial hydrological conditions (ICs) and meteorological forcings. The meteorological drought forecasting system was based on bias-corrected and downscaled 88-member North American Multimodel Ensemble (NMME)

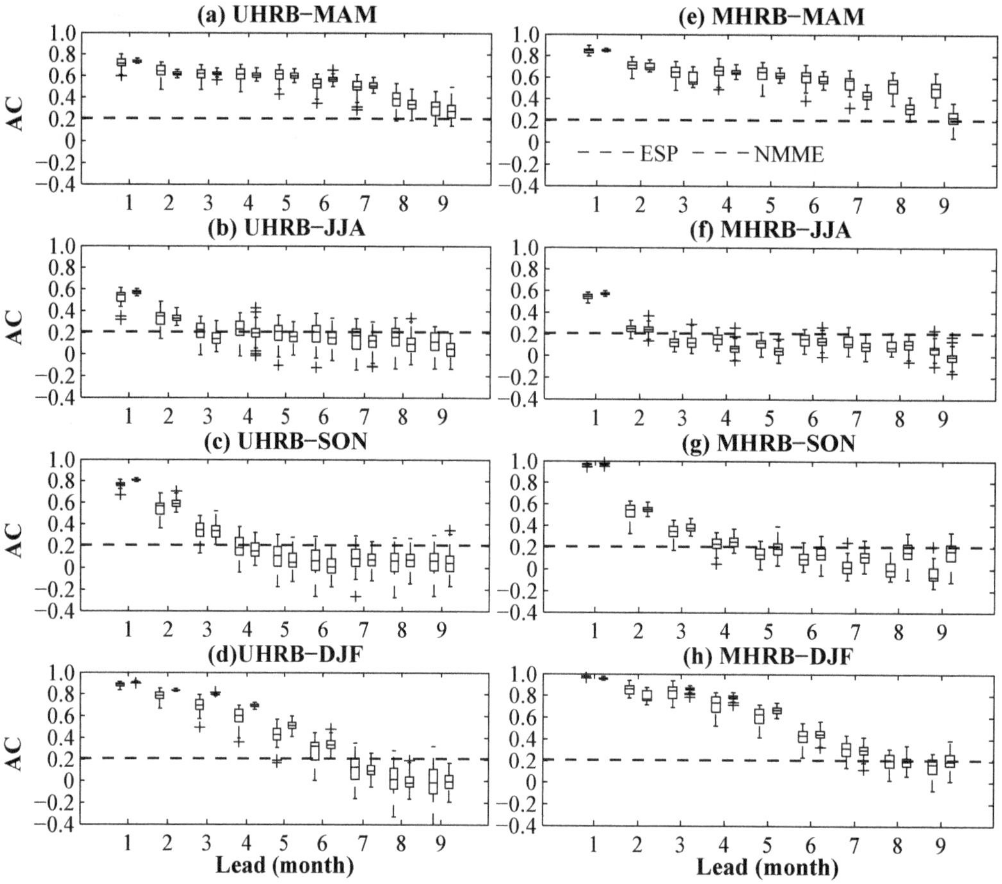

Figure 9. Anomaly correlation (AC) of the forecast of seasonal SSI1. The red box plots show the spread of AC of each member from NMME, and the blue box plots show that from ESP. The blue (red) crosses show the outliers for NMME (ESP) forecast skill. The dashed black line indicates the threshold (AC = 0.21) of 95 % confidence intervals calculated from the t test. The predictions and simulations are carried out with the human activities module switched off.

climate hindcasts during 1982–2010, and the hydrological drought forecasting system was established using the Distributed Time-Variant Gain Hydrological Model (DTVGM) driven by the post-processed predictions. The NMME-based hydrological predictions were compared with the ESP-type predictions, verified by offline simulations to ignore hydrological model structural errors. The DTVGM, with human activities modules (i.e., reservoir module and water use module), has been calibrated well over 140 subbasins in the HRB based on observed streamflow at three mainstream gauges and meteorological forcings for the period of 1981–2000. The Nash–Sutcliffe efficiency (NSE) values at a monthly scale were greater than 0.86 in the upstream during the calibration and validation periods. Given extensive irrigation and water diversions in the midstream, the NSE was greater than 0.52, indicating a reasonable human activities module in the DTVGM hydrological model.

For meteorological drought predictions, the upstream and midstream show higher meteorological drought predictability than forecast skill in terms of anomaly correlation (AC) and Brier score (BS). The forecast skill of the overall NMME mean is higher than that of most individual members. The NMME climate predictions show statistically significant predictability and forecast skills for meteorological drought in the first 2 lead months, with higher values in autumn and winter. For the hydrological drought predictions, the upstream shows higher skill than the midstream in terms of AC and Brier skill score (BSS), due to more complex hydrological processes and human activities in the midstream. The highest forecast skill occurs during winter and autumn, while the lowest skill exists during summer. During spring, the NMME-based hydrological predictions outperform ESP-type predictions up to 7 lead months, in spite of poor precipitation predictions. This may be due to more seasonal snowmelt having contributed to the streamflow in spring, which relies on the temperature predictions. The NMME-based hydrological predictions show some improvement against ESP up to 2–3 months lead during autumn. The highest forecast skill in winter could continue up to 6 months, probably due to the long memory of hydrological conditions.

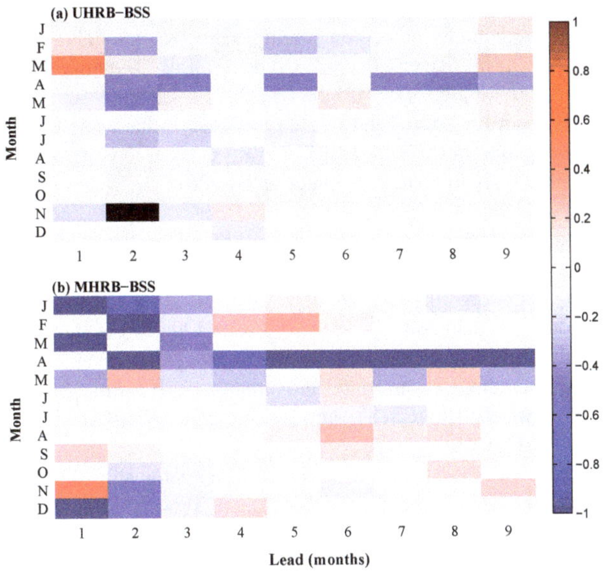

Figure 10. Brier skill score (BSS) of NMME forecast for hydrological drought events. Here, a hydrological drought event happens when the SSI1 value is below −1. The reference forecasts are simulations from ESP experiment. The predictions and simulations are carried out with the human activities module switched off.

Besides the ESP experiment, the reverse ESP was also conducted to investigate the relative role of initial conditions (ICs) in hydrological forecasting. The role of ICs could be more significant during the cold and dry seasons; for example, ICs prevail over the meteorological forcings up to 5–7 months during October–December and 2–3 months during January–March. However, the meteorological forcings outweigh the ICs in all lead months during warm and wet seasons (April–September). That reasonably explains the season-dependent hydrological forecast skill and inconsistency between meteorology and hydrology. In addition, a comparative experiment was conducted to explore the effect of human activities on hydrological predictability, via removing the human activities module from the DTVGM. Results show that human activities actually reduced the hydrological variability and increased the hydrological predictability during wet seasons (May and July–September) in the midstream. Therefore, the improvement of simulation of human activities could increase the hydrological drought forecast skill over a human-dominant basin.

Although the NMME-based forecasting system shows certain skill for meteorological and hydrological drought predictions, more efforts are needed to tackle issues in the following areas: (1) the physical mechanisms that caused climate anomalies need to be considered to improve climate models and the meteorological forecasting skill, as meteorological forcings play a dominant role in hydrological predictability during wet seasons. (2) The NSE value for the MHRB is greater than 0.52, which is still unsatisfactory. An unrefined human activities module in the hydrology model can lead to some uncertainties in the simulated streamflow and hydrological drought and thus the performance evaluation. For example, inaccurate calculations of irrigation water requirements and groundwater can increase errors in river flow and uncertainties in the influence of human activities on hydrological droughts. Therefore, refining the human activities' processes in the hydrological forecasting system could facilitate the understanding of the hydrological predictions over the regions with vast human activities. (3) Data assimilation and observation need to be improved for initial hydrological conditions (e.g., snow and soil moisture), which could promote the development of high-precision hydrological predictions. This requires more collaborations between scientists from different disciplines, including climate science, hydrology, agriculture, ecology and social economy.

Author contributions. FM, LL, AY and QD designed the experiments and FM carried them out. AY developed the DTVGM model code and FM performed the simulations. FM prepared the manuscript with contributions from all co-authors.

Competing interests. The authors declare that they have no conflict of interest.

Acknowledgements. This work was supported by the Strategic Priority Research Program of the Chinese Academy of Sciences (no. XDA19070104, XDA20060401), the Natural Science Foundation of China (no. 51879009, 41475093) and the State Key Laboratory of Earth Surface Processes and Resource Ecology Open Research Program (no. 2017-KF-17). This work is also supported by the Water Science Network WaterCube Program at Michigan State University.

Edited by: Lixin Wang

References

Barker, L. J., Hannaford, J., Chiverton, A., and Svensson, C.: From meteorological to hydrological drought using standardised indicators, Hydrol. Earth Syst. Sci., 20, 2483–2505, 2016.

Becker, E., Van Den Dool, H., and Zhang, Q.: Predictability and forecast skill in NMME, J. Climate, 27, 5891–5906, 2014.

Below, R., Grover-Kopec, E., and Dilley, M.: Documenting drought-related disasters. A global reassessment, J. Environ. Dev., 16, 328–344, https://doi.org/10.1177/1070496507306222, 2007.

Farahmand, A. and AghaKouchak, A.: A generalized framework for deriving nonparametric standardized drought indicators, Adv. Water Resour., 76, 140–145, https://doi.org/10.1016/j.advwatres.2014.11.012, 2015.

Gringorten, II.: A plotting rule for extreme probability paper, J. Geophys. Res., 68, 813–814, 1963.

Gustard, A., Bullock, A., and Dixon, J. M.: Low flow estimation in the United Kingdom, Institute of Hydrology, Wallingford, 88 pp., 1992.

Hao, Z., Singh, V. P., and Xia, Y.: Seasonal drought prediction: Advances, challenges, and future prospects, Rev. Geophys., 56, 108–141, https://doi.org/10.1002/2016RG000549, 2018.

Huang, J., Ji, M., Xie, Y., Wang, S., He, Y., and Ran, J.: Global semi-arid climate change over last 60 years, Clim. Dynam, 46, 1131–1150, 2016a.

Huang, J., Yu, H., Guan, X., Wang, G., and Guo, R.: Accelerated dryland expansion under climate change, Nat. Clim. Change, 6, 166–171, 2016b.

Huang, S., Li, P., Huang, Q., Leng, G., Hou, B., and Ma, L.: The propagation from meteorological to hydrological drought and its potential influence factors, J. Hydrol., 547, 184–195, 2017.

Hutchinson, M. F.: Interpolation of rainfall data with Thin Plate Smoothing Splines – Part I: two dimensional smoothing of data with short range correlation, J. Geogr. Inform. Decis. Anal., 2, 139–151, 1998a.

Hutchinson, M. F.: Interpolation of rainfall data with Thin Plate Smoothing Splines – Part II: analysis of topographic dependence, J. Geogr. Inform. Decis. Anal., 2, 152–167, 1998b.

Kirtman, B. P., Min, D., Infanti, J. M., Kinter, III, J. L., Paolino, D. A., Zhang, Q., Van den Dool, H., Saha, S., Mendez, M. P., Becker, E., Peng, P., Tripp, P., Huang, J., Dewitt, D. G., Tippett, M. K., Barnston, A. G., Li, S., Rosati, A., Schubert, S. D., Rienecker, M., Suarez, M., Li Z., Marshak, J., Lim, Y. K., Tribbia, J., Pegion K., Merryfield, W. J., Denis, B., and Wood, E. F.: The North American Multimodel Ensemble: Phase-1 Seasonal-to-Interannual Prediction; Phase-2 toward Developing Intraseasonal Prediction, Bull. Am. Meteorol. Soc., 95, 585–601, 2014.

Li, X., Cheng, G., Ge, Y., Li, H., Han, F., Hu, X., Tian, W., Tian Y., Pan X., Nian, Y., Zhang, Y., Ran, Y., Zheng, Y., Gao, B., Yang, D., Zheng, C., Wang X., Liu, S., and Cai, X.: Hydrological Cycle in the Heihe River Basin and Its Implication for Water Resource Management in Endorheic Basins, J. Geophys. Res.-Atmos., 123, 890–914, 2018.

Liu, Y., Ma, K., and Lin, Z.: Potential predictability of monthly precipitation over China, J. Meteor. Res., 14, 316–329, 2000.

Luo, L. and Wood, E. F.: Assessing the idealized predictability of precipitation and temperature in the NCEP Climate Forecast System, Geophys. Res. Lett., 33, L04708, https://doi.org/10.1029/2005GL025292, 2006.

Luo, L. and Wood, E. F.: Monitoring and predicting the 2007 U.S. drought, Geophys. Res. Lett., 34, L22702, https://doi.org/10.1029/2007GL031673, 2007.

Luo, L. and Wood, E. F.: Use of Bayesian Merging Techniques in a Multimodel Seasonal Hydrologic Ensemble Prediction System for the Eastern United States, J. Hydrometeorol., 9, 866–884, 2008.

Luo, L., Sheffield, J., and Wood, E. F.: Towards a global drought monitoring and forecasting predictability, NWS Science & Technology Infusion Climate Bulletin, 2008.

Ma, M. G. and Frank, V.: Interannual variability of vegetation cover in the Chinese Heihe River Basin and itsrelation to meteorological parameters, Int. J. Remote Sens., 27, 3473–3486, 2006.

Ma, F., Yuan, X., and Ye, A.: Seasonal Drought Predictability and Forecast Skill over China, J. Geophys. Res.-Atmos., 120, 8264–8275, 2015.

Ma, F., Ye, A., Deng, X., Zhou, Z., Liu, X., Duan, Q., Xu, J., Miao, C., Di, Z., and Gong, W.: Evaluating the skill of NMME seasonal precipitation ensemble predictions for 17 hydroclimatic regions in continental China, Int. J. Climatol., 36, 132–144, 2016.

Ma, F., Ye, A., and Duan, Q.: Seasonal drought ensemble predictions based on multiple climate models in the upper Han River Basin, China, Clim. Dynam., 1–14, https://doi.org/10.1007/s00382-017-3577-1, 2017.

Ma, F., Luo, L., Ye, A., and Duan, Q.: Drought characteristics and propagation in the semi-arid Heihe River basin in Northwestern China, J. Hydrometeorol., under review, 2018.

Mao, Y., Ye, A., Liu, X., Ma, F., Deng, X., and Zhou, Z.: High-resolution simulation of the spatial pattern of water use in continental China, Hydrolog. Sci. J., 61, 2626–2638, https://doi.org/10.1080/02626667.2016.1153102, 2016.

Mariotti, A., Schubert, S., Mo, K., Peters-Lidard, C., Wood, A., Pulwarty, R., Huang, J., and Barrie, D.: Advancing drought understanding, monitoring, and prediction, B. Am. Meteorol. Soc., 94, ES186–ES188, https://doi.org/10.1175/BAMS-D-12-00248.1, 2013.

McKee, T., Doesken, N., and Kleist, J.: The relationship of drought frequency and duration to time scales, in: Proceedings of the 8th conference of applied climatology on American Meteorological Society, Anaheim, CA, 17–22 January 1993, 179–184, 1993.

Mishra, A. K. and Singh, V. P.: Drought modeling – A review, J. Hydrol., 403, 157—175, https://doi.org/10.1016/j.jhydrol.2011.03.049, 2011.

Mo, K. and Lettenmaier, D. P.: Hydrologic prediction over the conterminous United States using the National Multi-Model Ensemble, J. Hydrometeorol., 15, 1457–1472, 2014.

Mo, K. and Lyon, B.: Global meteorological drought prediction using the North American Multi-Model Ensemble, J. Hydrometeorol., 16, 1409–1424, 2016.

Pan, M., Yuan, X., and Wood, E. F.: A probabilistic framework for assessing drought recovery, Geophys. Res. Lett., 40, 3637–3642, https://doi.org/10.1002/grl.50728, 2013.

Pozzi, W., Sheffield, J., Stefanski, R., Cripe, D., Pulwarty, R., Vogt, J. V., Heim Jr, R. R., Brewer, M. J., Svoboda, M., Westerhoff, R., van Dijk, A. I.J.M., Lloyd-Hughes, B., Pappenberger, F., Werner, M., Dutra, E., Wetterhall, F., Wagner, W., Schubert, S., Mo, K., Nicholson, M., Bettio, L., Nunez, L., van Beek, R., Bierkens, M., Gustavo Goncalves de Goncalves, L., Gerd Zell de Mattos, J., and Lawford, R.: Towards global drought early warning capability: Expanding international cooperation for the development of a framework for global drought monitoring and forecasting, B. Am. Meteorol. Soc., 94, 776–785, https://doi.org/10.1175/BAMS-D-11-00176.1, 2013.

Schepen, A., Wang, Q. J., and Everingham, Y.: Calibration, Bridging, and Merging to Improve GCM Seasonal Temperature Forecasts in Australia, Mon. Weather Rev., 144, 2421–2441, 2016.

Sheffield, J. and Wood, E. F.: Drought: Past Problems and Future Scenarios, Earthscan, London, Washington, D.C., 2012.

Shi, P., Yang, T., Zhang, K., Tang, Q., Yu, Z., and Zhou, X.: Large-scale climate patterns and precipitation in an arid endorheic region: linkage and underlying mechanism, Environ. Res. Lett., 11, 044006, https://doi.org/10.1088/1748-9326/11/4/044006, 2016.

Shukla, S. and Lettenmaier, D. P.: Seasonal hydrologic prediction in the United States: understanding the role of initial hydrologic conditions and seasonal climate forecast skill, Hydrol. Earth Syst. Sci., 15, 3529–3538, https://doi.org/10.5194/hess-15-3529-2011, 2011.

Shukla, S., Sheffield, J., Wood, E. F., and Lettenmaier, D. P.: On the sources of global land surface hydrologic predictability, Hydrol. Earth Syst. Sci., 17, 2781–2796, https://doi.org/10.5194/hess-17-2781-2013, 2013.

Shukla, S., Roberts, J., Hoell, A., Funk, C. C., Robertson, F., and Kirtman, B.: Assessing North American multimodel ensemble (NMME) seasonal forecast skill to assist in the early warning of anomalous hydrometeorological events over East Africa, Clim. Dynam, 1–17, https://doi.org/10.1007/s00382-016-3296-z, 2016.

Smith, A. B. and Katz, R. W.: US billion-dollar weather and climate disasters: Data sources, trends, accuracy and biases, Nat. Hazards, 67, 387–410, https://doi.org/10.1007/s11069-013-0566-5, 2013.

Sun, J., Jiang, H., Wang, K., Xu, J., Li, Z., Zhu, Q., and Luo, X.: The Analysis of Atmospheric Circulation and Water Vapor Conditions of Precipitation over the Heihe River Basin, 2011 International Conference on Computer Distributed Control and Intelligent Environmental Monitoring, Changsha, China, 19–20 February, 2271–2275, 2011.

Thober, S. and Kumar, R.: Seasonal soil moisture drought prediction over Europe using the North American Multi-Model Ensemble (NMME), J. Hydrometeorol., 16, 2329–2344, 2015.

Van Loon, A. F., Gleeson, T., Clark, J., Van Dijk, A. I., Stahl, K., Hannaford, J., Baldassarre G. D., Teuling A. J., Tallaksen, L. M., Uijlenhoet, R., Hannah, D. M., Sheffield, J., Svoboda, M., Verbeiren, B., Wagener, T., Rangecroft, S., Wanders, N., and Van Lanen, H. A. J.: Drought in the Anthropocene, Nat. Geosci., 9, 89–91, https://doi.org/10.1038/ngeo2646, 2016a.

Van Loon, A. F., Stahl, K., Di Baldassarre, G., Clark, J., Rangecroft, S., Wanders, N., Gleeson, T., Van Dijk, A. I. J. M., Tallaksen, L. M., Hannaford, J., Uijlenhoet, R., Teuling, A. J., Hannah, D. M., Sheffield, J., Svoboda, M., Verbeiren, B., Wagener, T., and Van Lanen, H. A. J.: Drought in a human-modified world: reframing drought definitions, understanding, and analysis approaches, Hydrol. Earth Syst. Sci., 20, 3631–3650, https://doi.org/10.5194/hess-20-3631-2016, 2016b.

Vicente-Serrano, S. M., López-Moreno, J. I., Beguería, S., Lorenzo-Lacruz, J., Azorin-Molina, C., and Morán-Tejeda, E.: Accurate computation of a streamflow drought index, J. Hydrol. Eng., 17, 318–332, 2012.

Wang, J. and Li, W.: Establishing simulated model of snowmelt runoff for large scale basin in western China, J. Glaciol. Geocryol., 21, 264–268, 1999.

Wang, S., Kang, S., Zhang, L., and Li, F.: Modelling hydrological response to different land-use and climate change scenarios in the Zamu River basin of northwest China, Hydrol. Proc., 22, 2502–2510, 2008.

Wang, S. and Yuan, X.: Extending seasonal predictability of Yangtze River summer floods, Hydrol. Earth Syst. Sci., 22, 4201–4211, https://doi.org/10.5194/hess-22-4201-2018, 2018.

Wilks, D. S.: Statistical methods in the atmospheric sciences, 3 Edn., International Geophysics Series, Vol. 100, edited by: Dmowaska, R., Hartmann, D., and Rossby, H. T., Academic Press, Kidlington, UK, 704 pp, 2011.

Wood, A. W. and Lettenmaier, D. P.: An ensemble approach for attribution of hydrologic prediction uncertainty, Geophys. Res. Lett., 35, L14401, https://doi.org/10.1029/2008GL034648, 2008.

Wood, A. W., Mauer, E. P., Kumar, A., and Lettenmaier, D. P.: Long-range experimental hydrologic forecasting for the eastern United States, J. Geophys. Res. Atmos., 107, 4429, https://doi.org/10.1029/2001JD000659, 2002.

Ye, A., Duan, Q., Zeng, H., Li, L., and Wang, C.: A Distributed Time-Variant Gain Hydrological Model based on remote sensing, J. Resour. Ecol., 1, 222–230, 2010.

Ye, A., Duan, Q., Zhan, C., Liu, Z., and Mao, Y.: Improving kinematic wave routing scheme in Community Land Model, Hydrol. Res., 44, 886–903, 2013.

Yuan, X.: An experimental seasonal hydrological forecasting system over the Yellow River basin – Part 2: The added value from climate forecast models, Hydrol. Earth Syst. Sci., 20, 2453–2466, https://doi.org/10.5194/hess-20-2453-2016, 2016.

Yuan, X. and Wood, E. F.: Multimodel seasonal forecasting of global drought onset, Geophys. Res. Lett., 40, 4900–4905, 2013.

Yuan, X., Zhang, M., Wang, L., and Zhou, T.: Understanding and seasonal forecasting of hydrological drought in the Anthropocene, Hydrol. Earth Syst. Sci., 21, 5477–5492, https://doi.org/10.5194/hess-21-5477-2017, 2017.

Zhang, A., Liu, W., Yin, Z., Fu, G., and Zheng, C.: How will climate change affect the water availability in the Heihe River Basin, northwest China?, J. Hydrometeorol., 17, 1517–1542, 2016.

Zhang, Y., Fu, G., Sun, B., Zhang, S., and Men, B.: Simulation and classification of the impacts of projected climate change on flow regimes in the arid Hexi Corridor of northwest China, J. Geophys. Res.-Atmos., 120, 7429–7453, 2015.

Zhao, Y. and Zhu, J.: Assessing quality of grid daily precipitation datasets in China in recent 50 years, Plateau Meteorology (in Chinese), 34, 50–58, 2015.

10

Reconstruction of droughts in India using multiple land-surface models (1951–2015)

Vimal Mishra[1], **Reepal Shah**[1], **Syed Azhar**[1], **Harsh Shah**[1], **Parth Modi**[1], **and Rohini Kumar**[2]

[1]Civil Engineering, Indian Institute of Technology (IIT) Gandhinagar, Gujarat, 382355, India

[2]UFZ-Helmholtz Centre for Environmental Research, Leipzig, Germany

Correspondence: Vimal Mishra (vmishra@iitgn.ac.in)

Abstract. India has witnessed some of the most severe historical droughts in the current decade, and severity, frequency, and areal extent of droughts have been increasing. As a large part of the population of India is dependent on agriculture, soil moisture drought affecting agricultural activities (crop yields) has significant impacts on socioeconomic conditions. Due to limited observations, soil moisture is generally simulated using land-surface hydrological models (LSMs); however, these LSM outputs have uncertainty due to many factors, including errors in forcing data and model parameterization. Here we reconstruct agricultural drought events over India during the period of 1951–2015 based on simulated soil moisture from three LSMs, the Variable Infiltration Capacity (VIC), the Noah, and the Community Land Model (CLM). Based on simulations from the three LSMs, we find that major drought events occurred in 1987, 2002, and 2015 during the monsoon season (June through September). During the Rabi season (November through February), major soil moisture droughts occurred in 1966, 1973, 2001, and 2003. Soil moisture droughts estimated from the three LSMs are comparable in terms of their spatial coverage; however, differences are found in drought severity. Moreover, we find a higher uncertainty in simulated drought characteristics over a large part of India during the major crop-growing season (Rabi season, November to February: NDJF) compared to those of the monsoon season (June to September: JJAS). Furthermore, uncertainty in drought estimates is higher for severe and localized droughts. Higher uncertainty in the soil moisture droughts is largely due to the difference in model parameterizations (especially soil depth), resulting in different persistence of soil moisture simulated by the three LSMs. Our study highlights the importance of accounting for the LSMs' uncertainty and consideration of the multi-model ensemble system for the real-time monitoring and prediction of drought over India.

1 Introduction

Drought is among the top natural disasters that affect food and fresh water security. The 2014–2015 drought in India affected more than 3.3 million people and resulted in the loss of INR 6 50 000 crore (Indian express, 11 May 2016). Drought characteristics such as frequency, areal extent, and intensity have increased in India, which can be attributed to erratic summer monsoon as well as an increase in air temperature (Shah and Mishra, 2014). Moreover, the frequency of severe and widespread multi-year droughts has also increased during recent decades (Mishra et al., 2016). For instance, India experienced 10 major droughts between 1950 and 1989, while 5 droughts occurred after 2000 (Pai et al., 2017). The drought of 2015 was among the most severe during the period of 1901–2015, which caused enormous damage to crops and affected various sectors of society (Mishra et al., 2016). Precipitation deficit during the monsoon (rainy) season not only affects water availability during that season, but also affects water availability in the post-monsoon (dry) season.

Despite an increase in irrigation infrastructure during the last few decades, about 66 % of Indian agriculture remains rain-fed and largely reliant on the monsoon season rainfall, which accounts for about 80 % of the total annual rainfall. Precipitation deficit during the monsoon season leads to a deficit in root-zone soil moisture during the post-monsoon crop-growing season. This deficit in the soil moisture can be amplified by positive temperature anomalies during the growing season. Due to lack of long-term observations of soil moisture, the impacts of climate variability and climate change on soil moisture drought are often studied using land-surface (hydrologic) models (LSMs, Mishra et al., 2014; Sheffield and Wood, 2008; Samaniego et al., 2013). However, these LSMs have differences in model parameterization and representation of hydrological processes (Mishra et al., 2017; Wang et al., 2009), which point to uncertainty related to model structural error deficits. Moreover, soil depths specified in the LSMs vary depending on an individual model configuration, which can lead to differences in soil moisture persistence (Wang et al., 2009). Soil moisture persistence is important for understanding the dynamics of soil moisture in response to meteorological forcing. For instance, LSMs with low soil moisture persistence may show higher sensitivity to temperature and/or precipitation anomalies. Differences in soil moisture persistence in LSMs can lead to uncertainty in drought monitoring and assessment.

There have been several projects on the inter-comparison of soil moisture and other hydrologic fluxes from different LSMs. For instance, the Global Land Data Assimilation System (GLDAS; Rodell et al., 2004), the earthH2Observe project (Beck et al., 2017), the Project for Intercomparison of Land Surface Parameterization Schemes (PILPS; Bowling et al., 2003), and the Global Soil Wetness Project (Dirmeyer et al., 1999, 2006) provide useful insights into the differences in soil moisture simulations from the LSMs and hydrological models. Our aim here is to understand the uncertainty in soil moisture drought characteristics simulated using three LSMs over India. We use observed gridded meteorological data to force the calibrated VIC, Noah, and CLM land-surface models. We estimate drought indices based on precipitation and 60 cm depth (as a surrogate for root-zone depth) soil moisture from the three LSMs to reconstruct the major drought events that occurred during the period of 1951–2015. We selected the top 60 cm to analyze soil moisture drought because, for many crops, effective root-zone depth falls in this region (Jalota and Arora, 2002). Also, we consider this depth for all three models in our drought assessment so as to reduce uncertainty due to specification of different root-zone depth (based on respective vegetation parameters) and soil layer thicknesses in three LSMs. We recognize that there exist different sources of uncertainty in model estimates (of soil moisture), including those arising from errors in input variables (e.g., meteorological forcings, surface and sub-surface characteristics); however, here, our aim is not to quantify uncertainty due to all the sources.

Rather, we limit ourselves to understanding the uncertainty in historical reconstructions of soil moisture droughts over India due to structural differences among different LSMs.

2 Methodology

2.1 Data

The three LSMs (VIC, Noah, and CLM) used in this study were forced with a common meteorological dataset that comprises daily precipitation, maximum and minimum temperatures, and wind speed. We used a 0.25° daily gridded precipitation product available for the period 1901–2015 from the India Meteorological Department (hereafter IMD; Pai et al., 2015), which was developed by IMD using data from 6995 gauge stations across India and an inverse distance weighting scheme (Shepard, 1984). In the gridded precipitation data from IMD, orographic and topographic features of precipitation are well captured along with the spatial variability associated with the Indian summer monsoon. We used 1° daily gridded maximum and minimum air temperatures from IMD (Srivastava et al., 2009), which were developed using the data from 395 observation stations across India. We re-gridded air temperatures from 1 to 0.25° using the method described in Maurer et al. (2002), which is based on a temperature lapse rate of 6.5 °C km^{-1} rise in elevation and the SYMAP algorithm. In re-gridding of air temperature, we used a 0.25° digital elevation model (DEM) that was resampled from the original 30 m elevation data from the Shuttle Radar Topography Mission (SRTM). The gridded precipitation and air temperature products have been used in many previous studies on drought and heat waves (Shah et al., 2017; Shah and Mishra, 2014; R. D. Shah and Mishra, 2016, 2015; Mishra et al., 2016).

2.2 Land-surface models

We used simulated soil moisture from the three LSMs, the VIC, the Noah, and the CLM, to assess uncertainty in root-zone (60 cm) depth soil moisture and retrospective drought assessment. These three LSMs have been widely used for producing land-surface fluxes at global and regional scales (Rodell et al., 2004; Shah and Mishra, 2015; Unnikrishnan et al., 2013; see also Table S1 in the Supplement for a brief description of major hydrological processes). All three LSMs were forced with the same meteorological forcing from IMD at 0.25° resolution, and additional (radiation-related) forcing variables for the Noah and CLM were derived from the MT-CLIM algorithm integrated in the VIC model (see Bohn et al., 2013). This will keep basic forcing data consistent across models. In all LSMs routines are there which disaggregate daily precipitation uniformly to a sub-daily timescale, while temperature and radiation are temporally disaggregated following the diurnal cycle. Regarding the usage of land-surface datasets, we note here that each model used a slightly different set of input datasets. Soil textural properties and resulting hydrologic parameters (like field capacity, wilting points, and available water) are mostly derived based on the Harmonized World Soil Database (HWSD) and the Food and Agriculture Organization (FAO) based soil maps. However, since land-surface models use soil parameters from two different sources that can lead to differences in available water (Fig. S2), we evaluated the difference in the simulated soil moisture anomalies from the VIC model using the soil parameters based on the FAO and HWSD datasets (Fig. S3). We do not find a significant difference in the simulated soil moisture anomalies based on the soil parameters from the two sources. It is worth noting that the underlying (soil-textural) dataset for the HWSD is mostly derived based on the FAO dataset. The different products may use different pedo-transfer functions to derive soil-related parameters. Nevertheless the derived (static) soil parameters do not induce significant differences in the temporal dynamics of simulated soil moisture anomalies (Figs. S2 and S3).

Vegetation characteristics specified within LSMs were derived from the Advanced Very High-Resolution Radiometer (AVHRR) and Moderate Resolution Imaging Spectroradiometer (MODIS) datasets. We evaluated the sensitivity of vegetation parameters derived from AVHRR and MODIS to simulated soil moisture using the Noah model (Fig. S4). In this case too we do not find any substantial differences in the dynamics of the simulated soil moisture anomalies due to differences in the underlying vegetation parameters estimated from the AVHRR and MODIS datasets (Fig. S4). We note that the analysis conducted here is at a 0.25° spatial resolution – as a result the differences among the model simulations due to fine-scale soil and vegetation characteristics may not be observed at a coarse scale.

All three LSMs were first spun up using 65 years (1951–2015) of data to establish initial conditions for the modeled states and fluxes. All three LSMs were manually calibrated to match simulated monthly streamflow with observed streamflow data obtained from the India-WRIS (http://www.india-wris.gov.in) at the gauging stations (H. L. Shah and Mishra, 2016) that are least affected by human interventions related to water diversion and water withdrawal for irrigation (see the Supplement, Fig. S1 and Table S2, for the geographical location of these stations). We identified the calibration parameter for each LSM based on prior studies (Cai et al., 2014; Hogue et al., 2005; Nijssen et al., 2001) and by performing a simple (one parameter at a time) sensitivity analysis. We used soil thickness also as calibration parameters following the success of calibrating the VIC using soil-layer thickness (Nijssen et al., 2001; R. D. Shah and Mishra, 2016). The calibration parameters were manually adjusted so as to match observed streamflow (see Table S2). Further, we evaluated the model skill by comparing simulated soil moisture with station and satellite-based soil moisture, and also by comparing the total column soil moisture changes with terrestrial water change based on GRACE products (see Sect. 3.1 for more details).

2.2.1 The Variable Infiltration Capacity (VIC) model

We used VIC v4.2.a (Liang et al., 1994), which is a semi-distributed, physically based hydrologic model in water balance mode at a daily time step. The VIC model simulates water and energy fluxes in each grid cell considering soil and vegetation parameters, and meteorological forcing as input. The model estimates total evapotranspiration as a sum of the canopy and bare soil evaporation and transpiration from vegetation mosaics. Any number of vegetation types can be represented within a grid cell to represent sub-grid variability in vegetation cover. Infiltration is estimated using a variable infiltration capacity curve. The VIC model has three soil layers and the top two layers respond quickly to rainfall, and diffusion is allowed from the middle to top layers when the middle (second) layer is wet. Baseflow from the bottom (third) layer is estimated using the Arno model formulation (Franchini and Pacciani, 1991). The bottom layer responds slowly to depict seasonal soil moisture behavior. We calibrated the VIC model parameters, which include depth of the soil layers, infiltration curve parameters, and parameters related to baseflow, following Nijssen et al. (2001). Vegetation and soil texture used in the VIC model were developed using the 1 km Advanced Very High-Resolution Radiometer (AVHRR) and Harmonized World Soil Database (HWSD), respectively, as described in Table S1. The VIC model requires soil parameters like field capacity and wilting point, which were derived by first identifying soil class based on the United States Department of Agriculture (USDA) classification and then applying the pedo-transfer functions of Cosby et al. (1984). Soil texture specified at 0.25° for deriving soil parameters is shown in Fig. S2a. More detailed information on the VIC model calibration can be obtained from the previous studies (Mishra et al., 2010; Nijssen et al., 2001; H. L. Shah and Mishra, 2016, R. D. Shah and Mishra, 2016).

2.2.2 The Noah model

We used the one dimensional Noah model (version 3.1; Mitchell, 2004; Schaake et al., 1996) which solves water and energy balance in each grid cell. The Noah model has four soil layers. The model uses the modified Penman–Monteith equation to represent the diurnal variation of the atmospheric resistance coefficient (Chen et al., 1996; Mahrt and Ek, 1984). In the Noah model, spatial variability of precipitation and infiltration is considered to estimate surface runoff, which is based on the exponential distribution of infiltration capacity. Baseflow is proportional to soil moisture storage. Vegetation parameters used in the Noah model were derived from the MODIS dataset and classified based on the Modified International Geosphere Biosphere Programme (IGBP) scheme. The MODIS-based IGBP product has 20 categories of land use/land cover data, which were derived during the observation period of 2001–2005. The vegetation parameters of the Noah model consist of vegetation fraction, stomatal resistance, minimum and maximum values of LAI, albedo, and roughness length. The major land cover classes are Forest, Shrubs, Savannas, Tundra, Grasslands, Croplands, Wetlands, Built-up, Ice, and Water. We used soil textures derived from a digital soil map (Fig. S2c) developed by the Food and Agriculture Organization (FAO). The forcing parameters required for the Noah model are daily precipitation, air temperatures (maximum and minimum), wind speed, surface pressure, relative humidity, long-wave radiation, and surface downward solar radiation. Daily meteorological forcing in the Noah model was internally disaggregated using a uniform distribution for precipitation and the diurnal cycle for other variables. We calibrated the Noah model parameters that include the depth of four soil layers, the Zilintikevich coefficient, a surface runoff parameter, and a bare soil evaporation component. The Zilintikevich coefficient controls the ratio of the roughness length for heat to the momentum by representing the aerodynamic resistance.

2.2.3 The Community Land Model (CLM)

The CLM is a land-surface component of the community-developed global climate system model version 3.0 (CCSM v3.0), which was developed by the National Centre for Atmospheric Research (NCAR). The CLM has 10 soil layers and, similarly to the VIC and Noah, simulates both water and energy fluxes in each grid cell. Surface runoff in the CLM is parameterized based on the TOPMODEL concept (Beven and Kirkby, 1979). Soil moisture storage in the CLM is modeled after removing surface runoff, infiltration, and evaporation from surface storage. The basic difference in the CLM from the VIC and Noah is that the CLM has a representation of a groundwater table which is updated dynamically (Niu et al., 2007). The atmospheric forcings required for the CLM are daily precipitation, air temperatures (maximum and minimum), wind speed, specific humidity, incident solar radiation, long-wave radiation, and surface pressure. Land cover used in the CLM is represented by 17 plant functional types (PFTs), which were derived from MODIS and are classified using the IGBP scheme similar to Bonan et al. (2002), while the soil textures (Fig. S2) used in the CLM are derived from FAO datasets. We calibrated the soil thickness parameter for the CLM model similarly to the VIC model. A detailed comparison of input parameters is provided in Table S1. All three LSMs were run without considering irrigation and groundwater extraction as our aim was to understand the role of atmospheric forcing in root-zone (60 cm depth) soil moisture drought uncertainty.

2.3 Drought indices

We used the Standardized Precipitation Index (hereafter SPI; McKee et al., 1993) and the Standardized Soil moisture Index (SSI; Hao and AghaKouchak, 2013) to represent meteorological and soil moisture (agriculture) droughts, respectively. We used 60 cm soil depth as a representative of root-zone soil moisture (Shah and Mishra, 2015). Since depths of root-zone and soil layers are different in all three LSMs (Table S1), we estimated 60 cm soil moisture for each grid cell and for each LSM, separately. A parametric (Gamma) distribution was fitted to precipitation and root-zone soil moisture to estimate SPI and SSI, respectively. For both SPI and SSI, the cumulative distribution functions obtained by fitting the Gamma distribution were mapped onto the normal distribution functions to represent a dimensionless index and to derive drought indices (see Shah and Mishra, 2015, and Appendix A for more details). We note that there are other approaches for estimating the soil moisture drought index – for example a non-parametric percentile-based drought index (Samaniego et al., 2013), but in this study we used a parametric (Gamma) distribution for estimating SSI so as to be consistent with the precipitation-based drought index (SPI).

2.4 Intensity–areal extent–frequency curves

Intensity–areal extent–frequency (IAF) curves for drought events were constructed to understand the frequency and severity of droughts in India for the period 1951–2015. The IAF curves were estimated using the root-zone soil moisture from the three LSMs (i.e. VIC, Noah, and CLM). We estimated drought severity using the 4-month SSI at the end of the monsoon and Rabi seasons (so as to represent the entire season) for the whole of India and for the Indo-Gangetic Plain region (longitude 75–90° E and latitude 23–30° N; Fig. S1) considering the climatological period of 1951–2015. The method to construct IAF curves has been described in detail in Mishra et al. (2016) and Mishra and Cherkauer (2010). IAF curves were estimated using the following steps: (i) for each year, the mean 4-month SSI value was estimated for all the grids for areal extents of 2, 5, 10, 20, 30, 40, 50, 60, 70, 80, 90, and 100 %; (ii) for each threshold of areal extent, mean severity of root-zone soil moisture drought was estimated for each year during the 1951–2015 period; (iii) the Generalized Extreme Value (GEV) distribution was fitted to the mean severity for the selected areal extents and parameters (shape, location, and scale) were estimated using the maximum likelihood method; (iv) drought severity was estimated for the selected return periods of 2, 5, 10, 20, 25, 50, 100, 200, and 500 years for each areal-extent threshold to construct IAF. Using IAF curves, mean intensity of drought can be estimated or for a given areal extent and frequency of drought. We evaluated the goodness of fit of the GEV distribution using QQ plots and a Chi-square goodness of fit test (Figs. S15–S17 and Tables S11–S13).

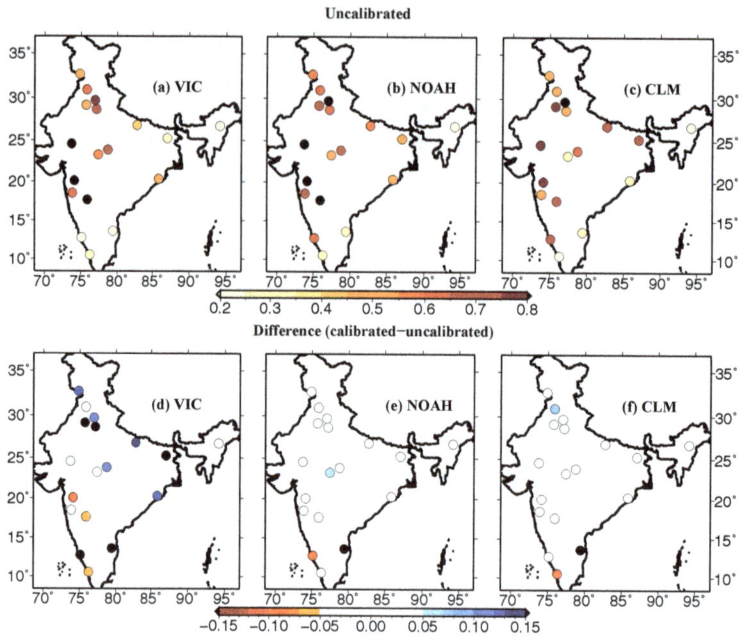

Figure 1. Correlation of weekly 60 cm simulated soil moisture with IMD gauge-based soil moisture (~ 60 cm) during the monsoon season, 2009–2013. **(a–c)** Correlation for control (default or uncalibrated) setup for the VIC, NOAH, and CLM, respectively. **(d–f)** show differences (calibrated–uncalibrated) in the correlation coefficient.

3 Results

3.1 Calibration and evaluation of land-surface models (LSMs)

The three land-surface model (VIC, Noah, and CLM) parameterizations were manually constrained (calibrated) against observed streamflow across a set of 18 major river basins covering approximately the entire landmass of India (see Fig. S1 and Table S2). The performance of the LSMs in capturing the temporal dynamics of monthly streamflow during calibration and validation periods is quite satisfactory for most of the river basins (Table S2). The median correlation r (and Nash–Sutcliffe efficiency, NS) values estimated across these basins during the calibration period are around 0.91 (0.78), 0.90 (0.70), and 0.90 (0.70) for the VIC, Noah, and CLM, respectively. A similar level of (median) skill is also observed during the validation period (Table S2). The skill of the multi-model averaged streamflow of the three LSMs is comparatively better than that of individual models – with an overall median r (and NS) value estimated across all basins of 0.91 (0.80) and 0.94 (0.77) during the calibration and validation periods, respectively. The ensemble mean of simulated streamflow from the three LSMs performed better against the observations. We also notice a relatively poor skill for all three LSMs and the ensemble mean in the coastal basins (e.g., Cauvery and eastern coast basins – Table S2), which could be attributed to a number of factors including errors in forcing data and model parameterizations. Nevertheless, considering the wide range of hydro-climatic gradients across India, the efficiency of the three LSMs for capturing the observed streamflow can be considered reasonable.

Next we evaluated the skill of each model in capturing the observed dynamics of near-surface and 60 cm soil moisture (Figs. 1 and S3–S4). We used three different sources of soil moisture observations for this comparison purpose. The first set consisted of the weekly soil moisture observations taken at 18 IMD-based stations during the monsoon (JJAS) season for the period 2009–2013 (Unnikrishnan et al., 2013). The model simulated 60 cm soil moisture dynamics were compared against observations, which generally revealed a good skill for all three models (Fig. 1). Model simulated soil moisture showed a relatively higher correlation with observations in the northern and western regions as compared to those located in the southern coastal belt. Among models, the Noah simulated soil moisture exhibited higher correlation as compared to other two LSMs. For this setup, we also compared the calibrated vs. uncalibrated model runs to understand what improvements (if any) could be achieved by the parameter calibration of simulation of soil moisture anomalies. We find limited benefits of the model calibration in this case – only the VIC model benefited from the model calibration, mainly in the northern region locations and a few of the southern locations.

The second set of evaluations considered the continuous soil moisture observation datasets at an IIT Kanpur site available from the International Soil Moisture Network (ISMN: Dorigo et al., 2011). Although all three models exhibited a general bias in capturing absolute values of the observed soil moisture, their daily variability observed over the course of the year is well captured by all three models (Fig. S5). Moreover, the Noah and CLM models show improvements in terms of reducing overall bias as a result of model calibration.

Finally, our third set of model evaluations considered an assessment of the model skill for capturing the remote-sensing-based soil moisture available from the ESA-CCI product (Dorigo et al., 2012). Here we used the modeled top-layer (10–30 cm) annual soil moisture over the period 1979–2012 for the comparison; however, the ESA-CCI soil moisture product is more reliable after 2001 (Dorigo et al., 2017) (Fig. S6). Despite the limitation that the ESA-CCI soil moisture inference is for the top few centimeters of the earth's surface, we find a positive correlation with modeled soil moisture for all three models across a large part of India. A relatively higher correlation (more than 0.6) can be noticed for regions in the northwestern and southern peninsular parts of India.

We also evaluated the skills of LSMs for the terrestrial water storage (TWS) anomalies from the Gravity Recovery and Climate Experiment (GRACE – release v5.0; Landerer and Swenson, 2012) derived products (1° × 1°) for the period 2002–2015. We used the ensemble mean of three available GRACE-TWS products (GeoForschungsZentrum, GFZ, Potsdam, Germany, Centre for Space Research at the University of Texas at Austin, USA, and Jet Propulsion Laboratory, USA) to reduce the noise (and scatter) among different TWS products. We compared ensemble mean GRACE-TWS against the monthly anomalies of modeled total column soil moisture from each of the three LSMs and their ensemble mean. The modeled total column soil moisture was aggregated to 1° spatial resolution to match the (coarse) resolution of the GRACE-TWS product. Overall, all three LSMs are able to capture the temporal dynamics of GRACE-TWS anomalies reasonably well across a large part of India (Fig. S7). The median correlation estimated across the modeled grid cells is more than 0.6 for all three LSMs – and the ensemble mean of simulated total column soil moisture anomalies showed an overall best (median) skill. All three LSMs (and the ensemble mean) exhibited a systematically lower performance in the northwestern part of India (Fig. S7), which is most probably related to groundwater pumping effects that are not modeled in LSMs but are captured in GRACE datasets (Asoka et al., 2017). Each LSM shows a slightly different area (grid cells) with the best skill score that motivates the use of a multi-model ensemble mean to capture the (GRACE-based) water storage anomalies across a large part of India.

3.2 Multi-model ensemble droughts in India

We estimated areal extent of severe to exceptional droughts (SPI < −1.3) based on the 4-month SPI at the end of the monsoon season (representing accumulated precipitation for the June to September period) for the period of 1951–2015 (Fig. 2a). The top five monsoon season (JJAS) drought events occurred in 1987 (areal extent of severe to exceptional droughts: 35 %), 2002 (33.5 %), 1979 (27.7 %), 1972 (26.3 %), and 2009 (24.6 %) at the all-India level. The monsoon season droughts of 2015 (with an areal extent of 17.4 %) and 2014 (14.4 %) ranked 8th and 10th during the period of 1951–2015. Mishra et al. (2016) reported that the 2014–2015 monsoon season drought in the Indo-Gangetic Plain was the most severe during the history of 116 years, with a return period of 542 years.

We find that uncertainty in simulated areal extent of soil moisture drought (4-month SSI at the end of the monsoon season) estimated based on three LSMs is moderate during the monsoon season and has a year-to-year variability (Fig. 2a). For example, the 1 standard deviation representing the uncertainty in the simulated areal drought extent is on average estimated to be around 1.6 %. However, during the 1972 and 1979 monsoon seasons, uncertainty in areal extent of drought is 8 and 14 %, respectively (Fig. 2a; see also Table S3). We estimated the ensemble mean (ENS-SSI) areal extent of 60 cm soil moisture drought from the three LSMs for the monsoon season and found that 1987 (36.7 %), 2002 (35.9 %), 2009 (31.5 %), 1972 (29.8 %), and

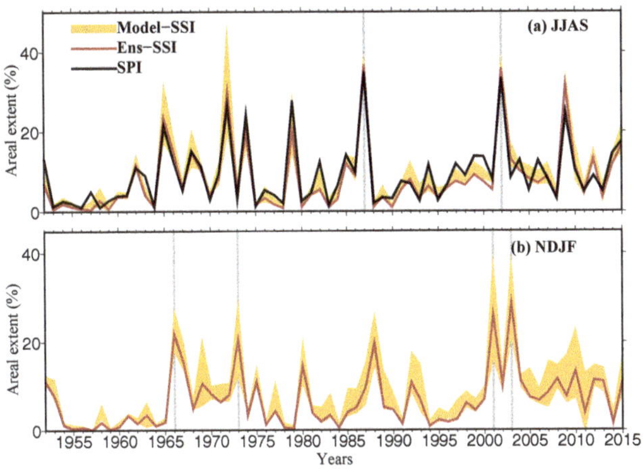

Figure 2. Uncertainty in areal extent (%) of 60 cm soil moisture drought simulated using the three LSMs (i.e. VIC, Noah, and CLM). **(a)** Multi-model ensemble (brown) mean 4-month Standardized Soil Moisture Index (SSI) and inter-model variation (shaded) estimated as 1 standard deviation for the monsoon season. The black line in **(a)** shows the 4-month Standardized Precipitation Index (SPI) at the end of the monsoon season (June through September), **(b)** and multi-model ensemble mean and uncertainty in the 4-month SSI estimated using the three LSMs for the Rabi season (November through February). The light brown shaded area shows uncertainty in severe-to-exceptional drought based on the model simulated SSI (SSI < −1.3). The dark brown line shows areal extent estimated based on the ensemble mean SSI for the three LSMs. The grey line marks the top drought years based on area under drought.

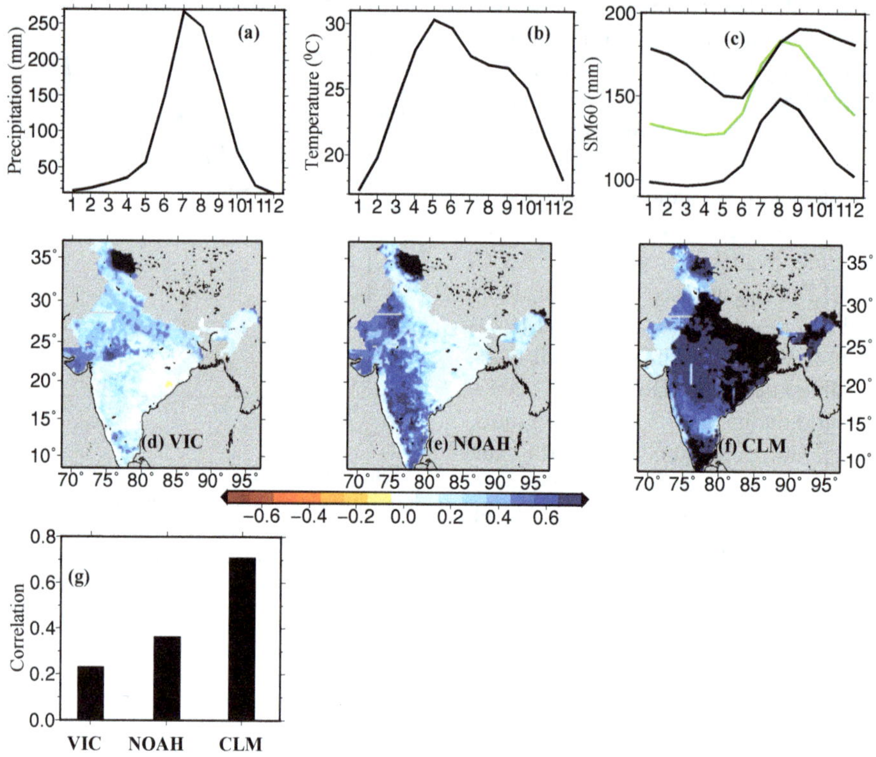

Figure 3. Uncertainty in persistence in root-zone soil moisture (60 cm). Seasonal cycle of all-India averaged **(a)** precipitation, **(b)** mean air temperature, and **(c)** 60 cm soil moisture simulated using the VIC (blue), the Noah (green), and the CLM (red). **(d–f)** Autocorrelation in 60 cm soil moisture at a 4-month lag simulated using the VIC, Noah, and CLM, respectively. **(g)** All-India median autocorrelation (4-month lag) in the 60 cm soil moisture from the three LSMs.

1965 (23.4 %) are the top 5 drought years during the period of 1951–2015 (Fig. 2a). We notice that 1979 (areal extent: 20 %) ranked 6th on record, while 2015 (16.5 %) and 2014 (7.13 %) ranked 8th and 15th, respectively, based on ENS-SSI areal extent of the simulated 60 cm soil moisture drought during the monsoon season.

Since the Rabi season (NDJF) is the key crop-growing season in India, we estimated areal extents of 60 cm soil moisture drought from the three LSMs for the period of 1951–2015 (Fig. 2b). While year-to-year variability in the uncertainty of areal extent of droughts was found, the uncertainty in 60 cm soil moisture based on the areal extent of droughts was substantially higher (∼ 5 %) during the last decade of 1951–2015, which might be associated with frequent drought events during this period. We found a high uncertainty (∼ 7 %) in the areal extent of soil moisture drought in 2011 (Fig. 2b). Based on the ENS-SSI areal extent of 60 cm soil moisture drought, the top 5 drought years were 2003 (areal extent 30 %, uncertainty 9.6 %), 2001 (27.6, 12.0 %), 1966 (22.6, 4.7 %), 1973 (20.7, 8.9 %), and 1988 (20.6, 6.1 %). Droughts in the Rabi season can be driven by both monsoon season precipitation deficit and positive temperature anomalies during this season. Uncertainty in the top five drought events in the Rabi season was substantial (5–12 %), which underscores the need for multi-model drought assessment in the growing season.

The VIC, Noah, and CLM show an all-India median autocorrelation of 0.23, 0.37, and 0.71, respectively, at a 4-month lag (Fig. 3), indicating that the CLM has the highest persistence in the 60 cm soil moisture. Spatial differences in soil moisture persistence were also observed (Fig. 3). Our results are in agreement with the findings of Wang et al. (2009), who reported higher persistence for CLM modeled soil moisture, which can be attributed to its higher water holding capacity and thicker soil column (Fig. S2). Figure S2 shows that though there is not much difference in soil texture provided as input to three LSMs, the available water in the total column is much higher for the CLM. We find that soil layer thickness is strongly related to soil moisture persistence regardless of model calibration (Fig. S8). The same thing can be noted even considering the first three soil layers which cover a 60 cm soil column for all three LSMs, the soil-layer thickness of the CLM dominates over another two LSMs (Fig. S9).

Furthermore we find that all-India averaged mean monthly soil moisture is highest during July in the VIC and Noah LSMs, which is consistent with the seasonal cycle of all-India averaged precipitation (Fig. 3a–c). However, all-India averaged mean monthly 60 cm soil moisture reaches the highest level in the month of August in the CLM model (Fig. 3a–c). The 1-month lag between peak precipitation and peak 60 cm soil moisture from the CLM can be due to a relatively deeper soil column (Fig. S8) and higher total column water holding capacity, a higher number of soil layers, and differences in processes related to soil hydrology as discussed in Wang et al. (2009) and Xia et al. (2012).

To understand the relationship between meteorological and agricultural droughts, lagged correlation analysis was performed between the 4-month SPI at the end of the monsoon season and the 4-month SSI (at the end of JJAS, JASO, ASON, and so on). We find that the 4-month SSI at the end of the monsoon season for the VIC model showed the highest correlation with the 4-month SPI (JJAS), while the 4-month SSI from the Noah model showed the lowest correlation (Fig. S10a). These results indicate that the 60 cm soil moisture from the VIC model responds faster to the monsoon season precipitation than the other two LSMs, which can be associated with soil moisture persistence and model parameterization (Van Loon et al., 2012; Wang et al., 2009; Xia et al., 2012). However, we notice that the correlation between the 4-month SPI at the end of the monsoon season and the 4-month SSI declines rapidly after October (ONDJ, NDJF, and so on) for the VIC and Noah models (Fig. S10a). On the other hand, the CLM shows substantially higher persistence even for the March–June 60 cm soil moisture, which can be attributed to a deeper soil column and differences in the other processes related to soil hydrology. These results also indicate that the anomalous precipitation during the monsoon season can last longer and have a substantial influence on agriculture drought estimated using the 60 cm soil moisture from the CLM as reflected by the strength of the relationship between the 12-month SPI and 12-month SSI in the CLM (Fig. S10b).

Areal extent and severity of agricultural droughts estimated using the 60 cm soil moisture show a considerable uncertainty mainly due to the differences in soil moisture persistence characteristics among three LSMs (Fig. 3). We estimated areal extent of agriculture drought from the three LSMs considering the period that showed maximum correlation against the monsoon season precipitation (Fig. S10c). For instance, 4-month SSI at the end of October (JASO) showed the highest correlation with 4-month SPI at the end of September (JJAS) for the VIC and Noah models (Fig. S10a). On the other hand, the 4-month SSI at the end of November (ASON) showed the highest correlation with the 4-month SPI at the end of September (JJAS) for the CLM. Therefore, we considered root-zone soil moisture for JASO, JASO, and ASON periods from the VIC, Noah, and CLM, respectively (Fig. S10c) to understand the response of the monsoon season deficit in precipitation on agricultural drought. We find that the uncertainty in the areal extent of agricultural drought is substantially reduced considering the lagged response of the monsoon season precipitation and soil moisture (Fig. S10c and Table S3), indicating that the major source of uncertainty in areal extent of agricultural droughts is soil moisture persistence in the LSMs.

3.3 Reconstruction of major droughts

We reconstructed major monsoon season drought events over India using 60 cm soil moisture from the three LSMs for the period of 1951–2015 (Fig. 4). The meteorological and agricultural droughts were represented using the 4-month SPI and 4-month SSI, respectively, at the end of the monsoon season. We estimated the ensemble mean 4-month SSI from the three LSMs to understand whether the individual LSMs show a larger difference from the ensemble mean. Based on the 4-month SPI at the end of the monsoon season, we selected the top two most widespread drought events that occurred in 1987 and 2002 (Figs. 2a and 4). Moreover, we also selected a recent drought event (2015) that caused an enormous water crisis in the Indo-Gangetic Plain (Mishra et al., 2016). For all three major droughts (1987, 2002, and 2015) in the monsoon season, we compared areal extents of 60 cm soil moisture drought in the monsoon season estimated from the three LSMs.

Areal extents of the monsoon season droughts (meteorological and agricultural) in 1987, 2002, and 2015 show that droughts were mainly caused by the monsoon season precipitation deficits (Fig. 4). We notice positive air temperature anomalies in all 3 years (1987, 2002, and 2015); however, patterns of agricultural and meteorological droughts were largely similar (Fig. 4). Among the three drought events, the monsoon season air temperature anomaly (positive) was strongest in the 2015 monsoon season. Uncertainty in areal extent of agricultural droughts estimated using the 60 cm soil moisture is presented in supplemental Table S3. We notice large uncertainty in the areal extent of drought simulated from the three LSMs for the 2015 event (Table S3). The VIC model simulated areal extent of soil moisture drought was 14 % during the 2015 monsoon season, while the areal extent of drought simulated from Noah and the CLM was 21.2 and 18.1 %, respectively (Table S3).

Similar to the monsoon season droughts, we compared the spatial pattern of droughts simulated by the three LSMs for major droughts in the Rabi season, which occurred in 1966, 1973, 2001, and 2003 (Fig. S11). We notice that major droughts in the Rabi season were also largely driven by the precipitation deficit, and the role of positive air temperature anomalies was relatively minor (Fig. S11). Overall, the VIC model shows the lesser intensity of drought during the Rabi season as compared to the Noah and CLM for all the years, which can be attributed to differences in soil moisture persistence in the three LSMs (Fig. 3a). We find higher uncertainty in areal extent of drought during the 2001 Rabi season (Fig. S11) than other years of major drought events, which may be due to higher impacts of air temperature on drought during 2001. The overall uncertainty in the areal extent of

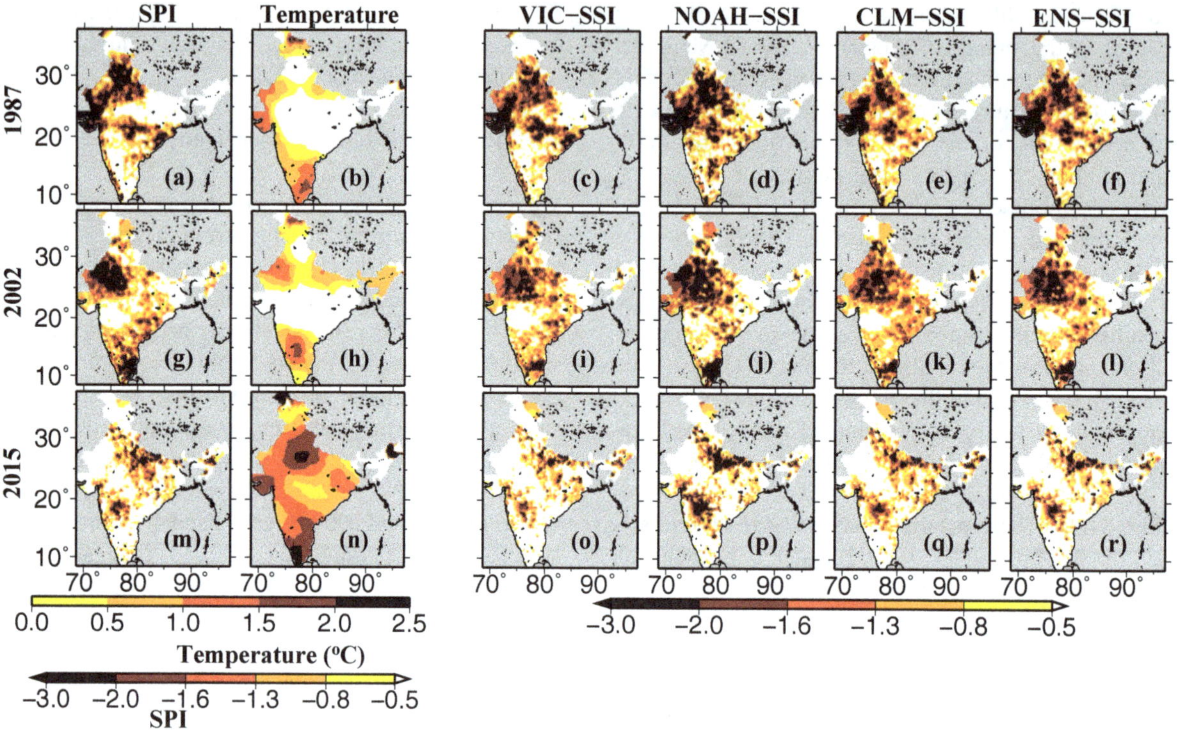

Figure 4. Reconstruction of monsoon season drought events of **(a–f)** 1987, **(g–l)** 2002, and **(m–r)** 2015, estimated based on **(a, g, k)** the 4-month SPI at the end of the monsoon season, **(c, i, o)** the 4-month SSI at the end of the monsoon season simulated using the VIC model, **(d, j, p)** same as **(c, i, o)** but for the Noah model, and **(e, k, q)** same as **(c, i, o)** but for the CLM. **(f, l, r)** Ensemble mean 4-month SSI simulated using the VIC, Noah, and CLM. **(b, h, n)** Air temperature anomaly during the monsoon season for the selected drought years.

agricultural droughts estimated using the 60 cm soil moisture estimated from the three models is presented in supplemental Table S4. In 2001, the areal extent of droughts simulated by the VIC, Noah, CLM, and their ensemble mean were 17.2, 40.7, 24.7, and 26.1 %, respectively (Table S4). Large differences in the areal extent of 60 cm soil moisture drought simulated from the three LSMs were also noted for the years 1966, 1973, 2001, and 2003 (Table S4).

3.4 Intensity–areal extent–frequency (IAF) of droughts

Uncertainty in the multi-model drought estimates was evaluated using IAF curves (Fig. 5). Drought intensity associated with the selected areal extent thresholds was estimated for the return periods of 10, 20, 50, 100, 200, and 500 years (Fig. 5). We find that multi-model-based drought intensity for a selected areal extent has a much larger uncertainty when the 95 % confidence interval of the GEV parameters was considered (Table S6). However, uncertainty in drought intensity was lower when only mean values of the GEV parameters for drought intensity and areal extents were considered. Uncertainty in drought intensity appeared to grow with an increase in the return period (Fig. 5). Based on the IAF curves, we find that the 2002 monsoon season drought has a return period of 50 years. Moreover, uncertainty in drought intensity from the three LSMs was larger for smaller areal extents (Fig. 5). For instance, a drought of 50 % areal extent and 50-year return period has intensities of −1.66, −1.91, and −1.53 simulated from the VIC, Noah, and CLM, respectively (Table S5). On the other hand, a drought of 5 % areal extent and 50-year return period has intensities of −2.89, −3.79, and −3.09 simulated from the VIC, Noah, and CLM, respectively (Table S5). Most of the return periods and areal-extent drought intensities were higher for the Noah model and lower for the VIC model (Table S5), which can be associated with the differences in 60 cm soil moisture persistence in the three LSMs.

A considerably higher uncertainty in IAF curves during the Rabi season was noticed (Fig. S12; Tables S7 and S8). For instance, a drought of 50 % areal extent and a 50-year return period can have intensities of −1.23, −1.62, and −1.50 for 60 cm soil moisture obtained from the VIC, Noah, and CLM, respectively (Fig. S12 and Table S7). A drought of 5 % areal extent and the 50-year return period have intensities of −2.17, −3.71, and −3.25 simulated from the VIC, Noah, and CLM, respectively (Table S7). Higher uncertainty in drought intensities during the Rabi season can be attributed to the response of soil moisture in the three LSMs to meteorological forcing (precipitation and air temperature).

As the Indo-Gangetic Plain is one of the most intensive crop-growing regions in the world, we evaluated the uncertainty in the IAF curves constructed using the 60 cm

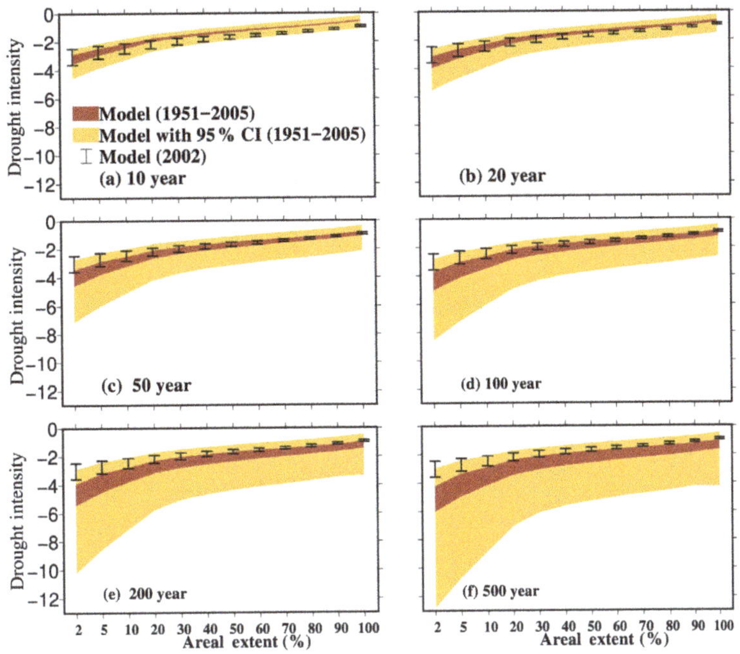

Figure 5. Uncertainty in intensity–areal extent–frequency (IAF) curves for the monsoon season 60 cm soil moisture drought estimated using the three LSMs. Dark brown color shade shows uncertainty in models without considering parameter uncertainty in the Generalized Extreme Value (GEV) distribution, while light brown color shows uncertainty considering the 95 % confidence interval of the GEV parameters for return periods **(a)** 10, **(b)** 20, **(c)** 50, **(d)** 100, **(e)** 200, and **(f)** 500 years. Black error bars indicate uncertainty for the 2002 monsoon season drought using three LSMs.

soil moisture from the three LSMs (Fig. S13). We used the 12-month SSI at the end of December based on the areal averaged mean annual 60 cm soil moisture over the Indo-Gangetic Plain. Similar to IAF curves for the all-India averaged 60 cm soil moisture, a large uncertainty was found due to differences in the GEV parameters (Fig. S13, Tables S9 and S10). Moreover, uncertainty in IAF curves of the Indo-Gangetic Plain increases with the return period and declines with an increase in areal extent of droughts. For instance, for an aerial extent of 5 % and a return period of 50 years, drought intensities were −2.59, −3.06, and −2.48 for the VIC, Noah, and CLM (Table S9). However, when areal extent increases to 50 %, drought intensities are −1.42, −1.49, and −1.45 for the VIC, Noah, and CLM, respectively (Table S9, Fig. S11). Overall, uncertainty in drought intensity is higher for localized droughts that have higher return periods. These results further indicate that 60 cm soil moisture drought characteristics can have large uncertainty arising from different LSMs, which can be reduced by considering the multi-model ensemble agricultural drought assessments.

3.5 Role of the monsoon season precipitation and Rabi season air temperature

We evaluated the differences in the coupling of 60 cm soil moisture (SSI) with monsoon season precipitation and air temperature (Figs. 6 and S14). We find that a 4-month SSI at the end of the monsoon season from the VIC model is strongly coupled (correlation coefficient = 0.90) with the monsoon season precipitation over India. On the other hand, the 60 cm SSI showed correlation coefficients of 0.79 and 0.74 against the monsoon season precipitation for the Noah and CLM, respectively (Fig. 6c and e). These results show differences in the response of 60 cm SSI against changes in the monsoon season precipitation. However, the all-India averaged 60 cm SSI showed stronger coupling with the monsoon season air temperature for the Noah and CLM models (correlation = −0.65 and −0.67) than that of the VIC model (correlation = −0.53) (Fig. 6b, d and f). Interestingly, the coupling between the 60 cm SSI for the Rabi and monsoon season precipitation is stronger for the CLM (correlation = 0.76) and Noah (correlation = 0.66) than that of the VIC model (correlation = 0.55). These results indicate that the monsoon season precipitation deficit can have a larger influence on the Rabi season drought in the CLM and Noah models compared to that of the VIC model. Similarly, the Rabi season air temperature showed a stronger relationship with the 60 cm SSI for the CLM (correlation = −0.51) and Noah (−0.37) than that of the VIC model (−0.31). Similar differences in the Rabi and monsoon season 60 cm SSI for the Indo-Gangetic Plain were observed for the three LSMs (Fig. S14), indicating that the drought indices based on the 60 cm soil moisture may show different sensitivity to at-

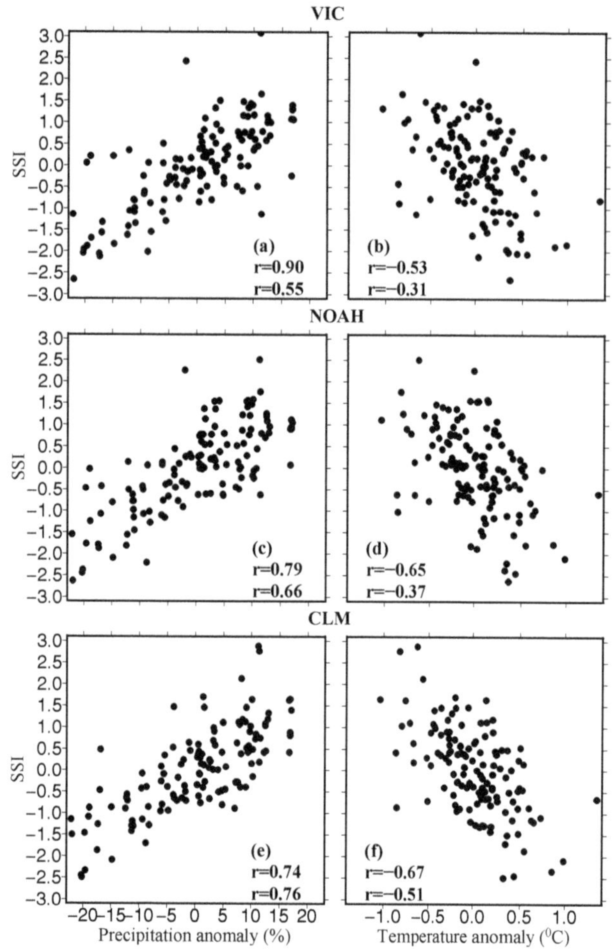

Figure 6. (a, c, e) Relationship between monsoon season precipitation anomaly (%) and 4-month SSI at the end of the monsoon season, and **(b, d, f)** same as **(a, c, e)** but for the relationship between 4-month SSI and the air temperature anomaly of the monsoon season. Correlation coefficients are shown for all-India SSI (blue) and 4-month SSI over the Indo-Gangetic Plain (red).

mospheric forcing, which can show a substantial variation across regions and seasons.

4 Discussion

We find that the three LSMs show major differences in agricultural droughts during the monsoon and Rabi seasons. Uncertainty in the intensity of droughts is higher in the Rabi season than that of the monsoon season, which can be associated with the role of air temperature in soil moisture. Also, we found differences in soil moisture drought simulated in response to precipitation and temperature deficits (Figs. 6 and S12). For instance, soil moisture from the VIC model shows a quick response to precipitation deficit, whereas Noah and the CLM show a delayed response. Moreover, localized droughts have more uncertainty than that of widespread droughts over India and the Indo-Gangetic Plain. The primary cause of the uncertainty in 60 cm soil moisture and droughts is related to soil moisture persistence (Fig. S16), which is associated with the soil water holding capacity as reported in Wang et al. (2009). We found that persistence in soil moisture is strongly linked with soil-layer thickness (see Fig. S8), which in turn affects the soil water holding capacity (available water, Fig. S16). Apart from the soil moisture persistence, there can be several other factors that can introduce uncertainty in 60 cm soil moisture simulations. For instance, all three LSMs have different calibration parameters, which do not cover the entire range of uncertainty due to manual calibration (De Lannoy et al., 2006; Samaniego et al., 2013). However, our results show that the model calibration has little impact on soil moisture anomalies, which are largely driven by the climate forcing. Moreover, during drought, there is a high degree of non-linearity; therefore, calibration parameters estimated through global optimization may also not yield the best results (De Lannoy et al., 2006). The differences in vegetation parameters in the three LSMs can also be attributed to uncertainty in 60 cm soil moisture (Peters-Lidard et al., 2008). For instance, the Noah model does not account for the sub-grid variability of vegetation, unlike the CLM and VIC models, which use a mosaic-based representation of vegetation. However, since we were interested in 60 cm soil moisture droughts, we assume that the major uncertainty in soil moisture simulations is due to soil hydraulic properties and different soil thickness in LSMs (Peters-Lidard et al., 2008; Teuling et al., 2009). With respect to the LSMs, we would like to note that the drought uncertainty assessments conducted here are limited to only three LSMs, which is comparatively a smaller size.

Disparities in soil moisture persistence in the three LSMs can have implications for real-time drought monitoring and forecast. For instance, Shukla et al. (2013) reported that hydrologic initial conditions play a major role in hydrological prediction skills at a global scale. Similar findings were noted by Shah et al. (2017), who found that hydrological initial conditions play a vital role in prediction skills of soil moisture droughts over India. Hydrologic prediction at short to seasonal scales can be influenced by soil moisture persistence and the LSMs with higher persistence can have more skill contributed by the initial hydrologic conditions. This further highlights a need for multi-model-based real-time drought monitoring and prediction systems over India, as shown in Wang et al. (2009). We found that the multi-model ensemble mean performs better than individual LSMs for streamflow and terrestrial water storage (TWS) from GRACE. Bohn et al. (2010) reported that the multi-model ensemble average may not always yield a higher prediction skill at seasonal scales; however, at shorter lead times, it can provide better confidence in prediction of soil moisture droughts due to higher skill from hydrologic initial conditions (Shukla et al., 2013; Shah et al., 2017). We also find disparities in coupling between monsoon/Rabi season 60 cm soil moisture and precipitation/air temperature. The differences in soil moisture sensitivity to atmospheric forcings can have implications for future projections of droughts from multiple LSMs. For instance, Prudhomme et al. (2014) reported that uncertainty in drought projections can be large, especially due to models that simulate the dynamic response of plants to climate. Overall, we find that 60 cm soil moisture droughts have uncertainty associated with their areal extent and severity. The uncertainty in drought estimates is largely due to differences in the soil moisture persistence. Uncertainty in drought estimates during the crop-growing season can be reduced using the multi-model ensemble mean, which can assist decision makers in India.

5 Conclusions

India has witnessed some of the most severe meteorological and agricultural droughts during the period of 1951–2015. The most wide-spread meteorological droughts during the monsoon season occurred in 1987, 2002, 1972, 1979, and 2009. During the Rabi season, the most wide-spread agricultural droughts occurred in 2003, 2001, 1966, 1973, and 1988. All three LSMs, as well as their ensemble mean, identified major 60 cm soil moisture droughts between 1951 and 2015.

The three LSMs (e.g. VIC, Noah, and CLM) showed differences in persistence of 60 cm soil moisture over India, which was largely associated with soil water holding capacity. The CLM showed the highest soil moisture persistence among the three LSMs. Due to differences in the soil moisture persistence, areal extent and intensity of droughts calculated by the three LSMs showed uncertainty. Using the IAF curves, we found that the uncertainty in intensity was higher for the localized droughts (with less areal extent). Uncertainty increases with the return period of droughts, indicating that localized and rare drought events are more different among the three LSMs.

All three LSMs showed differences in the coupling between 60 cm soil moisture and precipitation/air temperature, suggesting that LSMs have disparities in soil moisture sensitivity to precipitation and temperature anomalies in the monsoon and Rabi seasons. Considering the differences in drought characteristics simulated by the three models, multi-model ensemble mean can be a better estimate of agricultural droughts over India as demonstrated for streamflow and terrestrial water storage. Uncertainty in intensity and areal extent can be reduced substantially for the severe and localized droughts that can affect agricultural production. Future studies should consider soil moisture simulations from a large number of LSMs as well as other sources of uncertainty in the historical reconstruction of agricultural droughts over India. Moreover, including the uncertainty due to choice of different precipitation datasets and other meteorological forcing datasets can be important for regional drought impact assessments.

All three LSMs showed differences in the coupling between 60 cm soil moisture and precipitation/air temperature, suggesting that LSMs have disparities in soil moisture sensitivity to precipitation and temperature anomalies in the monsoon and Rabi seasons. Considering the differences in drought characteristics simulated by the three models, multi-model ensemble mean can be a better estimate of agricultural droughts over India as demonstrated for streamflow and terrestrial water storage. Uncertainty in intensity and areal extent can be reduced substantially for the severe and localized droughts that can affect agricultural production. Future studies should consider soil moisture simulations from a large number of LSMs as well as other sources of uncertainty in the historical reconstruction of agricultural droughts over India. Moreover, including the uncertainty due to choice of different precipitation datasets and other meteorological forcing datasets can be important for regional drought impact assessments.

Appendix A: Estimation of the Standardized Precipitation Index and the Standardized Soil Moisture Index

To estimate the Standardized Precipitation Index (SPI), the gamma probability density function is fitted to precipitation data accumulated to different timescales (e.g. 1-month, 4-month, 12-month, and 24-month) in the same way as Edwards and Mckee (1997) as shown in Eq. (A1):

$$g(x) = \frac{1}{\beta^{\alpha}\Gamma(\alpha)} x^{\alpha-1} e^{-x/\beta} \text{ for } x > 0, \tag{A1}$$

where α is a shape parameter ($\alpha > 0$), β is a scale parameter ($\beta > 0$), x is the precipitation amount ($x > 0$), and

$$\Gamma(\alpha) = \int_0^{\infty} y^{\alpha-1} e^{-y} \mathrm{d}y$$

where $\Gamma(\alpha)$ denotes the gamma function.

The shape parameter α and scale parameter β are estimated using maximum likelihood following Edward and Mckee (1997) for each grid and each timescale of interest as follows:

$$\hat{\alpha} = \frac{1}{4A}\left(1 + \sqrt{1 + \frac{4A}{3}}\right)$$

$$\hat{\beta} = \frac{\overline{x}}{\hat{\alpha}}$$

where $A = \ln(\overline{x}) - [\sum \ln(x)]/n$, with n being the number of precipitation events and $\overline{x}$ is mean precipitation.

After fitting the probability density function and identifying shape and scale parameters, the cumulative density function $\mathrm{cdf}(x) = \int_0^P x\mathrm{pdf}(t) \cdot \mathrm{d}t$ is fitted.

Cumulative probability for a given rainfall amount, $H(\mathrm{PRE})$, is then estimated from the following mixed distribution:

$$H(\mathrm{PRE}) = (m/n) + [1 - (m/n)]\mathrm{cdf}(\mathrm{PRE}),$$

where m is the number of zeros in a dataset and n is the sample size.

SPI is then estimated by applying equi-probability transformation of the $H(\mathrm{PRE})$ as described in Wu et al. (2007) following Abramowitz and Stegun (1965).

$$\mathrm{SPI} = -\left[t - \frac{c_0 + c_1 t + c_2 t^2}{1 + d_1 t + d_2 t^2 + d_3 t^3}\right] \text{ for } 0 < H(\mathrm{PRE}) \leq 0.5,$$

$$\mathrm{SPI} = +\left[t - \frac{c_0 + c_1 t + c_2 t^2}{1 + d_1 t + d_2 t^2 + d_3 t^3}\right] \text{ for } 0.5 < H(\mathrm{PRE}) < 1,$$

where

$$t = \sqrt{\left(\ln\left(\frac{1}{(H(\mathrm{PRE}))^2}\right)\right)} \text{ for } 0 < H(\mathrm{PRE}) \leq 0.5,$$

$$t = \sqrt{\left(\ln\left(\frac{1}{(1 - H(\mathrm{PRE}))^2}\right)\right)} \text{ for } 0.5 < H(\mathrm{PRE}) < 1,$$

$$c_0 = 2.515517;\ c_1 = 0.802853;\ c_2 = 0.010328,$$

$$d_1 = 1.432788;\ d_2 = 0.189269;\ d_3 = 0.001308.$$

For more details, refer to Mckee et al. (1993) and Lloyd-Hughes and Sanders (2002). SSI was estimated in the same way as SPI but using 60 cm soil moisture instead of precipitation.

Competing interests. The authors declare that they have no conflict of interest.

Acknowledgements. The authors acknowledge the funding from the ITRA-Water program. Comments and suggestions from the editor and three reviewers are greatly appreciated. Authors would like to thank C. K. Unnikrishnan (NCESS) for sharing the in situ soil moisture observations.

Edited by: Harrie-Jan Hendricks Franssen

References

Abramowitz, M. and Stegun, I. A.: Handbook of mathematical function, Editoral, Dover, p. 1046, 1965.

Asoka, A., Gleeson, T., Wada, Y., and Mishra, V.: Relative contribution of monsoon precipitation and pumping to groundwater storage changes in India, Nat. Geosci., 10, 109–117, 2017.

Beck, H. E., van Dijk, A. I. J. M., de Roo, A., Dutra, E., Fink, G., Orth, R., and Schellekens, J.: Global evaluation of runoff from 10 state-of-the-art hydrological models, Hydrol. Earth Syst. Sci., 21, 2881–2903, https://doi.org/10.5194/hess-21-2881-2017, 2017.

Beven, K. J. and Kirkby, M. J.: A physically based, variable contributing area model of basin hydrology/Un modèle à base physique de zone d'appel variable de l'hydrologie du bassin versant, Hydrolog. Sci. J., 24, 43–69, 1979.

Bohn, T. J., Sonessa, M. Y., and Lettenmaier, D. P.: Seasonal Hydrologic Forecasting: Do Multimodel Ensemble Averages Always Yield Improvements in Forecast Skill?, J. Hydrometeorol., 11, 1358–1372, https://doi.org/10.1175/2010JHM1267.1, 2010.

Bohn, T. J., Livneh, B., Oyler, J. W., Running, S. W., Nijssen, B., and Lettenmaier, D. P.: Global evaluation of MTCLIM and related algorithms for forcing of ecological and hydrological models, Agr. Forest Meteorol., 176, 38–49, 2013.

Bonan, G. B., Levis, S., Kergoat, L., and Oleson, K. W.: Landscapes as patches of plant functional types: An integrating concept for climate and ecosystem models, Global Biogeochem. Cy., 16, 5-1–5-23, https://doi.org/10.1029/2000GB001360, 2002.

Bowling, L. C., Lettenmaier, D. P., Nijssen, B., Graham, L. P., Clark, D. B., El Maayar, M., Essery, R., Goers, S., Gusev, Y. M., Habets, F., van den Hurk, B., Jin, J., Kahan, D., Lohmann, D., Ma, X., Mahanama, S., Mocko, D., Nasonova, O., Niu, G.-Y., Samuelsson, P., Shmakin, A. B., Takata, K., Verseghy, D., Viterbo, P., Xia, Y., Xue, Y., and Yang, Z.-L.: Simulation of high-latitude hydrological processes in the Torne–Kalix basin: PILPS Phase 2(e), Global Planet. Change, 38, 1–30, https://doi.org/10.1016/S0921-8181(03)00003-1, 2003.

Cai, X., Yang, Z.-L., David, C. H., Niu, G.-Y., and Rodell, M.: Hydrological evaluation of the Noah-MP land surface model for the Mississippi River Basin, J. Geophys. Res.-Atmos., 119, 23–38, https://doi.org/10.1002/2013JD020792, 2014.

Chen, F., Mitchell, K., Schaake, J., Xue, Y., Pan, H.-L., Victor, K., Duan, Q. Y., Michael, E., and Alan, B.: Modeling of land surface evaporation by four schemes and comparison with FIFE observations, J. Geophys. Res., 101, 7251–7268, 1996.

Cosby, B. J., Hornberger, G. M., Clapp, R. B., and Ginn, T.: A statistical exploration of the relationships of soil moisture characteristics to the physical properties of soils, Water Resour. Res., 20, 682–690, 1984.

De Lannoy, G. J., Houser, P. R., Pauwels, V., and Verhoest, N. E.: Assessment of model uncertainty for soil moisture through ensemble verification, J. Geophys. Res.-Atmos., 111, D10101, https://doi.org/10.1029/2005JD006367, 2006.

Dirmeyer, P. A., Dolman, A. J., and Sato, N.: The pilot phase of the global soil wetness project, B. Am. Meteorol. Soc., 80, 851–878, 1999.

Dirmeyer, P. A., Gao, X., Zhao, M., Guo, Z., Oki, T., and Hanasaki, N.: GSWP-2: Multimodel Analysis and Implications for Our Perception of the Land Surface, B. Am. Meteorol. Soc., 87, 1381–1397, https://doi.org/10.1175/BAMS-87-10-1381, 2006.

Dorigo, W. A., Wagner, W., Hohensinn, R., Hahn, S., Paulik, C., Xaver, A., Gruber, A., Drusch, M., Mecklenburg, S., van Oevelen, P., Robock, A., and Jackson, T.: The International Soil Moisture Network: a data hosting facility for global in situ soil moisture measurements, Hydrol. Earth Syst. Sci., 15, 1675–1698, https://doi.org/10.5194/hess-15-1675-2011, 2011.

Dorigo, W., de Jeu, R., Chung, D., Parinussa, R., Liu, Y., Wagner, W., and Fernández-Prieto, D.: Evaluating global trends (1988–2010) in harmonized multi-satellite surface soil moisture, Geophys. Res. Lett., 39, L18405, https://doi.org/10.1029/2012GL052988, 2012.

Dorigo, W., Wagner, W., Albergel, C., Albrecht, F., Balsamo, G., Brocca, L., Chung, D., Ertl, M., Forkel, M., Gruber, A., Haas, E., Hamer, P. D., Hirschi, M., Ikonen, J., de Jeu, R., Kidd, R., Lahoz, W., Liu, Y. Y., Miralles, D., Mistelbauer, T., Nicolai-Shaw, N., Parinussa, R., Pratola, C., Reimer, C., van der Schalie, R., Seneviratne, S. I., Smolander, T., and Lecomte, P.: ESA CCI soil moisture for improved Earth system understanding: State-of-the art and future directions, Remote Sens. Environ., 203, 185–215, https://doi.org/10.1016/j.rse.2017.07.001 10.1016/j.rse.2017.07, 2017.

Edwards, D. C. and McKee, T. B.: Characteristics of 20th century drought in the United States at multiple time scales, Climatology Report No. 97-2, Colorado State Univ., Ft. Collins, CO, 1997.

Franchini, M. and Pacciani, M.: Comparative analysis of several conceptual rainfall–runoff models, J. Hydrol., 122, 161–219, 1991.

Hao, Z. and AghaKouchak, A.: Multivariate Standardized Drought Index: A parametric multi-index model, Adv. Water Resour., 57, 12–18, https://doi.org/10.1016/j.advwatres.2013.03.009, 2013.

Hogue, T. S., Bastidas, L., Gupta, H., Sorooshian, S., Mitchell, K., and Emmerich, W.: Evaluation and transferability of the Noah land surface model in semiarid environments, J. Hydrometeorol., 6, 68–84, 2005.

Jalota, S. and Arora, V.: Model-based assessment of water balance components under different cropping systems in north-west India, Agr. Water Manage., 57, 75–87, https://doi.org/10.1016/S0378-3774(02)00049-5, 2002.

Landerer, F. W. and Swenson, S. C.: Accuracy of scaled GRACE terrestrial water storage estimates, Water Resour. Res., 48, W04531, https://doi.org/10.1029/2011WR011453, 2012.

Liang, X., Lettenmaier, D. P., Wood, E. F., and Burges, S.: A simple hydrologically based model of land surface water and energy fluxes for general circulation models, J. Geophys. Res., 99, 14415–14428, https://doi.org/10.1029/94JD00483, 1994.

Lloyd-Hughes, B. and Saunders, M. A.: A drought climatology for Europe, Int. J. Climatol., 22, 1571–1592, https://doi.org/10.1002/joc.846, 2002.

Mahrt, L. and Ek, M.: The influence of atmospheric stability on potential evaporation, J. Clim. Appl. Meteorol., 23, 222–234, 1984.

Maurer, E. P., Wood, A. W., Adam, J. C., and Lettenmaier, D. P.: A long-term hydrologically based dataset of land surface fluxes and states for the conterminous United States, J. Climate, 15, 3237–3251, https://doi.org/10.1175/1520-0442(2002)015,3237:ALTHBD.2.0.CO;2, 2002.

McKee, T. B., Doesken, N. J., and Kleist, J.: The relationship of drought frequency and duration to time scales, in: Proceedings of the 8th Conference on Applied Climatology, vol. 17, 179–183, American Meteorological Society, Boston, MA, available at: http://ccc.atmos.colostate.edu/relationshipofdroughtfrequency.pd (last access: 22 March 2017), 1993.

Mishra, V. and Cherkauer, K. A.: Retrospective droughts in the crop growing season: Implications to corn and soybean yield in the Midwestern United States, Agr. Forest Meteorol., 150, 1030–1045, 2010.

Mishra, V., Cherkauer, K. A., and Bowling, L. C.: Parameterization of Lakes and Wetlands for Energy and Water Balance Studies in the Great Lakes Region, J. Hydrometeorol., 11, 1057–1082, https://doi.org/10.1175/2010JHM1207.1, 2010.

Mishra, V., Shah, R., and Thrasher, B.: Soil Moisture Droughts under the Retrospective and Projected Climate in India, J. Hydrometeorol., 15, 2267–2292, https://doi.org/10.1175/JHM-D-13-0177.1, 2014.

Mishra, V., Aadhar, S., Akarsh, A., Pai, S., and Kumar, R.: On the frequency of the 2015 monsoon season drought in the Indo-Gangetic Plain, Geophys. Res. Lett., 43, 102–112, https://doi.org/10.1002/2016GL071407, 2016.

Mishra, V., Kumar, R., Shah, H. L., Samaniego, L., Eisner, S., and Yang, T.: Multimodel assessment of sensitivity and uncertainty of evapotranspiration and a proxy for available water resources under climate change, Climatic Change, 141, 451–465, 2017.

Mitchell, K. E.: The multi-institution North American Land Data Assimilation System (NLDAS): Utilizing multiple GCIP products and partners in a continental distributed hydrological modeling system, J. Geophys. Res., 109, D07S90, https://doi.org/10.1029/2003JD003823, 2004.

Nijssen, B., O'donnell, G. M., Hamlet, A. F., and Lettenmaier, D. P.: Hydrologic sensitivity of global rivers to climate change, Climatic Change, 50, 143–175, 2001.

Niu, G.-Y., Yang, Z.-L., Dickinson, R. E., Gulden, L. E., and Su, H.: Development of a simple groundwater model for use in climate models and evaluation with Gravity Recovery and Climate Experiment data, J. Geophys. Res.-Atmos., 112, D07103, https://doi.org/10.1029/2006JD007522, 2007.

Pai, D. S., Sridhar, L., Badwaik, M. R., and Rajeevan, M.: Analysis of the daily rainfall events over India using a new long period (1901–2010) high resolution ($0.25° \times 0.25°$) gridded rainfall data set, Clim. Dynam., 45, 755–776, https://doi.org/10.1007/s00382-014-2307-1, 2015.

Pai, D. S., Guhathakurta, P., Kulkarni, A., and Rajeevan, M. N.: Variability of Meteorological Droughts Over India, in: Observed Climate Variability and Change over the Indian Region, edited by: Rajeevan, M. N. and Nayak, S., Springer, Singapore, 73–87, 2017.

Peters-Lidard, C. D., Mocko, D. M., Garcia, M., Santanello, J. A., Tischler, M. A., Moran, M. S., and Wu, Y.: Role of precipitation uncertainty in the estimation of hydrologic soil properties using remotely sensed soil moisture in a semiarid environment, Water Resour. Res., 44, W05S18, https://doi.org/10.1029/2007WR005884, 2008.

Prudhomme, C., Giuntoli, I., Robinson, E. L., Clark, D. B., Arnell, N. W., Dankers, R., Fekete, B. M., Franssen, W., Gerten, D., Gosling, S. N., Hagemann, S., Hannah, D. M., Kim, H., Masaki, Y., Satoh, Y.,Stacke, T., Wada, Y., and Wisser, D.: Hydrological droughts in the 21st century, hotspots and uncertainties from a global multimodel ensemble experiment, P. Natl. Acad. Sci. USA, 111, 3262–3267, 2014.

Rodell, M., Houser, P. R., Jambor, U., Gottschalck, J., Mitchell, K., Meng, C.-J., Arsenault, K., Cosgrove, B., Radakovich, J., Bosilovich, M., Entin, J. K., Walker, J. P., Lohmann, D., and Toll, D.: The Global Land Data Assimilation System, B. Am. Meteorol. Soc., 85, 381–394, https://doi.org/10.1175/BAMS-85-3-381, 2004.

Samaniego, L., Kumar, R., and Zink, M.: Implications of Parameter Uncertainty on Soil Moisture Drought Analysis in Germany, J. Hydrometeorol., 14, 47–68, https://doi.org/10.1175/JHM-D-12-075.1, 2013.

Schaake, J., Koren, V., Duan, Q. Y., Mitchell, K., and Chen, F.: Simple iwater balance model for estimating runoff at different spatial and temporal scales, J. Geophys. Res., 101, 7461–7475, https://doi.org/10.1029/95JD02892, 1996.

Shah, H. L. and Mishra, V.: Hydrologic Changes in Indian Sub-Continental River Basins (1901–2012), J. Hydrometeorol., https://doi.org/10.1175/JHM-D-15-0231.1, 2016.

Shah, R. and Mishra, V.: Evaluation of the Reanalysis Products for the Monsoon Season Droughts in India, J. Hydrometeorol., 15, 1575–1591, https://doi.org/10.1175/JHM-D-13-0103.1, 2014.

Shah, R., Sahai, A. K., and Mishra, V.: Short to sub-seasonal hydrologic forecast to manage water and agricultural resources in India, Hydrol. Earth Syst. Sci., 21, 707–720, https://doi.org/10.5194/hess-21-707-2017, 2017.

Shah, R. D. and Mishra, V.: Development of an Experimental Near-Real-Time Drought Monitor for India, J. Hydrometeorol., 16, 327–345, https://doi.org/10.1175/JHM-D-14-0041.1, 2015.

Shah, R. D. and Mishra, V.: Utility of Global Ensemble Forecast System (GEFS) Reforecast for Medium-Range Drought Prediction in India, J. Hydrometeorol., 17, 1781–1800, https://doi.org/10.1175/JHM-D-15-0050.1, 2016.

Sheffield, J. and Wood, E. F.: Projected changes in drought occurrence under future global warming from multi-model, multi-scenario, IPCC AR4 simulations, Clim. Dynam., 31, 79–105, https://doi.org/10.1007/s00382-007-0340-z, 2008.

Shepard, D. S.: Computer Mapping: The SYMAP Interpolation Algorithm, in: Spatial Statistics and Models, edited by: Gaile, G. L. and Willmott, C. J., Springer Netherlands, Dordrecht, 133–145, https://doi.org/10.1007/978-94-017-3048-8_7, 1984.

Shukla, S., Sheffield, J., Wood, E. F., and Lettenmaier, D. P.: On the sources of global land surface hydrologic predictability, Hydrol. Earth Syst. Sci., 17, 2781–2796, https://doi.org/10.5194/hess-17-2781-2013, 2013.

Srivastava, A. K., Rajeevan, M., and Kshirsagar, S. R.: Development of a high resolution daily gridded temperature data set (1969–2005) for the Indian region, Atmos. Sci. Lett., 10, 249–254, https://doi.org/10.1002/asl.232, 2009.

Teuling, A. J., Uijlenhoet, R., van den Hurk, B., and Seneviratne, S. I.: Parameter sensitivity in LSMs: An analysis using stochastic soil moisture models and ELDAS soil parameters, J. Hydrometeorol., 10, 751–765, 2009.

Unnikrishnan, C. K., Rajeevan, M., Rao, S. V. B., and Kumar, M.: Development of a high resolution land surface dataset for the South Asian monsoon region, Curr. Sci., 105, 1235–1246, 2013.

Van Loon, A. F., Van Huijgevoort, M. H. J., and Van Lanen, H. A. J.: Evaluation of drought propagation in an ensemble mean of large-scale hydrological models, Hydrol. Earth Syst. Sci., 16, 4057–4078, https://doi.org/10.5194/hess-16-4057-2012, 2012.

Wang, A., Bohn, T. J., Mahanama, S. P., Koster, R. D., and Lettenmaier, D. P.: Multimodel Ensemble Reconstruction of Drought over the Continental United States, J. Climate, 22, 2694–2712, https://doi.org/10.1175/2008JCLI2586.1, 2009.

Wu, H., Svoboda, M. D., Hayes, M. J., Wilhite, D. A. and Wen, F.: Appropriate application of the Standardized Precipitation Index in arid locations and dry seasons, Int. J. Climatol., 27, 65–79, https://doi.org/10.1002/joc.1371, 2007.

Xia, Y., Mitchell, K., Ek, M., Sheffield, J., Cosgrove, B., Wood, E., Luo, L., Alonge, C., Wei, H., Meng, J., Livneh, B., Lettenmaier, D., Koren, V., Duan, Q., Mo, K., Fan, Y., and Mocko, D.: Continental-scale water and energy flux analysis and validation for the North American Land Data Assimilation System project phase 2 (NLDAS-2): 1. Intercomparison and application of model products, J. Geophys. Res.-Atmos., 117, D03109, https://doi.org/10.1029/2011JD016051, 2012.

11

Fourier approach for estimating the thermal attributes of streams

Masahiro Ryo[1,2,3], **Marie Leys**[1,3], **and Christopher T. Robinson**[1,3]

[1]Department of Aquatic Ecology, EAWAG, 8600 Duebendorf, Switzerland
[2]Department of Civil Engineering, Tokyo Institute of Technology, 2-12-1 Ookayama, Meguro-ku, 152-0033 Tokyo, Japan
[3]Institute of Integrative Biology, ETH-Zürich, 8092 Zürich, Switzerland

Correspondence to: Masahiro Ryo (masahiroryo@gmail.com)

Abstract. Temperature models that directly predict ecologically important thermal attributes across spatiotemporal scales are still poorly developed. This study developed an analytical method based on Fourier analysis to estimate seasonal and diel periodicities, as well as irregularities in stream temperature, at data-poor sites. The method extrapolates thermal attributes from highly resolved temperature data at a reference site to the data-poor sites on the assumption of spatial autocorrelation. We first quantified the thermal attributes of a glacier-fed stream in the Swiss Alps using 2 years of hourly recorded temperature. Our approach decomposed stream temperature into its average temperature of 3.8 °C, a diel periodicity of 4.9 °C, seasonal periodicity spanning 7.5 °C, and the remaining irregularity (variance) with an average of 0.0 °C but spanning 9.7 °C. These attributes were used to estimate thermal characteristics at upstream sites where temperatures were measured monthly, and we found that a diel periodicity and the variance strongly contributed to the variability at the sites. We evaluated the performance of our predictive mechanism and found that our approach can reasonably estimate periodic components and extremes. We could also estimate the variability in irregularity, which cannot be represented by other techniques that assume a linear relationship in temperature variabilities between sites. The results confirm that spatially extrapolating thermal attributes based on Fourier analysis can predict thermal characteristics at a data-poor site. The R scripts used in this study are available in the Supplement.

1 Introduction

Temperature is a fundamental determinant of physical and biogeochemical patterns and processes in ecosystems. Organisms respond to temperature via different adaptations and distinct life history strategies across various spatial and temporal dimensions (Cossins and Bowler, 1987). More broadly, global biodiversity has tracked external temperature transitions over millennia (Mayhew et al., 2012). In the Anthropocene, unprecedented rapid transitions in thermal regimes owing to changes in land use and climate are a global concern (IPCC, 2014). Consequently, adequate characterization of thermal regimes is an important prerequisite for understanding the role of temperature in ecosystems and for ecological conservation.

Thermal attributes (seasonal and diel periodicity patterns and irregular extremes), as representations of variability, have attracted a growing interest in recent climate-relevant ecological studies (e.g., Thompson et al., 2013). The combination of seasonal and diel periodicity cycles in temperature promotes diverse behaviors and spatial distributions of exothermic organisms according to respective life history strategies and life cycle stages (e.g., Vannote et al., 1980). In contrast, irregular extremes such as heatwaves can induce physiological exposure and vulnerability (Paaijmans et al., 2013) as well as causing abrupt shifts in biogeochemical processes (Frank et al., 2015) and ecological assemblages (Thompson et al., 2013). Hence, the variability in thermal periodicities and irregular extremes need to be distinguished from average temperatures.

In freshwaters, ecological responses to thermal regime shifts may be less understood than in marine and terrestrial ecosystems (Thompson et al., 2013), even though freshwater biodiversity has experienced major decreases in recent decades (WWF, 2014). Progress in understanding response patterns has been delayed partially because the quantification of thermal attributes is difficult for running waters. Physical disturbance in streams can be considerable due to periodic flow pulses (Poff et al., 1997). Particularly where flashy flows carry debris such as driftwoods or mobilize gravel riverbeds, e.g., braided rivers, the installation of data loggers for long-term time series monitoring can be logistically difficult. For natural floodplains acting as biodiversity hot spots (Tockner and Stanford, 2002), frequent time series temperature data (daily or sub-daily) recorded throughout a year or longer are difficult to acquire because the floodplain mosaic changes frequently and dramatically (Van der Nat et al., 2003). In these cases, researchers often rely on spot measures of temperature or the use of reference temperature time series at stations along the streamline. Both approaches have caveats (a lack of time series information or likely bias in the data) when estimating the thermal attributes at a data-poor site, thereby limiting understanding of the ecological consequences in freshwaters. Estimating thermal stream attributes using both spot measurements at study sites and time series measures at the nearest hydrological station would likely be a more robust approach.

There is still much room for improvement of statistical models for stream temperature (see review in Benyahya et al., 2007). Regression models employ correlative relationships with air temperature (e.g., Pilgrim et al., 1998) and streamflow (Webb et al., 2003), but correlative approaches considering water temperature at a nearby hydrological station along the streamline have not been implemented to date. Autoregressive models take into account the autocorrelation structure within water temperature time series data and also the correlation with external variables (Kothandaraman, 1971; Cluis, 1972; Long, 1972). To deal with seasonally changing parameters in regression and autoregressive models, periodic autoregressive models were introduced (Benyahya et al., 2007). However, temperature patterns at multiple temporal scales, particularly the combination of seasonal and diel periodicity patterns, are still rarely considered (Steel and Lange, 2007). Moreover, temperature models that directly predict ecologically important thermal attributes as response variables do not exist presently, as many models focus solely on temperature numeric values at a given time. Considerable error and bias in ecologically relevant attributes can arise if they are calculated using a modeled time series environmental condition (e.g., hydrologic indices calculated from a simulated river discharge; Ryo et al., 2015). Directly estimating ecologically relevant thermal attributes, therefore, is required for reliably predicting associated ecological responses.

Fourier analysis (Fourier, 1878) is well suited for analyzing combined multi-temporal patterns and predicting thermal attributes. Unfortunately, the use of Fourier analysis for assessing stream temperature patterns has slowed since its early emergence in 1970 (Kothandaraman, 1971; Cluis, 1972; Long, 1972; but see Maheu et al., 2015). This is surprising given the high potential for Fourier techniques to detect and describe periodicity at multiple scales in time series data, and water temperature data in particular. Here, we investigate the application of Fourier analysis to the assessment of thermal patterns in running waters.

We developed an analytical method to estimating average, seasonal, and diel periodicities in stream temperature, as well as irregular extremes at data-poor sites (i.e., spot measures) using Fourier analysis. We first quantified these thermal attributes with 2 years of hourly recorded time series temperature data from an Alpine glacial-fed stream in Val Roseg, Switzerland (see Ward and Uehlinger (2003) for a synthesis of research conducted in this catchment). Using those results, we predicted thermal patterns at sites along the same stream, where monthly spot measures of temperature were taken during the same 2 years. We compared the performance of the method with that of a linear regression model to underscore the advantages.

2 Methods

2.1 Compositional variables in stream temperature

We assume that hourly stream temperature $T(t)$ at a given time t is composed of its long-term average value $\bar{T}$, a seasonal periodicity pattern $S(t)$, and diel periodicity pattern $D(t)$. The seasonal and diel periodicity patterns are driven by meteorological (e.g., solar radiation and precipitation) and hydrological (e.g., discharge and snowmelt/ice melt) conditions. The remaining unexplained component of variance in the temperature time series results from multiple external factors such as sub-daily changes in weather conditions and a week-long heat waves: we call this component an irregularity $\varepsilon(t)$. A distinctive high/low value in the irregularity indicates thermal extremes that strongly disturb the periodicity patterns in temperature. Consequently, hourly stream temperature is expressed as

$$T(t) = \bar{T} + S(t) + D(t) + \varepsilon(t). \tag{1}$$

2.2 Fourier analysis for temperature decomposition

Fourier transformation converts a function of time $T(t)$ into a function of frequency $G(f)$ by transforming time series data into a sum of trigonometric curves. Both forward (Eq. 2) and backward transformations (Eq. 3) are identically reversible:

$$G(f) = \int_{-\infty}^{\infty} T(t)e^{-i2\pi ft}\mathrm{d}t, \tag{2}$$

$$T(t) = \int_{-\infty}^{\infty} G(f) e^{i2\pi ft} \mathrm{d}f. \tag{3}$$

As measured stream temperature is a discrete variable, we used the fast Fourier transformation algorithm to perform the transforms. The algorithm searches for a single solution to identically explain a time series variable $T(t)$ by summing the trigonometric curves ($e^{i2\pi ft}$) of different frequency f amplified by a corresponding spectral intensity $G(f)$:

$$G(f) = \sum_{t=0}^{N-1} T(t) e^{-i2\pi ft}, \tag{4}$$

$$T(t) = \frac{1}{N} \sum_{t=0}^{N-1} G(f) e^{i2\pi ft}, \tag{5}$$

where N is the length of the time series (rounded to the nearest power of 2) and equivalent to the number of summed curves. As seen in Eq. (5), the time series variable, hourly stream temperature $T(t)$, is formulated with summed trigonometric curves (with frequencies $f = 0, 1, 2, \ldots, N-1$) with magnitudes amplified by a corresponding spectral intensity $G(f)$. Frequencies with high spectral intensities are the most important contributors to variability of the time series. Stable patterns often dominate the variance of the time series and result in strong spectral intensities at frequencies close to 24 h and 1 year.

As the time series temperature $T(t)$ is formulated by Eqs. (1) and (5), we consider that all of the terms in Eq. (1) result from a sum of a subset of the terms in Eq. (5). Although our intention is to explain time series temperature with the four components above, Fourier analysis decomposes it into N components based on a series of trigonometric curves. Terms in Eq. (5), therefore, need to be "classified" as belonging to the terms in Eq. (1). The mismatch in the number of components requires summing some of the trigonometric curves to best represent seasonal and diel periodicity patterns. However, it is unknown which curves at what frequencies and spectral intensities are required to sufficiently express these periodicity patterns. For selecting curves to be summed, we considered the condition that the seasonal and diel periodicities consist of their period lengths (inverse of frequency, $1/f$) within the ranges of 1–365 days and 1–24 h, respectively. Moreover, we introduced a threshold value in spectral intensity in order to select only the dominant components as well as to avoid mixing noise with the periodicity patterns. The irregular component accounts for all frequencies with a spectral intensity below the (a) threshold value. Consequently, the shape of functions $S(t)$ and $D(t)$ depends on a spectral threshold value. We set the threshold value at 0.1 °C, a minimal unit of temperature measurement in the case study system (see Sect. 2.4). The analysis was performed using the fast Fourier transform (fft) function of the "stats" library in R 3.1.2 (R Core Team, 2014).

2.3 Extrapolation to spot-measured temperature data

Assuming that thermal attributes are autocorrelated in space along the river continuum, we can extrapolate time series data from reference locations to spot-measured data at sites along the same stream. For simplicity, we do not include external information (e.g., discharge and air temperature). After decomposing the time series temperature T_0 as Eq. (1) at a reference site, the decomposed factors are used to estimate temperature T_a at spot-measured site A along the same stream network. We assume that the four temperature components in Eq. (1) are linearly correlated between sites when other factors (e.g., major tributary or groundwater inputs) affecting stream temperature along the network are low enough to maintain the spatial-autocorrelation temperature patterns between sites. The temperature at site A is formulated as

$$T_a(t) = \bar{T}_a + S_a(t) + D_a(t) + \varepsilon_a(t) \tag{6}$$

$$= \beta_1 \bar{T}_0 + \beta_2 S_0(t) + \beta_3 D_0(t) + \beta_4 \varepsilon_0(t) + \beta_5, \tag{7}$$

where coefficients β_{1-4} are the weighting parameters for each component (>0) and β_5 is a parameter to adjust systematic bias. The following procedures are performed to estimate each parameter. First, linear regression between T_a and T_0 corresponding to temperatures at the measurement time of T_a is conducted to estimate β_1 (slope) and β_5 (intercept). Second, β_2 and β_3 are estimated by minimizing the mean square error value based on the linear regression between an estimated $\beta_4\varepsilon_0 = T_a - \{\beta_1 \bar{T}_0 + \beta_2 S_0(t) + \beta_3 D_0(t) + \beta_5\}$ and ε_a at the corresponding measurement times. Third, β_4 is numerically estimated similarly to the second step based on T_a and the estimated T_a. Note that this approach will require a reasonably high density of spot measurements, covering the diel range – ideally including minimum and maximum – in different seasons.

To highlight the benefits of the extrapolation method, we compared the component extrapolation approach to a linear regression model that simply extrapolates time series temperature based on the linear regression between T_a and T_0. Importantly, if the coefficients β_{1-4} have the same value as the slope of the linear regression and β_5 is the intercept, the Fourier approach is equivalent to the linear regression model. The R scripts used in this study are available in the Supplement.

2.4 The case study: an Alpine glacier-fed stream (Val Roseg, Switzerland)

The method was applied in the Roseg catchment, an Alpine valley located in the Bernina Massif of the Swiss Alps (Fig. 1). The catchment area is 66.5 km^2 and ca. 30 % glaciated (Swiss National Hydrological and Geological Survey (OFEV); year of record 2010). Elevations range from 1766 to 4049 m a.s.l. The Roseg River is fed by meltwaters of the Tschierva and Roseg glaciers. The Roseg glacial runoff first drains into the pro-glacial lake Roseg before merging

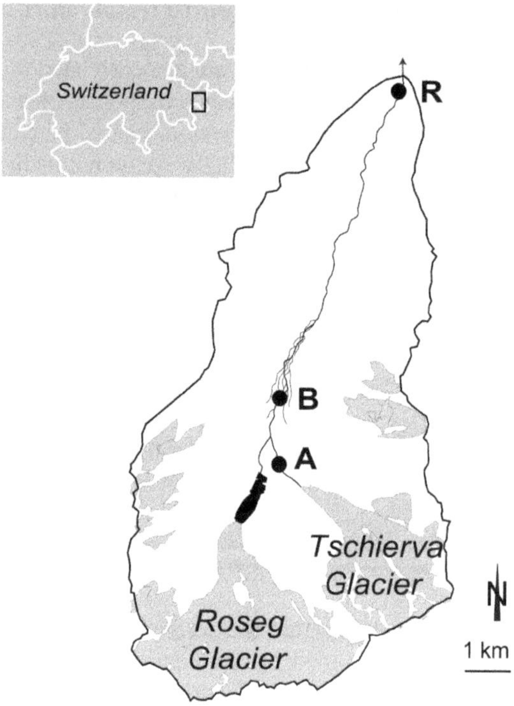

Figure 1. Location of the Val Roseg catchment in Switzerland. Study and reference (i.e., Pontresina hydrological station) sites are indicated by black dots and labeled A and B, and R, respectively. Glaciers are shown as shaded areas with the pro-glacial lake shown below the Roseg Glacier (modified from Uehlinger et al., 2003).

with the flow from the Tschierva Glacier (Fig. 1). The thermal attributes of the glacial meltwaters and runoff from the lake strongly influence the seasonal and diel periodic thermal patterns in this river. Mean annual discharge at the end of the catchment was 2.8 $m^3 s^{-1}$ (discharge record averaged for 1955–2013), and daily discharge ranged from 3.3 to 19.6 $m^3 s^{-1}$ in July and August and from 0.2 to 2.0 $m^3 s^{-1}$ between November and March 2013 (OFEV). The study system comprises (i) a long proglacial reach below the Tschierva Glacier, exhibiting extremely low temperatures due to glacial runoff (kryal); (ii) a single-thread channel downstream of the confluence of the proglacial reach and the Roseg lake outlet; (iii) a complex braided floodplain; and (iv) a constrained reach extending to the end of the catchment where the Pontresina hydrological station is located (see Tockner et al. (1997) and Uehlinger et al. (2003) for a detailed description).

Hourly time series temperature was recorded at a reference site, the Pontresina hydrological station (site R in Fig. 1: 46°29′23.6″ N, 09°53′53.3″ E; 1766 m a.s.l.) in 2012 and 2013 (Fig. 2a: provided by OFEV). Spot-measured water temperature was taken monthly at two sites, one located within the proglacial reach (site A: 46°24′38.3″ N, 09°51′31.2″ E; 2106 m a.s.l.) and one below the lake outlet confluence reach (site B: 46°25′05.9″ N, 09°51′27.1″ E; 2054 m a.s.l.) (Fig. 1). For both sites, stream temperature was measured using a conductivity meter (WTW LF323, Weilheim, Germany) at different daily times on each visit from April to October 2012 and 2013 (in total 14 times).

Table 1. Frequency f, period $1/f$, and spectral intensity $G(f)$ of trigonometric curves (threshold of $G(f) > 0.1$) composing the following thermal attributes: average $\bar{T}$, seasonal periodicity pattern $S(t)$, and diel periodicity pattern $D(t)$.

Attribute	f [Hz]	$1/f$ [h]	[days]	$G(f)$ [°C]
$\bar{T}$	0	–		3.78
$S(t)$	2	8760	365	2.01
	4	4380	183	0.11
	6	2920	122	0.19
$D(t)$	728	24.1	1	0.27
	730	24	1	0.59
	732	23.9	1	0.28
	734	23.9	1	0.10
	1458	12.0	0.5	0.12
	1460	12	0.5	0.21

3 Results

3.1 Thermal attributes at the reference site

By converting the time series temperature data at the reference site R (Fig. 2a) to the frequency domain, we identified frequency ranges with high spectral power (Fig. 3). Nine trigonometric curves exceeded the threshold of 0.1 °C (Table 1). Based on period length (Table 1), these curves were allocated to seasonal and diel components: three curves for seasonal periodicity patterns (Fig. 2b) and six curves for diel periodicity patterns (Fig. 2c). The time series temperature at the reference site R (Fig. 2a) was thus decomposed into an average of 3.8 °C, a seasonal cycle spanning 7.5 °C (Fig. 2b), a diel cycle spanning 4.9 °C (Fig. 2c), and an irregularity spanning 9.7 °C with an average of 0.0 °C and standard deviation of 0.92 °C (Fig. 2d). Hourly stream temperatures excluding the irregularity (i.e., $\bar{T} + S(t) + D(t)$; Fig. 2e) explained 92 % of the data variability (r^2), indicating a successful decomposition of the time series data and a high reliability of the approach to characterize stream temperature components at the reference site. The irregularity had a normal distribution (no inferred bias), indicating that the seasonal and diel periodicity patterns were extracted accurately from the original time series data.

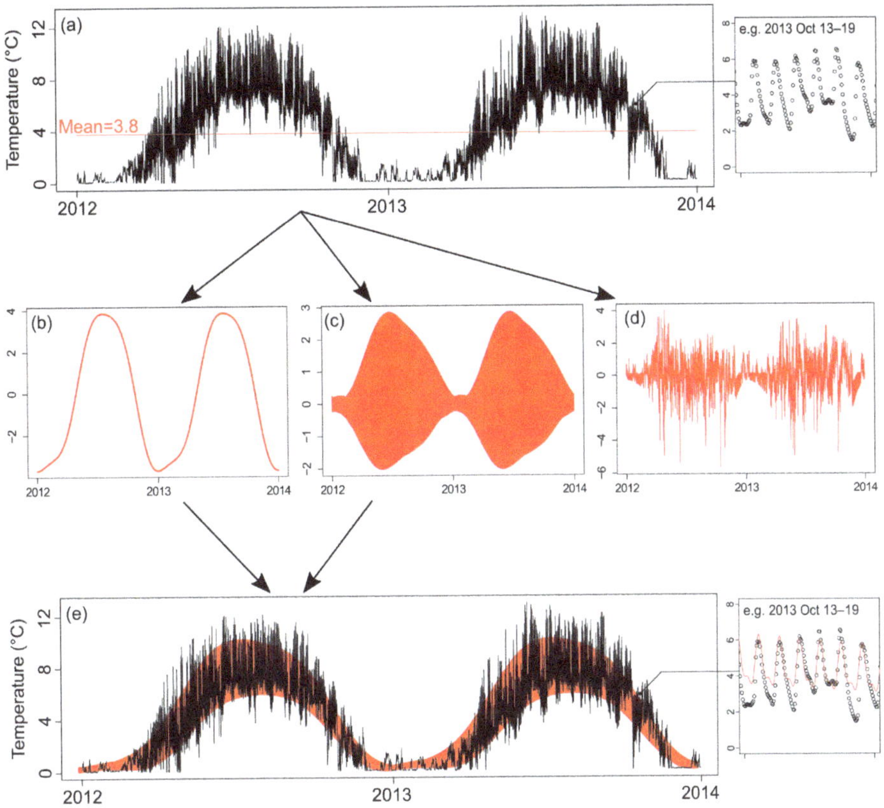

Figure 2. Hourly time series temperature at the reference site decomposed using Fourier decomposition analysis. **(a)** The observed record T_a in 2012–2013 is decomposed into an average value **(a)**, seasonal periodicity pattern S_a **(b)**, diel periodicity pattern D_a **(c)**, and the irregularity ε_a **(d)**. The comparison between the observed and measured temperature excluding ε_a at the reference site **(e)**.

3.2 Estimating time series temperature for spot-measured stream sites

Temperature at site A was approximately twofold lower than at site B during the study period. Temperature at site A had an average of 2.3 °C and spanned 0.4–4.8 °C; temperature at site B had an average of 5.4 °C and spanned 0.8–9.2 °C. Temperatures at site A and B were linearly correlated to temperature at the reference site R ($r^2 = 0.60$ and 0.92, respectively; Fig. 4). This result indicates that spatial autocorrelation of the thermal attributes exists between them, even though a thermal influence from the pro-glacial lake differentiates the thermal patterns between sites A and B (Fig. 1).

For Eq. (7), the proposed method estimated that the weighting parameters at site A ($\beta_{1-4} = \{0.32, 0.32, 0.50, 0.70\}$) showed higher heterogeneity in the relative contributions of the components than those at site B ($\beta_{1-4} = \{0.97, 0.97, 0.99, 0.93\}$). For site A, the relative contribution of diel variability and irregularity to the estimated temperature ($\beta_3 = 0.50$, $\beta_4 = 0.70$) was higher than the average and seasonal variability ($\beta_1 = \beta_2 = 0.32$). This result indicates that our approach can accurately estimate periodic components and extremes, including the variability in irregularity that cannot be represented by linear regression focusing on an average estimate. At site B, the parameter composition was less variable (0.93–0.99), and therefore the performance of the decomposition approach was similar to the performance of the linear model. Using the developed method, the temperature at site A was estimated better than the estimated temperature using linear regression ($r^2 = 0.66$ and 0.60, respectively). For site B, both approaches performed equivalently ($r^2 = 0.92$). These differences between sites A and B indicate that the thermal attributes at site A are more different from the reference site R than those at site B. The thermal attributes at site A are not affected by a thermal effluence from the pro-glacial lake (Fig. 1).

Based on the obtained parameters, time series temperatures at sites A and B were inferred (Fig. 5). At site A, due to relatively high contributions of diel cycles and irregularities ($\beta_3 = 0.50$, $\beta_4 = 0.70$), higher hourly variability was estimated by the proposed approach than the linear regression (Fig. 5a). The maximum values estimated by the proposed approach and linear regression were clearly distinct at 5.9 and 4.0 °C, respectively, while an average of 1.1 °C was equivalent for both estimates. The linear model clearly underestimated (<4.0 °C) the high temperatures recorded at 4.4 °C on 30 May 2012 and 4.8 °C on 5 June 2013. In con-

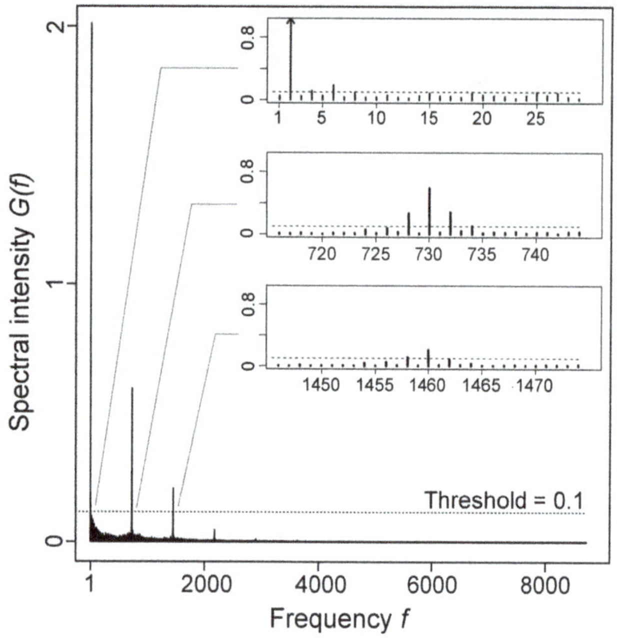

Figure 3. Frequency and spectral intensity of trigonometric curves converted from the 2 years of temperature data at the reference site. Trigonometric curves whose spectral intensity was higher than the threshold value are found in Table 1.

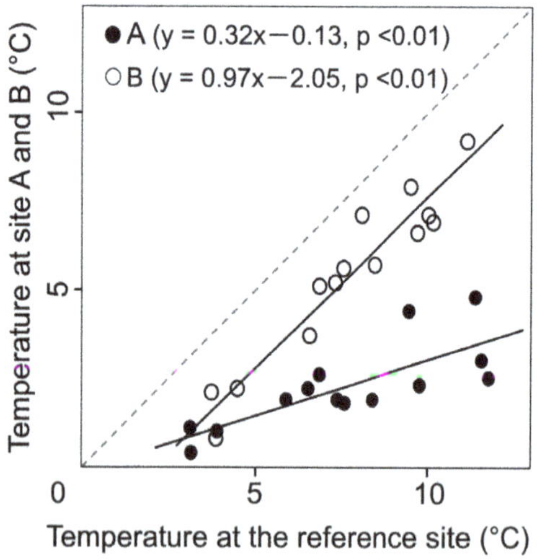

Figure 4. Linear regression of stream temperature at site A and B with the reference site temperature R.

trast, the proposed approach inferred a possibility of temperature reaching 5.9 °C, having a 97.5 % percentile value of 4.0 °C. This difference between approaches indicates the ability of the proposed approach to estimate extreme thermal pulses and their occurrence probability.

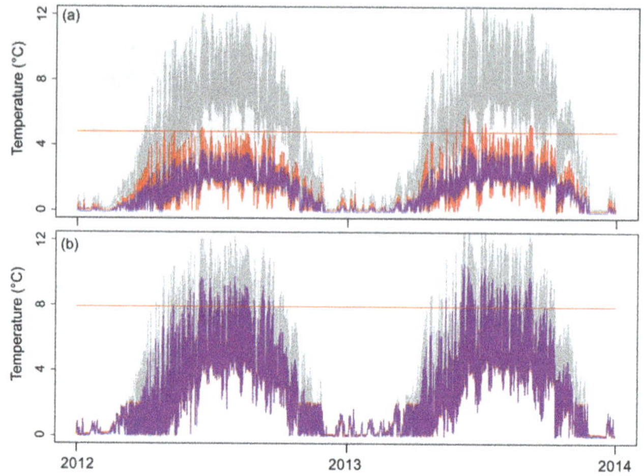

Figure 5. Hourly stream temperatures estimated using Fourier decomposition (red) and linear regression (purple) at site A and B (plots **(a)** and **(b)**, respectively) and temperature at the reference site (grey). Horizontal red line represents the maximum temperature recorded from the monthly spot measures (4.8 °C on 5 June 2013 at site A, and 7.9 °C on 9 July 2013 at site B).

4 Discussion

Currently, no stream temperature models explicitly predict ecologically important thermal attributes (seasonal and diel periodicity patterns and irregular extremes) because of the difficulty in capturing the combined patterns at multiple temporal scales in stream environments (Benyahya et al., 2007). This study developed a regression approach that predicts these thermal attributes at data-poor sites based on the pre-analysis of time series temperature data at a data-rich reference site along the stream. The method merges a Fourier transformation technique into a linear regression model to better represent periodic patterns at multiple temporal scales. The approach could estimate the relative composition of thermal attributes from a limited number of spot-measured data (see Eq. 7), while linear regression weighted the composition equally. The results emphasized the significance of developing further ecologically based thermal prediction models, aiming at deeper understandings in ecological responses to thermal attributes (Paaijmans et al., 2013; Thompson et al., 2013; Frank et al., 2015).

The developed prediction method confirmed its potential to evaluate the relative contribution of thermal attributes at a data-poor site. The method is somewhat comparable with multiple linear regressions ($Y = \beta_0 + \beta_1 X_1 + \ldots$) in terms of the assumption that the thermal attributes are independent of each other and autocorrelated between sites, and their relative contributions change according to location. The unique difference is that the developed method can directly use the periodicity patterns and irregularities for prediction as Eq. (7). This type of model was not addressed in a recent review on temperature models (Benyahya et al., 2007). As the statistical expression of our approach is linear (Eqs. 1 and 6), it can be easily coupled with approaches in the review; i.e., using other regression models employing air temperature (e.g., Pilgrim et al., 1998) and streamflow (Webb et al., 2003). Adding such information has the potential to increase accuracy, especially if these factors contain unique information that is unexplained by the spatial correlation of water temperature between sites. For example, if discharge represents a volume of snowmelt water that can influence the correlative relationship of water temperature between sites, inclusion of discharge into the model's structure would increase accuracy.

The advantage of Fourier decomposition analysis at multiple temporal scales is the ability to evaluate the possible range maxima for each thermal attribute of time series data (Fig. 2). Fourier analysis clearly detected strong periodicity patterns in time series temperature data (Fig. 3) and represented them with a small number of trigonometric curves (Table 1). By doing so, we found that the irregularity had a broader range (9.7 °C) than the spanning ranges of seasonal and diel periodicity patterns (7.5 and 4.9 °C, respectively) at the study stream. This high irregularity may be an important thermal attribute in glacier-fed streams. Further, our results clearly showed that seasonal trends existed in the amplitudes of diel periodicities and irregularities, both having lower variability in winter and greater variability in spring (Fig. 2c, d) (Hopkins, 1971). Importantly, these characteristics are difficult to detect without coupling the periodicity patterns at multiple temporal scales. By decomposing the time series temperature data at the reference site as a pre-analysis for prediction, this decomposition method (see Sects. 2.1 and 2.2) can be useful for quantifying thermal attributes in ecological studies.

The proposed method is highly promising towards evaluating potential changes in thermal variability due to climate change or anthropogenic thermal effluents in rivers and streams (Caissie, 2006; Webb et al., 2008). For instance, environmental change could modify each component of stream temperature with a different degree of severity such as increasing the average, shifting the peak timing in seasonality, reducing diel variability, or amplifying the irregularity. In addition, as thermal regimes play a key role in aquatic ecosystem structure and functioning (e.g., environmental niches determining community composition, biogeochemical cycles; Vannote et al., 1980), each component of stream temperature can be used to better understand ecosystem dynamics and responses (Thompson et al., 2013).

While demonstrating the broad applicability of this approach, we caution over some of its limitations. First, the analysis connecting Eqs. (1) and (7) is threshold dependent. We first estimated an appropriate range of threshold value visually so as to capture a handful of trigonometric curves (i.e., the three major peaks were shown in Fig. 3). Then, we compared the patterns modeled based on some threshold values. Too-low threshold value results in high sensitivity to noise, while too-high threshold value results in high insensitivity to periodic patterns. We compared the performance of the models based on a set of threshold values (0.05, 0.1, 0.2 °C, etc.) and determined the threshold graphically as the value 0.1 °C clearly separated periodicity patterns and irregularities (Fig. S1; Supplement). Therefore, a threshold value must be carefully chosen, and it needs to be evaluated whether the irregularity attribute is unbiased. Second, although the concept is feasible, complete validation was limited due to the low number of sample sites. To obtain more robust estimation, the model needs additional validation as well as calibration. Third, if a systematic shift in thermal pattern is observed during a target period, the method needs adjustment. Our target 2-year period was short enough to neglect long-term trends in temperature. For example, some studies using long-term records (>20 years) detected an interannual increasing trend in mean temperature due to anthropogenic thermal discharge and land-use change (Beschta and Taylor, 1988; Hostetler, 1991). In such cases, the proposed equation in Eq. (1) must be modified because it assumes constant average temperature over a target period for analysis; i.e., $\bar{T}$ should be a function of time as $\bar{T}(t)$. Lastly, the assumption of linearly related components between the reference site and other sites is not always met in all cases. For instance, glacier-fed and groundwater-fed streams can be composed of different seasonal and diel periodicity patterns (Brown and Hannah, 2008). In our target system, flow from the Roseg pro-glacial lake differentiated thermal patterns of sites A and B (Fig. 1).

Even though our analysis confirmed that the linear assumption was apparent in this system ($r^2 = 0.60$; Fig. 4), incorporation of the thermal pattern at the lake outlet into the analysis would certainly increase estimation accuracy.

Acknowledgements. The study was done as part of a fellowship in the "Young Researchers Exchange Program between Japan and Switzerland" under the "Japanese-Swiss Science and Technology Programme" (EG 11-2015) and supported by the Japan Society for the Promotion of Science (26-11771). The data used in this study were collected for a Swiss National Science Foundation project (31003A_152815). We thank the Swiss National Hydrological and Geological Survey from the Federal Office for the Environment (OFEV) for kindly providing water temperature data and S. Blaser for field assistance. We are indebted to the anonymous reviewers for their constructive reviews and comments. We thank India Mansour for proofing the language.

Edited by: S. Thompson

References

Benyahya, L., Caissie, D., St-Hilaire, A., Ouarda, T. B. M. J., and Bobée, B.: A review of statistical water temperature models, Can. Water Resour. J., 32, 179–192, 2007.

Beschta, R. L. and Taylor, R. L.: Stream temperature increases and land use in a forested Oregon watershed, Water Resour. Bull., 24, 19–25, 1988.

Brown, L. E. and Hannah, D. M.: Spatial heterogeneity of water temperature across an alpine river basin, Hydrol. Process., 22, 954–967, 2008.

Caissie, D.: The thermal regime of rivers: a review, Freshwater Biol., 51, 1389–1406, 2006.

Cluis, D. A.: Relationship between stream water temperature and ambient air temperature – a simple autoregressive model for mean daily stream water temperature fluctuations, Nord. Hydrol., 3, 65–71, 1972.

Cossins, A. R. and Bowler, K.: Temperature biology of animals, Springer Netherlands, 1987.

Fourier, J.: The analytical theory of heat, Cambridge University Press, 1878.

Frank, D., Reichstein, M., Bahn, M., Thonicke, K., Frank, D., Mahecha, M. D., Smith, P., van der Velde, M., Vicca, S., Babst, F., Beer, C., Buchmann, N., Canadell, J. G., Ciais, P., Cramer, W., Ibrom, A., Miglietta, F., Poulter, B., Rammig, A., Seneviratne, S. I., Walz, A., Wattenbach, M., Zavala, M. A., and Zscheischler, J.: Effects of climate extremes on the terrestrial carbon cycle: Concepts, processes and potential future impacts, Glob. Change Biol., 21, 2861–2880, 2015.

Hopkins, C. L.: The annual temperature regime of a small stream in New Zealand, Hydrobiologia, 37, 397–408, 1971.

Hostetler, S. W.: Analysis and modeling of long-term stream temperatures on the Steamboat Creek basin, Oregon: implication for land use and fish habitat, Water Resour. Bull., 27, 637–647, 1991.

Intergovernmental Panel on Climate Change (IPCC): Summary for policymakers, in: climate change 2014: Impacts, adaptation, and vulnerability, Part A: Global and sectoral aspects. Contribution of working group II to the fifth assessment report of the Intergovernmental Panel on Climate Change, 1–32, 2014.

Kothandaraman, V.: Analysis of water temperature variations in large river, J. Sanit. Eng. Div., 97, 19–31, 1971.

Long, L. L.: Mathematical modeling of river water temperatures, Univ. Missouri-Roll., Missouri University of Science and Technology, 1972.

Maheu, A., Poff, N. L., and St-Hilaire, A.: A classification of stream water temperature regimes in the conterminous USA, River Res. Appl., 2015.

Mayhew, P. J., Bell, M. A., Benton, T. G., and McGowan, A. J.: Biodiversity tracks temperature over time, P. Natl. Acad. Sci. USA, 109, 15141–15145, 2012.

Paaijmans, K. P., Heinig, R. L., Seliga, R. A., Blanford, J. I., Blanford, S., Murdock, C. C., and Thomas, M. B.: Temperature variation makes ectotherms more sensitive to climate change, Glob. Change Biol., 19, 2373–2380, 2013.

Pilgrim, J. M., Fang, X., and Stefan, H. G.: Stream temperature correlations with air temperatures in Minnesota: Implications for climate warming, J. Am. Water Resour. Assoc., 34, 1109–1121, 1998.

Poff, N. L., Allan, J. D., Bain, M. B., Karr, J. R., Prestegaard, K. L., Richter, B. D., Sparks, R. E., and Stromberg, J. C.: The natural flow regime: A paradigm for river conservation and restoration, Bioscience, 47, 769–784, 1997.

R Core Team, R: A language and environment for statistical computing, 2014.

Ryo, M., Iwasaki, Y., Yoshimura, C., and Saavedra, O. C. V.: Evaluation of spatial pattern of altered flow regimes on a river network using a distributed hydrological model, PLoS ONE, 10, 1–16, 2015.

Steel, E. A. and Lange, I. A.: Using wavelet analysis to detect changes in water temeprature regimes at multiple scales: Effects of multi-purpose dams in the Willamette River Basin, River Res. Appl., 23, 351–359, 2007.

Thompson, R. M., Beardall, J., Beringer, J., Grace, M., and Sardina, P.: Means and extremes: Building variability into community-level climate change experiments, Ecol. Lett., 16, 799–806, 2013.

Tockner, K. and Stanford, J. A.: Riverine flood plains: present state and future trends, Environ. Conserv., 29, 308–330, 2002.

Tockner, K., Malard, F., Burgherr, P., Robinson, C. T., Uehlinger, U., Zah, R., and Ward, J. V.: Physico-chemical characterization of channel types in a glacial floodplain ecosystem (Val Roseg, Switzerland), Arch. Hydrobiol., 140, 433–463, 1997.

Uehlinger, U., Malard, F., and Ward, J. V.: Thermal patterns in the surface waters of a glacial river corridor (Val Roseg, Switzerland), Freshwater Biol., 48, 284–300, 2003.

Van der Nat, D., Tockner, K., Edwards, P. J., Ward, J. V., and Gurnell, A. M.: Habitat change in braided flood plains (Tagliamento, NE-Italy), Freshwater Biol., 48, 1799–1812, 2003.

Vannote, R. L., Minshall, G. W., Cummins, K. W., Sedell, J. R., and Cushing, C. E.: The river continuum concept, Can. J. Fish. Aquat. Sci., 37, 130–137, 1980.

Ward, J. V. and Uehlinger, U.: Ecology of a glacial flood plain, Springer Netherlands, 2003.

Webb, B. W., Clack, P. D., and Walling, D. E.: Water-air temperature relationships in a Devon river system and the role of flow, Hydrol. Process., 17, 3069–3084, 2003.

Webb, B. W., Hannah, D. M., Moore, R. D., Brown, L. E., and Nobilis, F.: Recent advances in stream and river temperature research, Hydrol. Process., 22, 902–918, 2008.

World Wildlife Fund (WWF): Living Planet Report 2014, 2014.

12

On the skill of raw and post-processed ensemble seasonal meteorological forecasts in Denmark

Diana Lucatero[1], **Henrik Madsen**[2], **Jens C. Refsgaard**[3], **Jacob Kidmose**[3], **and Karsten H. Jensen**[1]

[1]Department of Geosciences and Natural Resource Management, University of Copenhagen, Copenhagen, Denmark
[2]DHI, Hørsholm, Denmark
[3]Geological Survey of Denmark and Greenland (GEUS), Copenhagen, Denmark

Correspondence: Diana Lucatero (diana.lucatero@ign.ku.dk)

Abstract. This study analyzes the quality of the raw and post-processed seasonal forecasts of the European Centre for Medium-Range Weather Forecasts (ECMWF) System 4. The focus is given to Denmark, located in a region where seasonal forecasting is of special difficulty. The extent to which there are improvements after post-processing is investigated. We make use of two techniques, namely linear scaling or delta change (LS) and quantile mapping (QM), to daily bias correct seasonal ensemble predictions of hydrologically relevant variables such as precipitation, temperature and reference evapotranspiration (E_{T0}). Qualities of importance in this study are the reduction of bias and the improvement in accuracy and sharpness over ensemble climatology. Statistical consistency and its improvement is also examined. Raw forecasts exhibit biases in the mean that have a spatiotemporal variability more pronounced for precipitation and temperature. This variability is more stable for E_{T0} with a consistent positive bias. Accuracy is higher than ensemble climatology for some months at the first month lead time only and, in general, ECMWF System 4 forecasts tend to be sharper. E_{T0} also exhibits an underdispersion issue, i.e., forecasts are narrower than their true uncertainty level. After correction, reductions in the mean are seen. This, however, is not enough to ensure an overall higher level of skill in terms of accuracy, although modest improvements are seen for temperature and E_{T0}, mainly at the first month lead time. QM is better suited to improve statistical consistency of forecasts that exhibit dispersion issues, i.e., when forecasts are consistently overconfident. Furthermore, it also enhances the accuracy of the monthly number of dry days to a higher extent than LS. Caution is advised when applying a multiplicative factor to bias correct variables such as precipitation. It may overestimate the ability that LS has in improving sharpness when a positive bias in the mean exists.

1 Introduction

Seasonal forecasting has gained increasing attention during the last three decades due to high societal impacts of extreme meteorological events that affect a plethora of weather-related sectors such as agriculture, environment, health, transport and energy, and tourism (Dessai and Soares, 2013). Information on weather-related hazards months ahead is important for protection against extremes for these sectors.

General circulation models (GCMs) have become the state-of-the-art technology for issuing meteorological forecasts at different timescales. GCM-based seasonal forecasting is possible due to signals, which can be extracted from slowly changing systems such as the ocean and, to a lesser extent, land, which then translate into a signal in the atmospheric patterns (Weisheimer and Palmer, 2014; Doblas-Reyes et al., 2013). El Niño–Southern Oscillation, ENSO, is the strongest of these signals, and its influence on seasonal forecasting is higher near the tropics (Weisheimer and Palmer, 2014).

Seasonal ensemble forecasts have been operational in Europe since the late 1990s, provided by the European Centre for Medium-Range Weather Forecast (ECMWF) (Molteni et al., 2011), and in the US since August 2004, provided by the National Center of Environmental Prediction (Saha et al., 2013). Other examples of operational seasonal forecasts include the ones generated by the Met Office in the UK (Maclachlan et al., 2015), the Australian Bureau of Meteorology (Hudson et al., 2013), the Beijing Climate Center (Liu et al., 2015) and the Hydrometeorological Center of Russia (Tolstykh et al., 2014).

ECMWF is a leading center for weather and climate predictions and its seasonal forecasting system is often regarded as the best (Weisheimer and Palmer, 2014). Research on the quality of the seasonal GCM forecasts has been done for different system versions (Molteni et al., 2011; Weisheimer et al., 2011). The system has also been compared to other GCMs (Kim et al., 2012a, b; Doblas-Reyes et al., 2013) or statistical (van Oldenborgh et al., 2005) seasonal forecasting systems.

Despite the efforts mentioned above and the documented improvements on forecasting skill of meteorological parameters, specially over the tropics (Molteni et al., 2011), several issues remain. The main one, and specific to forecasting in Europe and North America, is that the signal of the main driver of seasonal predictability, the ENSO, has been found to be weak or nonexistent (Molteni et al., 2011; Saha et al., 2013) in these regions, leading to poor skill of atmospheric variables such as precipitation. For example, Weisheimer and Palmer (2014) studied the reliability (consistency between the forecasted probabilities and their observed frequencies) and ranked forecasts using five categories from "dangerous" (1) to "perfect" (5) for two regimes of precipitation (wet or dry) and temperature (cold or warm). For the northern European region, they found dry (wet) forecasts during summer (started in May) to be "dangerous" ("marginally useful") and dry (wet) forecasts during winter (started in November) to be "not useful" ("marginally useful"). For temperature, results were less variable among the different categories with winter cold or warm and summer warm forecasts found to be marginally useful and summer cold temperatures forecasts found to be in category (5) for perfect. Moreover, Molteni et al. (2011) found weak anomaly correlations of precipitation and temperature during the summer for most of the regions located in northern Europe.

Due to the issues stated above, the need for post-processing the raw forecasts in the hope of improvements has gained importance in the scientific literature. A plethora of methods for statistical post-processing exist for a range of temporal scales. These methods consist of transfer functions, computed on the basis of reforecasts, or past records of forecast–observation pairs (Hamill et al., 2004) whose goal is to match forecast values with observed ones. The choice of a post-processing method is determined by the availability of reforecast data and the application at hand. Although in principle any method could be used for seasonal forecasts, this temporal scale represents a special difficulty due to the fact that initial condition skill is mostly gone and there is little detectable signal behind a large amount of chaotic error.

In particular, for the post-processing of ECMWF System 4 seasonal forecasts, a number of studies have been carried out: Crochemore et al. (2016), Peng et al. (2014), Trambauer et al. (2015) and Wetterhall et al. (2015). The most used methods are linear scaling (LS) and quantile mapping (QM), although Peng et al. (2014) used a Bayesian merging technique. In general, the studies are successful in improving the values of the forecast qualities the authors considered important. For example, Wetterhall et al. (2015) reported higher skill of forecasts of the frequency and duration of dry spells once an empirical quantile mapping has been applied to daily values of precipitation. Crochemore et al. (2016) analyzed the effect different implementations of the linear scaling and quantile mapping methods had on streamflow forecasting, concluding that the empirical quantile mapping improves the statistical consistency of the precipitation forecasts for different catchments throughout France.

The aforementioned studies have been made only for precipitation and/or mainly large areas. For hydrological applications seasonal forecasting skill of instantaneous values of precipitation, temperature and reference evapotranspiration (E_{T0}) at the catchment scale (100–1000 km^2) are, however, more important. Therefore, we analyze the bias, skill and statistical consistency of the ECMWF System 4 for Denmark focusing on precipitation, temperature and E_{T0} of relevance for seasonal streamflow forecasting at the catchment scale. We make use of the two most used methods for post-processing, namely linear scaling and quantile mapping (Zhao et al. 2017), applied to daily values. We focus on the skill of monthly aggregated values of gridded data throughout Denmark for both the raw and the corrected forecasts. We investigate the following questions:

1. What is the longest lead time for which an acceptable forecast is achieved?
2. Is it possible to extend the acceptable forecast lead time with different post-processing techniques?

In this study we argue that an acceptable forecast needs to have consistency between the observed probability distribution and the predictive one. This is what we call statistical consistency throughout the paper. A statistically consistent forecast system has low (or nonexistent) bias in both mean and variance. Secondly, we argue that the forecast to be used must be better than climatology, having a higher skill both in terms of accuracy and sharpness and giving priority to the former. These characteristics for an "acceptable forecast" follow the principle that the purpose of post-processing is to maximize sharpness subject to statistical consistency as discussed by Gneiting et al. (2007).

2 Data and methods

2.1 Ensemble prediction system and observational grid

Seasonal reforecasts of the ECMWF System 4 for the years 1990–2013 are used in the present study. The system is comprised of 15 members (for January, March, April, June, July, September, October and December) and 51 members (for February, May, August and November) with a spatial resolution of 0.7° and are run for 7 months with daily output. The increase in ensemble size for February, May, August and November attempts to aid in improving forecasts for the seasons with a higher predictability. Precipitation, temperature and E_{T0} are the variables under study. For the computation of E_{T0}, we make use of the Makkink equation (Hendriks, 2010), which takes as inputs temperature and incoming shortwave solar radiation from ECMWF System 4.

Observed daily values for precipitation, temperature and E_{T0} from the Danish Meteorological Institute (DMI) are used (Scharling and Kern-Hansen, 2012). The spatial scale for precipitation and temperature, E_{T0}, is 10 and 20 km, respectively. However, we assume temperature and E_{T0} to be equally distributed within the 20 km and set the same values of the 20 km to the 10 km grid. Then, in total there are 662 (for precipitation) and 724 (for temperature and E_{T0}) grid points, which cover the 43 000 km^2 area of Denmark. Moreover, precipitation is corrected for under-catch errors as explained in Stisen et al. (2011) and (2012). The time and spatial variations of the variables can be seen in Fig. 1. Values are monthly accumulations for precipitation and E_{T0} and monthly averages for temperature, averaged over the observed record (1990–2013). Danish weather is mainly driven by its proximity to the sea. There is a modest spatial precipitation gradient from west to east, which is more pronounced during autumn and winter. The driest month in terms of precipitation is April and the wettest is October. E_{T0} also shows a modest spatial variability during spring and summer, with larger values in eastern Denmark.

2.2 Post-processing strategy

Given the fact that the spatial resolution of the ensemble data differs from the resolution of the observed data, first the ensemble forecasts were interpolated to match the 10 km grid of observed values using an inverse distance weighting (Shepard, 1968), where the values at a given point on the higher-resolution grid (10 km) are computed using a weighted average of the four surrounding nodes of the lower-resolution forecast grid (70 km). The weights are computed as the inverse of the Euclidean distances between the observed grid node and the forecast nodes. Forecasts are then post-processed for each grid point, time of forecast (month) and lead time (month) for each variable separately. Moreover, the computation is done in a leave-one-out cross-validation mode (Wilks, 2011, and Mason and Baddour, 2008) such that the year being corrected is withdrawn from the sample. This is to ensure independence between training and validation data. Then, for example, for precipitation, 662 × 12 × 7 × 24 (no. of grid points, no. of months, no. of lead times and no. of years in the sample, respectively) correction models are computed.

2.3 Post-processing methods

2.3.1 Delta method – linear scaling (LS)

The linear scaling approach operates under the assumption that forecast values and observations will agree in their monthly mean once a scale or shift factor has been applied (Teutschbein and Seibert, 2012). LS is the simplest possible post-processing method as it only corrects for biases in the mean. The factor is commonly computed differently for precipitation, E_{T0} and temperature due to the different nature of the variables, as precipitation and E_{T0} cannot be negative.

For precipitation and E_{T0},

$$f^*_{k,i} = \frac{\sum_{j=1}^{N-1} y_j}{\sum_{j=1}^{N-1} \overline{f}_j} f_{k,i} \text{ for } i \neq j. \tag{1}$$

For temperature,

$$f^*_{k,i} = f_{k,i} - \frac{1}{N-1}\left[\sum_{j=1}^{N-1} \overline{f}_j - \sum_{j=1}^{N-1} y_j\right] \text{ for } i \neq j, \tag{2}$$

where $f_{k,i}$denotes ensemble member k for $k = 1, \ldots, M$ of forecast–observation pair $i = 1, \ldots, N$; M denotes the number of members (15 or 51) and N is the number of forecast–observation pairs; $\overline{f}_j$ denotes the ensemble mean; and y_j denotes the verifying observation. Note that, as stated in Sect. 2.3, both the means of $\overline{f}_j$ and y_j are computed with the sample that withdraws forecast and observation pair i. Finally, $f^*_{k,i}$ represents the corrected ensemble member. Note that for precipitation, before applying the correction factor, we set all values of daily precipitation below a specific threshold to zero to remove the "drizzle effect" (Wetterhall et al., 2015). The threshold was chosen so that the number of dry days on a given forecast month matches the number of observed dry days. This threshold varies according to the month and the year and spatially, with an average value of 1.5 mm day^{-1} over Denmark.

2.3.2 Quantile mapping (QM)

QM relies on the idea of Panofsky and Brier (1968). This method matches the quantiles of the predictive and observed distribution functions in the following way:

$$f^*_{k,i} = G_i^{-1}\left(F_i\left(f_{k,i}\right)\right), \tag{3}$$

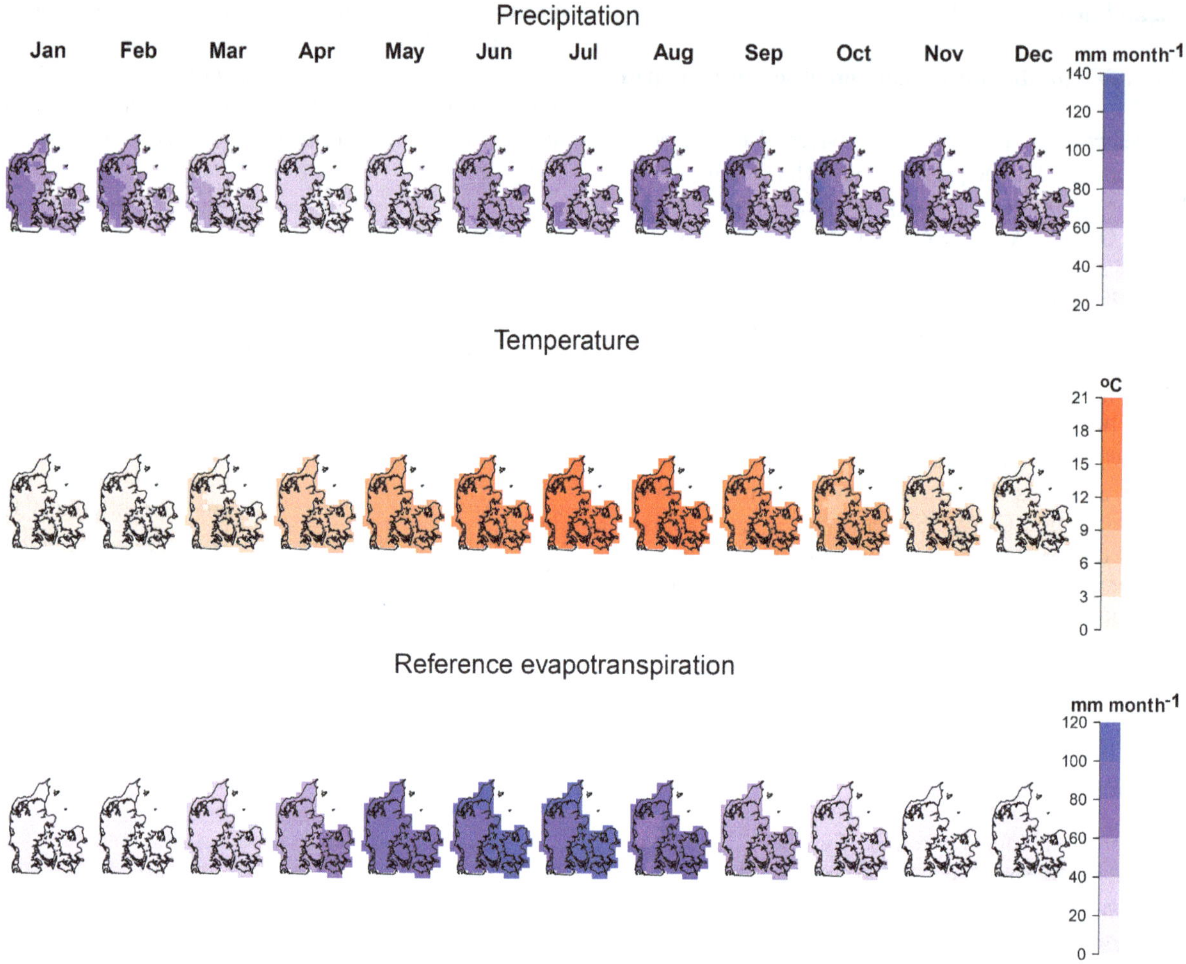

Figure 1. Spatiotemporal variability of precipitation (mm month^{-1}), temperature (°C) and reference evapotranspiration (mm m^{-1}) of monthly aggregated values (P, E_{T0}) and monthly averages (T) of the observation period (1990–2013).

where F_i represents the predictive cumulative distribution function (CDF) for forecast–observation pair i, and G_i represents the observed CDF. Again, note that, as stated in Sect. 2.3, both F_i and G_i are computed with a sample that withdraws forecast and observation pair i.

F_i is calculated as an empirical distribution function fitted with all ensemble members of daily values of a given month for a given lead time and grid point. For example, for a forecast of target month June initialized in May, F_i is fitted using a sample comprising 30 (days) × 23 (number of years in the reforecast minus the year to be corrected) × 51 (number of ensemble members). The same is done for G_i, except that the fitting sample is comprised of 30 × 23 values only. F and G are computed as an empirical CDF. Linear interpolation is needed to approximate the values between the bins of F and G. Extrapolation is then needed to map ensemble values and percentiles that are outside the fitting range. Note that other approaches to deal with values outside the sample range exist that are more suitable when the focus of the study is the extreme values. For example, Wood et al. (2002) fitted an extreme value distribution to extend the empirical distributions of the variables of interest. However, analyzing the effects of different fitting strategies is out of the scope of the present paper.

2.4 Verification metrics

In order to evaluate the raw forecasts and their improvement after post-processing, we analyzed four attributes of forecast quality: bias, skill in regards to accuracy and sharpness, and statistical consistency.

2.4.1 Bias

Bias is a measure of under- or overestimation of the mean of the ensemble in comparison with the observed mean:

$$\%\,\text{Bias} = \left(\frac{\sum_{i=1}^{N} \overline{f}_i}{\sum_{i=1}^{N} y_i} - 1 \right) \times 100 \tag{4}$$

for precipitation and E_{T0} and

$$\mathrm{Bias} = \frac{1}{N}\left[\sum_{i=1}^{N}\overline{f}_i - \sum_{i=1}^{N} y_i\right] \tag{5}$$

for temperature. f_i and y_i are the same as in Eq. (1).

If the bias is negative, the forecasting system exhibits a systematic underprediction. Conversely, if the amount is positive the system shows an average overprediction. Values closer to 0 are desirable.

2.4.2 Skill

The skill of a forecasting system is the improvement the system has, on average, with respect to a reference system, which could be used instead. For example, climatology for seasonal forecasts or persistence for short-range forecasts. The skill score is computed in the following manner:

$$\mathrm{Skill} = \frac{\mathrm{Score_{sys}} - \mathrm{Score_{ref}}}{\mathrm{Score_{per}} - \mathrm{Score_{ref}}}, \tag{6}$$

where $\mathrm{Score_{sys}}$, $\mathrm{Score_{ref}}$ and $\mathrm{Score_{per}}$ are the score value of the system to be evaluated, the reference system and the value of a perfect system, respectively. The range of the skill is from $-\infty$ to 1 and values closer to 1 are preferred. In this paper we calculate the skill with respect to accuracy and sharpness. We compute the continuous rank probability score (CRPS) (Hersbach, 2000), as a general measure of the accuracy of the forecast as it contains information on both forecast biases in the mean and spread. The computation of the score is as follows:

$$\mathrm{CRPS} = \frac{1}{N}\sum_{i=1}^{N}\int_{-\infty}^{\infty}\left[P_i(x) - H(x - y_i)\right]^2 \mathrm{d}x, \tag{7}$$

where $P_i(x)$ is the CDF of the ensemble forecast for pair i and $H(x - y_i)$ is the Heaviside function that takes the value 1 when $x > y_i$ and 0 otherwise; and y_i and N are, as in Eq. (1), the verifying observation for forecast–observation pair i and the number of forecast–observation pairs, respectively. We made use of the EnsCrps function of the R package SpecsVerification (Siegert, 2015) developed in R version 0.4-1. For the skill with respect to sharpness we use the average along $i = 1, \ldots, N$ of the differences between the 25th and the 75th percentiles of each of the ensemble CDFs, P_i.

In Eq. (6), our reference is ensemble climatology (1990–2013), where the year to be evaluated is withdrawn from the sample. Both the accuracy and sharpness score for a perfect system (see Eq. 6, $\mathrm{Score_{per}}$) is equal to 0 so the skill score can be then simplified as

$$\mathrm{Skill} = 1 - \frac{\mathrm{Score_{sys}}}{\mathrm{Score_{ref}}}, \tag{8}$$

which, once multiplied by 100, is the percentage of improvement (if positive) or worsening (if negative) over the reference forecast. Throughout the paper, the skill related to accuracy will be denoted as CRPSS whereas the skill due to sharpness will be denoted as SS.

Furthermore, to define the statistical significance of the differences between the skill of ensemble climatology and ECMWF System 4 forecasts, as well as the post-processed predictions, a Wilcoxon–Mann–Whitney test (WMW test; see Hollander et al., 2014) was carried out. The WMW test, unlike the most common t test, makes no assumptions about the underlying distributions of the samples. We applied the test for each grid point, target month and lead time.

2.4.3 Statistical consistency

We use the probability integral transform (PIT) diagram for a depiction of the statistical consistency of the system. The PIT diagram is the CDF of the z_i's defined as $z_i = P_i(X \leq y_i)$. Therefore, z_i is the value that the verifying observation y_i attains within the ensemble CDF, P_i. The diagram represents an easy check of the biases in the mean and dispersion of the forecasting system. For a forecasting system to be consistent, meaning that the observations can be seen as a draw of the forecast CDF, the CDF of z_i should be close to the CDF of a uniform distribution on the [0, 1] range. Deviations from the 1 : 1 diagonal represent bias issues in the ensemble mean and spread. The reader is referred to Laio and Tamea (2007) and Thyer et al. (2009) for an interpretation of the diagram. Similar to Laio and Tamea (2007), we make use of the Kolmogorov bands to have a proper graphical statistical test for uniformity. Finally, we make use of the Anderson–Darling test (Anderson and Darling, 1952) for a numerical test of the uniformity of the PIT diagrams. We carry out the test using the ADGofTest (Gil, 2011) R package (R Core Team, 2017). Here, the null hypothesis is that the PIT diagram follows a uniform distribution on the [0–1] range.

2.5 Accuracy of maximum monthly daily precipitation and number of dry days

For applications such as flooding and forecasting of low flows and droughts, water managers might be interested not only in the skill of monthly accumulated precipitation but also in the skill of other precipitation quantities. We will use a rather simple approach to check for deficiencies of the raw forecasts and whether the post-processing methods improve these deficiencies. We will analyze the improvement in the prediction of monthly maximum daily precipitation and number of dry days in each month. For this study, a dry day is defined as the day with observed zero-precipitation, and the comparison with the ensembles is made daily.

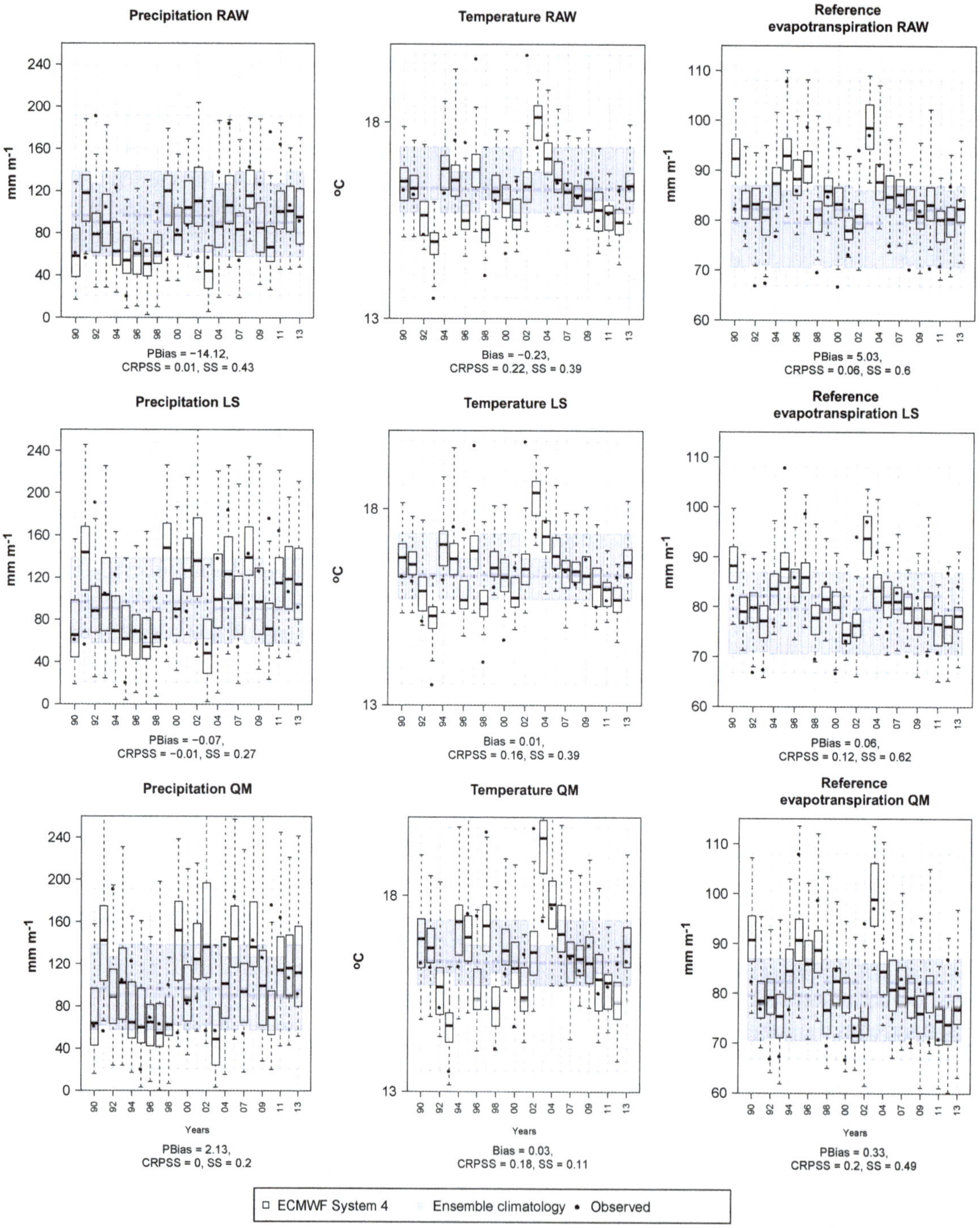

Figure 2. Example of a monthly forecast, valid for August and a lead time of 1 month for a grid point located in west-central Denmark. Blue box plots are the ensemble climatological forecasts. Black box plots are ECMWF System 4 raw or post-processed forecasts. Red dots represent observed values. Bias as in Eqs. (4) and (5) and skill scores for accuracy and sharpness (CRPSS and SS) as in Eq. (8). The first column corresponds to the raw forecast; the second and third columns correspond to the corrected forecasts with the linear scaling or delta change (LS) and quantile mapping (QM) methods, respectively.

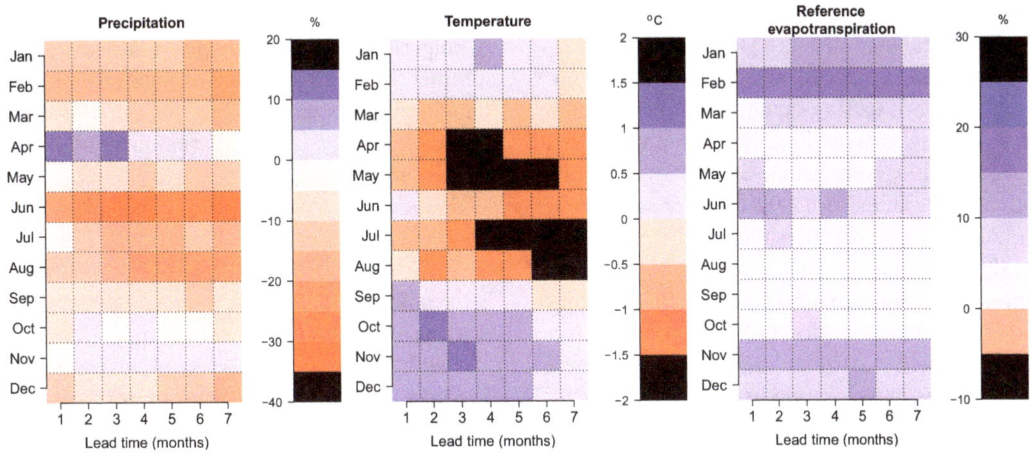

Figure 3. Percentage bias and absolute bias of monthly values of raw forecasts. The y axis represents the target month and the x axis represents the different lead times at which target months are forecasted. Values in blue range represent a positive bias and values in red represent a negative bias.

3 Results

3.1 Analysis of raw forecasts

The first row in Fig. 2 depicts the ECMWF System 4 forecast and the ensemble climatological forecast for August accumulated precipitation and E_{T0} as well as averaged temperature for one grid located in west-central Denmark for the first month lead time. The values for different forecast qualities for that grid point are also included. For a forecasting system to be useful, it has to be at least better than a climatological forecast. For the given example here, the reference forecast is wider than the ECMWF System 4 forecast. This is an example of a month where we have a slightly better skill than the ensemble climatological forecast for the three variables in question. For example, raw temperature predictions from ECMWF System 4 improve, on average, on the reference forecast by 22 % in terms of accuracy. This level of skill is attained due to the sharper forecasts that exhibit a low bias (−0.23 °C). On the other hand, sharpness is only a desirable property when biases are low. This is illustrated for precipitation forecasts attaining a high skill due to sharpness (0.43) but at the expense of a low skill due to accuracy (0.01). This is caused by the high negative bias (−14.12 %) where, for example for 1992 and 2010, the verifying observation lies outside the ensemble range contributing negatively to the CRPS in Eq. (7).

3.1.1 Bias

In an effort to summarize the results, a spatial average of the bias throughout Denmark was computed. Figure 3 shows the spatial average bias of precipitation, temperature and E_{T0} of the raw forecasts. The y axis represents the target month, for example April, and the x axis represents the forecast lead time – lead time 5 is the forecast for April initiated in December. As we can see from Fig. 3, bias depends on the target month and, to a lesser degree, on lead time. For precipitation, the lowest bias can be found throughout autumn to the beginning of winter, followed by a general underestimation of precipitation that is at its highest for June. April shows an overestimation which might be due to the drizzle effect in a month where dry days are slightly more common, in comparison to March and May (the percentage of dry days within the month is 50 %, 57 % and 56 % in March, April and May, respectively). The drizzle effect issue is a very well-known problem of GCMs and is related to the generation of small precipitation amounts, usually around 1.0–1.5 mm day^{-1}, where observed precipitation is not present (Wetterhall et al., 2015).

Temperature bias averaged over Denmark has a range that lies within [−2, 2] °C. The bias switches from positive to negative when temperatures start to increase in March and from negative to positive bias when temperatures start decreasing in August. This indicates that the forecast of temperature has a smaller annual amplitude than observed. Lowest biases are encountered during January and February with a bias of 0.5 °C, and it is higher during late spring and summer, with a negative bias of almost 2 °C. Finally, the bias range for E_{T0} is smaller than precipitation, taking values within [0 %–25 %] on average over Denmark. In general, there is a positive bias, which is at its highest during February.

However, averaging does not tell the whole story. We are also interested in the spatial variation of biases over Denmark. Figure 4 shows the spatial distribution of bias for the first month lead time and its evolution during summer (JJA). In general, there is an underestimation of precipitation throughout Denmark, which much more pronounced during June. Nevertheless, there also exists a positive bias in central Jutland (mid-west Denmark) and on the urban area of Copenhagen (mid-east Denmark) reaching a value around 10 %–

20 %. The positive bias area grows in July, occupying most of Jutland and North Zealand (mid-east Denmark).

Other seasons were also mapped and shown in the Supplement as Figs. S1 to S3. During autumn and winter there is also a general negative bias, which is more pronounced in central Jutland, reaching values of −30 %. Nevertheless, an overestimation exists in eastern Denmark for those seasons. For winter, this overestimation is present in the sea grid points. Finally, during spring the spatial variability changes. For example, most grid points exhibit a positive bias during April, except for the southeast region of Denmark that has a small negative bias between 0.0 % and 5.0 %. During May, a tendency of over-forecasting is present in central Jutland.

The spatial distribution of temperature bias during autumn and winter (Figs. S2 and S3, respectively) follows a similar pattern with a general positive bias reaching its highest values in the southeast region (from 1.5 to 2.0 °C). A negative bias is seen during spring (Fig. S1) and late summer across Denmark (Fig. 4). In June, a positive bias of [0–2 °C] is present in a large area of the Jutland peninsula (Fig. 4). Finally, the spatial variation of the bias of E_{T0} is less pronounced and, in general, positive. Nonetheless, exceptions exist. There is a negative bias in small regions located in the coastal areas or sea grid points, which ranges from −10 % to 0 %.

The results presented above are specifically for lead time 1, i.e., forecasts of accumulated precipitation and E_{T0} and average temperature for August initialized on 1 August. The spatial variation of the bias for other lead times was also mapped (not shown) and analyzed. In general, similar spatial patterns were found for all three variables, being the same along the target months and regardless of lead time.

3.1.2 Skill

Figure 5 summarizes the results for skill in the following manner. First, the values presented are, like in Fig. 3, the spatial average of skill for all of Denmark. Secondly, we evaluate the skill in terms of accuracy (CRPSS, first row) and sharpness (SS, second row). As seen in the evaluation of bias, skill appears to be dependent on the target month and, to a lesser degree, on lead time.

For precipitation, and looking at the first month lead time, ECMWF System 4 skill in accuracy is mildly better than that of climatology for February, March, April, July, August, November and December, with a CRPSS of 0.15 at most. In general, skill in accuracy decreases for lead time 2 onwards. April stands out for having a slightly positive skill in accuracy for almost all lead times, but comes at the expense of having a wider spread than ensemble climatology. For temperature, for the first month lead time, a positive skill exists in terms of accuracy for almost all months, except for late spring. Skill in terms of accuracy also decreases with lead time, but February stands out for having a mild skill for almost all lead times. Forecasts are also in general sharper than ensemble climatology, except for January and March at longer lead times. E_{T0} appears to have skill only for late summer and the beginning of autumn in terms of accuracy. This may be explained by the fact that forecasts are sharper than climatology, indicating that there could be an underdispersion issue.

Note that the computation of the skill score for accuracy was done using a $CRPS_{sys}$ of 15 or 51 ensemble members, while the $CRPS_{ref}$ is comprised of 23 members. The disparity in the number of ensemble members can cause forecasts to be at a disadvantage (advantage) compared to a reference forecast with larger (smaller) ensemble size. Ferro et al. (2008) provided estimates of unbiased skill scores that consider the differences in ensemble size. To remove this effect, we calculated the CRPSS using the unbiased estimator for $CRPS_{ref}$ in Eq. (22) in Ferro et al. (2008). In general, there was a mild increase in the CRPSS value for the months with 15 members, as expected (not shown). The opposite holds for the months with 51 members (February, May, August and November), where a mild decrease in the CRPSS values (not shown) was obtained. Because the changes in CRPSS are moderate, in the rest of the document we will report the results of the CRPSS using the original ensemble sizes ($CRPS_{sys}$ with 15 or 51 ensemble members, $CRPS_{ref}$ with 23 members).

Figure 6 shows the skill in accuracy for monthly values and for a lead time of 1 month mapped across Denmark and its monthly evolution during the summer (JJA). The other seasons are also mapped and analyzed and are included in the Supplement (Figs. S4 to S6). Higher skill is observed for temperature, for which ECMWF System 4 improves the ensemble climatological forecast with up to 50 %. Precipitation and E_{T0} have lower skill in comparison to temperature, reaching a value of 0.3 for specific months and regions in Denmark. The spatial variation of skill for precipitation seems scattered across Denmark and through the year. Some notable exceptions are the higher skill in accuracy that ECMWF System 4 has in western Jutland during November and December and the low skill attained in October across Denmark (Figs. S5 and S6). The spatial variation of skill in accuracy of monthly averaged temperature seems more pronounced during autumn and spring, with northern Denmark attaining the highest values of skill for these seasons. For the remaining seasons, skill over climatology is present across the country, except for late spring where eastern Denmark has the largest negative skill. Finally, the spatial variation of skill of E_{T0} is more pronounced for the months April to November with both positive and negative skill. In general, in this period, eastern Denmark attains positive, although mild for some months, values of skill, except for November.

In general, the areas with highest biases shown in Fig. 4 are associated with the lowest skill scores. For instance, for precipitation in October in southern central Jutland, the negative bias reaches values around 30 %–40 % (Fig. S2). This leads to values of CRPS almost 60 % smaller than that of ensemble climatology (Fig. S5). The opposite also holds –

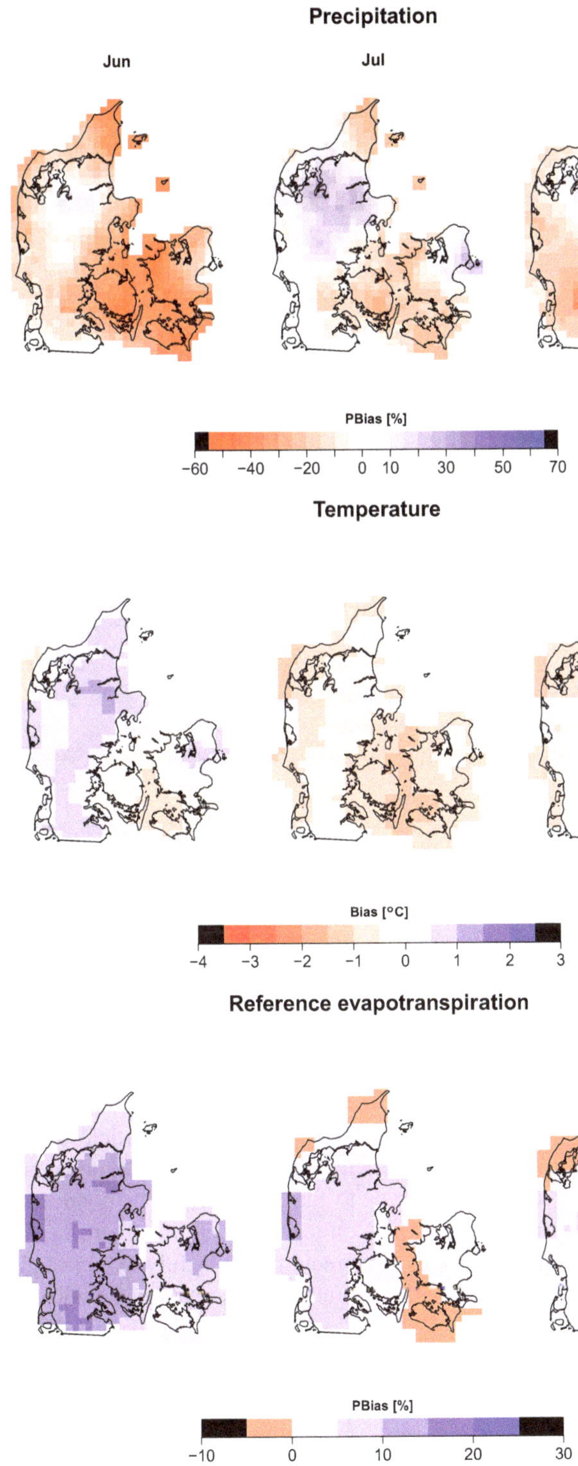

Figure 4. Percentage bias and absolute bias of monthly values of raw forecasts for the summer (JJA). Forecast lead time of 1 month.

areas where biases are lower tend to have the highest benefits over ensemble climatology, i.e., precipitation in March across Denmark or in November in western Jutland.

Skill related to accuracy was also mapped for lead times of 2–7 months (not shown). In general, the geographical regions having a statistically significant positive skill score for a lead time of 1 month disappear, except for some smaller regions where a slight positive skill between 0.0 and 0.1 is found, i.e., precipitation in the April forecast initiated in February (lead

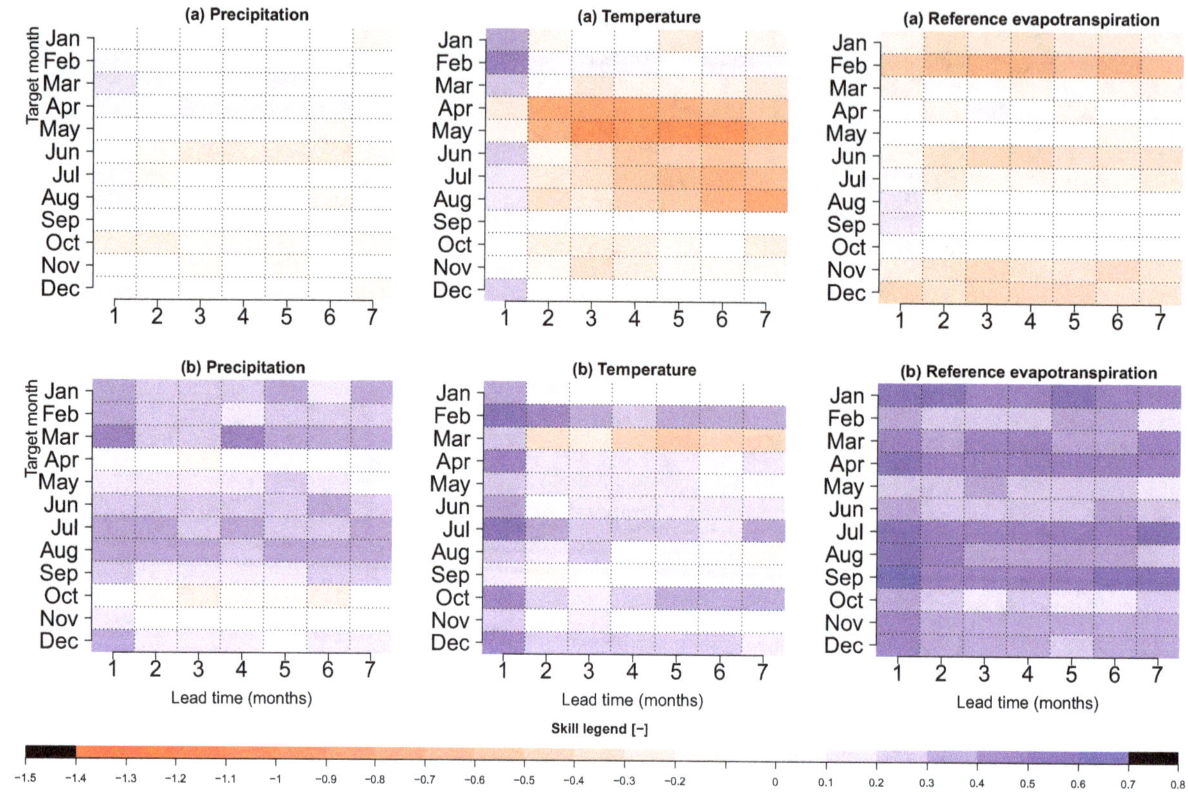

Figure 5. Skill in terms of accuracy **(a)** and sharpness **(b)** of monthly values of raw forecasts. The y axis represents the target month and the x axis represents the different lead times at which target month is forecasted.

time of 3 months), which contributes to the mild positive skill at longer lead times as seen in Fig. 5.

The skill related to sharpness was also mapped for all target months and lead times (not shown). In general, forecasts are sharper than ensemble climatology as seen in Fig. 5 across Denmark for all three variables under study. This situation persists, in general, along all lead times, except for precipitation in April and October, in addition to temperature in January, March and November (Fig. 5). For these months, the lack of sharpness is present throughout Denmark. Nevertheless, for precipitation in April, the region with the lack of sharpness is in southern Denmark, along all lead times.

3.1.3 Statistical consistency of monthly aggregated P, T and E_{T0}

The first row in Fig. 7 shows the PIT diagram for raw ECMWF System 4 summer forecasts for a lead time of 1 month. The observations and forecast of the diagram come from a grid point located in western Denmark (squared shape in Fig. 8). The remaining seasons can be seen in Figs. S7 to S9. Raw precipitation forecasts, for this grid point, exhibit an underprediction of the mean for winter and autumn and mixed results for the remaining seasons. For example, the underprediction bias is reduced for spring, except for April, where the system exhibits a positive bias. Raw temperature predictions of winter, October and November, in addition to June, exhibit an overprediction, which is lowest for January and February. Spring and summer temperatures exhibit an underprediction, which is highest for July (Figs. S7 and 7). Finally, raw forecasts of E_{T0} during all seasons exhibit an overprediction of the mean, in accordance with the results in Sect. 3.1.1. The statistical consistency at longer lead times for all variables (2–7 months, not shown) depends, similarly to bias, on the target month and, to a much lesser degree, on lead time. Temperature in March and, to a lesser degree, precipitation in July exhibit underdispersion issues; i.e., too often the observations lie outside the ensemble range.

3.2 Analysis of post-processed forecasts

The second and third rows in Fig. 2 show the corrected forecasts and the bias and skill scores after post-processing using the LS and the QM method, respectively. The results represent a particular grid point and forecast of August initialized on 1 August. After post-processing, the reduction of bias is evident for the three variables under study. Nevertheless, and contrary to what one should expect, this reduction of bias does not necessarily translate into an increase in skill in accuracy, at least for precipitation and temperature and for this month and grid point. The quantification of the reduction of bias and increase in accuracy after post-processing for the

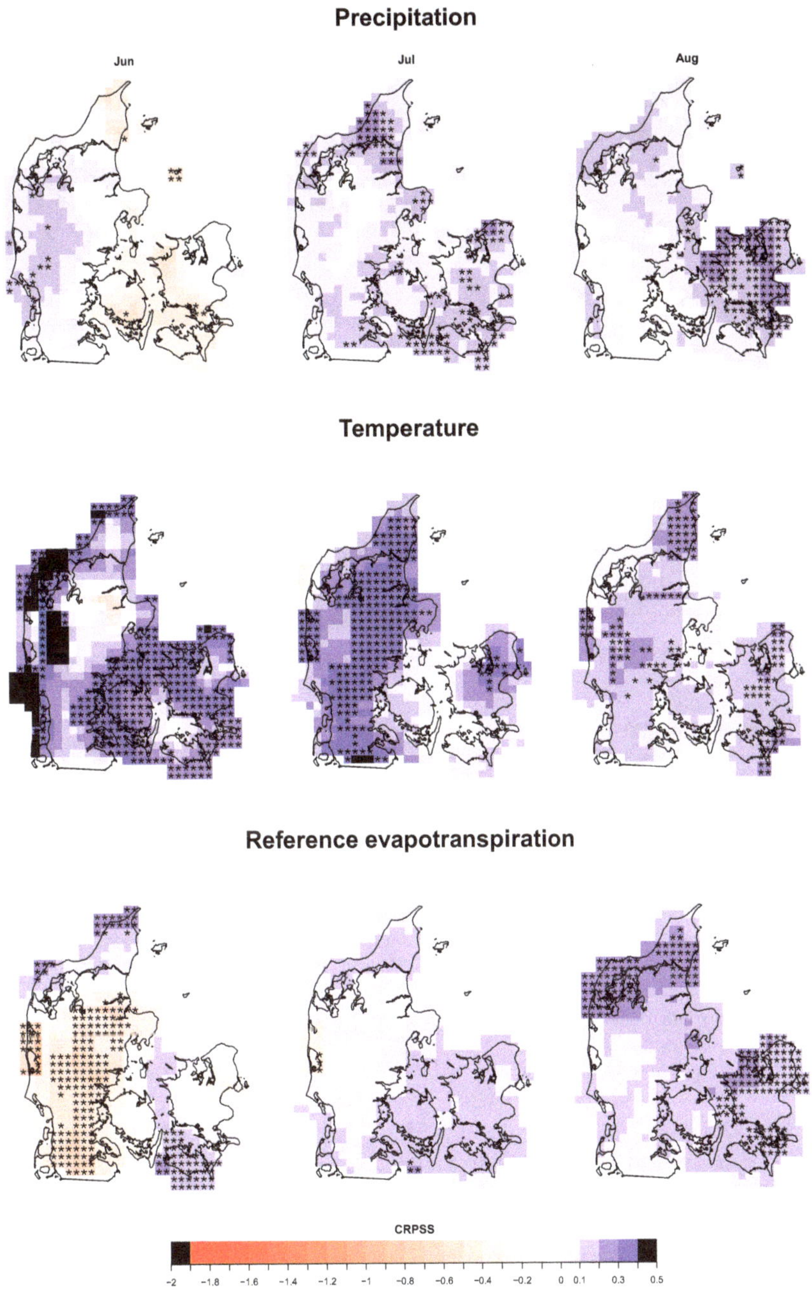

Figure 6. Spatial variability of skill in accuracy for summer (JJA) raw forecasts for a lead time of 1 month. The grids marked with "*" are points where the distribution of the accuracy for ensemble climatology differs from the accuracy distribution of the ECMWF System 4 forecasts at a 5 % significance using the WMW test.

whole of Denmark, through the year and for different lead times, is discussed in the Sect. 3.2.2 below.

3.2.1 Bias

Any post-processing technique used should be able to at least remove biases in the mean. This is accomplished using both techniques. Figure 8 shows the bias of precipitation, temperature and E_{T0} and its evolution through the year for a lead time of 1 month. Bias is shown for four locations scattered around Denmark. Figure 8 shows that the yearly variability of the bias is collapsed to almost 0 %, although for precipitation and winter E_{T0}, the LS method seems to be doing slightly better at removing the bias in comparison to QM. This comes as no surprise, as the LS method forces this bias to be zero.

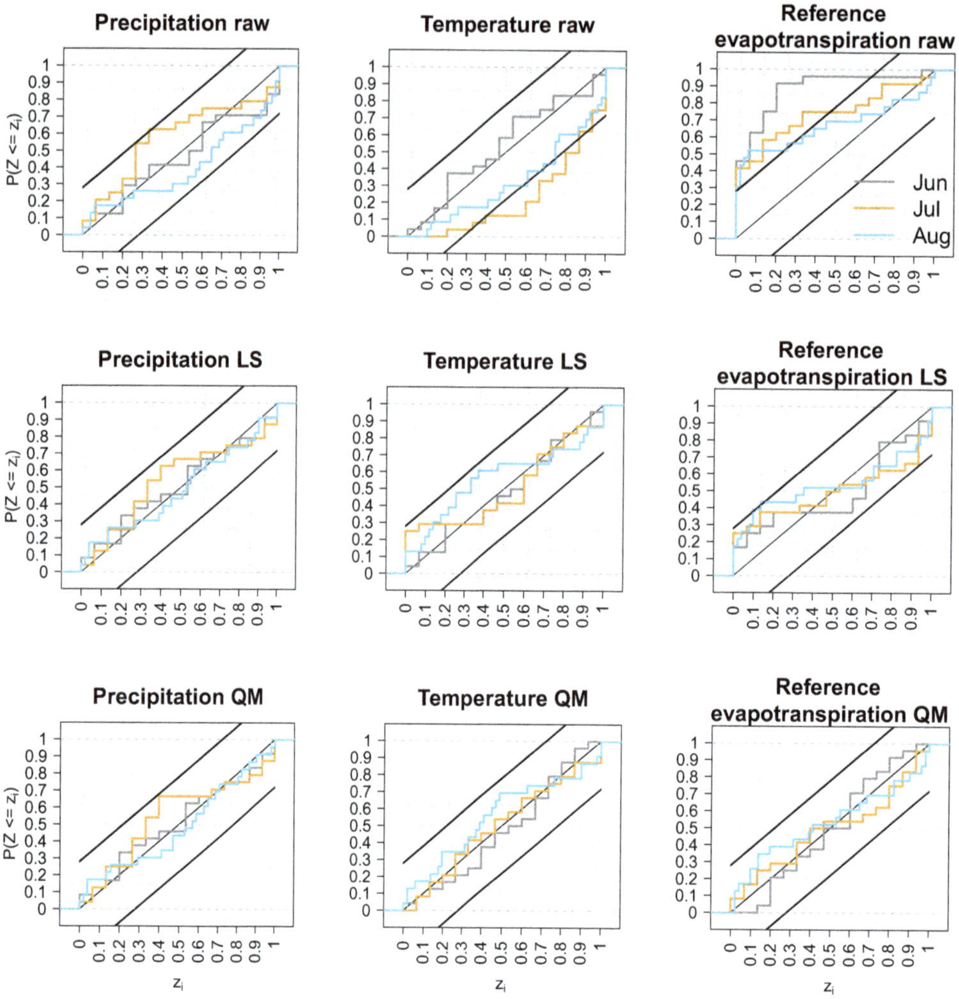

Figure 7. PIT diagrams of summer P, T and E_{T0} for the raw and post-processed forecasts for a lead time of 1 month for one grid point located in western Denmark.

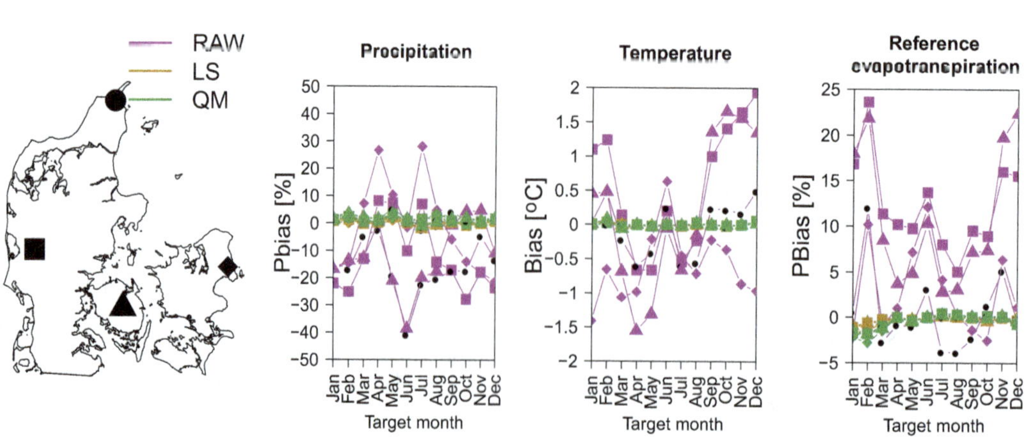

Figure 8. Biases of raw and post-processed precipitation, temperature and E_{T0} at four locations in Denmark. Biases are for the different target months and for a lead time of 1 month. Different locations are represented with different symbol shapes according to the map on the left, whereas the raw and the different post-processing techniques are represented with different colors.

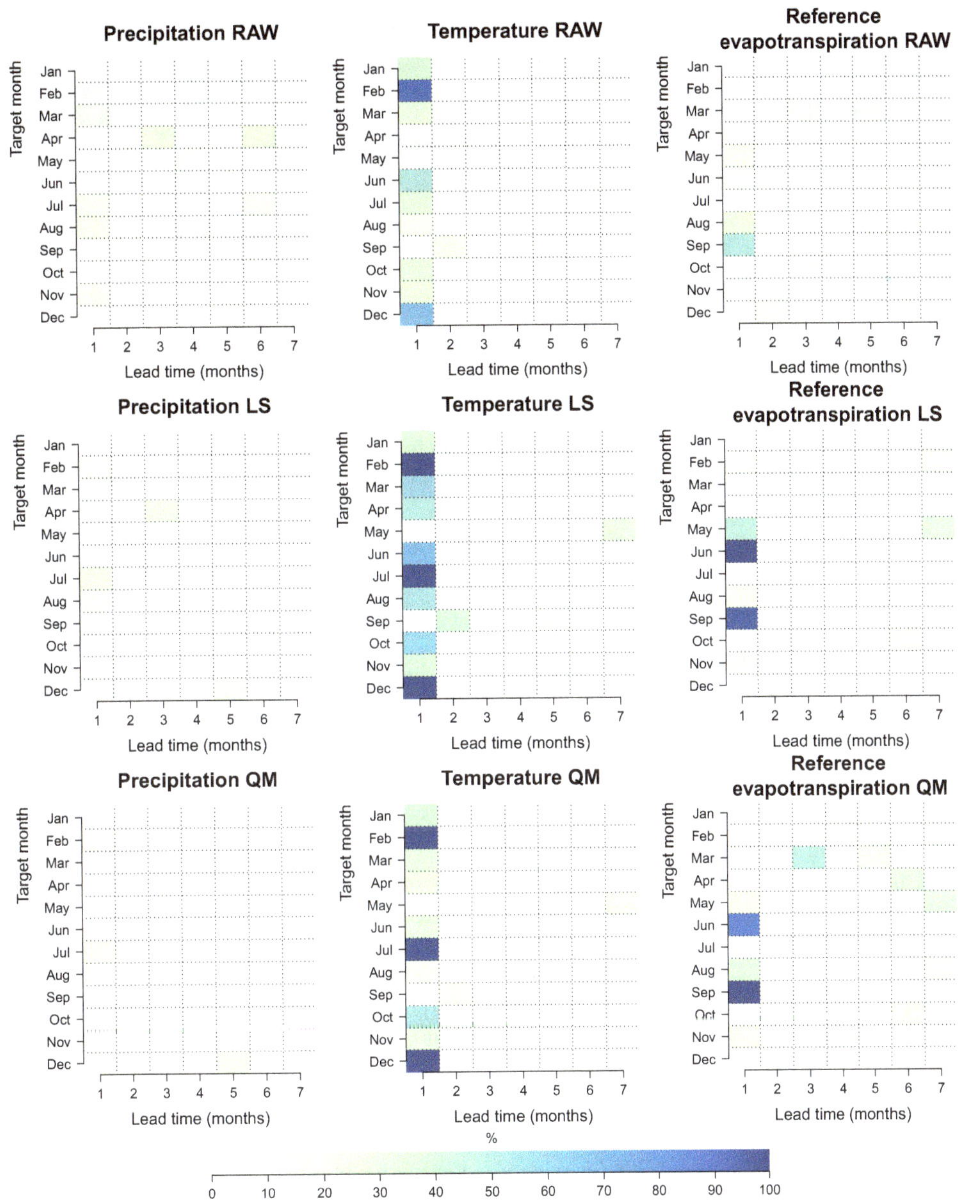

Figure 9. Percentage of grid points with statistically significant positive CRPSS (see Eq. 8).

3.2.2 Skill

In order to quantify the improvement over the raw forecast, we compared the number of grid points for which the skill score was positive and negative. Furthermore, the scores are only considered positive or negative if the differences in the distribution of the skill between ensemble climatology and ECMWF System 4 forecasts are statistically significant at the 0.05 level using the WMW test. Consequently, we introduced a third category for which there is no statistically significant difference in skill between climatology and ECMWF System 4 forecasts.

Figure 9 shows the percentage of grid cells with a statistically significant positive skill due to accuracy, Eq. (8), for the raw forecasts (first raw) and the post-processed forecasts (LS, second row; QM, third row). All target months and lead times are included. If a post-processing method is successful in increasing the regions with positive skill scores, then the box for that target month and lead time is bluer in comparison to the raw forecasts. For precipitation, there is no obvious increase in skill due to accuracy, except, perhaps, February and July forecasts for the first month lead time. There are, however, instances for which the percentage of positive skill grid points decreases. The most obvious cases are March and November (first month lead time) with a reduction of almost

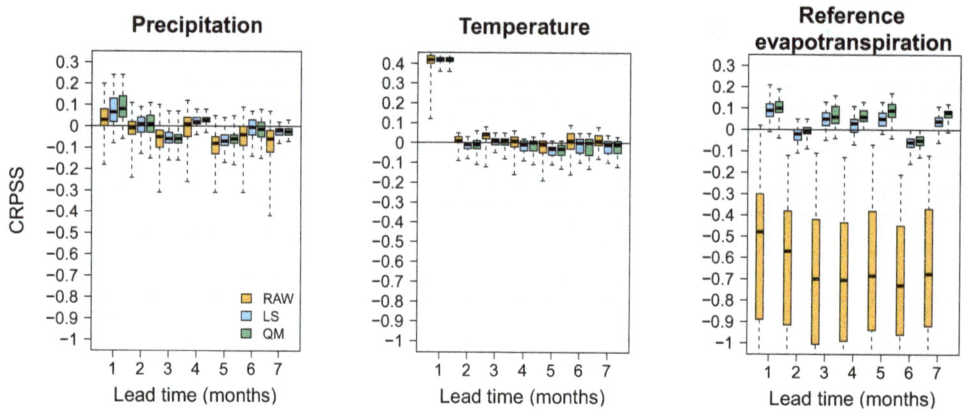

Figure 10. Spatial variability of skill in terms of accuracy for P, T and E_{T0} for the raw and post-processed forecasts of February as a target month at different lead times. Box plots represent the values of CRPSS, see Eq. (8) with climatology as reference, of all the 662 or 724 grid points covering Denmark.

half, i.e., from 13.6 % (raw) to 5.7 % (LS) for March. On the contrary, temperature and E_{T0} exhibit a greater improvement, at least on the first month lead time. For instance, the percentage of grid points with positive skill increases from 4.5 % to 50 % for temperature in April (LS) and from 30 % to 100 % (LS) for temperature in July. The biggest improvement for E_{T0} appears in June (first month lead time), reaching 90 % of positive grid cells after post-processing (LS).

In addition, the negative and equal categories were also plotted and included in the Supplement as Figs. S10 and S11. After post-processing, there are instances where a considerable amount of grid cells change from a statistically significant negative skill score to the third category (no significant differences between ensemble climatology and ECMWF System 4 score distributions, Fig. S11). This is true for temperature and E_{T0} at longer lead times. One of the obvious examples is February E_{T0} at lead time 6 (forecast initiated in September); the percentage of grid points with negative skill scores decreases from 80.5 % to 4.3 % after post-processing (Fig. S10). On the other hand, the percentage of grid points with no significant differences in skill increases from 20 % to 95.7 % after post-processing (Fig. S11) for this example.

To further illustrate the above situation, Fig. 10 shows the variability of the CRPSS when considering all grid points (662 or 724 grid points across Denmark). Figure 10 shows the CRPSS for the target month of February at all lead times and the raw and post-processed skill. The figure shows a reduction of the spatial variability of skill in accuracy and for this month, this reduction is more pronounced for E_{T0}. However, and as mentioned above, the reduction of spatial variability of accuracy is not enough to ensure statistically significant positive differences in skill.

Results for sharpness (Figs. S12 to S14), similar to Fig. 9, show that a loss of sharpness occurs after post-processing in comparison to raw forecasts for LS and QM applied to precipitation and QM applied to temperature and E_{T0}. Sharpness seems to be maintained for temperature and E_{T0} when we use the LS method. This can be explained by the fact that the correction factor applied to temperature forecasts is additive, which in turn changes the level of the ensemble members and has no effect in the spread of the forecasts, leaving the sharpness score equal to that of the raw forecasts. On the other hand, when the correction factor is multiplicative, as in Eq. (1) for precipitation and E_{T0}, not only the level but also the spread is affected. It will increase the spread when the correction factor is above 1 (which indicates an underprediction issue) and, conversely, reduce the spread when the correction factor is below 1 (indicating an overprediction issue). The larger the correction factor is, the larger effect it will have in the ensemble spread. For E_{T0}, where biases are in general lower than biases in precipitation, sharpness seems not to be affected. This effect is somewhat artificial and may lead to misleading evaluations of the power LS has in correcting for biases in spread.

3.2.3 Statistical consistency of post-processed monthly aggregated forecasts

The second and third rows in Figs. 7 and S7–S9 show the PIT diagrams of corrected forecasts for one grid point located in western Denmark. In general, the statistical consistency seems to be improved (points closer to the 1 : 1 diagonal in Fig. 7) to the same degree for both post-processing methods. Although, for E_{T0}, this consistency is better enhanced by QM. This fact may be explained by the more evident sharpness loss after correcting forecasts with the QM method (Figs. S11 to S13).

Figure 11. depicts the results of the AD test in the following manner. First, the first, second and third rows represent the results of the raw and post-processed forecasts with LS and QM methods, respectively. Secondly, as for Fig. 9, the x axis represents the lead time and the y axis the tar-

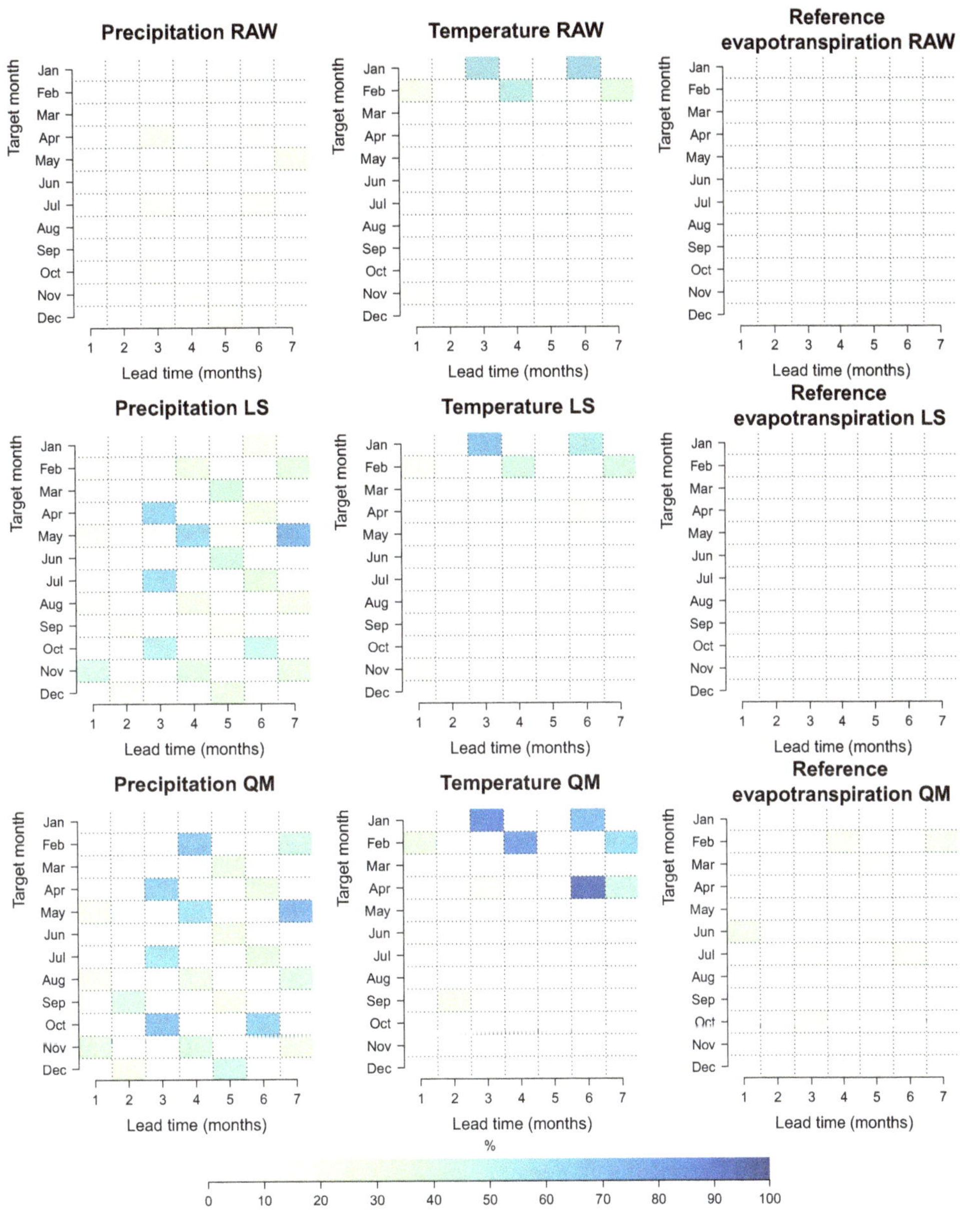

Figure 11. Percentage of grid points for which we fail to reject the null hypothesis of uniformity using the Anderson–Darling test at the 5 % significance level.

get month. Finally, the percentage shows the proportion of grid points for which the null hypothesis of uniformity at the 5 % significance level is accepted. A variety of results are found by the inspection of Fig. 11. First, the percentage of grid points for which the uniformity hypothesis is accepted is very low for raw forecasts. This conclusion holds except for temperature in January and February, with forecasts initialized in months with 51 ensemble members. Secondly, the percentage increases after post-processing for selected months and lead times, usually involving target months with 51 ensemble members. This increase is more visible in post-processed forecasts using the QM method. Finally, the statistical consistency of E_{T0} appears to remain low even after post-processing.

3.2.4 Accuracy of extreme precipitation and number of dry days

Figure 12 shows the skill in terms of accuracy for both monthly maximum precipitation and the monthly number of dry days. Box plots represent the range of the skill score of all 662 grid cells. Skill scores are for January forecasts for all seven lead times. Two features are highlighted. First, spatial variability of skill gets reduced after post-processing. Secondly, for the skill of the number of dry days, results show

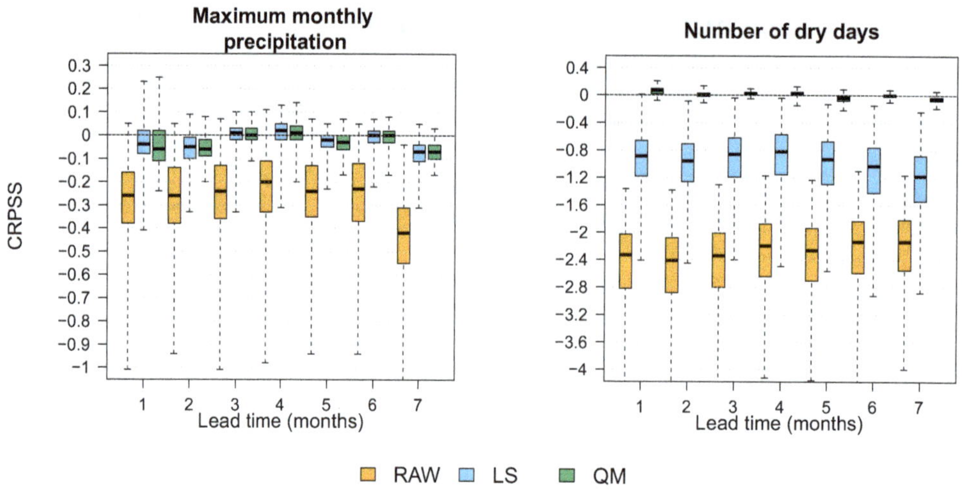

Figure 12. Skill of daily monthly maximum P and number of dry days for target month January for all 7-month lead times for the raw and post-processed forecasts. Box plots represent the values of CRPSS, of Eq. (8) with climatology as reference, of all the 662 or 724 grid points covering Denmark.

that QM performs significantly better than LS. This is not surprising as QM adjusts for biases in the whole range of percentiles of the distributions, whereas LS only focuses on the mean. Note that the apparent increase in skill after LS post-processing is a consequence of the drizzle effect removal before bias correction. Despite the reduction of the spatial variability and an increase, on average, in the skill of post-processed monthly maximum precipitation and number of dry days, results still fail to show an improvement over climatology, as the CRPSS is still negative, even after bias corrections are implemented.

4 Summary and conclusions

The present study had two objectives. The first was to analyze the bias and skill of the ECMWF System 4 in comparison to a climatological ensemble forecast, i.e., a forecast based on observed climatology over a period of 24 years. The second was to compare the statistical consistency between the predictive distribution and the distribution of its verifying observations. This analysis was done for hydrologically relevant variables: precipitation, temperature and E_{T0}. The conclusions of the first objective of the study and which answer the first question posed in Sect. 1 can be summarized as follows:

- Raw seasonal forecasts of precipitation, temperature and E_{T0} from ECMWF System 4 exhibit biases which depend on the target month and, to a lesser extent, on lead time. This result is also in accordance to what was found in Crochemore et al. (2016). There is a persistent over-forecasting issue for E_{T0}, which can be the result of a combination of biases originating in of both temperature and incoming shortwave solar radiation.
- In general, skill in terms of accuracy is only present during the first month lead time, which is basically the skill of the medium-range forecast. Crochemore et al. (2016) found a similar degree of skill of the raw ECMWF System 4 forecasts for mean areal precipitation in France.
- One seeming advantage ECMWF System 4 has over ensemble climatology is that forecasts are sharper. However, this overconfidence, combined with the biases in the mean, leads to lower levels of accuracy in comparison to the accuracy of the ensemble climatological forecasts. Using the PIT diagrams, we could confirm the results for the bias on the mean of precipitation, temperature and E_{T0}.

The second objective was to improve the forecasts using two relatively simple methods of post-processing: LS and QM. This was done having in mind the biases GCMs have in the mean and dispersion. Modest improvements were found and can be summarized as follows:

- Both methods perform equally well in removing biases in the mean.
- In terms of accuracy, mild improvements are seen on the first month lead time, especially for temperature and E_{T0}, where a higher portion of grid points over Denmark can reach a positive skill. Precipitation and longer lead times are still difficult to improve. This may be explained by the same situation as discussed in Zhao et al. (2017). QM performs better when there exists a linear relationship between ensemble mean and observations. This linear relationship may be absent at longer lead times reducing the effectiveness of these methods.
- Looking at the spatial distribution of skill in sharpness we see that for precipitation, both methods tend to de-

crease it, with a slight increase in QM over LS. For temperature and E_{T0}, LS seems to be able to keep the sharpness of the raw forecasts. This is not the case for QM, as it manages to make the areas where a slight positive skill is present disappear. Note, however, that sharpness using the LS method is improved when the correction factor is multiplicative and less than 1 (positive bias; i.e., E_{T0}). The opposite holds: sharpness is inflated when the multiplicative correction factor is greater than 1 (negative bias; i.e., precipitation). This has implications for the computation of the CRPS, as its value increases (decreases) for wider (narrower) ensemble forecasts, on top of the penalization for biased predictions. This situation may also explain why in Crochemore et al. (2016) LS has a better improvement in terms of sharpness than QM, at least for precipitation in spring.

– Statistical consistency is improved using QM. Moreover, QM also performs better in correcting biases of low values of precipitation. This is not a surprising result, as QM corrects for biases for the entire percentile range.

We are aware that our research has limitations. The first is that methods applied here were implemented on a grid-to-grid basis. This may fail to maintain spatiotemporal and intervariable dependencies. Spatial correction methods have been suggested such as the ones used by Feddersen and Andersen (2005) and Di Giuseppe et al. (2013). Another suggestion has been to recover these dependencies by adding a final post-processing step such as the methods proposed in Clark et al. (2004) or Schefzik et al. (2013). The second limitation is the exclusion of post-processing methods tailored to ensemble forecasts, which consider the joint distribution of forecasts and observations (Raftery et al., 2005; Zhao et al., 2017). Their inclusion would gain a deeper insight to the comparison presented here by increasing the complexity of the correction methods and the evaluation of their added value in comparison to simpler approaches.

Post-processing for seasonal forecasting is still a subject at its infancy and, although one could argue that advances in seasonal forecasting will make post-processing unnecessary in the future, there are still opportunities for enhancement. GCMs suffer from several issues as discussed here; however, we still encourage their use. They are physically based, sharper than climatological forecasts. We believe that once bias issues are fixed by means of a more realistic representation of coupled and subgrid processes and/or a better integration of observational data using data assimilation (Weisheimer and Palmer, 2014; Doblas-Reyes et al., 2013), GCMs will be able to provide valuable information at longer lead times for sector applications such as water management.

Author contributions. DL conceived, designed and conducted the study and co-wrote the manuscript; HM, JCR, JK and KHJ conceived and designed the study and contributed to the manuscript.

Competing interests. The authors declare that they have no conflict of interest.

Special issue statement. This article is part of the special issue "Sub-seasonal to seasonal hydrological forecasting". It is not associated with a conference.

Acknowledgements. This study was supported by the project HydroCast – Hydrological Forecasting and Data Assimilation, contract no. 0603-00466B (http://hydrocast.dhigroup.com/, last access: 16 December 2018), funded by the Innovation Fund Denmark. Special thanks to Florian Pappenberger for providing the ECMWF System 4 reforecast, to Andy Wood and Pablo Mendoza for hosting the first author at NCAR, and to Louise Crochemore (SMHI) for aiding in the implementation of the unbiased CRPSS. Finally, the authors would like to thank three anonymous reviewers for providing feedback that certainly increased the quality of the work.

Edited by: Maria-Helena Ramos

References

Anderson, T. W. and Darling, D. A.: Asymptotic theory of certain "goodness of fit" criteria based on stochastic processes, The annals of mathematical statistics, 23, 193–212, 1952.

Clark, M., Gangopadhyay, S., Hay, L., Rajagopalan, B., and Wilby, R.: The Schaake Shuffle: A Method for Reconstructing Space–Time Variability in Forecasted Precipitation and Temperature Fields, J. Hydrometeorol., 5, 243–262, https://doi.org/10.1175/1525-7541(2004)005<0243:TSSAMF>2.0.CO;2, 2004.

Crochemore, L., Ramos, M.-H., and Pappenberger, F.: Bias correcting precipitation forecasts to improve the skill of seasonal streamflow forecasts, Hydrol. Earth Syst. Sci., 20, 3601–3618, https://doi.org/10.5194/hess-20-3601-2016, 2016.

Dessai, S. and Soares, M. B.: European Provision Of Regional Impact Assessment on a Seasonal-to-decadal timescale. Deliverable 12.1 Systematic literature review on the use of seasonal to decadal climate and climate impacts predictions across European sectors, Euporias, 12, 1–26, available at: http://www.euporias.eu/system/files/D12.1_Final.pdf (last access: 1 June 2017), 2013.

Di Giuseppe, F., Molteni, F., and Tompkins, A. M.: A rainfall calibration methodology for impacts modelling based on spatial mapping, Q. J. Roy. Meteor. Soc., 139, 1389–1401, https://doi.org/10.1002/qj.2019, 2013.

Doblas-Reyes, F. J., García-Serrano, J., Lienert, F., Biescas, A. P., and Rodrigues, L. R. L.: Seasonal climate predictability and forecasting: Status and prospects, Wiley Interdiscip. Rev. Clim. Chang., 4, 245–268, https://doi.org/10.1002/wcc.217, 2013.

Feddersen, H. and Andersen, U.: A method for statistical downscaling of seasonal ensemble predictions, Tellus A, 57, 398–408, https://doi.org/10.1111/j.1600-0870.2005.00102.x, 2005.

Ferro, C. A. T., Richardson, D. S., and Weigel, A. P.: On the effect of ensemble size on the discrete and continuous ranked probability scores, Meteorol. Appl., 15, 19–24, https://doi.org/10.1002/met.45, 2008.

Gil, C.: ADGofTest, Anderson-Darling GoF test, R package version 0.3, available at: https://CRAN.R-project.org/package=ADGofTest (last access: 31 July 2018), 2011.

Gneiting, T., Balabdaoui, F., and Raftery, A. E.: Probabilistic Forecasts, Calibration and Sharpness, J. R. Stat. Soc. B, 69, 243–268, 2007.

Hamill, T. M., Whitaker, J. S., and Wei, X.: Ensemble re-forecasting: Improving medium-range forecast skill using retrospective forecasts, B. Am. Meteorol. Soc., 3825–3830, https://doi.org/10.1175/1520-0493(2004)132<1434:ERIMFS>2.0.CO;2, 2004.

Hendriks, M.: Introduction to Physical Hydrology, Oxford University Press, 2010.

Hersbach, H.: Decomposition of the Continuous Ranked Probability Score for Ensemble Prediction Systems, Weather Forecast., 15, 559–570, https://doi.org/10.1175/1520-0434(2000)015<0559:DOTCRP>2.0.CO;2, 2000.

Hollander, M., Wolfe, D. A., and Chicken, E.: Nonparametric statistical methods, 3rd edn., Wiley Series in Probability and Statistics, 2014.

Hudson, D., Marshall, A. G., Yin, Y., Alves, O., and Hendon, H. H.: Improving Intraseasonal Prediction with a New Ensemble Generation Strategy, Mon. Weather Rev., 141, 4429–4449, https://doi.org/10.1175/MWR-D-13-00059.1, 2013.

Kim, H. M., Webster, P. J., Curry, J. A., and Toma, V. E.: Asian summer monsoon prediction in ECMWF System 4 and NCEP CFSv2 retrospective seasonal forecast, Clim. Dynam., 39, 2957–2991, https://doi.org/10.1007/s00382-012-1470-5, 2012a.

Kim, H. M., Webster, P. J., and Curry, J. A.: Seasonal prediction skill of ECMWF System 4 and NCEP CFSv2 retrospective forecast for the Northern Hemisphere Winter, Clim. Dynam., 39, 2957–2973, https://doi.org/10.1007/s00382-012-1364-6, 2012b.

Laio, F. and Tamea, S.: Verification tools for probabilistic forecasts of continuous hydrological variables, Hydrol. Earth Syst. Sci., 11, 1267–1277, https://doi.org/10.5194/hess-11-1267-2007, 2007.

Liu, X., Wu, T., Yang, S., Jie, W., Nie, S., Li, Q., Cheng, Y., and Liang, X.: Performance of the seasonal forecasting of the Asian summer monsoon by BCC_CSM1.1(m), Adv. Atmos. Sci., 32, 1156–1172, https://doi.org/10.1007/s00376-015-4194-8, 2015.

Maclachlan, C., Arribas, A., Peterson, K. A., Maidens, A., Fereday, D., Scaife, A. A., Gordon, M., Vellinga, M., Williams, A., Comer, R. E., Camp, J., Xavier, P., and Madec, G.: Global Seasonal forecast system version 5 (GloSea5): A high-resolution seasonal forecast system, Q. J. Roy. Meteor. Soc., 141, 1072–1084, https://doi.org/10.1002/qj.2396, 2015.

Mason, S. J. and Baddour, O.: Statistical Modelling, in: Seasonal Climate: Forecasting and Managing Risk, Springer, 82, 167–206, 2008.

Molteni, F., Stockdale, T., Balmaseda, M., Balsamo, G., Buizza, R., Ferranti, L., Magnusson, L., Mogensen, K., Palmer, T., and Vitart, F.: The new ECMWF seasonal forecast system (System 4), ECMWF Tech. Memo., 656, 49 pp., available at: https://www.ecmwf.int/sites/default/files/elibrary/2011/11209-new-ecmwf-seasonal-forecast-system-system-4.pdf (last access: 1 June 2017), 2011.

Panofsky, H. W. and Brier, G.W.: Some Applications of Statistics to Meteorology, The Pennsylvania State University Press, Philadelphia, US, 1968.

Peng, Z., Wang, Q. J., Bennett, J. C., Schepen, A., Pappenberger, F., Pokhrel, P., and Wang, Z.:Statistical Calibration and Bridging of ECMWF System4 outputs for forecasting seasonal precipitation over China, J. Geophys. Res.-Atmos., 119, 7116–7135, https://doi.org/10.1002/2013JD021162, 2014.

Raftery, A. E., Gneiting, T., Balabdaoui, F., and Polakowski, M.: Using Bayesian Model Averaging to Calibrate Forecast Ensembles, Mon. Weather Rev., 133, 1155–1174, https://doi.org/10.1175/MWR2906.1, 2005.

R Core Team R: A language and environment for statistical computing. R Foundation for Statistical Computing, Vienna, Austria, available at: https://www.R-project.org/ (last access: 31 July 2018), 2017.

Saha, S., Moorthi, S., Wu, X., Wang, J., Nadiga, S., Tripp, P., Behringer, D., Hou, Y.-T., Chuang, H., Iredell, M., Ek, M., Meng, J., Yang, R., Mendez, M. P., van den Dool, H., Zhang, Q., Wang, W., Chen, M., and Becker, E.: The NCEP Climate Forecast System Version 2, J. Climate, 27, 2185–2208, https://doi.org/10.1175/JCLI-D-12-00823.1, 2013.

Scharling, M. and Kern-Hansen, C.: Climate Grid Denmark - Dateset for use in research and education, DMI Tech. Rep., 10, 1–12, available at: https://www.dmi.dk/vejr/arkiver/vejrarkiv/ (last access: 16 December 2018), 2012.

Schefzik, R., Thorarinsdottir, T. L., and Gneiting, T.: Uncertainty Quantification in Complex Simulation Models Using Ensemble Copula Coupling, Stat. Sci., 28, 616–640, https://doi.org/10.1214/13-STS443, 2013.

Shepard, D. S.: A two dimensional interpolation function for irregularity spaced data. Proceedings of the 23rd Associations for Computing Machinery Conference, ACM, 517–524, 1968.

Siegert, S.: SpecsVerification: Forecast verification routines for the SPECS FP7 project, R package version 0.4-1, available at: https://cran.r-project.org/web/packages/SpecsVerification/index.html, (last access: 12 June 2017), 2015.

Stisen, S., Sonnenborg, T. O., Højberg, A. L., Troldborg, L., and Refsgaard, J. C.: Evaluation of Climate Input Biases and Water Balance Issues Using a Coupled Surface–Subsurface Model, Vadose Zone J., 10, 37–53, https://doi.org/10.2136/vzj2010.0001, 2011.

Stisen, S., Højberg, A. L., Troldborg, L., Refsgaard, J. C., Christensen, B. S. B., Olsen, M., and Henriksen, H. J.: On the importance of appropriate precipitation gauge catch correction for hydrological modelling at mid to high latitudes, Hydrol. Earth Syst. Sci., 16, 4157–4176, https://doi.org/10.5194/hess-16-4157-2012, 2012.

Teutschbein, C. and Seibert, J.: Bias correction of regional climate model simulations for hydrological climate-change impact studies: Review and evaluation of different methods, J. Hydrol., 456–457, 12–29, https://doi.org/10.1016/j.jhydrol.2012.05.052, 2012.

Thyer, M., Renard, B., Kavetski, D., Kuczera, G., Franks, S. W., and Srikanthan, S.: Critical evaluation of parameter consistency and predictive uncertainty in hydrological modeling: A case study using Bayesian total error analysis, Water Resour. Res., 45, W00B14, https://doi.org/10.1029/2008WR006825, 2009.

Tolstykh, M. A., Diansky, N. A., Gusev, A. V., and Kiktev, D. B.: Simulation of Seasonal Anomalies of Atmospheric Circulation Using Coupled Atmosphere–Ocean Model, Izv, Atmos. Ocean. Phys., 50, 131–142, https://doi.org/10.7868/S0002351514020126, 2014.

Trambauer, P., Werner, M., Winsemius, H. C., Maskey, S., Dutra, E., and Uhlenbrook, S.: Hydrological drought forecasting and skill assessment for the Limpopo River basin, southern Africa, Hydrol. Earth Syst. Sci., 19, 1695–1711, https://doi.org/10.5194/hess-19-1695-2015, 2015.

van Oldenborgh, G. J., Balmaseda, M. A., Ferranti, L., Stockdale, T. N., and Anderson, D. L. T.: Evaluation of atmospheric fields from the ECMWF seasonal forecasts over a 15-year period, J. Climate, 18, 3250–3269, https://doi.org/10.1175/JCLI3421.1, 2005.

Weisheimer, A. and Palmer, T. N.: On the reliability of seasonal climate forecasts, J. R. Soc. Interface, 11, 20131162, https://doi.org/10.1098/rsif.2013.1162, 2014.

Weisheimer, A., Doblas-Reyes, F. J., Jung, T., and Palmer, T. N.: On the predictability of the extreme summer 2003 over Europe, Geophys. Res. Lett., 38, 1–5, https://doi.org/10.1029/2010GL046455, 2011.

Wetterhall, F., Winsemius, H. C., Dutra, E., Werner, M., and Pappenberger, E.: Seasonal predictions of agro-meteorological drought indicators for the Limpopo basin, Hydrol. Earth Syst. Sci., 19, 2577–2586, https://doi.org/10.5194/hess-19-2577-2015, 2015.

Wilks, D. S.: Statistical methods in the atmospheric sciences, 3rd edn., Elsevier, 2011.

Wood, A. W., Maurer, E. P., Kumar, A., and Lettenmaier, D.: Long-range experimental hydrologic forecasting for the eastern United States, J. Geophys. Res., 107, 4429, https://doi.org/10.1029/2001JD000659, 2002.

Zhao, T., Bennett, J., Wang, Q. J., Schepen, A., Wood, A., Robertson, D., and Ramos, M.-H.: How suitable is quantile mapping for post-processing GCM precipitation forecasts?, J. Climate, 30, 3185–3196, https://doi.org/10.1175/JCLI-D-16-0652.1, 2017.

13

Historical drought patterns over Canada and their teleconnections with large-scale climate signals

Zilefac Elvis Asong[1], **Howard Simon Wheater**[1], **Barrie Bonsal**[2], **Saman Razavi**[1], **and Sopan Kurkute**[1]

[1]Global Institute for Water Security and School of Environment and Sustainability, University of Saskatchewan, 11 Innovation Blvd, Saskatoon, SK, Canada S7N 3H5
[2]Environment and Climate Change Canada, 11 Innovation Blvd, Saskatoon, SK, Canada S7N 3H5

Correspondence: Zilefac Elvis Asong (elvis.asong@usask.ca)

Abstract. Drought is a recurring extreme climate event and among the most costly natural disasters in the world. This is particularly true over Canada, where drought is both a frequent and damaging phenomenon with impacts on regional water resources, agriculture, industry, aquatic ecosystems, and health. However, nationwide drought assessments are currently lacking and impacted by limited ground-based observations. This study provides a comprehensive analysis of historical droughts over the whole of Canada, including the role of large-scale teleconnections. Drought events are characterized by the Standardized Precipitation Evapotranspiration Index (SPEI) over various temporal scales (1, 3, 6, and 12 consecutive months, 6 months from April to September, and 12 months from October to September) applied to different gridded monthly data sets for the period 1950–2013. The Mann–Kendall test, rotated empirical orthogonal function, continuous wavelet transform, and wavelet coherence analyses are used, respectively, to investigate the trend, spatio-temporal patterns, periodicity, and teleconnectivity of drought events. Results indicate that southern (northern) parts of the country experienced significant trends towards drier (wetter) conditions although substantial variability exists. Two spatially well-defined regions with different temporal evolution of droughts were identified – the Canadian Prairies and northern central Canada. The analyses also revealed the presence of a dominant periodicity of between 8 and 32 months in the Prairie region and between 8 and 40 months in the northern central region. These cycles of low-frequency variability are found to be associated principally with the Pacific–North American (PNA) and Multivariate El Niño/Southern Oscillation Index (MEI) relative to other considered large-scale climate indices. This study is the first of its kind to identify dominant periodicities in drought variability over the whole of Canada in terms of when the drought events occur, their duration, and how often they occur.

1 Introduction

Drought is a naturally occurring environmental phenomenon and a major natural hazard that can have devastating impacts on regional water resources, agriculture, industry, and other social-ecological systems, with far-reaching impacts in an increasingly globalized and uncertain world (IPCC, 2013; Sternberg, 2011). Although still among the least understood extreme weather events affecting larger areas, droughts have proved to be the costliest and most widespread of natural disasters (Bryant, 2005; Wilhite, 2000b). This is primarily due to their usually lengthy duration, severity, and large spatial extent, sometimes reaching continental scales and lasting for many years (Sheffield et al., 2009). Generally, droughts can affect all components of the hydrological cycle, from its origin as a deficit in precipitation (P; Dai, 2011; Palmer, 1965; McKee et al., 1993), to its combination with high evapotranspiration losses that can lead to a deficit in soil moisture and subsequent manifestation into a hydrological drought (Tallaksen and Stahl, 2014).

A review of key drought concepts (e.g. classification and indices) and the relation between droughts and large-scale climate indices has been carried out by Mishra and Singh (2010). However, due to the wide variety of sectors affected by droughts, their diverse spatial and temporal structures, the interdependence across climatic, hydrologic, geomorphic, ecological, and societal variables, and the demand placed on water supply by different users, there is no universal definition of droughts and associated impacts. The most used drought classification is that initially proposed by Dracup et al. (1980) and later integrated by Wilhite and Glantz (1985), and Wilhite (2000a). Based on the degree of water deficit, droughts are often classified into three types including (1) meteorological, (2) agricultural, and (3) hydrological. Further details on drought classification and definitions are found in Mishra and Singh (2010), Dai (2011), and Van Loon et al. (2016).

Studies on regional drought characteristics are important and should be incorporated in water resource management efforts (Mishra and Singh, 2011; Wheater and Gober, 2013). Of particular interest is the analysis of drought occurrence over Canada, a country in which drought is among the most costly natural hazards, particularly in the interior Prairie region; see e.g. Bonsal et al. (2011a). During the period 1950–2010, nationwide annual mean surface air temperature, T, increased by 1.5 °C (Vincent et al., 2012). Being the second largest country in the world and with a large continental interior, this rapid warming has been accompanied by significant changes in many other hydroclimatic elements in different parts of the country, including increases in P (Mekis and Vincent, 2011), decreases in the duration of snow cover (Brown and Braaten, 1998), and decreases in annual streamflow (Zhang et al., 2001). Climate projections also indicate that many regions of Canada will likely experience increasing drought risk by the end of the 21st century (Masud et al., 2017; Bonsal et al., 2013; Dibike et al., 2017).

Historically, most areas of Canada have experienced periodic droughts with different durations, severities, and marked spatial extent, but the agricultural belt of the Canadian Prairies has tended to be highly susceptible to droughts due in part to its location in the lee of the Rocky Mountains and its strong dependence on rain-fed agriculture (Shabbar and Skinner, 2004). In particular, devastating drought events over western Canada during the 1890s, 1910s, 1930s, 1960s, 1980s, 1999–2005, and most recently in 2015 have been identified by using a variety of drought indicators at various scales (Bonsal et al., 2011b; Bonsal and Regier, 2007; Szeto et al., 2016). The 1961 drought (the worst single year drought on the Prairies, with about 50 % of normal growing season precipitation) led to a total net farm income drop of 48 % (USD 300 million) compared with the previous year (Bonsal et al., 1999). The drought of 1988 had many impacts on the agricultural sectors of Canada, including wind erosion, livestock, incomes, farm management, crop production, and prices. In addition, the sparse snow cover and high spring Ts resulted in little or no spring runoff from Prairie watersheds in 1988, such that the mean runoff volume was 60 to 70 % of the normal amount (Wheaton et al., 1992). Furthermore, the 1999–2005 drought which was at its most severe between 2001 and 2002 was felt across Canada but was concentrated in the Prairies, and cost the regional economy an estimated USD 3.6 billion in lost agricultural output (Council of Canadian Academies, 2013).

The uncertainty of drought characterization in Canada in an era of changing climate and increasing pressure from competing water users poses a major challenge to sustainable water management. A better understanding of the spatial distribution of drought, and its frequency, intensity, and duration is thus required. Increased knowledge of these drought characteristics and their relationship to large-scale ocean–atmosphere forcing is necessary for predicting seasonal drought severity, as well as for planning for impacts due to future climate change. Previous studies have documented significant links between low-frequency internal climate variability and the Canadian hydroclimate. For example, positive phases of the Pacific Decadal Oscillation (PDO) and El Niño–Southern Oscillation (ENSO) have been associated with warm winter T in western and central Canada (Bonsal et al., 2001; Shabbar and Yu, 2012; Shabbar and Khandekar, 1996) and a reduction of snow cover in western Canada (Brown and Braaten, 1998). A review of the association of large-scale variability and low streamflows over Canada was conducted by Bonsal and Shabbar (2008). They found a higher frequency of low-flow events to coincide with warmer/drier conditions during El Niño events and positive phases of the PDO and the Pacific–North American (PNA) pattern. Nazemi et al. (2017) investigated the major drivers of annual streamflow variability in the headwaters of the Canadian Prairies during the 20th century and found the PDO to significantly determine monotonic trends and shifts in the central tendency of annual mean streamflow.

Fleming and Quilty (2006) quantified the effects of organized modes of climate variability upon groundwater resources, by examining the influence of ENSO on water levels in shallow aquifers in British Columbia. They found water levels to be above average during La Niña years and below average during El Niño years, an indication of variability in winter and spring P that recharges the aquifer systems. Similarly, Tremblay et al. (2011) analyzed the variability of groundwater systems for three different regions across Canada and their linkage to the North Atlantic Oscillation (NAO), the Arctic Oscillation (AO), the PNA, and ENSO. Their findings indicated that groundwater variability in the Prince Edward Island region is mostly modulated by the NAO and AO, in Manitoba it is influenced by the PNA, while for Vancouver Island NAO, AO, and ENSO showed the highest influence. Perez-Valdivia et al. (2012) found variability in groundwater levels in the Canadian Prairies in the 2–7-year and 7–10-year bands to be highly influenced by ENSO while oscillation modes in the 18–22-year band reflected a negative correlation with the PDO index.

The development of a comprehensive drought monitoring system capable of providing early warning of a drought's onset, severity, persistence, and spatial extent in a timely man-

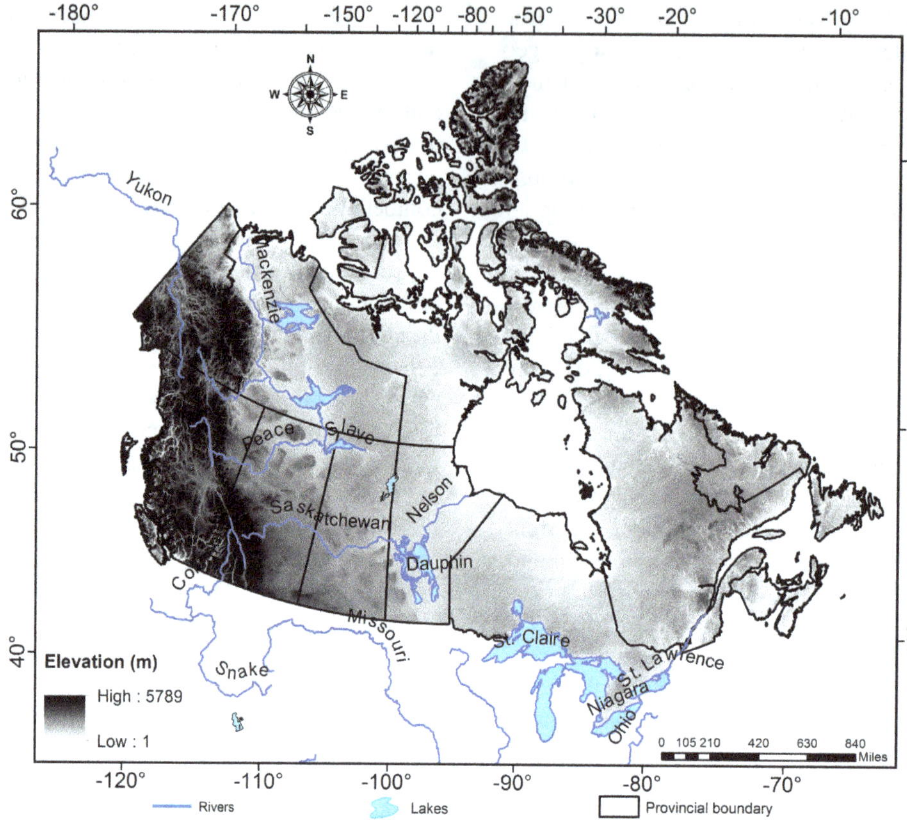

Figure 1. Study area showing topographic and hydrographic features of Canada.

ner is a critical component in establishing a national drought policy or strategy. However, such nationwide drought assessments in Canada are hampered partly by observational uncertainties. The paucity and heterogeneous distribution of P and T estimates are an important limitation of drought characterization. Ground-based measurements (e.g. gauges) are limited, especially over the Rocky Mountains and north of the 60th parallel, and suffer from inaccuracies associated with cold climate processes (Wang and Lin, 2015; Wong et al., 2016; Asong et al., 2017; Asong et al., 2016). For this purpose, it is worthwhile to study the long-term time series of P and T regarding their non-homogeneous climatic and hydrological properties. It is also important to identify homogeneous regions within Canada with distinct drought features for improved drought risk assessment and for a more efficient water resource management at the regional level. So far, most studies on droughts have been limited to Canada south of 60° N (Masud et al., 2015; Bonsal et al., 2013; Dibike et al., 2017). Nevertheless, we have not come across studies that attempt to establish the link between nationwide drought characteristics (e.g. spatial structure, temporal patterns, periodicities) and the large-scale ocean–atmospheric modes of variability in a comprehensive manner.

This study aims to fill these gaps by providing a comprehensive analysis of historical droughts over the whole of Canada. Drought events are characterized by the Standardized Precipitation Evapotranspiration Index (SPEI; Vicente-Serrano et al., 2010) over various temporal scales (1, 3, 6, and 12 consecutive months, 6 months from April to September, and 12 months from October to September). First, trends in the SPEI are investigated by means of the Mann–Kendall test. Major patterns of long-term change and periodicity of drought events are then characterized using the rotated empirical orthogonal function and continuous wavelet transform techniques, respectively. In addition, potential key drivers of drought are investigated using wavelet coherence analysis, with a special emphasis on the role played by large-scale modes of climate variability. Finally, due to the uncertainty associated with climate variables, especially in the northern and mountainous regions where ground-based measurements are inevitably limited (Zhang et al., 2000), this study utilizes and compares two common Canada-wide gridded data sets (monthly P and T) for the period 1950–2013.

The paper is organized as follows. Section 2 provides a description of the data and analysis methods. Section 3 discusses the detailed characteristics of drought over different regions of Canada by applying the different aforementioned statistical analyses to the drought index. This section also discusses the physical and dynamical mechanisms driving the observed dry episodes in the country. Finally, a summary and conclusions are given in Sect. 4.

2 Materials and methods

2.1 Study area

The study area comprises the entire Canadian landmass (Fig. 1). The region includes several major river systems including the Great Lakes–St. Lawrence River system, which is one of the largest freshwater resources globally. Topography plays an important role in shaping regional climates, ranging from wet maritime on the coasts to dry continental across the Prairies and Boreal Plain. Snowfall is restricted to winter months (approximately October to April depending on region). The occurrence, intensity, and timing of seasonal P greatly influence the functioning of ecosystems in various terrestrial ecozones in this region (Hogg et al., 2000). Based on the period 1950–2013, mean annual P varied from more than 2460 mm on the west and 1260 mm on the east coastal regions to less than 360 mm in the interior Prairie (southern central) and northern (above 60° N) regions (Fig. 2). The long-term monthly (January–December) minimum and maximum Ts ranged from −30 to +15 °C (see Sect. 2.2 below for data sources). Characterized by a highly variable hydroclimate and diminishing water resources, southern parts of Canada are home to cities with the highest population densities and support a vibrant agro-based economy that was hard-hit by the most severe and prolonged droughts of 1988, 1999–2005, and 2015, as well as severe floods of 2011, 2013, and 2014 (Wheater and Gober, 2015; Pomeroy et al., 2016).

2.2 Data sources

2.2.1 Gridded observations – ANUSPLIN

The Australian National University Spline (ANUSPLIN) implementation of trivariate thin plate smoothing splines (Hutchinson et al., 2009) has been used to provide gridded climate data over continental Canada, available on a 0.0833 grid spacing (∼ 10 km). Variables include monthly minimum T (Tmin), maximum T (Tmax), and P amounts. Station data from Environment and Climate Change Canada observing sites were interpolated onto the high-resolution grid using the ANUSPLIN smoothing splines, with longitude, latitude, and elevation as interpolation predictors (McKenney et al., 2011). Prior to interpolation, observed station data (Fig. S1 in the Supplement) were quality-controlled and corrected for station relocation, changes in the definition of the climate day, and trace P amounts. Hopkinson et al. (2011) showed that annual mean absolute interpolation errors in ANUSPLIN were limited to 1.0 °C for Tmax, 1.3 °C for Tmin, and about 9 % for annual P over southern Canada.

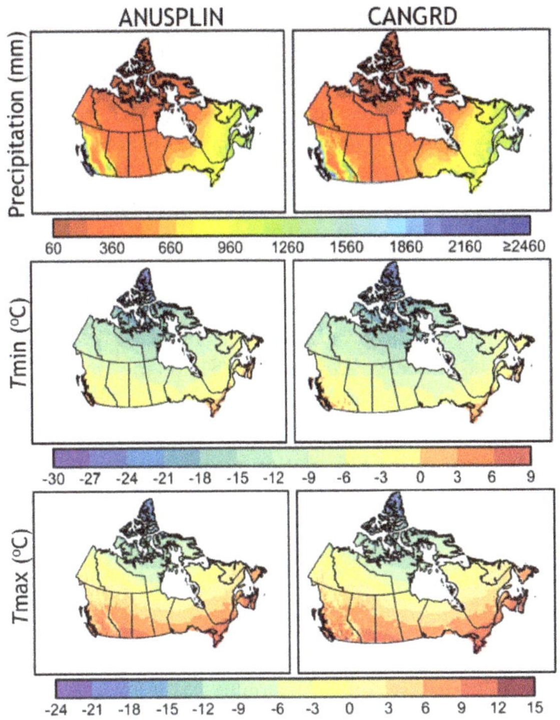

Figure 2. Mean annual climatology of precipitation, minimum temperature (Tmin), and maximum temperature (Tmax) for the period 1950–2013.

2.2.2 Gridded observations – CANGRD

The Canadian gridded (CANGRD) data originate from the second generation of Adjusted and Homogenized Canadian Climate Data – Daily Temperature and Precipitation (http://open.canada.ca/data/en/dataset/d6813de6-b20a-46cc-8990-01862ae15c5f, last access: 25 April 2018), with over 330 locations (Fig. S2) for T and 460 for total P (note that these numbers are not constant over time). These data have been quality-controlled and adjusted to account for known changes in measurement practices. In particular, records from stations separated by less than 10 km were merged so that correlations between stations would be small. See Mekis and Vincent (2011) for a detailed discussion on merging techniques and trends in the mean climatologies of these data. For CANGRD, these data were interpolated to evenly spaced (50 km) grids using Gandin's optimal interpolation (Gandin, 1966) technique. As in the case of ANUSPLIN, monthly P, Tmax, and Tmin values were extracted from 1950–2013 and used in the analyses.

2.2.3 Teleconnection indices

To analyze the key drivers of drought events over Canada, six large-scale climate anomalies that have been linked to hydroclimatic variability over Canada (Shabbar and Khandekar, 1996; Asong et al., 2015; Zhao et al., 2013; Perez-

Valdivia et al., 2012; Bonsal and Shabbar, 2008; Fleming and Quilty, 2006; Nazemi et al., 2017) and/or North America (Ropelewski and Halpert, 1986) are investigated. They include the Pacific Decadal Oscillation (PDO; Mantua and Hare, 2002), North Atlantic Oscillation (NAO; Hurrell and Van Loon, 1997), Pacific–North American (PNA; Barnston and Livezey, 1987), Arctic Oscillation (AO; Zhou et al., 2001), Atlantic Multidecadal Oscillation (AMO; Enfield et al., 2001), and Multivariate El Niño/Southern Oscillation Index (MEI; Wolter, 1987; Wolter and Timlin, 2011). Monthly values of all indices are sourced from https://www.esrl.noaa.gov/psd/ (last access: 25 April 2018) for the period 1950–2013.

2.3 Drought index calculation and drought identification

Many quantitative metrics have been developed and used for identification and monitoring of droughts (Mishra and Singh, 2011; Raible et al., 2017). A variety of these indices measure, in most cases, how much P and T for a given period deviate from historical averages. Examples include the Palmer drought severity index (PDSI; Palmer, 1965), Palmer hydrological drought index (Palmer, 1965), self-calibrated Palmer drought severity index (Wells et al., 2004), standardized precipitation index (SPI; McKee et al., 1993), the SPEI (Vicente-Serrano et al., 2010), and multivariate standardized drought index (Hao and AghaKouchak, 2013). In Canada, most drought analyses (Dibike et al., 2016; Masud et al., 2017) have utilized climate-based indices since the T and P variables are readily available for longer periods and span larger areas, compared to hydrologic variables such as soil moisture and streamflow. In this study, we make use of the SPEI as a meteorological proxy for drought quantification.

As detailed in Vicente-Serrano et al. (2010), the SPEI is a multi-scalar drought index based on a water balance approach that uses the monthly difference between P and potential evapotranspiration (PET) to analyze wet/dry spells over multiple timescales. SPEI involves calculating monthly PET and then subtracting this from the corresponding monthly P to obtain the climatic water balance. Several derivations have been put forth for calculating PET, including the widely used Penman (Penman, 1948), Thornthwaite (Thornthwaite, 1948), Priestley–Taylor (Priestley and Taylor, 1972), and Hargreaves (Hargreaves et al., 1985) methods. However, most of these approaches require long records for solar radiation, Ts, wind speed, and air pressure, which are not readily available in Canada and many regions of the world. Reanalysis products are an alternative data set for historical drought analysis in Canada and could lead to robust estimates of PET based on the Penman–Monteith algorithm (Maidment, 1993). However, they have been found to be uncertain compared to observations (Wong et al., 2017). A more suitable product with close performance to observations is the Global Environmental Multiscale (GEM) numerical weather prediction model output (Côté et al., 1998). However, it is limited in record length (2002–present). Therefore, the Hargreaves method (Hargreaves, 1994) which simply uses Tmin and Tmax for estimating PET is employed in this study. Once PET is calculated, the difference between P and PET for the month j is calculated as in Eq. (1):

Table 1. Categories of dryness/wetness degree according to the SPEI values (McKee et al., 1993).

Categories	SPEI classifications
Extremely dry	≤ -2.00
Severely dry	-1.99 to -1.50
Moderately dry	-1.49 to -1.00
Near normal	-0.99 to 0.99
Moderately wet	1.00 to 1.49
Severely wet	1.50 to 1.99
Extremely wet	≥ 2.00

$$Q_{\mathrm{j}} = P_{\mathrm{j}} - \mathrm{PET}_{\mathrm{j}}, \tag{1}$$

where Q_{j}values represent monthly water surplus or deficit.

To compute SPEI, the monthly Q_{j}values are first standardized with respect to the long-term monthly mean values. The 1-month SPEI is generally representative of meteorological drought, while timescales between 3 and 6 months are considered as an agricultural drought index. Longer scales such as 6 and 12 months are used to represent hydrological drought and are useful for monitoring surface water resources (Beguería et al., 2014; Hayes et al., 2011). To ascertain the variability of spatio-temporal patterns for different types of droughts, the SPEI was used at different timescales, namely, at 1 (SPEI1), 3 (SPEI3), 6 (SPEI6), and 12 (SPEI12) consecutive months from January–December, at 6 months during the warm season (April–September, SPEI6$_{\mathrm{Apr-Sep}}$), and at 12 months during the hydrologic year (October–September, SPEI12$_{\mathrm{Oct-Sep}}$). SPEI1, SPEI3, SPEI6, and SPEI12 account for the sub-annual variability of droughts, while SPEI6$_{\mathrm{Apr-Sep}}$ and SPEI12$_{\mathrm{Oct-Sep}}$ account for the inter-annual variability. A drought event occurs any time the SPEI is continuously negative and reaches an intensity of -1.0 or less, and it ends when SPEI becomes positive. Whenever a drought event has been detected with a start and termination month, drought properties such as duration, magnitude, and frequency can be determined. SPEI categories are shown in Table 1. Unless indicated, we focus mostly on other SPEI timescales except SPEI1, which at 1 month is mainly a meteorological drought index.

2.4 Trend analysis

Long-term trends in drought intensity and variability (interannual) are analyzed using the SPEI time series at each grid point. This enables investigation of the percentage of grid points with increasing/decreasing trends during the period 1950–2013 based on the ANUSPLIN- and CANGRD-derived SPEI values. Pre-whitening as described in Yue et al. (2002) is first applied to the monthly SPEI anomalies to remove lag-1 autocorrelation, since serial correlation is generally recognized to influence trends in auto-correlated time series, which may distort the power of the Mann–Kendall test. Pre-whitening is limited to a low-order autoregressive model, i.e. AR(1), since it can falsify the structure of variability in time series across timescales (particularly with higher order models) (Razavi and Vogel, 2018). The pre-whitened SPEI values on different timescales are then used for detecting statistically significant ($p<0.05$) trends on a pixel basis. Further details on Mann–Kendall trends can also be found in Hamed and Rao (1998).

2.5 Empirical orthogonal function analysis

Hydroclimatological data are often characterized by nonlinearity and high dimensionality. Thus, the challenging task is to find ways to reduce the dimensionality of the system to as few modes as possible by expressing the data in such a way as to highlight their similarities and differences (Hannachi et al., 2007). The empirical orthogonal function (EOF) analysis (also known as principal component analysis, PCA) is among the most widely and extensively used method in hydroclimate sciences for accomplishing such tasks (Richman, 1986; Wilks, 2011; Preisendorfer and Mobley, 1988). EOF is employed to find hydroclimate subregions that experienced similar drought features during the period 1950–2013. The EOF consists of computing the covariance matrix of the SPEI series with the corresponding eigenvalues and eigenvectors (Uvo, 2003).

Following the rule by North et al. (1982), the sampling errors at 95 % confidence level of the eigenvalues associated with the leading components were estimated in order to establish how many modes to retain for rotation. To achieve more stable and physically explainable patterns, a rotation of the retained components with the varimax procedure (Richman, 1986) was applied. The patterns defined in this way are referred to as rotated empirical orthogonal functions (REOFs). In summary, the REOF (spatial patterns) values indicate the spatial representativeness of each rotated temporal component (RPC; temporal patterns). Subsequently, we obtained the most revealing patterns of drought evolution across Canada and determined the spatial extent of each component series by mapping the factorial matrix values (i.e. correlation between each REOF and the original SPEI series). Finally, the time variability of the selected RPCs of SPEI was examined for possible trends in the identified subregions using linear regression. The slopes of the trends were computed by applying the method of least squares linear fitting to the time series.

2.6 Wavelet analysis

After delimitating hydroclimate subregions that experienced similar drought features, the RPC time series corresponding to the REOFs was then analyzed in a time–frequency domain (to reveal dominant oscillations) by means of the continuous wavelet transform (CWT). Subsequently, by utilizing the wavelet coherence (WCO) technique (Grinsted et al., 2004; Torrence and Compo, 1998; Addison, 2016), the relationships between the dominant oscillations and large-scale climate indices that have possibly modulated historical drought patterns across Canada are investigated. The CWT projects the spectral-temporal characteristics of a time series onto a time–frequency plane from which the dominant periodicities and their duration can be distinguished (Fugal, 2009; Grinsted et al., 2004).

The wavelet power spectrum (WPS) is defined as the squared modulus of the CWT (Jiang et al., 2014; Hao et al., 2012), and represents the signal energy at a specific scale (period) and time. From the WPS, the various periodicities and the time intervals of their occurrence can be determined. For a given time series $\{y_n\}$, with $n = 1, 2, 3, \ldots, N$, the CWT is given by

$$W_n(s) = \sum_{n'=1}^{N} \left(\frac{\delta t}{s}\right)^{0.5} y_{n'} \psi^* \left[\frac{(n'-n)\,\delta t}{s}\right], \qquad (2)$$

where $W_n(s)$ are the wavelet coefficients, n is the time index describing the location of the wavelet in time, s is the wavelet scale, and δt is the sampling interval. The function ψ is the mother wavelet, and the asterisk (*) denotes its complex conjugate. The Morlet wavelet was used since it provides a good balance between time and frequency domains and is suitable when the purpose is to extract dominant signals (Grinsted et al., 2004). The statistical significance of the CWT was assessed against a red noise background at a 95 % confidence interval. The CWT function creates a cone of influence (COI) that delimitates a region of the WPS beyond which edge effects become important and the power could be suppressed (Torrence and Compo, 1998).

Wavelet analysis can also be used to identify the covariance between two time series. This can be done using the concept of wavelet coherence (Grinsted et al., 2004). Wavelet coherence (WCO) reveals local similarities between two time series and may be considered a local correlation coefficient in the time–frequency (time period) plane. For climatological time series, WCO can be used to identify their possible teleconnection with large-scale atmospheric drivers. The WCO between two time series can be computed by normalizing and smoothing their cross wavelet spectrum:

$$W_n^{XY}(s) = W_n^X(s)\, W_n^{Y^*}(s), \qquad (3)$$

where $W_n^X(s)$ and $W_n^Y(s)$ represent the WPS of the time series $\{x_i\}$ and $\{y_i\}$, respectively. The statistical significance

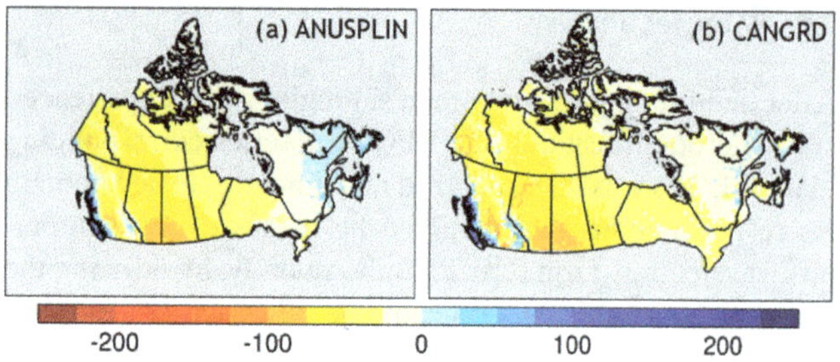

Figure 3. Spatial structure of long-term mean monthly water surplus/deficit (mm), i.e. *P*-PET, derived from ANUSPLIN **(a)** and CANGRD **(b)**.

Table 2. Percentage of grids with decreasing (Dec.) and increasing (Inc.) trends for the SPEI.

	ANUSPLIN		CANGRD	
	Inc.	Dec.	Inc.	Dec.
SPEI1	8.9	14.7	8.6	5.9
SPEI3	4.1	10.6	5.4	3.0
SPEI6	2.8	9.6	3.7	2.3
SPEI12	1.4	10.0	2.9	4.0
$SPEI6_{Apr-Sep}$	0.9	6.2	2.2	4.0
$SPEI12_{Oct-Sep}$	1.2	10.0	2.7	3.1

($p<0.05$) of the WCO is estimated using Monte Carlo methods with respect to a red-noise spectrum, resulting in significant periodicities of coherence delineated by significance contours. As in the CWT, regions outside of the COI should be interpreted with caution (Torrence and Compo, 1998).

3 Results and discussion

3.1 Spatial structure of long-term climatic water balance

Figure 3 shows the average monthly (January–December) water surplus/deficit (mm) as defined in Eq. (1) for ANUSPLIN and CANGRD. It is evident that, apart from the Pacific/Atlantic maritime, most of the continental Canadian interior experienced moisture deficits, with the Prairie being the most affected region (this region also has the highest climatological *T*min and *T*max, and some of the least mean annual *P* during the study period – Fig. 2). Other than the coastal areas, there is a general east to west moisture deficit pattern, mostly determined by the *P* and *T* pattern. For comparison, 5.8 % (33.3 %) of CANGRD points experienced moisture surplus (deficit), and 4.3 % (41.6 %) for ANUSPLIN. This implies that ANUSPLIN showed a relatively higher tendency towards dryness relative to CANGRD.

Table 3. Percentage of variance explained by the first two varimax rotated loadings (REOFs) of the SPEI at various timescales, computed using observations from ANUSPLIN and CANGRD data sets.

	ANUSPLIN		CANGRD	
	REOF1	REOF2	REOF1	REOF2
SPEI1	15.0	13.2	19.5	11.9
SPEI3	15.8	13.8	18.9	12.9
SPEI6	17.9	14.4	19.1	12.7
SPEI12	20.6	13.3	19.1	13.0
$SPEI6_{Apr-Sep}$	19.5	12.6	16.5	11.7
$SPEI12_{Oct-Sep}$	20.5	13.4	19.3	13.1

3.2 Long-term trends

Figure 4 depicts the spatial structure of long-term (1950–2013) SPEI trends at various timescales. Only grid points with significant trends are shown. It is noteworthy that significant trends largely occur in spatially isolated blocks. Decreasing trends are observable in the southern parts of the Prairies, the foothills of the Rocky Mountains, and Pacific maritime regions, whereas increasing trends are limited to a small area in the north and parts of the Atlantic coast to the east, similar to the water balance shown in Fig. 3.

Table 2 shows the percentage of grids with decreasing and increasing trends for ANUSPLIN and CANGRD. For both data sets, the percentage generally decreases with increasing SPEI timescale. ANUSPLIN has a higher tendency towards dryness (decreasing trends) unlike CANGRD. Using SPEI12 as an example, 10 % of ANUSPLIN pixels experienced decreasing trends compared to 4 % in the case of CANGRD. Conversely, apart from SPEI1, CANGRD showed a strong tendency towards wetness (increasing trends) relative to ANUSPLIN. These differences can be attributed partly to the inputs (e.g. the number of stations considered for gridding ANUSPLIN is larger than the number of stations of the CANGRD dataset), estimation methods, and spatial resolution, which are different for both data sets (see Sect. 2.2).

3.3 Spatial patterns of drought

Concerning the EOF analysis, and taking into account the percentage of variance explained by each rotated component (REOF), two main patterns were identified for subsequent discussion and analysis. Table 3 presents the explained variances of varimax rotated components relative to the considered SPEI timescales and data sets. The first two components (for both data sets) explain about 28 to 33.9 % of the total variance depending on the timescale, with the minimum and maximum variances observed for SPEI1 and SPEI12, respectively, in the case of ANUSPLIN. For CANGRD, $SPEI6_{Apr-Sep}$ and SPEI12 explained the minimum and maximum variances, respectively.

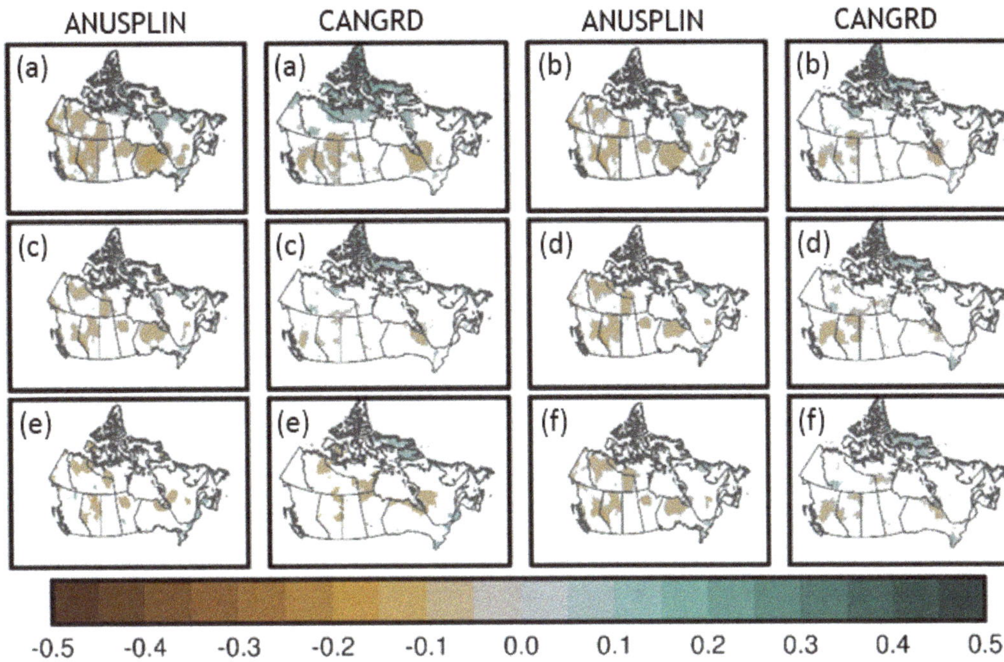

Figure 4. Trends in SPEI at each grid point for ANUSPLIN and CANGRD. Only significant values are shown on the map. Brown indicates decreasing and green is increasing. **(a)** SPEI at 1 month, **(b)** SPEI at 3 months, **(c)** SPEI at 6 months, **(d)** SPEI at 12 months, **(e)** SPEI at 6 months from Apr–Sep, and **(f)** SPEI at 12 months of the water/hydrologic year, i.e. Oct–Sep. Trends are significant at 95 % confidence level.

Figure 5 shows that between the two main components (REOF), the regions with higher correlations (>0.7) do not overlap, with clearly spatially disjunctive structure. For all SPEI timescales, the loading patterns of the first component (REOF1) with a maximum explained temporal variance of 20.6 % (SPEI12, ANUSPLIN) and 19.5 % (SPEI1, CANGRD) highlights mainly the drought evolution on the interior Prairie ecozone of Canada. This semi-arid region is relatively low-lying and characterized by high hydroclimate variability and some of the least annual mean P. The second component (REOF2) explains a maximum variance of around 13.8 % (SPEI3) in the case of ANUSPLIN and 13.1 % ($SPEI12_{Oct-Sep}$) for CANGRD. REOF2 is mainly representative of the northern central part of Canada (with the least annual mean P), including the Taiga Shield, Taiga Plains, Southern Arctic, and Taiga Cordillera ecozones, with positive correlations across all timescales and data sets. In summary, the foregoing analyses suggest that the Prairies and northern central regions are the leading droughts modes for Canada, in that they are well captured by the two data sets, with high positive loadings at all investigated timescales.

3.4 Temporal drought characteristics

The RPCs of REOF1 and REOF2 for ANUSPLIN and CANGRD, as well as correlation coefficients between the data sets, are shown in Fig. 6. It is evident that the RPCs from the two data sets are strongly correlated (cor > 0.75) for all SPEI timescales, except for RPC2 of SPEI3 where cor = 0.45 (the centroids of SPEI3 REOF2 in Fig. 5 are slightly different for both data sets). The primary focus is on $SPEI6_{Apr-Sep}$ and $SPEI12_{Oct-Sep}$, which reflect the agricultural season (when droughts are most critical for rain-fed and irrigated agriculture) and hydrological year in Canada, respectively. Fig. 6 reveals that the Prairies (REOF1 and RPC1) experienced moderate to extreme (Table 1) drought episodes starting in the mid-1950s, early and late 1960s, the 1970s and 1980s, the period from mid-1997–2005, and the summer of 2009. Some of the drought episodes were extremely dry and severe as they were prolonged in time such as the ones in the late 1970s, mid-1980s, and 1997–2005. These identified drought events correspond well with the findings of Bonsal et al. (2011a), who identified large-scale Prairie droughts in 1961, 1967, 1988, and 2001 using ANUSPLIN and CANGRD data. In the northern central region (REOF2 and RPC2), although fewer droughts were detected for the hydrological year ($SPEI12_{Oct-Sep}$), with the most intense occurring in 2000 (based on the ANUSPLIN data set), the seasonal SPEI series ($SPEI6_{Apr-Sep}$) detected several drought events in this region. Extremely dry episodes in this region occurred in the early 1960s, 1980s, early 1990s, and 2000s.

Linear trends of the RPCs for the two subregions are depicted in Fig. 7 and Table 4. For the Prairie subregion, both ANUSPLIN and CANGRD showed insignificant decreasing trends for SPEI6, $SPEI6_{Apr-Sep}$, and $SPEI12_{Oct-Sep}$. However, in the northern central region, the trend direction

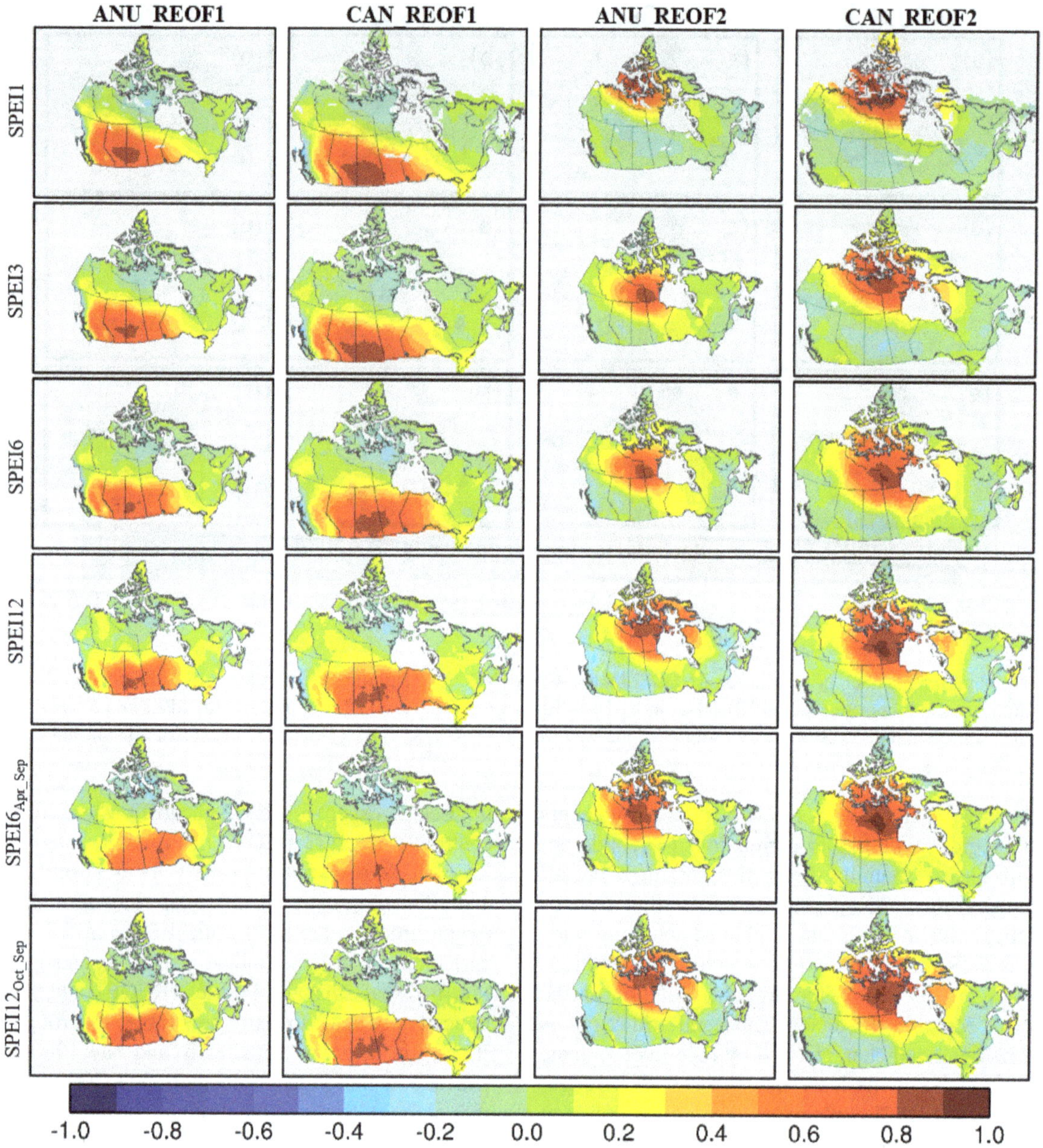

Figure 5. Spatial patterns of the first two REOFs of SPEI at various timescales. The spatial extent of the first two REOFs was characterized by mapping the values of the factorial matrix. See Table 3 for information on variances explained by each REOF pattern.

is not the same for both data sets at all SPEI timescales. For ANUSPLIN (CANGRD), decreasing (increasing) trends are found in $\mathrm{SPEI6}_{\mathrm{Apr-Sep}}$. Conversely, CANGRD (ANUSPLIN) shows decreasing (increasing) trends in SPEI3. Regionally, the SPEI at different timescales tends to display more significant trends in the northern central relative to the Prairies. Most of the trends are significant in the case of CANGRD compared to ANUSPLIN. The apparent differences in trends between the two data sets and subregions may be attributed to the low station density in areas above 60° N which can introduce higher uncertainty in the gridded *P* and *T* fields (Vincent et al., 2015).

3.5 Frequency estimation/periodicity of drought

To detect the dominant frequencies in the RPCs of the SPEI series in Fig. 6, the time series were further analyzed using a CWT. The wavelet power spectrum (WPS) of the RPCs is shown in Fig. 8. Periodicities of drought will be identified for $\mathrm{SPEI6}_{\mathrm{Apr-Sep}}$ and $\mathrm{SPEI12}_{\mathrm{Oct-Sep}}$ and their relationship to teleconnection indices examined. These two timescales correspond to the warm season and water year, respectively, and represent periods when moisture shortages are most critical for various sectors in Canada. Other SPEI timescales which explain sub-annual variability are included in Fig. 8

Table 4. Long-term (1950–2013) trends of monthly time series (RPCs) for the first two REOFs of SPEI at various timescales. The slope year^{-1} (a) and p value (b) is indicated, with significant ($p < 0.05$) trends shown in bold.

	ANUSPLIN				CANGRD			
	REOF1		REOF2		REOF1		REOF2	
	(a)	(b)	(a)	(b)	(a)	(b)	(a)	(b)
SPEI3	1.3×10^{-4}	0.432	1.6×10^{-4}	0.334	1.3×10^{-4}	0.410	-1.0×10^{-4}	0.539
SPEI6	-2.3×10^{-4}	0.157	-1.3×10^{-5}	0.935	-4.5×10^{-5}	0.787	-5.1×10^{-4}	**0.002**
SPEI12	-1.8×10^{-5}	0.909	5.4×10^{-4}	**0.001**	2.6×10^{-5}	0.873	6.4×10^{-4}	$\mathbf{1 \times 10^{-4}}$
$SPEI6_{Apr-Sep}$	-8.5×10^{-5}	0.855	-3.2×10^{-4}	0.492	-1.7×10^{-3}	$\mathbf{1 \times 10^{-4}}$	1×10^{-3}	**0.026**
$SPEI12_{Oct-Sep}$	-1.2×10^{-4}	0.456	-3.7×10^{-4}	**0.029**	-4.6×10^{-5}	0.785	-4.6×10^{-4}	**0.006**

for the interested reader. For $SPEI6_{Apr-Sep}$, from the WPS of RPC1 in Fig. 8, significant inter-annual variability of between 8 and 32 months is evident throughout much of the entire lengths of the SPEI time series. However, it is concentrated most heavily during the mid-1950s, early and late 1960s, the 1970s and 1980s, and the period from mid-1997–2010. For $SPEI12_{Oct-Sep}$, a strong frequency band is centred mostly around 16–66 months (~ 4 years) and is concentrated during the mid-1950s and late 1960s. Moreover, 32–64-month frequencies are dominant in the mid-1980s and the years between 1990 and 2010.

In the northern central region (RPC2), for $SPEI6_{Apr-Sep}$, a significant periodicity of a cycle of between 8 and 40 months as a dominant period of variability is evident. It is concentrated in the early 1960s, 1980s, late 1990s, and 2000s. For $SPEI12_{Oct-Sep}$, the WPS indicates a significant high power for relatively low-frequency (16–100 months, i.e. ~ 7 years) signals, concentrated over the period 1956–1975 and 1995–2013. The foregoing analysis reveals that the Prairie region (REOF1) of Canada is dominated by high-frequency power signals with high cycles of oscillation for both $SPEI6_{Apr-Sep}$ and $SPEI12_{Oct-Sep}$, while the northern central region (REOF2) is dominated by relatively low-frequency power signals and low cycles of oscillation. The analysis further indicates that significant inter-annual periodicities (<10 years) dominate drought variability over the two identified subregions across Canada. The dominant periodicities and the intervals during which they occurred are summarized in Table 5.

3.6 Coherence between drought and large-scale climate drivers

The WCO technique was used to identify both frequency bands and time intervals at which $SPEI6_{Apr-Sep}$ and $SPEI12_{Oct-Sep}$, and large-scale climate indices are co-varying (Torrence and Webster, 1999). The results of WCO coefficients between the RPCs of $SPEI6_{Apr-Sep}$ and $SPEI12_{Oct-Sep}$ and teleconnection indices are shown in Figs. 9–11. In these plots, the coloured shading represents the magnitude in the coherence as shown in the colour bar, which varies from 0 to 1 and indicates the timescale variability in the correlation between the two time series. As in Fig. 8, the black contours represent the significant regions. The relative phase relationships are shown as arrows, with in phase pointing right (positive correlation) and antiphase pointing left (negative correlation), whereas a vertical upward (downward) arrow indicates that the second time series lags (leads) the first in phase by 90°. If an association exists between two time series, a slowly varying phase lag would be expected, and the phenomenon would be phase-locked; i.e. the phase arrows point only in one direction for a given wavelength (Grinsted et al., 2004; Gobena and Gan, 2006). For this study, phase angle associations were noted strictly as either being in-phase-locked or antiphase-locked. To simplify and limit the length of the paper, only results for large-scale climate indices with the strongest correlations with frequencies of drought are shown. Other results are included as supplementary material.

Figure 9 illustrates the squared WCO between the temporal patterns of drought and MEI. It is apparent that, for $SPEI6_{Apr-Sep}$, the MEI had significant coherence with PC1 mainly concentrated in the 8–16-month band and from 1960–1980 in the Prairie region, indicating that the two time series are phase-locked over this time interval. Also, the strongest coherence between PC1 and MEI occurred over the 32–50-month scale, spanning the period 1986–2002. In the case of $SPEI6_{Apr-Sep}$ PC2 (northern central region), discontinuous in-phase coherence patterns were detected in the 2–32-month bands. The dominant high energy coherence occurred in the 16–32-month scale, spanning the period 1985–2005. Similarly, for the hydrological year timescale ($SPEI12_{Oct-Sep}$), drought in the Prairies ($SPEI12_{Oct-Sep}$PC1) experienced significant high power with the MEI at the 16–32-month scale from 1980 to 2005 based on ANUSPLIN (1986–2005 based on CANGRD), whereas in the northern central region ($SPEI12_{Oct-Sep}$PC2), significant in-phase cross power between drought and the MEI occurred in the 32–64-month period over the years 1978–2000 (in the case of ANUSPLIN). As in Fig. 8, it can be seen that the Prairie region has more short-term periodicities compared to the northern central region. It is also evident that although the MEI co-varied with

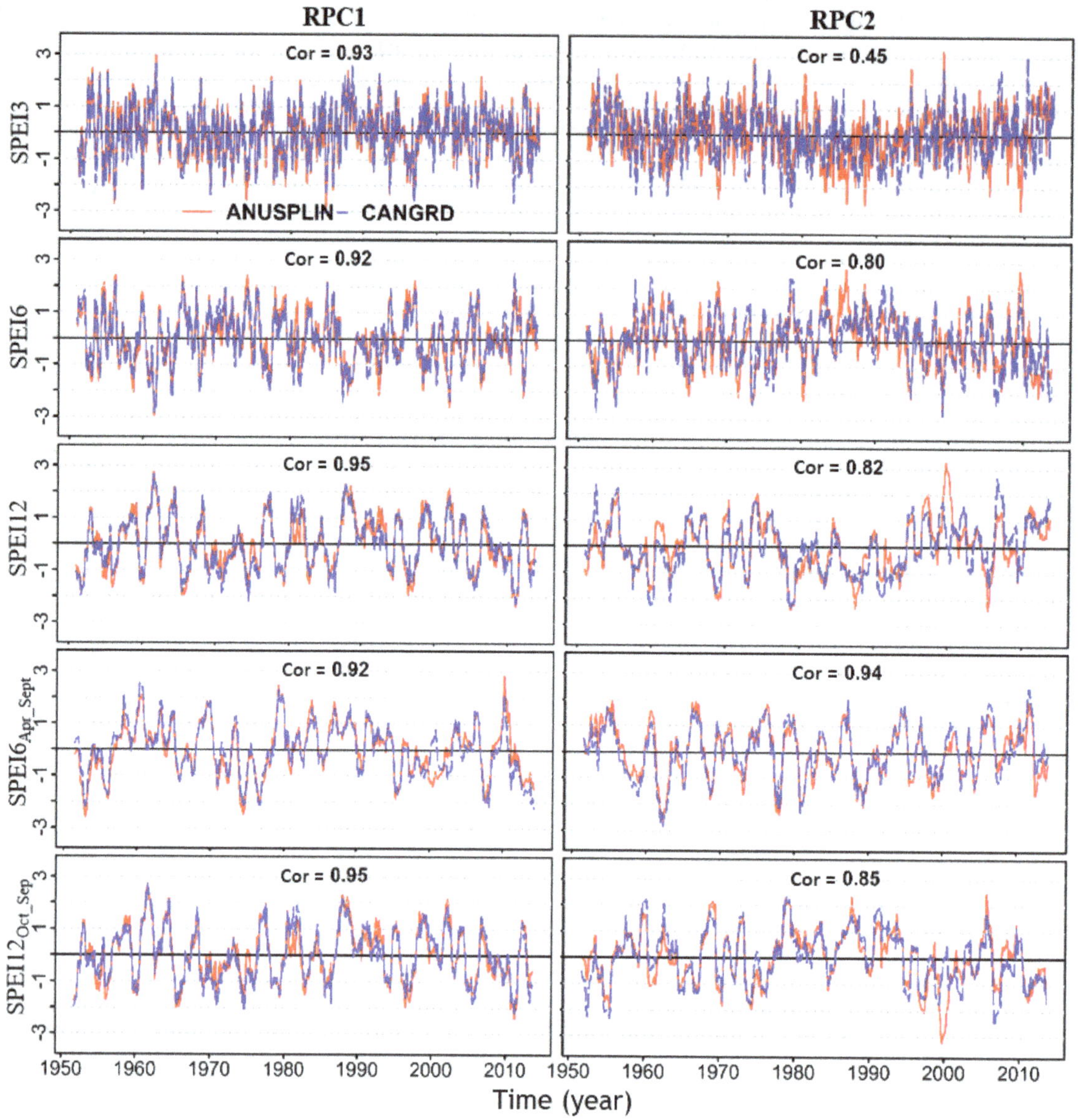

Figure 6. Temporal patterns (RPCs) of the first two rotated principal components (REOFs) of SPEI at various timescales. Indicated within each box is the pattern correlation between ANUSPLIN and CANGRD.

Table 5. Dominant periods and the intervals of significant variance for SPEI6$_{Apr-Sep}$ and SPEI12$_{Oct-Sep}$ during 1950–2013.

		Dominant periods (months)	Intervals of variance
SPEI6$_{Apr-Sep}$	REOF1 (Prairie)	8–32	1955–2001; 2002–2013
	REOF2 (Northern)	8–40	1955–2000
SPEI12$_{Oct-Sep}$	REOF1 (Prairie)	16–60; 32–64	1955–1968; 1970–2002
	REOF2 (Northern)	16–100	1956–1975; 1995–2013

drought events, the frequency is higher but shorter lived in the Prairies relative to the northern central region.

Figure 10 shows that for both the seasonal (SPEI6$_{Apr-Sep}$) and annual (SPEI12$_{Oct-Sep}$) drought time series, sporadic but significant coherence is observed intermittently from year to year with the PDO. For SPEI6$_{Apr-Sep}$, dominant strong coherence occurred between 16 and 32 months over the period 1988–2003 for the Prairies region, whereas fluctuations were intermittently observed between 1955 and 2000 for SPEI12$_{Oct-Sep}$ over the Prairie and northern central re-

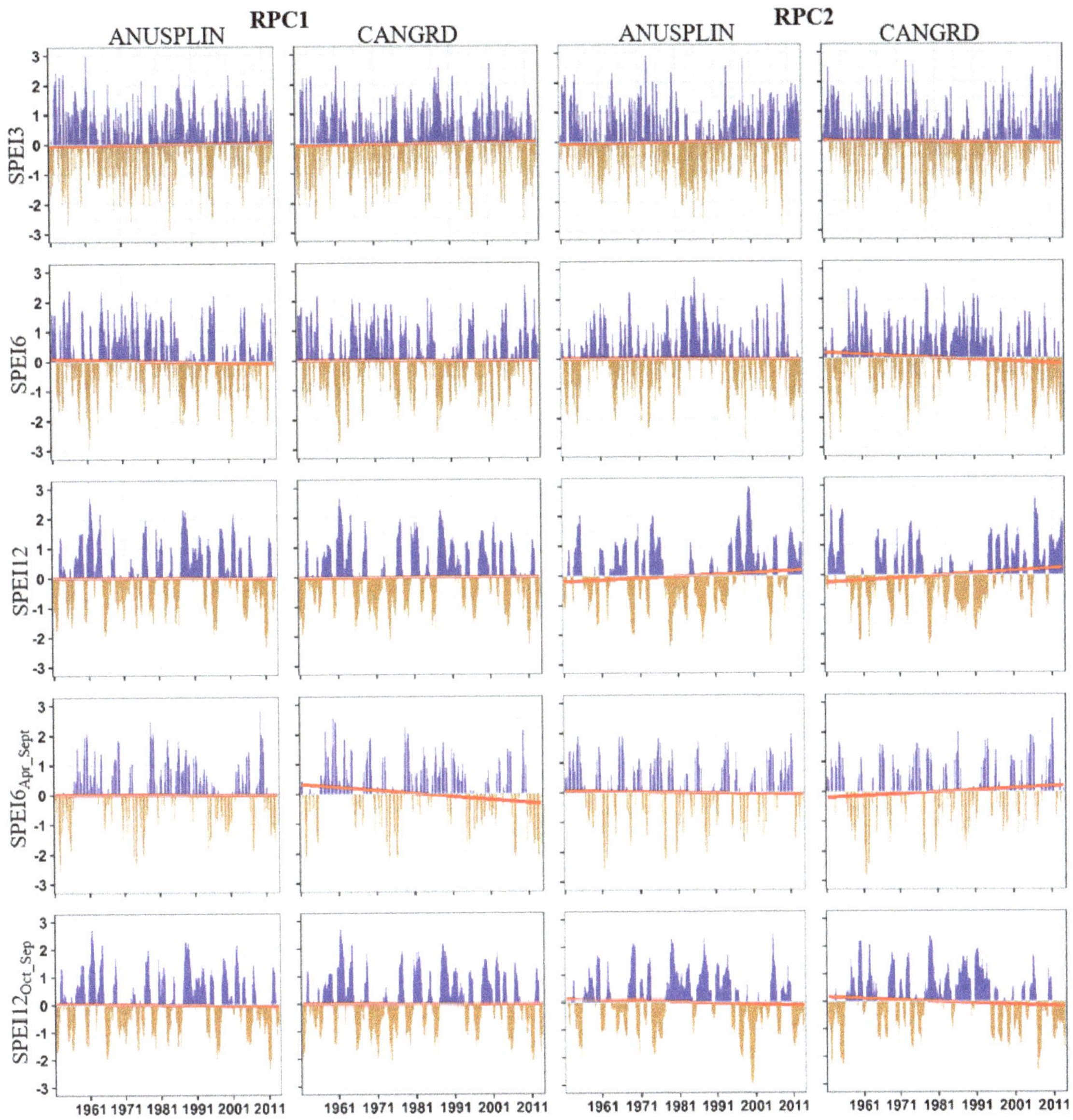

Figure 7. Long-term (1950–2013) trends (red line) of the RPCs for each drought subregion and data set.

gions ranging from 16 to 64 months. Unlike the PDO, the PNA showed a strong in-phase relationship with drought over Canada (Fig. 11). It is apparent that the PNA co-varied significantly with $SPEI6_{Apr-Sep}PC1$ over most of the years during the study period. Oscillations in the PNA are manifested in the $SPEI6_{Apr-Sep}PC1$ on wavelengths varying from 2 to 108 months ($\sim$ 9 years), suggesting that the PNA actively mirrors $SPEI6_{Apr-Sep}PC1$. Particularly, the PNA was phase-locked with drought during the period 1980–2001 at the 16–32-month scale. Apart from the late 1990s and early 2000s at the 8–16 month scale, no significant coherence was found between the PNA and $SPEI6_{Apr-Sep}PC2$. Although with less coherence relative to $SPEI6_{Apr-Sep}$, the PNA also co-varied with $SPEI12_{Oct-Sep}$, more so for the northern Central region compared to the Prairie region. In the Prairie region, the strongest coherence between $SPEI12_{Oct-Sep}PC1$ and the PNA occurred over the 16–32-month scale, while for the northern central region, significant cross power between the PNA and $SPEI12Oct_{Apr-Sep}PC2$ occurred over the 32–64-month scale, spanning the periods 1960–1973 and 1985–2000.

The relationship between drought and the AMO is shown in Fig. S3. It is clear that the AMO did not co-vary with $SPEI6_{Apr-Sep}$ over both homogenous drought regions. However, it is important to note that Shabbar and Skinner (2004) found a significant correlation pattern between the winter AMO index and following summer PDSI in the north of the Prairies province. Here, we only made use of the AMO index for the months April–September of each year. For $SPEI12_{Oct-Sep}$, significant and high en-

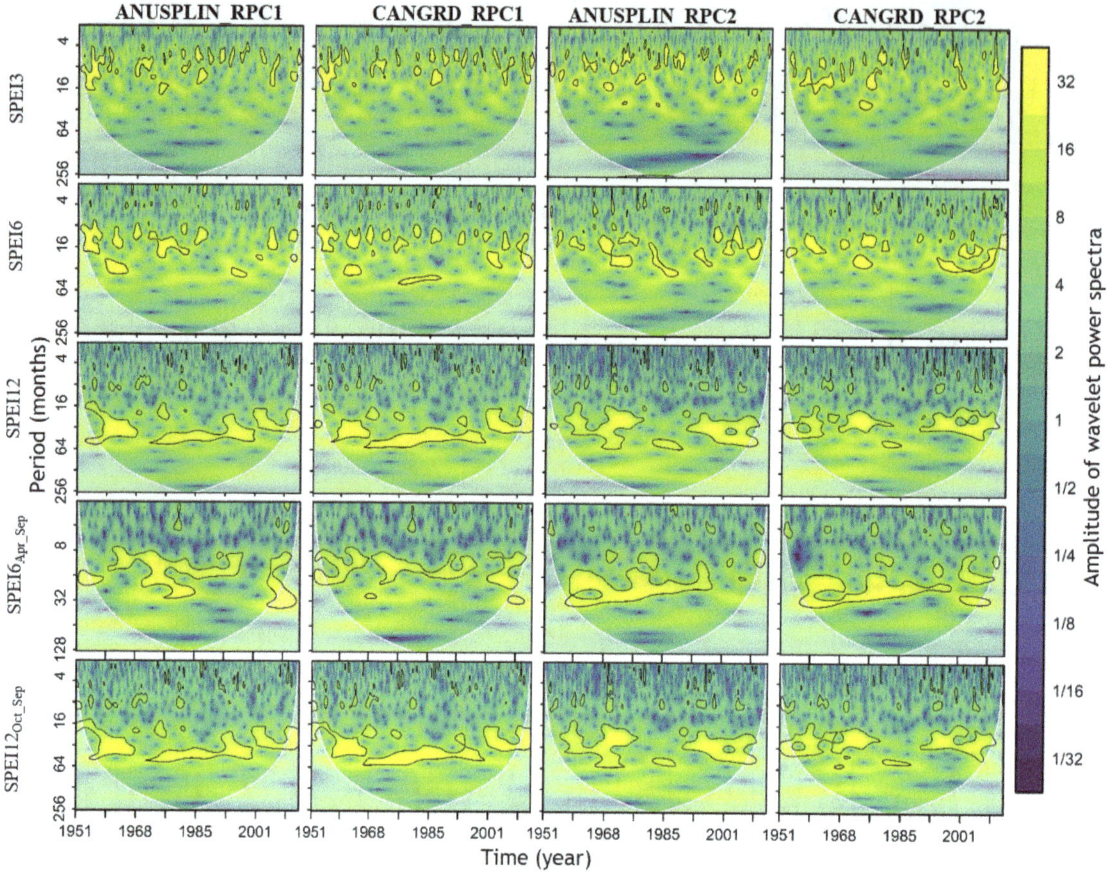

Figure 8. Wavelet power spectrum of the time series (RPCs) shown in Fig. 6. The black contour designates the 95 % confidence level against red noise, and the cone of influence (COI), where edge effects might distort the picture, is shown as a lighter grey shade.

ergy existed for the Prairie region only ($SPEI12_{Oct-Sep}PC1$), mainly distributed in the 8–64-month (~ 5 years) band and spanning the period 1970–2005. For the AO (Fig. S4), in-phase significant coherence existed with $SPEI6_{Apr-Sep}$ PC2 and $SPEI12_{Oct-Sep}PC1$. In the northern central region ($SPEI6_{Apr-Sep}$ PC2), 16–32-month high-energy regions were found over the period 1963–1978, and 8–16-month significant coherence also existed from 1994 to 2002. In the Prairie region ($SPEI12_{Oct-Sep}PC1$), drought was in phase with the AO especially from 1981 to 2003, where significant cross power and coherence were mainly concentrated in the 64–128-month band (based on CANGRD). The results further indicate that the AO was in antiphase with $SPEI6_{Apr-Sep}$ PC1 and $SPEI12_{Oct-Sep}PC2$ during the study period from 1950 to 2013. In terms of the NAO (Fig. S5), sporadic significant coherence is noticed with the seasonal $SPEI6_{Apr-Sep}$ at a higher frequency, ranging between 4 and 20 months and mainly concentrated over the period 1975–1990. There is also statistically significant in-phase coherence between $SPEI12_{Oct-Sep}PC1$ and the NAO in the 70–128-month band in the late 1970s–late 1990s, based on CANGRD.

The foregoing analysis has shown significant covariance between drought variability over Canada and large-scale climate indices, especially the MEI, PNA and PDO. This is in line with previous studies (e.g. Bonsal and Shabbar, 2008, Fleming and Quilty, 2006, Tremblay et al., 2011, and Perez-Valdivia et al., 2012) that have established links between Canadian hydroclimate and teleconnection indices. Furthermore, one should expect most of the climate indices to yield similar findings given that they appear to be interrelated at several timescales (Sheridan, 2003; Gan et al., 2007; Ng and Chan, 2012). We recommend that future studies examine the degree to which such interrelations can affect the findings reported here by eliminating, for example, the influence of the PDO on the MEI via partial wavelet coherence analysis (Ng and Chan, 2012). Also, a lead/lag response of the identified drought frequencies as well their correlations to positive and negative phases of various teleconnections constitute an area for future research. In addition, only one PET estimation method was used in this study and its selection was constrained by data availability during the study period. We recommend the use of other simple or complex methods to calculate PET and assess their impact on drought analysis over Canada since the Hargreaves method has been known to underestimate PET relative to the Penman–Monteith method (McMahon et al., 2013). Further-

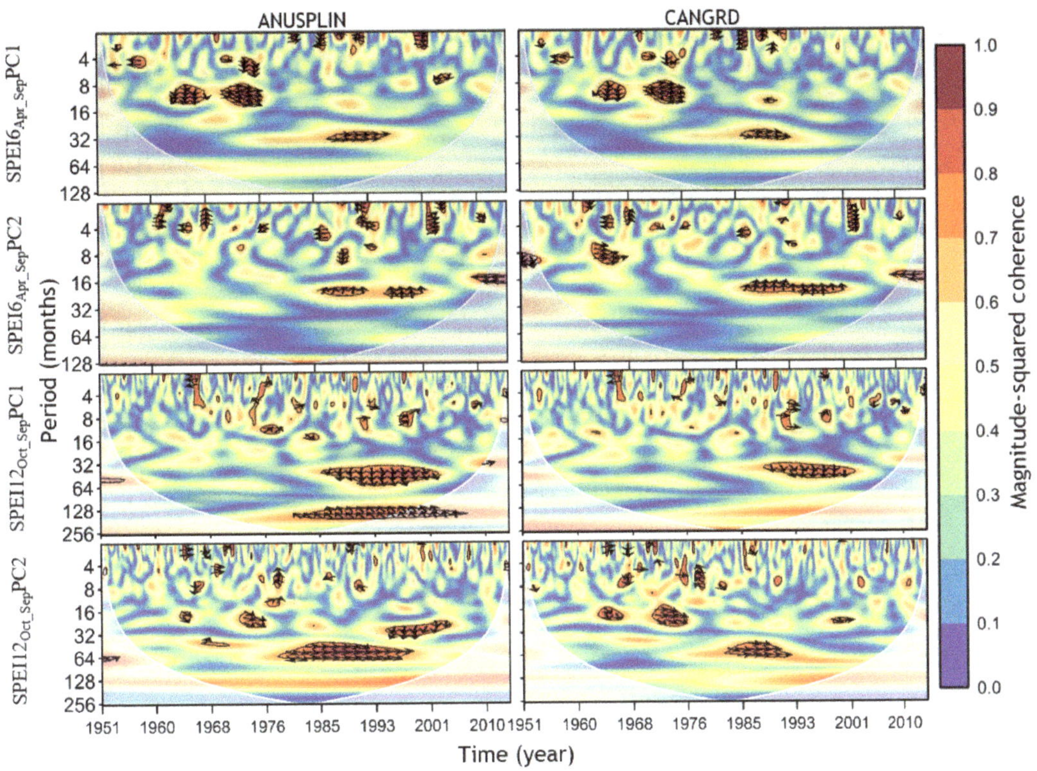

Figure 9. Squared wavelet coherence between the MEI and the temporal patterns of drought ($SPEI6_{Apr-Sep}$ and $SPEI12_{Oct-Sep}$). Phase arrows pointing right indicate signals are in phase, whereas left-pointing arrows indicate an antiphase relationship. Arrows deviating from the horizontal are indicative of lead–lag relationships between the two signals. The black contour designates the 95 % confidence level against red noise, and the cone of influence (COI), where edge effects might distort the picture, is shown as a lighter grey shade.

more, the Hargreaves method responds only to changes in temperature and can lead to misleading results under global warming. For example, Sheffield et al. (2012) found little change in global drought over the past 60 years (1950–2008). In terms of data sources, the findings reported here should be validated against other data sets such as long-term satellite products as they become available.

4 Summary and conclusions

This study performs a comprehensive analysis of historical droughts over the whole of Canada, considering the role of teleconnections by analyzing different monthly P and T products for the period 1950–2013. SPEI, a climatological drought index, is applied over various temporal scales to evaluate various drought characteristics such as trends, spatio-temporal patterns of long-term change, inter/intra-annual variability, and periodicity/frequency. In addition, potential prominent modes of low-frequency variability such as the PNA, AO, MEI, PDO, AMO, and NAO which are well established to influence the hydroclimate of Canada and North America are investigated as precursors to historical drought occurrence. The main conclusions from the analyses are given in the following.

Apart from the Pacific/Atlantic maritime regions, most of the continental Canadian interior experienced moisture deficits, with the Prairie region being the most affected region between 1950 and 2013. Based on a trend analysis derived from the water balance, significant decreasing trends in SPEI values are observed in the southern Prairies, the foothills of the Rocky Mountains, and Pacific maritime regions, whereas increasing trends are limited to a small area in the north and parts of the Atlantic coast to the east. Therefore, southern parts of the country showed a trend towards drier conditions and vice versa for the northern regions. Note that the northern region (above 60° N) has lower station density and as such higher uncertainty in the gridded P and T fields.

EOF identifies two main spatially disjunctive subregions of drought variability over Canada – the Prairies and northern central Canada. Based on both seasonal ($SPEI6_{Apr-Sep}$) and annual ($SPEI12_{Oct-Sep}$) SPEI values, the Prairie subregion experienced moderate to extreme droughts episodes starting in the mid-1950s, early and late 1960s, the 1970s and 1980s, the period from mid-1997–2005, and the summer of 2009. Some of the drought episodes were extremely dry and severe as they were prolonged in time such as the ones in the late 1970s, mid-1980s, and 1997–2005. In the northern central region, although fewer (likely due to below 0 tempera-

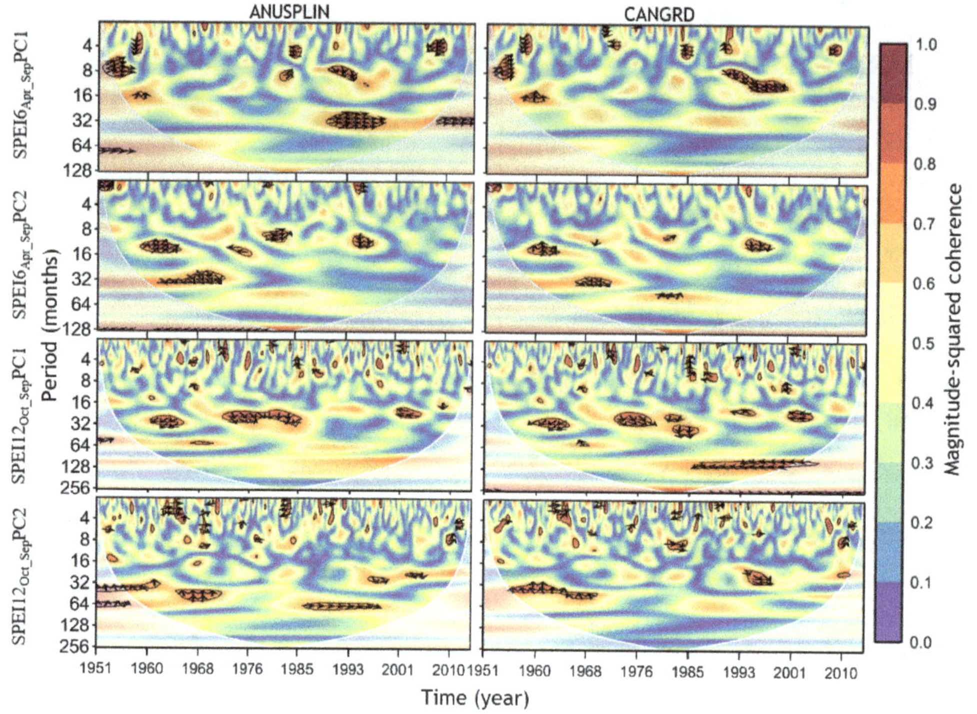

Figure 10. Squared wavelet coherence between the PDO and the temporal patterns of drought ($SPEI6_{Apr-Sep}$ and $SPEI12_{Oct-Sep}$). Phase arrows pointing right indicate signals are in phase, whereas left-pointing arrows indicate an antiphase relationship. Arrows deviating from the horizontal are indicative of lead–lag relationships between the two signals. The black contour designates the 95 % confidence level against red noise, and the cone of influence (COI), where edge effects might distort the picture, is shown as a lighter grey shade.

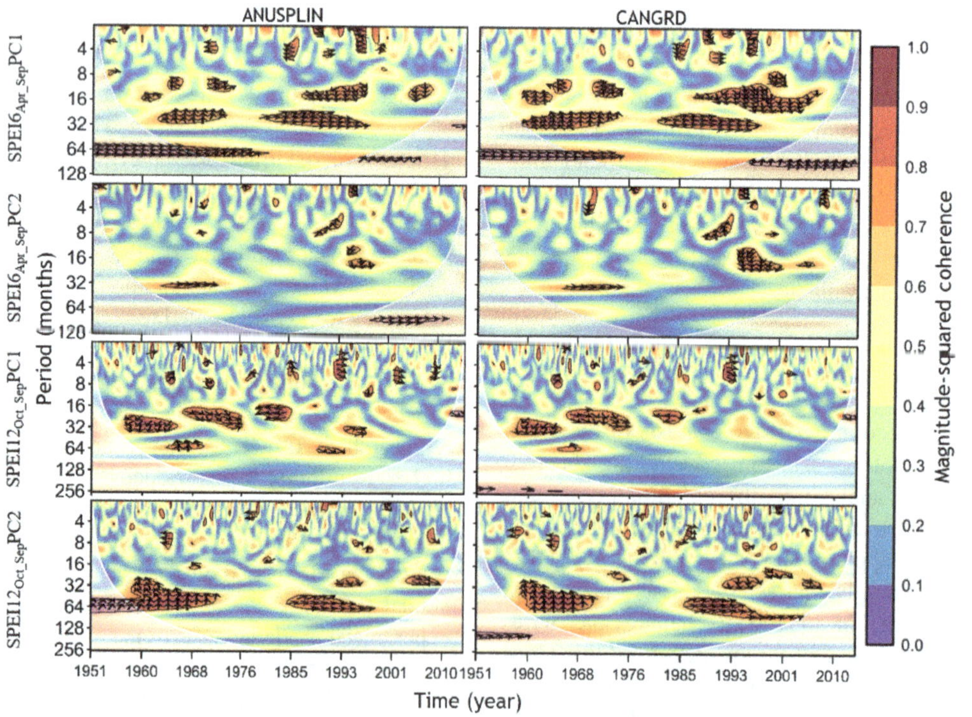

Figure 11. Squared wavelet coherence between the PNA and the temporal patterns of drought ($SPEI6_{Apr-Sep}$ and $SPEI12_{Oct-Sep}$). Phase arrows pointing right indicate signals are in phase, whereas left-pointing arrows indicate an antiphase relationship. Arrows deviating from the horizontal are indicative of lead–lag relationships between the two signals. The black contour designates the 95 % confidence level against red noise, and the cone of influence (COI), where edge effects might distort the picture, is shown as a lighter grey shade.

tures for most of the cold season) droughts were detected for the hydrological year ($SPEI12_{Oct-Sep}$), with the most intense occurring in 2000, the seasonal SPEI series ($SPEI6_{Apr-Sep}$) detected extremely dry episodes in this region in the early 1960s, 1980s, early 1990s, and 2000s. However, drought variability in the Prairies in particular experienced largely insignificant trends, a finding similar to numerous previous studies.

Wavelet analysis was particularly useful to detect periodical signals in the SPEI time series patterns for each subregion. For $SPEI6_{Apr-Sep}$, the analysis reveals clearly the presence of a dominant periodicity of between 8 and 32 months, persisting approximately from 1955 to 2001 in the Prairie region, while in the North central region, a significant periodicity of a cycle between 8 and 40 months as a dominant period of variability spanning the years 1955–2000 is apparent. For $SPEI12_{Oct-Sep}$, a strong power frequency band over the Prairie region, centred mostly around 16–66 months (~ 4 years) and spanning the period 1955–1968 is found. Moreover, 32–64-month periodic high-power signals are dominant during the years 1970–2002. In the northern central region, a significant high power for relatively low-frequency (16–100 months, i.e. ~ 7 years), spanning the period 1956–1975 and 1995–2013, is detected. Therefore, the Prairie region of Canada is dominated by high-frequency (i.e. more frequent and shorter cycles of dry events) power signals for both $SPEI6_{Apr-Sep}$ and $SPEI12_{Oct-Sep}$, while the northern central region is dominated by low-frequency (i.e. less frequent cycles of dry events) power signals. The analysis further indicates that significant inter-annual periodicities (with a period of <10 years) dominate drought variability over the two identified major regions of drought variability across Canada.

The identified drought short-time (long-time) inter-annual periodicities in the Prairie (northern central region) are likely associated with the immediate and significant influence of the MEI and PNA in particular, as these large-scale climate indices have maximum regions with a 5 % significance level in the WCO plots. For the MEI index, 8–16 and 32–50 months were the most predominant and effective periods for drought occurrence in Canada, while for the PNA, the 2–108-month period was the most predominant over the Prairie region and for $SPEI6_{Apr-Sep}$ compared with the northern central region. For $SPEI12_{Oct-Sep}$, the PNA showed more coherence with drought at the 32–64-month scale.

The foregoing analysis has indicated the need to consider various observational data sets in drought characterization, given the uncertainty in data. In terms of trends, the ANUSPLIN data set indicated a higher tendency for drought over the study period relative to CANGRD. Furthermore, irrespective of the timescale of accumulation, ANUSPLIN tends to reveal more drought severity compared to CANGRD although the correlation between the time series of the two data sets from each of the homogenous drought subregions is very strong. Therefore, further applications using other gridded data sets to verify the role played by the spatial resolution of the input data on regional drought patterns are recommended.

The identification of these subregions with similar drought variability and characteristics can be useful for drought risk management at a regional scale in Canada. Two of the most important river basins are both in the Prairies region (Saskatchewan River basin) and northern region (MacKenzie River basin).

Lastly, this study is the first of its kind to identify dominant periodicities in drought variability over the whole of Canada in terms of when the drought events occur, their duration, and how often they do so over the Prairies and northern central regions.

Competing interests. The authors declare that they have no conflict of interest.

Special issue statement. This article is part of the special issue "Understanding and predicting Earth system and hydrological change in cold regions". It is not associated with a conference.

Acknowledgements. The authors are thankful to Environment and Climate Change Canada and Natural Resources Canada for making the datasets freely available to the scientific community. We would like to thank two anonymous referees and John Hanesiak, the editor, for their contributions that have led to a significantly improved manuscript.

Edited by: John Hanesiak

References

Addison, P. S.: The Illustrated Wavelet Transform Handbook: Introductory Theory and Applications in Science, Engineering, Medicine and Finance, Second Edition, CRC Press, Taylor & Francis Group, Boca Raton, Florida, USA, 2016.

Asong, Z. E., Khaliq, M. N., and Wheater, H. S.: Regionalization of precipitation characteristics in the Canadian Prairie Provinces using large-scale atmospheric covariates and geophysical attributes, Stoch. Env. Res. Risk A., 29, 875–892, https://doi.org/10.1007/s00477-014-0918-z, 2015.

Asong, Z. E., Khaliq, M. N., and Wheater, H. S.: Multisite multivariate modeling of daily precipitation and temperature in the Canadian Prairie Provinces using generalized linear models, Clim. Dynam., 47, 2901–2921, https://doi.org/10.1007/s00382-016-3004-z, 2016.

Asong, Z. E., Razavi, S., Wheater, H. S., and Wong, J. S.: Evaluation of Integrated Multi-satellitE Retrievals for GPM (IMERG) over Southern Canada against Ground Precipitation Observations: A Preliminary Assessment, J. Hydrometeorol., 18, 1033–1050, https://doi.org/10.1175/jhm-d-16-0187.1, 2017.

Barnston, A. G. and Livezey, R. E.: Classification, Seasonality and Persistence of Low-Frequency Atmospheric Circulation Patterns, Mon. Weather Rev., 115, 1083–1126, https://doi.org/10.1175/1520-0493(1987)115<1083:csapol>2.0.co;2, 1987.

Beguería, S., Vicente-Serrano, S. M., Reig, F., and Latorre, B.: Standardized precipitation evapotranspiration index (SPEI) revisited: parameter fitting, evapotranspiration models, tools, datasets and drought monitoring, Int. J. Climatol., 34, 3001–3023, 10.1002/joc.3887, 2014.

Bonsal, B. and Regier, M.: Historical comparison of the 2001/2002 drought in the Canadian Prairies, Climate Res., 33, 229–242, 2007.

Bonsal, B. and Shabbar, A.: Impacts of large-scale circulation variability on low streamflows over Canada: a review, Can. Water Resour. J., 33, 137–154, 2008.

Bonsal, B., Zhang, X., and Hogg, W.: Canadian Prairie growing season precipitation variability and associated atmospheric circulation, Climate Res., 11, 191–208, 1999.

Bonsal, B., Shabbar, A., and Higuchi, K.: Impacts of low frequency variability modes on Canadian winter temperature, Int. J. Climatol., 21, 95–108, https://doi.org/10.1002/joc.590, 2001.

Bonsal, B., Aider, R., Gachon, P., and Lapp, S.: An assessment of Canadian prairie drought: past, present, and future, Clim. Dynam., 41, 501–516, https://doi.org/10.1007/s00382-012-1422-0, 2013.

Bonsal, B. R., Wheaton, E. E., Meinert, A., and Siemens, E.: Characterizing the Surface Features of the 1999–2005 Canadian Prairie Drought in Relation to Previous Severe Twentieth Century Events, Atmos. Ocean., 49, 320–338, https://doi.org/10.1080/07055900.2011.594024, 2011a.

Bonsal, B., Wheaton, E., Meinert, A., and Siemens, E.: Characterizing the Surface Features of the 1999–2005 Canadian Prairie Drought in Relation to Previous Severe Twentieth Century Events, Atmos. Ocean., 49, 320–338, https://doi.org/10.1080/07055900.2011.594024, 2011b.

Brown, R. D. and Braaten, R. O.: Spatial and temporal variability of Canadian monthly snow depths, 1946–1995, Atmos. Ocean., 36, 37–54, https://doi.org/10.1080/07055900.1998.9649605, 1998.

Bryant, E. A.: Natural Hazards, University of Cambridge, Cambridge, UK, 2005.

Côté, J., Gravel, S., Méthot, A., Patoine, A., Roch, M., and Staniforth, A.: The Operational CMC–MRB Global Environmental Multiscale (GEM) Model. Part I: Design Considerations and Formulation, Mon. Weather Rev., 126, 1373–1395, https://doi.org/10.1175/1520-0493(1998)126<1373:tocmge>2.0.co;2, 1998.

Council of Canadian Academies: Water and Agriculture in Canada: Towards Sustainable Management of Water Resources. Chapter 6–Building the Foundation for Sustainable Management of Water in Agriculture, available at: http://www.scienceadvice.ca/en/assessments/completed/water-agri.aspx (last access: 25 April 2018), 2013.

Dai, A.: Drought under global warming: a review, Wiley Interdisciplinary Reviews: Climate Change, 2, 45–65, https://doi.org/10.1002/wcc.81, 2011.

Dibike, Y., Prowse, T., Bonsal, B., and O'Neil, H.: Implications of future climate on water availability in the western Canadian river basins, Int. J. Climatol., 37, 3247–3263, https://doi.org/10.1002/joc.4912, 2016.

Dibike, Y., Prowse, T., Bonsal, B., and O'Neil, H.: Implications of future climate on water availability in the western Canadian river basins, Int. J. Climatol., 37, 3247–3263, https://doi.org/10.1002/joc.4912, 2017.

Dracup, J. A., Lee, K. S., and Paulson, E. G.: On the definition of droughts, Water Resour. Res., 16, 297–302, https://doi.org/10.1029/WR016i002p00297, 1980.

Enfield, D. B., Mestas-Nuñez, A. M., and Trimble, P. J.: The Atlantic Multidecadal Oscillation and its relation to rainfall and river flows in the continental U.S, Geophys. Res. Let., 28, 2077–2080, https://doi.org/10.1029/2000GL012745, 2001.

Environment and Climate Change Canada: Canadian Gridded Temperature and Precipitation Anomalies CANGRD, available at: https://open.canada.ca/data/en/dataset/3d4b68a5-13bc-48bb-ad10-801128aa6604, last access: 25 April 2018.

Fleming, S. W. and Quilty, E. J.: Aquifer Responses to El Niño–Southern Oscillation, Southwest British Columbia, Ground Water, 44, 595–599, https://doi.org/10.1111/j.1745-6584.2006.00187.x, 2006.

Fugal, D. L.: Conceptual Wavelets in Digital Signal Processing: An In-depth, Practical Approach for the Non-mathematician, Space & Signals Technical Pub., University of California, USA, 2009.

Gan, T. Y., Gobena, A. K., and Wang, Q.: Precipitation of southwestern Canada: Wavelet, scaling, multifractal analysis, and teleconnection to climate anomalies, J. Geophys. Res.-Atmos., 112, D10110, https://doi.org/10.1029/2006JD007157, 2007.

Gandin, L. S.: Objective analysis of meteorological fields. By L. S. Gandin. Translated from the Russian. Jerusalem (Israel Program for Scientific Translations), 1965. Pp. vi, 242: 53 Figures; 28 Tables. £4 1s. 0d, Q. J. Roy. Meteor. Soc., 92, 447–447, https://doi.org/10.1002/qj.49709239320, 1966.

Gobena, A. K. and Gan, T. Y.: Low-frequency variability in Southwestern Canadian stream flow: links with large-scale climate anomalies, Int. J. Climatol., 26, 1843–1869, https://doi.org/10.1002/joc.1336, 2006.

Grinsted, A., Moore, J. C., and Jevrejeva, S.: Application of the cross wavelet transform and wavelet coherence to geophysical time series, Nonlin. Processes Geophys., 11, 561–566, https://doi.org/10.5194/npg-11-561-2004, 2004.

Hamed, K. H. and Rao, R. A.: A modified Mann-Kendall trend test for autocorrelated data, J. Hydrol., 204, 182–196, https://doi.org/10.1016/S0022-1694(97)00125-X, 1998.

Hannachi, A., Jolliffe, I. T., and Stephenson, D. B.: Empirical orthogonal functions and related techniques in atmospheric science: A review, Int. J. Climatol., 27, 1119–1152, https://doi.org/10.1002/joc.1499, 2007.

Hao, Z. and AghaKouchak, A.: Multivariate Standardized Drought Index: A parametric multi-index model, Adv. Water Resour., 57, 12–18, https://doi.org/10.1016/j.advwatres.2013.03.009, 2013.

Hao, Y., Liu, G., Li, H., Li, Z., Zhao, J., and J. Yeh, T.-C.: Investigation of karstic hydrological processes of Niangziguan Springs (North China) using wavelet analysis, Hydrol. Process., 26, 3062–3069, https://doi.org/10.1002/hyp.8265, 2012.

Hargreaves, G. H.: Defining and Using Reference Evapotranspiration, J. Irrig. Drain. E., 120, 1132–1139, https://doi.org/10.1061/(ASCE)0733-9437(1994)120:6(1132), 1994.

Hargreaves, G. L., Hargreaves, G. H., and Riley, J. P.: Agricultural Benefits for Senegal River Basin, J. Irrig. Drain. E., 111, 113–124, https://doi.org/10.1061/(ASCE)0733-9437(1985)111:2(113), 1985.

Hayes, M., Svoboda, M., Wall, N., and Widhalm, M.: The Lincoln Declaration on Drought Indices: Universal Meteorological Drought Index Recommended, B. Am. Meteorol. Soc., 92, 485–488, https://doi.org/10.1175/2010bams3103.1, 2011.

Hogg, E. H., Price, D. T., and Black, T. A.: Postulated Feedbacks of Deciduous Forest Phenology on Seasonal Climate Patterns in the Western Canadian Interior, J. Climate, 13, 4229–4243, https://doi.org/10.1175/1520-0442(2000)013<4229:pfodfp>2.0.co;2, 2000.

Hopkinson, R. F., McKenney, D. W., Milewska, E. J., Hutchinson, M. F., Papadopol, P., and Vincent, L. A.: Impact of Aligning Climatological Day on Gridding Daily Maximum–Minimum Temperature and Precipitation over Canada, J. Appl. Meteorol. Clim., 50, 1654–1665, https://doi.org/10.1175/2011jamc2684.1, 2011.

Hurrell, J. W. and Van Loon, H.: Decadal Variations In Climate Associated With The North Atlantic Oscillation, Climatic Change, 36, 301–326, https://doi.org/10.1023/a:1005314315270, 1997.

Hutchinson, M. F., McKenney, D. W., Lawrence, K., Pedlar, J. H., Hopkinson, R. F., Milewska, E., and Papadopol, P.: Development and Testing of Canada-Wide Interpolated Spatial Models of Daily Minimum–Maximum Temperature and Precipitation for 1961–2003, J. Appl. Meteorol. Clim., 48, 725–741, https://doi.org/10.1175/2008jamc1979.1, 2009.

IPCC: Climate Change 2013: The Physical Science Basis. Contribution of Working Group I to the Fifth Assessment Report of the Intergovernmental Panel on Climate Change, Cambridge University Press, Cambridge, UK and New York, NY, USA, 1535 pp., 2013.

Jiang, R., Gan, T. Y., Xie, J., and Wang, N.: Spatiotemporal variability of Alberta's seasonal precipitation, their teleconnection with large-scale climate anomalies and sea surface temperature, Int. J. Climatol., 34, 2899–2917, https://doi.org/10.1002/joc.3883, 2014.

Maidment, D. R.: Handbook of hydrology, McGraw-Hill, New York, 1993.

Mantua, N. J. and Hare, S. R.: The Pacific Decadal Oscillation, J. Oceanogr., 58, 35–44, https://doi.org/10.1023/a:1015820616384, 2002.

Masud, M. B., Khaliq, M. N., and Wheater, H. S.: Analysis of meteorological droughts for the Saskatchewan River Basin using univariate and bivariate approaches, J. Hydrol., 522, 452–466, https://doi.org/10.1016/j.jhydrol.2014.12.058, 2015.

Masud, M. B., Khaliq, M. N., and Wheater, H. S.: Future changes to drought characteristics over the Canadian Prairie Provinces based on NARCCAP multi-RCM ensemble, Clim. Dynam., 48, 2685–2705, https://doi.org/10.1007/s00382-016-3232-2, 2017.

McKee, T., Doesken, N., and Kleist, J.: The Relationship of Drought Frequency and Duration to Time Scales. Paper Presented at 8th Conference on Applied Climatology, American Meteorological Society: Anaheim, CA, 1993.

McKenney, D. W., Hutchinson, M. F., Papadopol, P., Lawrence, K., Pedlar, J., Campbell, K., Milewska, E., Hopkinson, R. F., Price, D., and Owen, T.: Customized Spatial Climate Models for North America, B. Am. Meteorol. Soc., 92, 1611–1622, https://doi.org/10.1175/2011bams3132.1, 2011.

McMahon, T. A., Peel, M. C., Lowe, L., Srikanthan, R., and McVicar, T. R.: Estimating actual, potential, reference crop and pan evaporation using standard meteorological data: a pragmatic synthesis, Hydrol. Earth Syst. Sci., 17, 1331–1363, https://doi.org/10.5194/hess-17-1331-2013, 2013.

Mekis, É. and Vincent, L. A.: An Overview of the Second Generation Adjusted Daily Precipitation Dataset for Trend Analysis in Canada, Atmos. Ocean., 49, 163–177, https://doi.org/10.1080/07055900.2011.583910, 2011.

Mishra, A. K. and Singh, V. P.: A review of drought concepts, J. Hydrol., 391, 202–216, https://doi.org/10.1016/j.jhydrol.2010.07.012, 2010.

Mishra, A. K. and Singh, V. P.: Drought modeling–A review, J. Hydrol., 403, 157–175, https://doi.org/10.1016/j.jhydrol.2011.03.049, 2011.

Natural Resources Canada: ANUSPLIN data, available at: https://cfs.nrcan.gc.ca/projects/3/3, last access: 25 April 2018.

Nazemi, A., Wheater, H. S., Chun, K. P., Bonsal, B., and Mekonnen, M.: Forms and drivers of annual streamflow variability in the headwaters of Canadian Prairies during the 20th century, Hydrol. Process., 31, 221–239, https://doi.org/10.1002/hyp.11036, 2017.

Ng, E. K. and Chan, J. C.: Geophysical applications of partial wavelet coherence and multiple wavelet coherence, J. Atmos. Ocean Tech., 29, 1845–1853, 2012.

North, G. R., Bell, T. L., Cahalan, R. F., and Moeng, F. J.: Sampling Errors in the Estimation of Empirical Orthogonal Functions, Mon. Weather Rev., 110, 699–706, https://doi.org/10.1175/1520-0493(1982)110<0699:seiteo>2.0.co;2, 1982.

Palmer, W.: Meteorological drought, Research Paper No. 45. U.S. Weather Bureau, NOAA Library and Information Services Division, Washington, D.C. 20852, 1–25, 1965.

Penman, H. L.: Natural evaporation from open water, bare soil and grass, P. Roy. Soc. Lon. A. Mat., 193, 120–145, https://doi.org/10.1098/rspa.1948.0037, 1948.

Perez-Valdivia, C., Sauchyn, D., and Vanstone, J.: Groundwater levels and teleconnection patterns in the Canadian Prairies, Water Resour. Res., 48, W07516, https://doi.org/10.1029/2011WR010930, 2012.

Pomeroy, J. W., Fang, X., and Marks, D. G.: The cold rain-on-snow event of June 2013 in the Canadian Rockies – characteristics and diagnosis, Hydrol. Process., 30, 2899–2914, https://doi.org/10.1002/hyp.10905, 2016.

Preisendorfer, R. W. and Mobley, C. D.: Principal component analysis in meteorology and oceanography, Elsevier, Amsterdam, 1988.

Priestley, C. H. B. and Taylor, R. J.: On the Assessment of Surface Heat Flux and Evaporation Using Large-Scale Parameters, Mon. Weather Rev., 100, 81–92, https://doi.org/10.1175/1520-0493(1972)100<0081:otaosh>2.3.co;2, 1972.

Raible, C. C., Bärenbold, O., and Gómez-navarro, J. J.: Drought indices revisited–improving and testing of drought indices in a simulation of the last two millennia for Europe, Tellus A, 69, 1287492, https://doi.org/10.1080/16000870.2017.1296226, 2017.

Razavi, S. and Vogel, R.: Prewhitening of hydroclimatic time series? Implications for inferred change and variability across time scales, J. Hydrol., 557, 109–115, https://doi.org/10.1016/j.jhydrol.2017.11.053, 2018.

Richman, M. B.: Rotation of principal components, J. Climatol., 6, 293–335, https://doi.org/10.1002/joc.3370060305, 1986.

Ropelewski, C. F. and Halpert, M. S.: North American Precipitation and Temperature Patterns Associated with the El Niño/Southern Oscillation (ENSO), Mon. Weather Rev., 114, 2352–2362, https://doi.org/10.1175/1520-0493(1986)114<2352:napatp>2.0.co;2, 1986.

Shabbar, A. and Khandekar, M.: The impact of el Nino-Southern oscillation on the temperature field over Canada: Research note, Atmos. Ocean., 34, 401–416, https://doi.org/10.1080/07055900.1996.9649570, 1996.

Shabbar, A. and Skinner, W.: Summer Drought Patterns in Canada and the Relationship toGlobal Sea Surface Temperatures, J. Climate, 17, 2866–2880, https://doi.org/10.1175/1520-0442(2004)017<2866:sdpica>2.0.co;2, 2004.

Shabbar, A. and Yu, B.: Intraseasonal Canadian Winter Temperature Responses to Interannual and Interdecadal Pacific SST Modulations, Atmos. Ocean., 50, 109–121, https://doi.org/10.1080/07055900.2012.657154, 2012.

Sheffield, J., Andreadis, K. M., Wood, E. F., and Lettenmaier, D. P.: Global and Continental Drought in the Second Half of the Twentieth Century: Severity–Area–Duration Analysis and Temporal Variability of Large-Scale Events, J. Climate, 22, 1962–1981, https://doi.org/10.1175/2008jcli2722.1, 2009.

Sheffield, J., Wood, E. F., and Roderick, M. L.: Little change in global drought over the past 60 years, Nature, 491, 435, https://doi.org/10.1038/nature11575, 2012.

Sheridan, S. C.: North American weather-type frequency and teleconnection indices, Int. J. Climatol., 23, 27–45, 2003.

Sternberg, T.: Regional drought has a global impact, Nature, 472, 169–169, 2011.

Szeto, K., Zhang, X., White, R. E., and Brimelow, J.: The 2015 Extreme Drought in Western Canada, B. Am. Meteorol. Soc., 97, S42–S46, https://doi.org/10.1175/bams-d-16-0147.1, 2016.

Tallaksen, L. M. and Stahl, K.: Spatial and temporal patterns of large-scale droughts in Europe: Model dispersion and performance, Geophys. Res. Let., 41, 429–434, https://doi.org/10.1002/2013GL058573, 2014.

Thornthwaite, C. W.: An Approach toward a Rational Classification of Climate, Geogr. Rev., 38, 55–94, https://doi.org/10.2307/210739, 1948.

Torrence, C. and Compo, G. P.: A Practical Guide to Wavelet Analysis, B. Am. Meteorol. Soc., 79, 61–78, https://doi.org/10.1175/1520-0477(1998)079<0061:apgtwa>2.0.co;2, 1998.

Torrence, C. and Webster, P. J.: Interdecadal Changes in the ENSO–Monsoon System, J. Climate, 12, 2679–2690, https://doi.org/10.1175/1520-0442(1999)012<2679:icitem>2.0.co;2, 1999.

Tremblay, L., Larocque, M., Anctil, F., and Rivard, C.: Teleconnections and interannual variability in Canadian groundwater levels, J. Hydrol., 410, 178–188, 2011.

U.S. Department of Commerce, National Oceanic & Atmospheric Administration, Earth System Research laboratory, Physical Sciences Division: Climate Indices: Monthly Atmospheric and Ocean Time Series, available at: https://www.esrl.noaa.gov/psd/, 25 April 2018.

Uvo, C. B.: Analysis and regionalization of northern European winter precipitation based on its relationship with the North Atlantic oscillation, Int. J. Climatol., 23, 1185–1194, https://doi.org/10.1002/joc.930, 2003.

Van Loon, A. F., Stahl, K., Di Baldassarre, G., Clark, J., Rangecroft, S., Wanders, N., Gleeson, T., Van Dijk, A. I. J. M., Tallaksen, L. M., Hannaford, J., Uijlenhoet, R., Teuling, A. J., Hannah, D. M., Sheffield, J., Svoboda, M., Verbeiren, B., Wagener, T., and Van Lanen, H. A. J.: Drought in a human-modified world: reframing drought definitions, understanding, and analysis approaches, Hydrol. Earth Syst. Sci., 20, 3631–3650, https://doi.org/10.5194/hess-20-3631-2016, 2016.

Vicente-Serrano, S. M., Beguería, S., and López-Moreno, J. I.: A Multiscalar Drought Index Sensitive to Global Warming: The Standardized Precipitation Evapotranspiration Index, J. Climate, 23, 1696–1718, https://doi.org/10.1175/2009jcli2909.1, 2010.

Vincent, L. A., Wang, X. L., Milewska, E. J., Wan, H., Yang, F., and Swail, V.: A second generation of homogenized Canadian monthly surface air temperature for climate trend analysis, J. Geophys. Res.-Atmos., 117, D18110, https://doi.org/10.1029/2012JD017859, 2012.

Vincent, L. A., Zhang, X., Brown, R. D., Feng, Y., Mekis, E., Milewska, E. J., Wan, H., and Wang, X. L.: Observed Trends in Canada's Climate and Influence of Low-Frequency Variability Modes, J. Climate, 28, 4545–4560, https://doi.org/10.1175/jcli-d-14-00697.1, 2015.

Wang, X. L. and Lin, A.: An algorithm for integrating satellite precipitation estimates with in situ precipitation data on a pentad time scale, J. Geophys. Res.-Atmos., 120, 3728–3744, https://doi.org/10.1002/2014JD022788, 2015.

Wells, N., Goddard, S., and Hayes, M. J.: A Self-Calibrating Palmer Drought Severity Index, J. Climate, 17, 2335–2351, https://doi.org/10.1175/1520-0442(2004)017<2335:aspdsi>2.0.co;2, 2004.

Wheater, H. and Gober, P.: Water security in the Canadian Prairies: science and management challenges, Philos. T. R. Soc. A., 371, A 2013 371 20120409, https://doi.org/10.1098/rsta.2012.0409, 2013.

Wheater, H. and Gober, P.: Water security and the science agenda, Water Resour. Res., 51, 5406–5424, https://doi.org/10.1002/2015WR016892, 2015.

Wheaton, E., Arthur, L., Chorney, B., Shewchuk, S., Thorpe, J., Whiting, J., and Wittrock, V.: The prairie drought of 1988, Climatological Bulletin, 26, 188–205, 1992.

Wilhite, D. A.: Drought as a natural hazard: concepts and definitions, 3–18, Routledge, London, 2000a.

Wilhite, D. A.: Drought as a natural hazard: Concepts and definitions. Drought: A Global Assessment, 3–18, Routledge, London, 2000b.

Wilhite, D. A. and Glantz, M. H.: Understanding: the Drought Phenomenon: The Role of Definitions, Water Int., 10, 111–120, https://doi.org/10.1080/02508068508686328, 1985.

Wilks, D. S.: Statistical methods in the atmospheric sciences, Academic press, Cambridge, Massachusetts, USA, 2011.

Wolter, K.: The Southern Oscillation in Surface Circulation and Climate over the Tropical Atlantic, Eastern Pacific, and Indian Oceans as Captured by Cluster Analysis, J. Clim. Appl. Meteorol., 26, 540–558, https://doi.org/10.1175/1520-0450(1987)026<0540:tsoisc>2.0.co;2, 1987.

Wolter, K. and Timlin, M. S.: El Niño/Southern Oscillation behaviour since 1871 as diagnosed in an extended multivariate ENSO index (MEI.ext), Int. J. Climatol., 31, 1074–1087, https://doi.org/10.1002/joc.2336, 2011.

Wong, J. S., Razavi, S., Bonsal, B. R., Wheater, H. S., and Asong, Z. E.: Inter-comparison of daily precipitation products for large-scale hydro-climatic applications over Canada, Hydrol. Earth Syst. Sci., 21, 2163–2185, https://doi.org/10.5194/hess-21-2163-2017, 2017.

Yue, S., Pilon, P., Phinney, B., and Cavadias, G.: The influence of autocorrelation on the ability to detect trend in hydrological series, Hydrol. Process., 16, 1807–1829, 2002.

Zhang, X., Vincent, L. A., Hogg, W. D., and Niitsoo, A.: Temperature and precipitation trends in Canada during the 20th century, Atmos. Ocean., 38, 395–429, https://doi.org/10.1080/07055900.2000.9649654, 2000.

Zhang, X., Harvey, K. D., Hogg, W. D., and Yuzyk, T. R.: Trends in Canadian streamflow, Water Resour. Res., 37, 987–998, https://doi.org/10.1029/2000WR900357, 2001.

Zhao, H., Higuchi, K., Waller, J., Auld, H., and Mote, T.: The impacts of the PNA and NAO on annual maximum snowpack over southern Canada during 1979–2009, Int. J. Climatol., 33, 388–395, https://doi.org/10.1002/joc.3431, 2013.

Zhou, S., Miller, A. J., Wang, J., and Angell, J. K.: Trends of NAO and AO and their associations with stratospheric processes, Geophys. Res. Let., 28, 4107–4110, https://doi.org/10.1029/2001GL013660, 2001.

14

A Bayesian modelling method for post-processing daily sub-seasonal to seasonal rainfall forecasts from global climate models and evaluation for 12 Australian catchments

Andrew Schepen[1], **Tongtiegang Zhao**[2], **Quan J. Wang**[2], **and David E. Robertson**[3]

[1]CSIRO Land and Water, Dutton Park 4102, Australia

[2]Department of Infrastructure Engineering, The University of Melbourne, Parkville 3010, Australia

[3]CSIRO Land and Water, Clayton 3168, Australia

Correspondence: Andrew Schepen (andrew.schepen@csiro.au)

Abstract. Rainfall forecasts are an integral part of hydrological forecasting systems at sub-seasonal to seasonal timescales. In seasonal forecasting, global climate models (GCMs) are now the go-to source for rainfall forecasts. For hydrological applications however, GCM forecasts are often biased and unreliable in uncertainty spread, and calibration is therefore required before use. There are sophisticated statistical techniques for calibrating monthly and seasonal aggregations of the forecasts. However, calibration of seasonal forecasts at the daily time step typically uses very simple statistical methods or climate analogue methods. These methods generally lack the sophistication to achieve unbiased, reliable and coherent forecasts of daily amounts and seasonal accumulated totals. In this study, we propose and evaluate a Rainfall Post Processing method for Seasonal forecasts (RPP-S), which is based on the Bayesian joint probability modelling approach for calibrating daily forecasts and the Schaake Shuffle for connecting the daily ensemble members of different lead times. We apply the method to post-process ACCESS-S forecasts for 12 perennial and ephemeral catchments across Australia and for 12 initialisation dates. RPP-S significantly reduces bias in raw forecasts and improves both skill and reliability. RPP-S forecasts are also more skilful and reliable than forecasts derived from ACCESS-S forecasts that have been post-processed using quantile mapping, especially for monthly and seasonal accumulations. Several opportunities to improve the robustness and skill of RPP-S are identified. The new RPP-S post-processed forecasts will be used in ensemble sub-seasonal to seasonal streamflow applications.

1 Introduction

Rainfall forecasts are an integral part of hydrological forecasting systems at sub-seasonal to seasonal timescales (Bennett et al., 2016; Crochemore et al., 2017; Wang et al., 2011). Inclusion of climate information in seasonal streamflow forecasts enhances streamflow predictability (Wood et al., 2016). One strategy for integrating climate information into hydrological models is to conditionally resample historical rainfall (e.g. Beckers et al., 2016; Wang et al., 2011). An alternative approach is to use rainfall forecasts from dynamical climate models.

On one hand, ensemble rainfall forecasts from GCMs (global climate models) are attractive for hydrological prediction in that they forecast multiple seasons ahead and have a well-established spatial and temporal forecast structure. On the other hand, a major issue with GCM forecasts at sub-seasonal to seasonal timescales is that the forecasts are often biased and lacking in predictability of local climate (e.g. Kim et al., 2012; Tian et al., 2017). It is therefore necessary to post-process GCM rainfall forecasts using statistical or dynamical methods before they can be used in hydrological models (Yuan et al., 2015).

Several conceptually simple statistical correction methods are used for directly post-processing daily GCM rainfall forecasts including: additive bias correction, multiplicative bias correction and quantile mapping (Ines and Hansen, 2006). Crochemore et al. (2016), for example, recently evaluated linear scaling and quantile mapping for post-processing ECMWF System 4 rainfall forecasts in France. Quantile mapping adjusts forecast means and ensemble spread, but it is not a full calibration method because it does not account for the correlation between forecasts and observations (Zhao et al., 2017). It is useful for bias correction of climate change projections where a full statistical calibration is inappropriate (Teutschbein and Seibert, 2012). Since additive and multiplicative bias correction and quantile mapping methods do not account for intrinsic GCM skill they are ineffective to use as post-processing tools when GCM forecasts are unskilful.

Post-processing methods that take model skill into account typically fall under the model output statistics (MOS; Glahn and Lowry, 1972) banner. MOS type approaches are well established in the weather forecasting community for short-term forecasting and modern variants are normally probabilistic. Wilks and Hamill (2007), for example, studied ensemble MOS approaches for post-processing global forecast system (GFS) forecasts of rainfall and temperature up to 14 days ahead.

MOS methods can also be thought of as full calibration methods. In this regard, several Bayesian calibration approaches are known to be effective at post-processing GCM rainfall totals aggregated to monthly and seasonal timescales (Hawthorne et al., 2013; Luo et al., 2007; Schepen et al., 2014). However, it is apparent that full calibration methods are not normally applied to the post-processing of daily GCM forecasts of rainfall in the sub-seasonal to seasonal period.

That said, some studies have explored more sophisticated methods for post-processing daily rainfall forecasts from GCMs. An example of such a study is that undertaken by Pineda and Willems (2016), which applied a non-homogenous hidden Markov model (NHHMM) to forecasts in the northwestern region of South America. Their method extracted information from GCM forecasts of rainfall fields and sea surface temperatures. The study also built on the NHHMM method originally developed by Robertson et al. (2004) for prediction of rainfall occurrence in Brazil. In Australia, Shao and Li (2013) and Charles et al. (2013) applied an analogue downscaling method (Timbal and McAvaney, 2001) to produce downscaled daily rainfall forecasts from POAMA (the Predictive Ocean Atmosphere Model for Australia). The NHHMM and analogue methods are not straightforward to apply operationally, as they require the identification of optimal climate predictors in different climatic regions. The methods do not, by design, lead to forecasts that are always reliable in ensemble spread and at least not worse than climatology forecasts.

Statistical post-processing of daily rainfall forecasts is a formidable challenge, perhaps explaining the lack of sophisticated methods. Barriers include: short GCM hindcast records, a high prevalence of zero rainfall amounts, seasonal variations in rainfall patterns, and intrinsically low GCM skill. Amplifying the challenge is that GCM skill decays rapidly as the lead time increases. Lavers et al. (2009), for example, examined temperature and rainfall forecasts using DEMETER and CFS GCMs, and found that, in an idealised scenario, skill in the first 30 days is primarily attributable to skill in the first 15 days and much less to skill over the next 15 days. Post-processing methods should therefore be designed to capture as much skill as possible in the first fortnight and take GCM skill into account when post-processing forecasts further ahead.

In this study, we seek to develop a new, more direct daily rainfall post-processing method that operates solely on rainfall output from GCMs and provides a full forecast calibration taking GCM skill into account. We build a new sub-seasonal to seasonal rainfall post-processor, which has some similarities to the rainfall forecast post-processor (RPP) developed by Robertson et al. (2013b) and Shrestha et al. (2015) for post-processing numerical weather prediction (NWP) forecasts for short-term streamflow forecasting. The new system is hereafter referred to as RPP-S, which stands for rainfall forecast post-processor – seasonal.

The proposed RPP-S method applies the Bayesian joint probability (BJP) modelling approach to post-process daily GCM forecasts of rainfall. BJP has never been used in this situation before and it is therefore important to fully evaluate the merits of BJP as a component of a daily forecast post-processing system. As GCMs can produce ensemble forecasts for over 100 days ahead, RPP-S is developed to generate daily rainfall amounts that aggregate to intra-seasonal and seasonal totals. To this end, the Schaake Shuffle (Clark et al., 2004) is also included as a component of RPP-S.

We apply RPP-S to post-process raw catchment rainfall forecasts from ACCESS-S, Australia's new seasonal forecasting GCM. We evaluate bias, reliability and skill for 12 ephemeral and perennial catchments across Australia. RPP-S catchment rainfall forecasts are compared to catchment rainfall forecasts derived from the Bureau of Meteorology's bias-corrected product for ACCESS-S. Opportunities to develop the method further are discussed.

2 Data and catchments

2.1 ACCESS-S rainfall forecasts

Raw, gridded GCM rainfall forecasts are obtained from the Bureau of Meteorology (BOM's) new ACCESS-S forecasting system. Raw catchment rainfall forecasts are derived through a process of area-weighted averaging of the ensemble mean.

ACCESS-S is a customised version of the UK Met Office's seasonal forecasting system. It contains a fully coupled model representing the interactions among the Earth's atmosphere, oceans and land surface, including sea ice. The current horizontal spatial resolution of the ACCESS-S atmospheric model is approximately 60 km in the mid-latitudes. Full details of the current system are provided by Hudson et al. (2017).

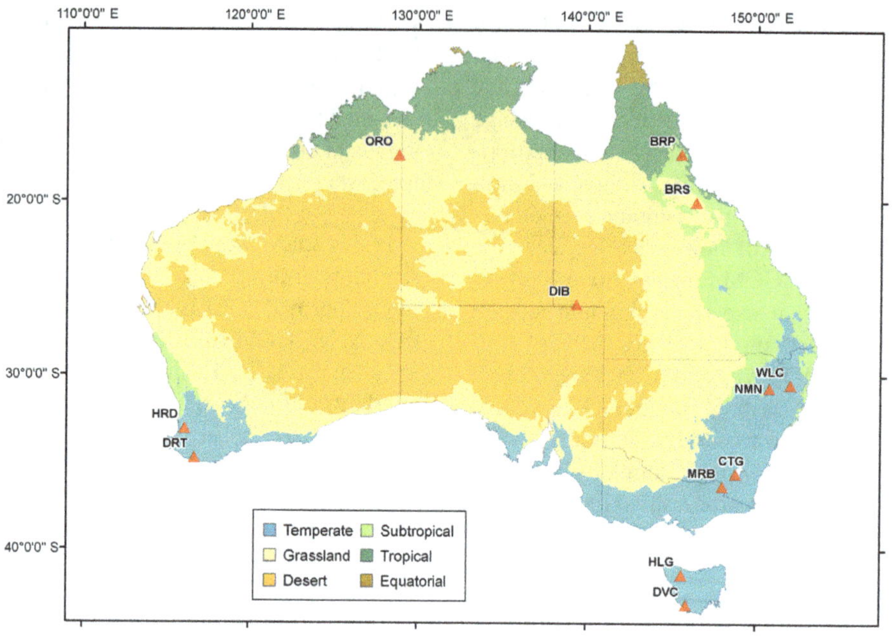

Figure 1. Map of Australian climate zones overlaid by gauging locations plotted as red triangles and labelled with catchment ID. Details of the catchments, including size, are presented in Table 1.

Available ACCESS-S hindcasts are initialised at midnight UTC on days 1 and 25 of each month (additional start dates will become available in the future). A burst ensemble comprising 11 ensemble members is generated. Hindcasts are available for the period 1990–2012 (23 years). A longer hindcast set is not possible because of a lack of suitable initial conditions. The next version of ACCESS-S will be delivered with more sets of initial conditions and longer hindcasts.

2.2 Observed rainfall

Observed rainfall is derived from the Australian Bureau of Meteorology's 5km daily rainfall analysis (AWAP). We make use of AWAP data from 1950 onwards. Catchment rainfall is derived through a process of area-weighted averaging.

An important note on time zones: Australian rainfall is recorded as 24 h totals to 09:00 LT (local time). Consequently, AWAP data are not perfectly synchronised with the ACCESS-S forecasts, which are initialised at midnight UTC. The asynchronism is compounded by time zone differences and daylight savings.

We align the GCM data and observed data as best we can. For east coast locations, an ACCESS-S forecast overlaps with the following day's AWAP rainfall analysis with a discrepancy of 1–2 h. The time discrepancy for west coast locations is approximately 2 h more.

2.3 Catchments

Twelve perennial and ephemeral catchments spread over Australia are selected for application and evaluation. Catchment information including name, gauge ID, regional location and size are shown in Table 1. Catchment locations and climate zones are mapped in Fig. 1. The catchments reside in highly distinct climate zones and vary markedly in size from 100 to 119 034 km^2. Evaluation across climate zones and for varying catchment sizes helps to comprehensively evaluate the effectiveness of the post-processing methods.

3 Methods

3.1 Bayesian joint probability models

Our post-processing method embeds the BJP modelling approach, which was originally designed for forecasting seasonal streamflow totals (Robertson et al., 2013a; Wang and Robertson, 2011; Wang et al., 2009). BJP has since been applied to calibrate hourly rainfall forecasts (Robertson et al., 2013b; Shrestha et al., 2015) and seasonal rainfall forecasts (Hawthorne et al., 2013; Khan et al., 2015; Peng et al., 2014; Schepen and Wang, 2014; Schepen et al., 2014). BJP was most recently adapted for sub-seasonal to seasonal streamflow forecasting (Schepen et al., 2016; Zhao et al., 2016).

BJP formulates a joint probability distribution to characterise the relationship between forecast ensemble means (predictors) and corresponding observations (predictands). The joint distribution is modelled as a bivariate normal distribution after transformation of the marginal distributions. Data transformation is handled using the flexible log – sinh transformation (Wang et al., 2012). The log – sinh transformation transforms y by

Table 1. Catchment ID, catchment name (river and gauging location), gauge ID, region and catchment size.

ID	Catchment name	Gauge ID	Region	Area (km^2)
BRP	Barron River above Picnic Crossing	110003A	Queensland	228
BRS	Burdekin River above Sellheim	120002	Queensland	36 260
DIB	Diamantina River at Birdsville	A0020101	Queensland	119 034
NMN	Namoi River above North Cuerindi	419005	New South Wales	2532
WLC	Wollomombi River above Coninside	206014	New South Wales	377
CTG	Cotter River above Gingera	410730	Murray-Darling Basin	148
MRB	Murray River above Biggara	401012	Murray-Darling Basin	1165
DVC	Davey River D/S Crossing River	473	Tasmania	698
HLG	Hellyer River above Guildford Junction	61	Tasmania	100
DRT	Deep River above Teds Pool	606001	Western Australia	474
HRD	Harvey River above Dingo Road	613002	Western Australia	148
ORO	Ord River at Old Ord Homestead	809316	Northern Australia	19 513

$$f(y) = \beta^{-1}\ln(\sinh(\alpha + \beta y)), \quad (1)$$

where α and β are transformation parameters. A predictor x is transformed to g. Likewise, a predictand y is transformed to h. The relationship between g and h is formulated by a bivariate normal distribution:

$$\begin{bmatrix} g \\ h \end{bmatrix} \sim N(\boldsymbol{\mu}, \boldsymbol{\Sigma}), \quad (2)$$

where

$$\boldsymbol{\mu} = \begin{bmatrix} \mu_g \\ \mu_h \end{bmatrix} \quad (3)$$

is the vector of means and

$$\boldsymbol{\Sigma} = \begin{bmatrix} \sigma_g^2 & \rho_{gh}\sigma_h\sigma_g \\ \rho_{gh}\sigma_h\sigma_g & \sigma_h^2 \end{bmatrix} \quad (4)$$

is the covariance matrix.

In forecasting mode, a BJP model is conditioned on new predictor values. For a particular parameter set, θ, and new transformed predictor value g_{new},

$$h_{\text{new}}|g_{\text{new}}, \theta \sim N\left(\mu_h + \rho_{gh}\frac{\sigma_h}{\sigma_g}(g_{\text{new}} - \mu_g), \sigma_h^2\left(1 - \rho_{gh}^2\right)\right). \quad (5)$$

3.2 Daily rainfall post-processing

We note some specifics of the BJP implementation used in this study. To allow the use of the continuous bivariate normal distribution, values of zero rainfall are treated as censored, meaning the true value is assumed to be less than or equal to zero. To simplify the inference of model parameters, data rescaling is used within the modelling process. To prevent undesirable extrapolation effects in forecasting mode, predictor values are limited to twice the maximum predictor value used to fit the model (future research will address this limitation).

Model parameters may be poorly estimated when there are insufficient non-zero data values used in data inference. We find that in very dry catchments, certain days of the year may have forecasts or observations with no non-zero values at all. In such cases, inference is not possible. To overcome this problem, predictor and predictand data sets are created by pooling data for nearby days and for multiple GCM initialisation dates. We choose to pool data within an 11-day sliding window to be consistent with the Bureau of Meteorology's bias correction scheme (see Sect. 3.4); for the first 5 days the window is fixed to the first 11 days. However, the RPP-S method places no restriction on the configuration, and alternative configurations are discussed in Sect. 5. We choose to pool data for ACCESS-S hindcasts initialised on day 1 of the month and day 25 of the month prior. Predictor and predictand data are subsequently paired according to the number of days since forecast initialisation. Consider, for example, post-processing of forecasts initialised on 1 February. Days 5–15 from the 1 February run and days 5–15 from the 25 January run are used to fit a model that is used to post-process forecasts for 10 February.

We note that pooled data are not completely independent, meaning there is potential to underestimate parameter uncertainty using BJP. Nevertheless, we expect the effect to be limited by the weak predictor–predictand relationships and weak persistence in daily rainfall.

We now describe the model fitting and forecasting steps. After pooling the data, model fitting proceeds as follows:

1. Rescale predictor and predictand data so that each series ranges within [0, 5].

2. Estimate the log − sinh transformation parameters for the predictors using the maximum a posteriori (MAP) estimation.

3. Estimate the log − sinh transformation parameters for the predictands in the same way.

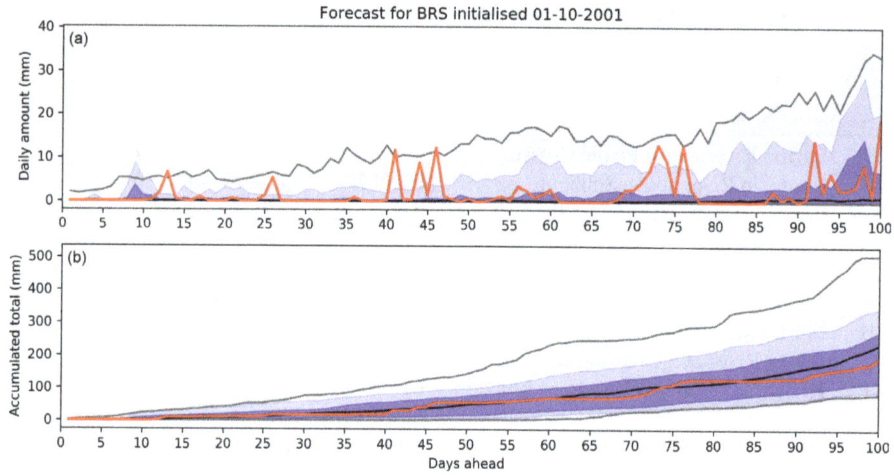

Figure 2. Example RPP-S rainfall forecast for the Burdekin River at Sellheim, initialised on the 1 October 2001 and forecasting 100 days ahead. Forecasts of daily amounts are shown in **(a)** and forecasts of accumulated totals are shown in **(b)**. Dark blue is the forecast [0.25, 0.75] quantile range; medium blue is the forecast [0.10, 0.90] quantile range; and light blue is the forecast [0.05, 0.95] quantile range. Grey lines are the climatological reference forecast [0.05, 0.95] quantile range. The black line is the climatological reference forecast median. The red line is the observation.

4. Apply the transformations to normalise the predictor and predictand data.
5. Apply the transformations to transform the predictor and predictand censoring thresholds.
6. Sample parameter sets representing the posterior distribution of the bivariate normal distribution parameters using Markov chain Monte Carlo sampling, the transformed data and transformed censoring thresholds.

Forecasting proceeds as follows:

1. Transform the predictor value using the log – sinh transformation for predictors.
2. Sample one h_{new} for each parameter set.
3. Back-transform the ensemble members using the transformation for predictands.
4. Rescale the ensemble members to the original space (opposite of step 1 in model fitting).
5. Set negative values to zero.

3.3 Forming ensemble time-series forecasts

BJP forecast ensemble members are initially random and are not linked across days. To deal with this problem, we apply the Schaake Shuffle (Clark et al., 2004), which uses historical data to link ensemble members and create sequences with realistic temporal patterns. The Schaake Shuffle works by imposing the rank correlation of observations on the forecast ensembles, using the trajectories of historical observations as a dependence template. Several variations on the Schaake Shuffle exist and choices can be made about how to construct a temporal dependence template. Schefzik (2016), for example, propose selecting historical observations using a similarity criterion. The steps of the Schaake Shuffle as implemented in this study are as follows:

1. Select a large number of years from the past with the same month and day as the forecast initialisation date (excluding the date of the current forecast).
2. Construct a two-dimensional dependence template by incrementing the dates by one day for each day ahead to be forecast.
3. Extract the observed data for each date in the dependence template.
4. For each day, reorder ensemble members according to the rank of the historical data for that day (e.g. if the second year selected has the smallest observed value, place the smallest ensemble member in second position).
5. Repeat steps (2)–(4) for blocks of ensemble members until the full ensemble has been shuffled.

After applying the Schaake Shuffle to all ensemble members across all lead times, the forecasts can be considered ensemble time-series forecasts.

An example RPP-S forecast for the BRS catchment is shown in Fig. 2. The top panel is the forecast of daily amounts and bottom panel is the forecast of accumulated totals. Hereafter, a daily amount is taken to mean a 24 h rainfall total on any given day. An accumulated total is the sum of daily rainfall amounts over a number of days. In this example, the forecast correctly predicts a dry beginning to the forecast period. The quantile ranges of the daily forecasts are reasonably consistent with the observed values. The accumulated forecast is somewhat narrower than the climatology reference forecast at monthly and seasonal timescales and the forecast is predicting a seasonal rainfall total less than the climatological median.

3.4 Verification

We use RPP-S to post-process all available ACCESS-S forecasts initialised on day 1 of each month for 100 days ahead (the sub-seasonal to seasonal forecast period). For water resources management, whilst it is important that daily amounts are realistic, it is vital that the accumulated totals are as reliable and as skilful as possible. Forecasts are verified against both daily amounts and accumulated totals.

RPP-S forecasts are generated using leave-one-year-out cross-validation. RPP-S forecasts are compared with the BOM's QM forecasts (see Sect. 3.5), which are also generated using leave-one-year-out cross-validation.

Bias is recognised as the correspondence between the mean of forecasts and the mean of observations. Bias is visually assessed by plotting the bias for a set of events against the average forecast for the same set of events. We calculate bias as the mean error

$$\mathrm{BIAS} := \frac{1}{T}\sum_{t=1}^{T}\left(\overline{y}_{\mathrm{fcst},t} - y_{\mathrm{obs},t}\right), \tag{6}$$

where $\overline{y}_{\mathrm{fcst},t}$ is the forecast ensemble mean for event t and $y_{\mathrm{obs},t}$ is the corresponding observation. Bias is calculated separately for each catchment, initialisation date and day. The bias is calculated across 23 events. For a given lead time, we also calculate the average absolute bias (AB) across all 12 catchments and 12 initialisation dates. Notwithstanding that the bias is scale dependent, the average absolute bias is used to compare the magnitude of biases for different model forecasts and for different lead times.

Reliability is the statistical consistency of forecasts and observations – a reliable forecasting system will accurately estimate the likelihood of an event. Reliability is checked by analysing the forecast probability integral transforms (PITs) of rainfall observations. The PIT for a forecast-observation pair is defined by $\pi = F(y_{\mathrm{obs}})$, where F is the forecast cumulative distribution function (CDF). In the case that $y_{\mathrm{obs}} = 0$, a pseudo-PIT value is sampled from a uniform distribution with a range $[0, \pi]$ (Wang and Robertson, 2011). If a forecasting system is reliable, π follows a standard uniform distribution. Reliability can be visually examined by plotting the set of π_t $(t = 1, 2, \ldots, T)$ with the corresponding theoretical quantile of the uniform distribution using the PIT uniform probability plot (or simply PIT plot). A perfectly reliable forecast follows the 1 : 1 line. In this study, we do not plot individual PIT diagrams. Instead, reliability is summarised using the α-index (Renard et al., 2010)

$$\alpha = 1.0 - \frac{2}{n}\sum_{t=1}^{n}\left|\pi_t^* - \frac{t}{n+1}\right|, \tag{7}$$

where π_t^* is the sorted π_t in increasing order. The α-index represents the total deviation of π_t^* from the corresponding uniform quantile (i.e. the tendency to deviate from the 1 : 1 line in PIT diagrams). The α-index ranges from 0 (worst reliability) to 1 (perfect reliability).

Forecast skill is evaluated using the continuous ranked probability score (CRPS; Matheson and Winkler, 1976). The CRPS for a given forecast and observation is defined as:

$$\mathrm{CRPS} = \int \left[F(y) - H(y - y_{\mathrm{obs}})\right]^2 \mathrm{d}y, \tag{8}$$

where y is the forecast variable; y_{obs} is the observed value; F is the forecast CDF; and H is the Heaviside step function, which equals 0 if $y < y_{\mathrm{obs}}$ and equals 1 otherwise. Model forecasts are compared to reference forecasts by calculating skill scores:

$$\text{CRPS skill score} = \frac{\overline{\mathrm{CRPS}_{\mathrm{ref}}} - \overline{\mathrm{CRPS}}}{\overline{\mathrm{CRPS}_{\mathrm{ref}}}} \times 100(\%), \tag{9}$$

where the overbar indicates averaging across a set of events. Reference forecasts for each day are produced using a BJP model fitted to observed data within an 11-day window. The Schaake Shuffle is also applied to instil correct temporal characteristics into the reference forecasts. The CRPS skill score is positively oriented (whereas CRPS is negatively oriented). As a percentage, a maximum score of 100 is indicative of perfectly sharp and accurate forecasts. A score of 0 indicates no overall improvement compared to the reference forecast. A negative score indicates poor quality forecasts in the sense that the naïve reference forecast is more skilful.

3.5 Comparison with forecasts post-processed using quantile mapping

We compare RPP-S forecasts with forecasts that have been post-processed at the ACCESS-S grid scale using quantile mapping (QM). The gridded QM forecasts are supplied by the Bureau of Meteorology and we derive the catchment forecasts through area-weighting. QM matches the statistical distribution of past forecasts to the distribution of observations to reduce errors in the forecast mean and improve forecast spread (Crochemore et al., 2016; Zhao et al., 2017). A post-processed forecast value is obtained by first working out the quantile fraction (cumulative probability) of the new forecast using the CDF of past forecasts, then inverting the quantile fraction using the CDF of observations. The Bureau of Meteorology applied a separate quantile mapping model to each day. The CDF of the past forecasts is formed using 11 ensemble members in an 11-day sliding window and 22 years of data in leave-one-year-out cross-validation. The statistical

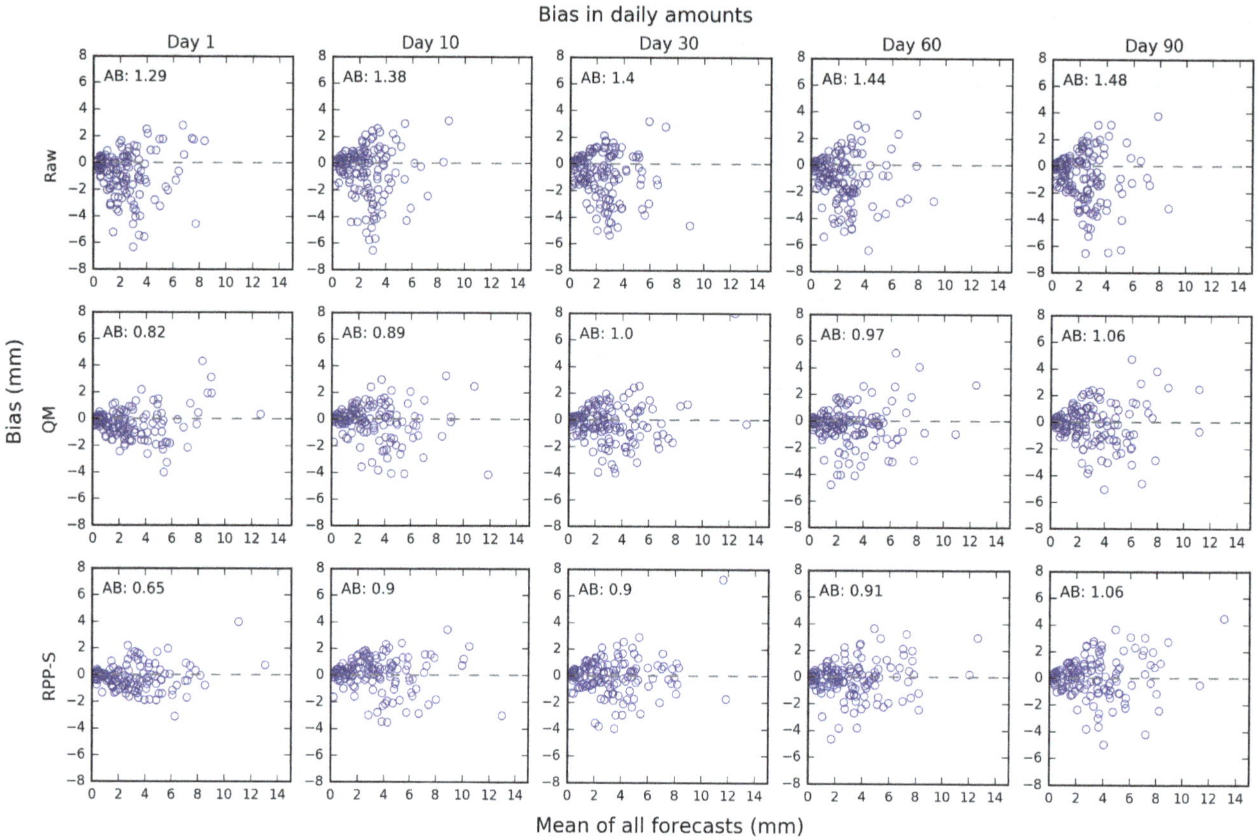

Figure 3. Bias in daily rainfall forecasts for raw, QM and RPP forecasts (rows) and selected days ahead/lead times (columns). The scatterplots are of forecast mean over all events (x axis) versus bias (y axis). There is one blue circle for each catchment and forecast initialisation time. The average absolute bias (AB) is printed in the top left corner.

distribution of the observations is formed using the observations in an 11-day sliding window and 22 years of data. For the first 5 days the window is fixed to the first 11 days. If the raw forecast ensemble member is above the previously known maximum forecast value, then the forecast value is instead linearly rescaled by $o_{\max}/f_{\max}$ where $o_{\max}$ is the maximum observed value and $f_{\max}$ is the maximum past forecast value.

4 Results

4.1 Bias in forecasts of daily amounts

Biases in forecasts of daily amounts are analysed for selected days using Fig. 3. Each circle represents the bias for a catchment and initialisation date. The bias is plotted against the average forecast (averaged over all events). As expected, raw forecasts are more biased than post-processed forecasts. The AB for raw forecasts ranges, for the examples given, from approximately 1.3 to 1.5 mm. The bias for raw forecasts tends to be negative, indicating that ACCESS-S has a propensity to underestimate daily rainfall amounts.

QM and RPP-S are similarly effective at reducing biases in daily amounts and both substantially reduce bias. After post-processing, some residual bias remains for any single day since the bias is corrected using pooled observations. The AB for QM forecasts ranges from roughly 0.8 to 1.1 mm. The AB for RPP-S forecasts ranges from about 0.7 to 1.1 mm. The differences in bias between QM and RPP-S for daily amounts are evidently insignificant. Visual examination of the QM and RPP-S scatter plots shows no wholesale difference in bias between the QM and RPP-S forecasts of daily amounts.

4.2 Bias in forecasts of accumulated totals

Biases in forecasts of accumulated totals are analysed using Fig. 4 after rescaling to mm day^{-1}. As with the results for daily amounts, the raw forecasts are more biased than post-processed forecasts and the bias tends to be negative. Visual examination of the scatterplots for the raw forecasts reveals, particularly for days 30, 60 and 90, that for some catchments and initialisation times, the raw forecasts are unbiased for sub-seasonal to seasonal rainfall totals. This suggests that the GCM performs well in some regions. Nevertheless, it

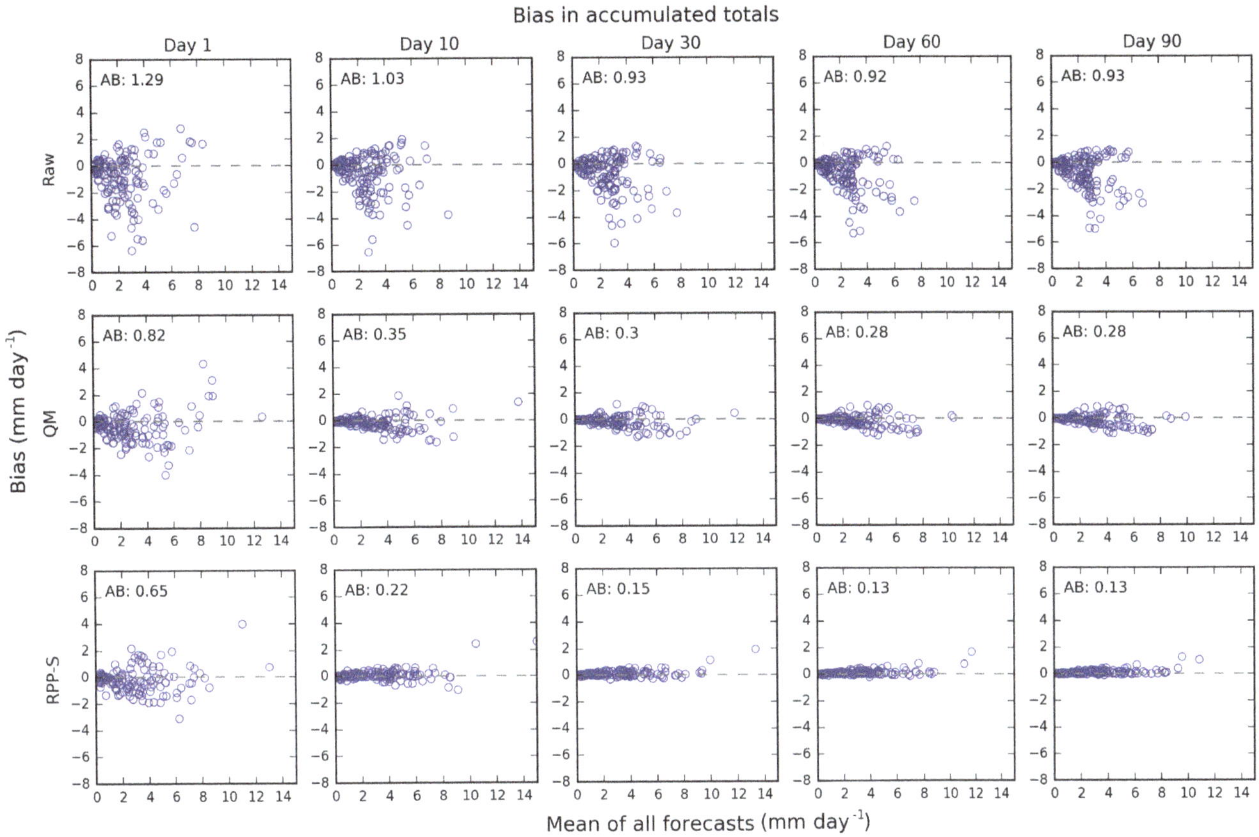

Figure 4. As for Fig. 3, except for accumulated totals.

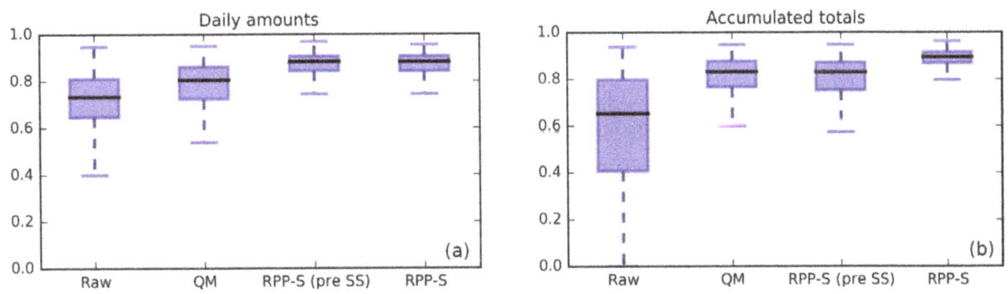

Figure 5. α-index of reliability for forecast daily rainfall amounts **(a)** and forecast accumulated rainfall totals **(b)**. Results for four types of forecasts are presented: raw, QM, RPP-S before Schaake Shuffle (pre SS) and RPP-S forecasts after the Schaake Shuffle. Higher α-index indicates better reliability. The box plots display the median as a black line. The box spans the interquartile range and the whiskers span the [0.1, 0.9] quantile range.

is evident that post-processing is necessary before using the ACCESS-S forecasts in hydrological forecasting.

In contrast to the results for daily amounts, QM and RPP-S have differing efficacy for reducing biases in accumulated totals, although both still reduce bias substantially. Visual examination of the QM and RPP-S scatter plots shows that the RPP-S points fit more tightly around the zero line than the QM points. The better ability of the RPP-S forecasts to reduce biases is reflected in the AB; for example, for day 90 the AB for QM forecasts is 0.28 mm day^{-1} and for RPP-S forecasts the AB is 0.13 mm day^{-1} (cf. the AB for raw forecasts of 0.93 mm day^{-1}).

4.3 Reliability

Reliability is analysed using Fig. 5, which presents box plots of α-index for forecasts of daily rainfall amounts (left panel) and accumulated totals (right panel). The box plots describe the distribution of α-index values for the same cases as we evaluated bias in Sect. 4.2 except that we omit the day 1 result from the accumulated totals analysis. Results are pre-

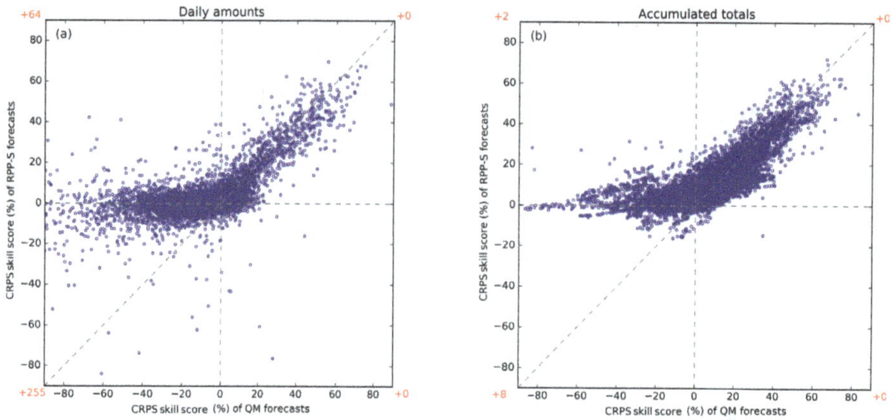

Figure 6. Scatterplots of CRPS skill scores for daily amounts **(a)** and accumulated totals **(b)**. Results for QM are on the horizontal axis and results for RPP-S are on the vertical axis. Higher CRPS skills scores reflect better forecast performance. Red text preceded by a "+" symbol indicates the number of points plotted outside the axis limits in the quadrant nearest the text.

sented for RPP-S before and after the Schaake Shuffle has been applied.

Raw forecasts have the poorest reliability for both daily amounts and accumulated totals. RPP-S has better reliability overall than QM for daily amounts. It is noted that the Schaake Shuffle has no effect on forecast reliability for daily amounts since it is plainly a reordering of already randomised ensemble members.

RPP-S forecasts are more reliable than raw forecasts and QM forecasts for accumulated totals. The RPP-S forecasts become significantly more reliable after applying the Schaake Shuffle and linking the ensemble members so that they have realistic temporal patterns (discussed further in Sect. 5).

4.4 Skill scores – overall performance

Skill scores for QM forecasts are plotted against skill scores for RPP-S forecasts in a scatterplot (Fig. 6). Skill scores for daily amounts are plotted in the left panel; skill scores for accumulated totals are plotted in the right panel. Accumulated totals are for two days or more. RPP-S forecasts and QM forecasts tend to be positively skilful for the same cases, although there is considerable variation in the magnitude of the skill scores. A striking feature of the scatterplots is that when QM forecasts are negatively skilful, the RPP-S forecasts tend to be neutrally skilful, particularly for the accumulated totals. It is evident that skill for daily amounts can be sharply negative. The skill for daily amounts can be difficult to estimate because of the small sample size and the inability to accurately forecast daily amounts beyond about 10 days. In contrast, the skill for accumulated totals is easier to estimate as it benefits from temporal averaging and the accumulation of skill from earlier periods.

4.5 Skill scores – detailed evaluation

Skill scores are partitioned according to catchment, lead time and initialisation date in Figs. 7–9, respectively. Since forecasts of accumulated totals are more informative for water resources management and forecast skill is generally known to be limited beyond the first week or two, we focus the remainder of the results on accumulated totals.

CRPS skill scores are plotted for each catchment in Fig. 7. Skill scores for RPP-S forecasts are vastly positive except for, most noticeably, some cases in ORO, NMN and HLG. Negative skill can be caused by insufficient information to fit a stable RPP-S model and can also be an artefact of cross-validation in the presence of extreme events. In several catchments, such as WLC and DRT, RPP-S forecasts are seen to significantly outperform QM forecasts by virtue of QM forecasts frequently being significantly negatively skilful and RPP-S forecasts rarely being negatively skilful. In several catchments, including BRP and CTG, both QM and RPP-S skill scores are predominantly positive, demonstrating that simple bias correction techniques can appear sufficient in localised studies.

CRPS skill scores are plotted for groups of lead times in Fig. 8. For days 2–10, the RPP-S and QM forecasts are similarly skilful and vastly positively skilful. There are some instances of negative skill, which are likely to be artefacts of cross-validation. For days 11–19 and 20–28, the positive relationship between QM and RPP-S forecasts gradually weakens. From day 29 onwards, the QM forecasts become negatively skilful in many instances whereas the RPP-S skill scores tend to level out around zero and are rarely negative to the same degree. For very long lead times, QM skill scores can be higher than RPP-S skill scores, although overall the skill scores for accumulated seasonal totals are quite low ($< 30\,\%$). Both these factors signal further improvement in the RPP-S forecasts is possible.

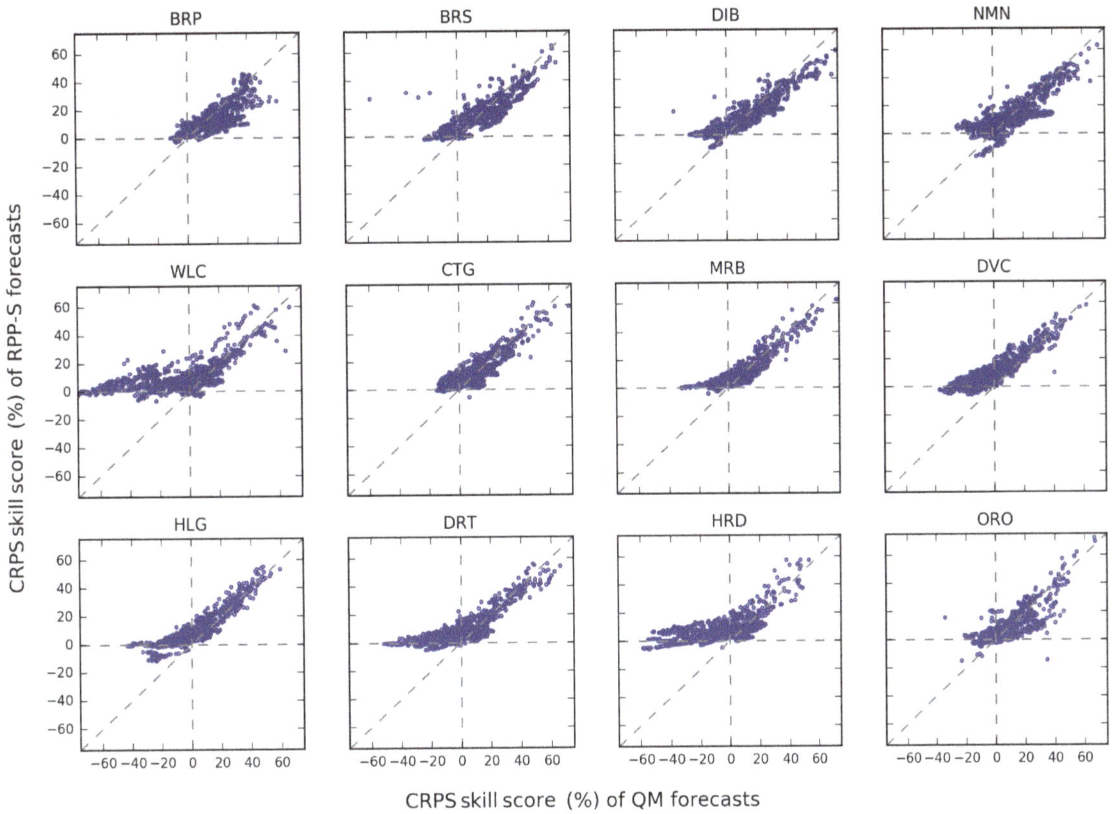

Figure 7. Scatterplots of CRPS skill scores for accumulated totals for each catchment. The scatterplots plot QM skill scores (horizontal axis) against RPP-S skill scores (vertical axis). Each scatterplot combines results for all initialisation dates and all lead times.

CRPS skill scores are plotted for each initialisation date in Fig. 9. Forecasts initialised from 1 September to 1 November have the highest proportion of cases where both QM and RPP-S forecasts are positively skilful, suggesting that ACCESS-S produces its best sub-seasonal to seasonal forecasts during the Austral spring and summer.

5 Discussion

We demonstrate that RPP-S is able to improve daily GCM rainfall forecasts by: reducing bias (Figs. 3–4); improving reliability (Fig. 5); and ensuring that forecasts are typically at least as skilful as a climatological reference forecast (Figs. 6–9). RPP-S forecasts are comprehensively compared with QM forecasts. RPP-S forecasts outperform QM forecasts, primarily because QM does not take into account the correlation between forecasts and observations nor corrects for autocorrelation problems. Since RPP-S is built upon the BJP modelling approach, it explicitly models the correlation between the forecasts and observations, and thus takes into account model skill in the calibration. Our results add to the findings of Zhao et al. (2017) who studied the post-processing of monthly rainfall forecasts from POAMA (Australia's GCM preceding ACCESS-S). While Zhao et al. (2017) demonstrated that QM is very effective for bias correction, they did not consider accumulated totals. In our study, we find that the RPP-S forecasts are less biased and more reliable than QM forecasts for accumulated totals.

Figure 5 illustrates the importance of the Schaake Shuffle for producing reliable forecasts using RPP-S. If the forecasts are not shuffled, then the forecasts of accumulated totals will tend to be too narrow in terms of ensemble spread, making the forecasts over-confident and less reliable. Related to this, RPP-S forecasts are more reliable for daily amounts than QM forecasts, yet QM and RPP-S forecasts do not exhibit any obvious differences in the magnitude of biases (Fig. 3) – evidence that the RPP-S forecasts have a more appropriate ensemble spread, since the α-index integrates information about forecast bias and ensemble spread.

We applied a consistent methodology to perennial and ephemeral catchments and established BJP models separately for each day, catchment and initialisation date. Alternative configurations are also possible, for example, we trialled establishing only six models by pooling data within week 1, week 2, weeks 3–4, and subsequent 4 week periods (not shown). The data in week 1 were used to fit model BJP1, the data in week 2 were used to fit model BJP2, and so on. Fewer days were pooled close to the initialisation date in an attempt to extract skill from the initial conditions and

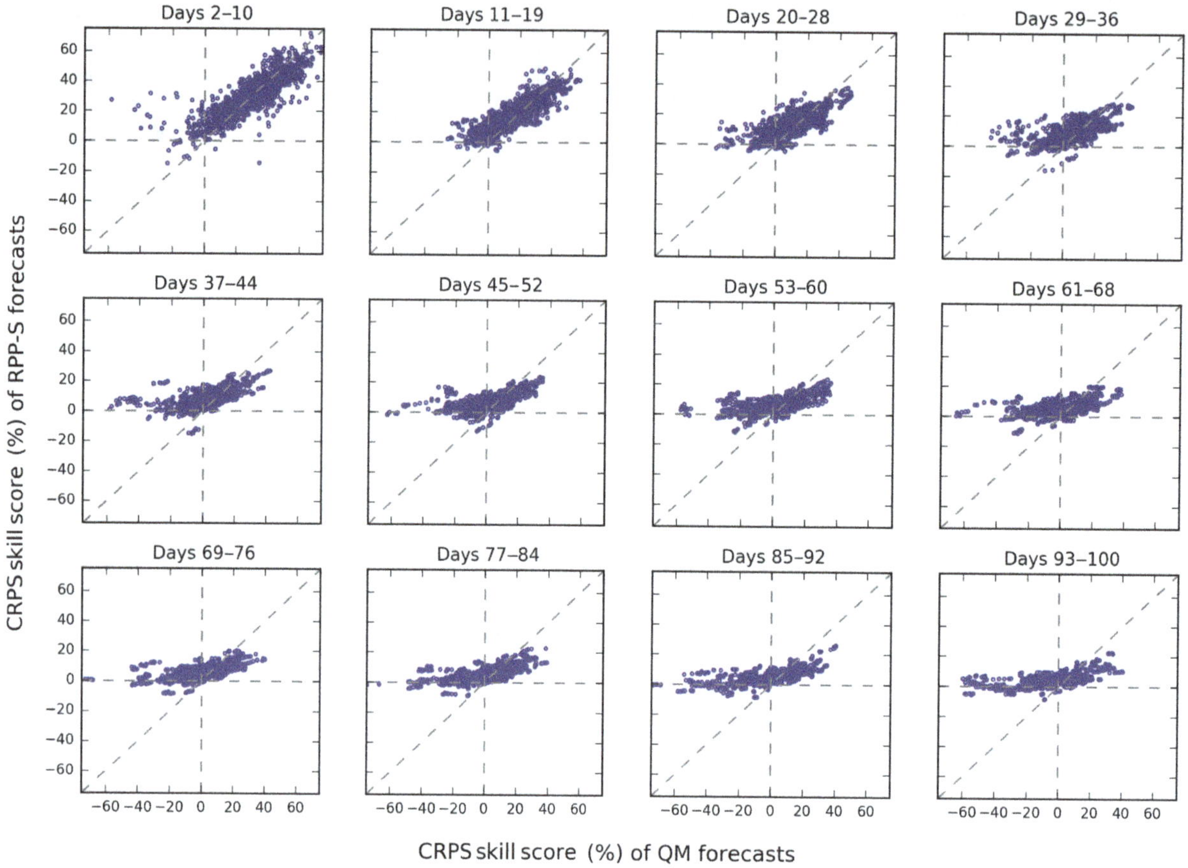

Figure 8. Scatterplots of CRPS skill scores for groups of days (lead times). The scatterplots plot QM skill scores (horizontal axis) against RPP-S skill scores (vertical axis). Each scatterplot combines results for all initialisation dates and all catchments.

more days were pooled for longer lead times to better approximate the climatological distribution. We found significant efficiency gains with little loss in performance, except for dry catchments, where a small increase in the frequency of negative skill scores occurred. Therefore, performance and efficiency need to be considered when establishing RPP-S models in new catchments.

The RPP-S method is sophisticated in that it is a full calibration approach, however, there are opportunities to improve the methodology. By pooling the data for many days in model parameter inference for example, it is assumed that rainfall from one day to the next is independent, which is an oversimplification. New inference methods that treat the rainfall data as conditionally independent therefore ought to be investigated. Future studies will seek to address the matters of independence and overlapping data in more sophisticated ways.

We make use of ACCESS-S runs initialised on day 1 of the month and day 25 of the previous month. These initialisation dates are only 4–7 days apart and therefore the climatology of daily rainfall is unlikely to change significantly over this period of time. It is technically possible to establish an RPP-S model using initialisation dates spanning several months. If temporally distant initialisation dates are included in model parameter inference, new strategies may be needed to ensure that the effects of seasonality are minimised. One possible approach is to standardise forecasts and observations prior to fitting the BJP model. In this way, the model transformation and climatological parameters will be allowed to vary by day of year. Such strategies for building more robust RPP-S models and coincidentally minimising the effect of seasonality will be investigated in follow up work.

RPP-S and QM CRPS scores are calculated using ensembles of different sizes and we consider the effect of ensemble size on our results. When a forecasting system is perfectly reliable, a larger ensemble should yield a better CRPS score (Ferro et al., 2008). However, CRPS has a weakness in that it discourages forecasting extremes (Fricker et al., 2013); indeed it is our experience that ensembles that are unbiased but too narrow (like the QM forecasts) can score overly well in terms of CRPS. Our position is that because the RPP-S and QM forecasts are not similarly reliable, we are unable to make meaningful adjustments to CRPS scores to allow for the effect of ensemble size. In any case, we understand that QM forecasts can be significantly negatively skilful and un-

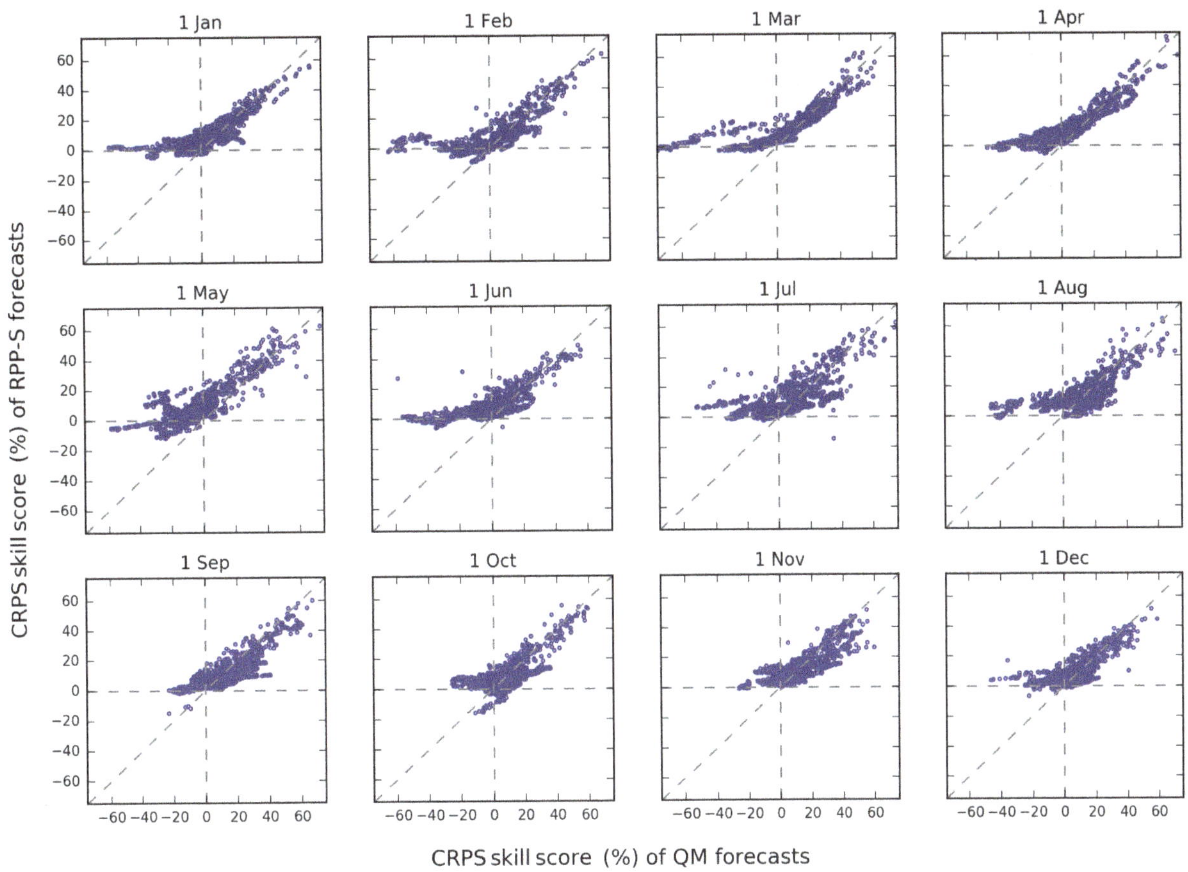

Figure 9. Scatterplots of CRPS skill scores for accumulated totals for each initialisation date. The scatterplots plot QM skill scores (horizontal axis) against RPP-S skill scores (vertical axis). Each scatterplot combines results for all catchments and all lead times.

reliable, and small adjustments to the CRPS of QM forecasts will not affect those conclusions.

RPP-S is designed to post-process forecasts of daily amounts. An alternative strategy is to post-process accumulated totals and subsequently disaggregate to daily amounts. BJP models may be applied to post-process monthly and seasonal totals, which are then disaggregated to daily timescales. Future work will investigate the relative merits of direct daily post-processing versus a seasonal calibration and disaggregation approach. This follow-up work is of particular interest as our study has shown daily post-processing skill is limited beyond 10-15 days. It is not clear how much of the seasonal forecasting skill is attributable to skill in the initial period and how much of the skill is attributable to seasonal climate signals in the GCM.

While our study has focused on rainfall, other variables are important in hydrology. Temperature forecasts, for example, are required in areas of snowmelt. Therefore, future research will investigate the extension of the RPP-S ideas to include other meteorological variables.

An alternative to statistical post-processing of GCM outputs is to run a regional climate model (RCM) to provide much more localised information than a global GCM. A review study by Xue et al. (2014) found that RCMs have limited downscaling ability for sub-seasonal to seasonal forecasts. In that regard, RCM outputs may also be statistically post-processed, which may lead to better forecasts in some regions. RCMs are suited to specialised studies and less suited to post-processing operational GCM forecasts in support of national scale hydrological forecasting services.

6 Conclusion

We have developed a novel method for post-processing daily rainfall amounts from seasonal forecasting GCMs. RPP-S is a full calibration approach that makes use of the BJP modelling approach to account for predictor–predictand skill relationships in post-processing. Reliable forecasts of sub-seasonal and seasonal accumulated totals are produced by linking ensemble members together using the Schaake Shuffle.

We applied RPP-S to 12 catchments across Australia in diverse climate zones. The method is robust in terms of being capable of post-processing forecasts in all cases, even in very dry catchments.

Compared to raw forecasts and QM post-processing (which does not account for predictor–predictand skill relationships), RPP-S performs significantly better in terms of correcting bias, reliability and skill. The only exception to this conclusion is that QM and RPP-S are similarly effective for correcting biases in daily amounts. RPP-S is particularly effective at delivering reliable, skilful, monthly and seasonal rainfall forecasts. Thus RPP-S forecasts are highly suitable for feeding into hydrological models for seasonal streamflow forecasting and other water resources management applications. Pooling multiple GCM runs and data for adjacent days in model parameter inference is a practical measure that can enable statistical post-processing across a range of perennial and ephemeral streams. There are many avenues of research that could significantly improve the robustness and performance of RPP-S forecasts. In the future, more effort could be devoted to applying RPP-S more widely and relating forecast performance to catchment characteristics, therefore yielding a better understanding of forecast bias, reliability and skill.

Competing interests. The authors declare that they have no conflict of interest.

Special issue statement. This article is part of the special issue "Sub-seasonal to seasonal hydrological forecasting". It is not associated with a conference.

Acknowledgements. This research was supported by the Water Information Research and Development Alliance (WIRADA), a partnership between CSIRO and the Bureau of Meteorology. We thank the Bureau of Meteorology for providing the ACCESS-S data, AWAP data and catchment information used in this study. We thank Durga Lal Shrestha for helpful comments on the manuscript and two anonymous reviewers for their insightful comments that helped improve the manuscript.

Edited by: Andy Wood

References

Beckers, J. V. L., Weerts, A. H., Tijdeman, E., and Welles, E.: ENSO-conditioned weather resampling method for seasonal ensemble streamflow prediction, Hydrol. Earth Syst. Sci., 20, 3277–3287, https://doi.org/10.5194/hess-20-3277-2016, 2016.

Bennett, J. C., Wang, Q. J., Li, M., Robertson, D. E., and Schepen, A.: Reliable long-range ensemble streamflow forecasts: Combining calibrated climate forecasts with a conceptual runoff model and a staged error model, Water Resour. Res., 52, 8238–8259, 2016.

Charles, A., Timbal, B., Fernandez, E., and Hendon, H.: Analog downscaling of seasonal rainfall forecasts in the Murray darling basin, Mon. Weather Rev., 141, 1099–1117, 2013.

Clark, M., Gangopadhyay, S., Hay, L., Rajagopalan, B., and Wilby, R.: The Schaake shuffle: A method for reconstructing space-time variability in forecasted precipitation and temperature fields, J. Hydrometeorol., 5, 243–262, 2004.

Crochemore, L., Ramos, M.-H., and Pappenberger, F.: Bias correcting precipitation forecasts to improve the skill of seasonal streamflow forecasts, Hydrol. Earth Syst. Sci., 20, 3601–3618, https://doi.org/10.5194/hess-20-3601-2016, 2016.

Crochemore, L., Ramos, M.-H., Pappenberger, F., and Perrin, C.: Seasonal streamflow forecasting by conditioning climatology with precipitation indices, Hydrol. Earth Syst. Sci., 21, 1573–1591, https://doi.org/10.5194/hess-21-1573-2017, 2017.

Ferro, C. A. T., Richardson, D. S., and Weigel, A. P.: On the effect of ensemble size on the discrete and continuous ranked probability scores, Meteorol. Appl., 15, 19–24, 2008.

Fricker, T. E., Ferro, C. A., and Stephenson, D. B.: Three recommendations for evaluating climate predictions, Meteorol. Appl., 20, 246–255, 2013.

Glahn, H. R. and Lowry, D. A.: The Use of Model Output Statistics (MOS) in Objective Weather Forecasting, J. Appl. Meteorol., 11, 1203–1211, 1972.

Hawthorne, S., Wang, Q., Schepen, A., and Robertson, D.: Effective use of general circulation model outputs for forecasting monthly rainfalls to long lead times, Water Resour. Res., 49, 5427–5436, 2013.

Hudson, D., Shi, L., Alves, O., Zhao, M., Hendon, H. H., and Young, G.: Performance of ACCESS-S1 for key horticultural regions, Bureau of Meteorology, Melbourne, 39 pp., 2017.

Ines, A. V. and Hansen, J. W.: Bias correction of daily GCM rainfall for crop simulation studies, Agr. Forest Meteorol., 138, 44–53, 2006.

Khan, M. Z. K., Sharma, A., Mehrotra, R., Schepen, A., and Wang, Q.: Does improved SSTA prediction ensure better seasonal rainfall forecasts?, Water Resour. Res., 51, 3370–3383, https://doi.org/10.1002/2014WR015997, 2015.

Kim, H.-M., Webster, P. J., and Curry, J. A.: Seasonal prediction skill of ECMWF System 4 and NCEP CFSv2 retrospective forecast for the Northern Hemisphere Winter, Clim. Dynam., 39, 2957–2973, 2012.

Lavers, D., Luo, L., and Wood, E. F.: A multiple model assessment of seasonal climate forecast skill for applications, Geophys. Res. Lett., 36, L23711, https://doi.org/10.1029/2009GL041365, 2009.

Luo, L., Wood, E. F., and Pan, M.: Bayesian merging of multiple climate model forecasts for seasonal hydrological predictions, J. Geophys. Res.-Atmos., 112, D10102, https://doi.org/10.1029/2006JD007655, 2007.

Matheson, J. E. and Winkler, R. L.: Scoring rules for continuous probability distributions, Management Science, 22, 1087–1096, 1976.

Peng, Z., Wang, Q., Bennett, J. C., Schepen, A., Pappenberger, F., Pokhrel, P., and Wang, Z.: Statistical calibration and bridging of ECMWF System4 outputs for forecasting seasonal precipitation over China, J. Geophys. Res.-Atmos., 119, 7116–7135, 2014.

Pineda, L. E. and Willems, P.: Multisite Downscaling of Seasonal Predictions to Daily Rainfall Characteristics over Pacific–Andean River Basins in Ecuador and Peru Using a Nonhomogeneous Hidden Markov Model, J. Hydrometeorol., 17, 481–498, 2016.

Renard, B., Kavetski, D., Kuczera, G., Thyer, M., and Franks, S. W.: Understanding predictive uncertainty in hydrologic modeling: The challenge of identifying input and structural errors, Water Resour. Res., 46, W05521, https://doi.org/10.1029/2009WR008328, 2010.

Robertson, A. W., Kirshner, S., and Smyth, P.: Downscaling of Daily Rainfall Occurrence over Northeast Brazil Using a Hidden Markov Model, J. Climate, 17, 4407–4424, 2004.

Robertson, D. E., Pokhrel, P., and Wang, Q. J.: Improving statistical forecasts of seasonal streamflows using hydrological model output, Hydrol. Earth Syst. Sci., 17, 579–593, https://doi.org/10.5194/hess-17-579-2013, 2013a.

Robertson, D. E., Shrestha, D. L., and Wang, Q. J.: Post-processing rainfall forecasts from numerical weather prediction models for short-term streamflow forecasting, Hydrol. Earth Syst. Sci., 17, 3587–3603, https://doi.org/10.5194/hess-17-3587-2013, 2013b.

Schefzik, R.: A similarity-based implementation of the Schaake shuffle, Mon. Weather Rev., 144, 1909–1921, 2016.

Schepen, A. and Wang, Q.: Ensemble forecasts of monthly catchment rainfall out to long lead times by post-processing coupled general circulation model output, J. Hydrol., 519, 2920–2931, 2014.

Schepen, A., Wang, Q., and Robertson, D. E.: Seasonal forecasts of Australian rainfall through calibration and bridging of coupled GCM outputs, Mon. Weather Rev., 142, 1758–1770, 2014.

Schepen, A., Zhao, T., Wang, Q. J., Zhou, S., and Feikema, P.: Optimising seasonal streamflow forecast lead time for operational decision making in Australia, Hydrol. Earth Syst. Sci., 20, 4117–4128, https://doi.org/10.5194/hess-20-4117-2016, 2016.

Shao, Q. and Li, M.: An improved statistical analogue downscaling procedure for seasonal precipitation forecast, Stoch. Environ. Res. Risk Assess., 27, 819–830, 2013.

Shrestha, D. L., Robertson, D. E., Bennett, J. C., and Wang, Q.: Improving Precipitation Forecasts by Generating Ensembles through Postprocessing, Mon. Weather Rev., 143, 3642–3663, 2015.

Teutschbein, C. and Seibert, J.: Bias correction of regional climate model simulations for hydrological climate-change impact studies: Review and evaluation of different methods, J. Hydrol., 456, 12–29, 2012.

Tian, D., Wood, E. F., and Yuan, X.: CFSv2-based sub-seasonal precipitation and temperature forecast skill over the contiguous United States, Hydrol. Earth Syst. Sci., 21, 1477–1490, https://doi.org/10.5194/hess-21-1477-2017, 2017.

Timbal, B. and McAvaney, B.: An analogue-based method to downscale surface air temperature: application for Australia, Clim. Dynam., 17, 947–963, 2001.

Wang, E., Zhang, Y., Luo, J., Chiew, F. H. S., and Wang, Q. J.: Monthly and seasonal streamflow forecasts using rainfall-runoff modeling and historical weather data, Water Resour. Res., 47, W05516, https://doi.org/10.1029/2010WR009922, 2011.

Wang, Q. and Robertson, D.: Multisite probabilistic forecasting of seasonal flows for streams with zero value occurrences, Water Resour. Res., 47, W02546, https://doi.org/10.1029/2010WR009333, 2011.

Wang, Q., Robertson, D., and Chiew, F.: A Bayesian joint probability modeling approach for seasonal forecasting of streamflows at multiple sites, Water Resour. Res., 45, W05407, https://doi.org/10.1029/2008WR007355, 2009.

Wang, Q. J., Shrestha, D. L., Robertson, D. E., and Pokhrel, P.: A log-sinh transformation for data normalization and variance stabilization, Water Resour. Res., 48, W05514, https://doi.org/10.1029/2011WR010973, 2012.

Wilks, D. S. and Hamill, T. M.: Comparison of Ensemble-MOS Methods Using GFS Reforecasts, Mon. Weather Rev., 135, 2379–2390, 2007.

Wood, A. W., Hopson, T., Newman, A., Brekke, L., Arnold, J., and Clark, M.: Quantifying Streamflow Forecast Skill Elasticity to Initial Condition and Climate Prediction Skill, J. Hydrometeorol., 17, 651–668, 2016.

Xue, Y., Janjic, Z., Dudhia, J., Vasic, R., and De Sales, F.: A review on regional dynamical downscaling in intraseasonal to seasonal simulation/prediction and major factors that affect downscaling ability, Atmos. Res., 147, 68–85, 2014.

Yuan, X., Wood, E. F., and Ma, Z.: A review on climate-model-based seasonal hydrologic forecasting: physical understanding and system development, Wiley Interdisciplin. Rev. Water, 2, 523–536, 2015.

Zhao, T., Schepen, A., and Wang, Q.: Ensemble forecasting of sub-seasonal to seasonal streamflow by a Bayesian joint probability modelling approach, J. Hydrol., 541, 839–849, https://doi.org/10.1016/j.jhydrol.2016.07.040, 2016.

Zhao, T., Bennett, J., Wang, Q. J., Schepen, A., Wood, A., Robertson, D., and Ramos, M.-H.: How suitable is quantile mapping for post-processing GCM precipitation forecasts?, J. Climate, 30, 3185–3196, https://doi.org/10.1175/JCLI-D-16-0652.1, 2017.

PERMISSIONS

All chapters in this book were first published in HESS, by Copernicus Publications; hereby published with permission under the Creative Commons Attribution License or equivalent. Every chapter published in this book has been scrutinized by our experts. Their significance has been extensively debated. The topics covered herein carry significant findings which will fuel the growth of the discipline. They may even be implemented as practical applications or may be referred to as a beginning point for another development.

The contributors of this book come from diverse backgrounds, making this book a truly international effort. This book will bring forth new frontiers with its revolutionizing research information and detailed analysis of the nascent developments around the world.

We would like to thank all the contributing authors for lending their expertise to make the book truly unique. They have played a crucial role in the development of this book. Without their invaluable contributions this book wouldn't have been possible. They have made vital efforts to compile up to date information on the varied aspects of this subject to make this book a valuable addition to the collection of many professionals and students.

This book was conceptualized with the vision of imparting up-to-date information and advanced data in this field. To ensure the same, a matchless editorial board was set up. Every individual on the board went through rigorous rounds of assessment to prove their worth. After which they invested a large part of their time researching and compiling the most relevant data for our readers.

The editorial board has been involved in producing this book since its inception. They have spent rigorous hours researching and exploring the diverse topics which have resulted in the successful publishing of this book. They have passed on their knowledge of decades through this book. To expedite this challenging task, the publisher supported the team at every step. A small team of assistant editors was also appointed to further simplify the editing procedure and attain best results for the readers.

Apart from the editorial board, the designing team has also invested a significant amount of their time in understanding the subject and creating the most relevant covers. They scrutinized every image to scout for the most suitable representation of the subject and create an appropriate cover for the book.

The publishing team has been an ardent support to the editorial, designing and production team. Their endless efforts to recruit the best for this project, has resulted in the accomplishment of this book. They are a veteran in the field of academics and their pool of knowledge is as vast as their experience in printing. Their expertise and guidance has proved useful at every step. Their uncompromising quality standards have made this book an exceptional effort. Their encouragement from time to time has been an inspiration for everyone.

The publisher and the editorial board hope that this book will prove to be a valuable piece of knowledge for researchers, students, practitioners and scholars across the globe.

LIST OF CONTRIBUTORS

F. M. Anayah
Utah Water Research Laboratory, Utah State University, Logan, UT 84322-8200, USA
Palestine Technical University, Kadoorie, Tulkarm, Palestine

J. J. Kaluarachchi
College of Engineering, Utah State University, Logan, UT 84322-4100, USA

Stephan Hülsmann
United Nations University Institute for Integrated Management of Material Fluxes and of Resources (UNU-FLORES), 01067 Dresden, Germany

Christian Bernhofer
Faculty of Environmental Sciences, Institute of Hydrology and Meteorology, Technische Universität Dresden, 01062 Dresden, Germany

Solomon Hailu Gebrechorkos
United Nations University Institute for Integrated Management of Material Fluxes and of Resources (UNU-FLORES), 01067 Dresden, Germany
Faculty of Environmental Sciences, Institute of Hydrology and Meteorology, Technische Universität Dresden, 01062 Dresden, Germany

Stefan Liersch, Julia Tecklenburg, Hagen Koch and Fred Fokko Hattermann
Potsdam Institute for Climate Impact Research (PIK), Telegraphenberg A31, 14473 Potsdam, Germany

Henning Rust, Andreas Dobler and Madlen Fischer
Free University of Berlin (FUB), Institute of Meteorology, Carl-Heinrich-Becker-Weg 6–10, 12165 Berlin, Germany

Tim Kruschke
GEOMAR Helmholtz Centre for Ocean Research Kiel, Wischhofstr. 1–3, 24148 Kiel, Germany

Yiheng Chen and Dawei Han
Department of Civil Engineering, University of Bristol, Bristol, BS8 1TR, UK

Zongxue Xu
College of Water Sciences, Beijing Normal University, Beijing, 100875, China
Beijing Key Laboratory of Urban Hydrological Cycle and Sponge City, Beijing, 100875, China

Qi Chu
College of Water Sciences, Beijing Normal University, Beijing, 100875, China
Beijing Key Laboratory of Urban Hydrological Cycle and Sponge City, Beijing, 100875, China
Department of Civil Engineering, University of Bristol, Bristol, BS8 1TR, UK

Sebastian A. Krogh and John W. Pomeroy
Centre for Hydrology, University of Saskatchewan, 121 Research Dr., Saskatoon, SK S7N 1K2, Canada

Bart Schilperoort, Miriam Coenders-Gerrits, Willem Luxemburg and Hubert Savenije
Delft University of Technology, Water Resources Section, Stevinweg 1, 2628 CN Delft, the Netherlands

César Jiménez Rodríguez
Delft University of Technology, Water Resources Section, Stevinweg 1, 2628 CN Delft, the Netherlands
Tecnológico de Costa Rica, Escuela de Ingeniería Forestal. 159-7050, Cartago, Costa Rica

César Cisneros Vaca
University of Twente, Faculty of Geo-Information Science and Earth Observation (ITC), Hengelosestraat 99, 7514 AE, Enschede, the Netherlands

Chongli Di, Tiejun Wang and Siliang Li
Institute of Surface-Earth System Science, Tianjin University, Tianjin, 300072, P. R. China

Xiaohua Yang
State Key Laboratory of Water Environment Simulation, School of Environment, Beijing Normal University, Beijing, 100875, P. R. China

Jana von Freyberg and James W. Kirchner
Department of Environmental Systems Science, ETH Zurich, Zurich, Switzerland
Swiss Federal Institute for Forest, Snow and Landscape Research (WSL), Birmensdorf, Switzerland

Scott T. Allen
Department of Environmental Systems Science, ETH Zurich, Zurich, Switzerland

Stefan Seeger and Markus Weiler
Faculty of Environment and Natural Resources, University of Freiburg, Freiburg im Breisgau, Germany

Aizhong Ye and Qingyun Duan
State Key Laboratory of Earth Surface and Ecological Resources, Faculty of Geographical Science, Beijing Normal University, Beijing 100875, China

Lifeng Luo
Department of Geography, Environment, and Spatial Sciences, Michigan State University, East Lansing, Michigan, USA

Feng Ma
State Key Laboratory of Earth Surface and Ecological Resources, Faculty of Geographical Science, Beijing Normal University, Beijing 100875, China
Department of Geography, Environment, and Spatial Sciences, Michigan State University, East Lansing, Michigan, USA

Vimal Mishra, Reepal Shah, Syed Azhar, Harsh Shah and Parth Modi
Civil Engineering, Indian Institute of Technology (IIT) Gandhinagar, Gujarat, 382355, India

Rohini Kumar
UFZ-Helmholtz Centre for Environmental Research, Leipzig, Germany

Masahiro Ryo
Department of Aquatic Ecology, EAWAG, 8600 Duebendorf, Switzerland
Department of Civil Engineering, Tokyo Institute of Technology, 2-12-1 Ookayama, Meguro-ku, 152-0033 Tokyo, Japan
Institute of Integrative Biology, ETH-Zürich, 8092 Zürich, Switzerland

Marie Leys and Christopher T. Robinson
Department of Aquatic Ecology, EAWAG, 8600 Duebendorf, Switzerland
Institute of Integrative Biology, ETH-Zürich, 8092 Zürich, Switzerland

Diana Lucatero and Karsten H. Jensen
Department of Geosciences and Natural Resource Management, University of Copenhagen, Copenhagen, Denmark

Henrik Madsen
DHI, Hørsholm, Denmark

Jens C. Refsgaard and Jacob Kidmose
Geological Survey of Denmark and Greenland (GEUS), Copenhagen, Denmark

Zilefac Elvis Asong, Howard Simon Wheater, Saman Razavi and Sopan Kurkute
Global Institute for Water Security and School of Environment and Sustainability, University of Saskatchewan, 11 Innovation Blvd, Saskatoon, SK, Canada S7N 3H5

Barrie Bonsal
Environment and Climate Change Canada, 11 Innovation Blvd, Saskatoon, SK, Canada S7N 3H5

Andrew Schepen
CSIRO Land and Water, Dutton Park 4102, Australia

Tongtiegang Zhao and Quan J. Wang
Department of Infrastructure Engineering, The University of Melbourne, Parkville 3010, Australia

David E. Robertson
CSIRO Land and Water, Clayton 3168, Australia

Index

S

T

U

W

Y

www.ingramcontent.com/pod-product-compliance
Lightning Source LLC
LaVergne TN
LVHW060532210926
848485LV00005B/1186